Macromolecular Symposia

Symposium Editor: H. R. Kricheldorf

199

pp. 1–521

September 2003

Macromolecular Symposia publishes lectures given at international symposia and is issued irregularly, with normally 14 volumes published per year. For each symposium volume, an Editor is appointed. The articles are peer-reviewed. The journal is produced by photo-offset lithography directly from the authors' typescripts.
Further information for authors can be obtained from:
Editorial office "Macromolecular Symposia"
Wiley-VCH Verlag GmbH & Co. KGaA,
Boschstrasse 12, 69469 Weinheim,
Germany
Tel. +49 (0) 62 01/6 06-2 38 or -5 81; Fax +49 (0) 62 01/6 06-3 09 or 5 10;
E-mail: macro-symp@wiley-vch.de
http://www.ms-journal.de
Suggestions or proposals for conferences or symposia to be covered in this series should also be sent to the Editorial office at the address above.

Macromolecular Symposia:
Annual subscription rates 2003 (print only or online only)*
Germany, Austria € 1318; Switzerland SFr 2168; other Europe € 1318; outside Europe US $ 1568.
Macromolecular Package, including Macromolecular Chemistry & Physics (18 issues), Macromolecular Bioscience (12 issues), Macromolecular Rapid Communications (18 issues), Macromolecular Theory & Simulations (9 issues) is also available. Details on request.
* For a 5 % premium in addition to **Print Only** or **Online Only**, Institutions can also choose both print and online access.
Packages including Macromolecular Symposia and Macromolecular Materials & Engineering are also available. Details on request.
Single issues and back copies are available. Please inquire for prices.

Orders may be placed through your bookseller or directly at the publishers:
WILEY-VCH Verlag GmbH & Co. KGaA, P. O. Box 10 11 61, 69451 Weinheim, Germany, Tel. +49 (0) 62 01/6 06-400, Fax +49 (0) 62 01/60 61 84. E-mail: service@wiley-vch.de

Macromolecular Symposia (ISSN 1022-1360) is published with 14 volumes per year by WILEY-VCH Verlag GmbH & Co. KGaA, P. O. Box 10 11 61, 69451 Weinheim, Germany. Air freight and mailing in the USA by Publications Expediting Inc., 200 Meacham Ave., Elmont, NY 11003, USA. Application to mail at Periodicals Postage rate is paid at Jamaica, NY 11431, USA. US POSTMASTER please send address changes to: Macromolecular Symposia, c/o Wiley-VCH, III River Street, Hoboken, NJ 07030, USA.

Printing: Strauss Offsetdruck, Mörlenbach. Binding: J. Schäffer, Grünstadt

Polycondensation 2002

Progress in Step-Growth Polymerization and
Structure-Property Relationships of Polycondensates

4th International Symposium on Polycondensates held in Hamburg, Germany
September 15–18, 2002

Symposium Editor (Chairman)

Professor Hans R. Kricheldorf

Institut für Technische und Makromolekulare Chemie
Bundesstrasse 45
20146 Hamburg
Germany

ISBN 3-527-30703-6

Contents of Macromolecular Symposia 199

Polycondensation 2002
Hamburg (Germany), 2002

* Asterisks indicate the name(s) of the author(s) to whom inquiries should be addressed.

Author Index

Preface

Polycondensation is a synthetic methodology which is particularly important for theoretical, practical and historical reasons. Firstly, because of the favorable thermodynamic aspects (reaction entropy close to zero), polycondensations allow a broader variation of monomer and polymer structure than any other synthetic approach. Secondly, numerous commercial polymers are produced via polycondensation processes. Thirdly, the first "man-made fibers" were polycondensates (nylons). Finally, all biopolymers are produced by living organisms via a special kind of polycondensation reaction. In order to create a platform allowing experts in the field of polycondensation chemistry to discuss their research activities and latest results, several colleagues and I have launched a series of international symposia dedicated to the synthesis, characterization and application of polycondensates. The first symposium of this series was organized in Paris in September 1996. The second meeting followed two years later in Anapolis, the third symposium in Tokyo in the year 2000 and the fourth symposium in Hamburg in 2002. Most results presented as oral contributions at the fourth meeting in Hamburg are summarized in the chapters of this symposium volume.

H. R. Kricheldorf

What Does Polycondensation Mean?

Hans R. Kricheldorf

Institut für Technische und Makromolekulare Chemie, Bundesstr. 45,
D-20146 Hamburg, Germany
Email: kricheld@chemie.uni-hamburg.de

Summary: This contribution has the function of an introduction to the entire volume. It deals with several fundamental definitions and classifications related to the chemistry of polycondensation processes, and it includes modifications of the classical theory of step-growth polymerizations.

Keywords: cyclic polymers, hyperbranched polymers, kinetic control, polycondensation, thermodynamic control

Introduction

The classical theory of step-growth polymerizations as it is presented in all textbooks of polymer science is mainly based on the experimental work of Carothers[1,2] and on the theoretical contributions of Flory.[3,4] One of the milestones in polymer science which has emerged from this work is the demonstration that the reactivity of endgroups is (in general) independent of the chain length (previously denied by Staudinger and other chemists). The classical theory of polycondensation describes the chain growth of linear monomers which are symbolized as "a-b" monomers, when different endgroups are present or "a-a" and "b-b" monomers when having identical endgroups. All reactive species, including linear oligomers and polymers can react with each other at any time and the chain lengths increase with the conversion according to the "Carothers equation" (1). The influence of chain terminators (frequently added in technical syntheses to regulate the average degree of polymerization, $\overline{DP}$) or an imbalance of the stoichiometry (which has the same effect) may be expressed by eq. (2). The frequency of chains characterized by an individual DP is given by eq. (3), and corresponding mass distribution by eq. (4). In the following paragraphs further definitions and classifications will be presented along with several modifications of the classical theory of step-growth polymerization.

 DOI: 10.1002/masy.200350901

$$\overline{DP} = \frac{1}{1-p} \tag{1}$$

$\overline{DP}$ = average degree of polymerization

p = conversion of functional groups

$$\overline{DP} = \frac{1+r}{2r(1-p)+1-r} = 1 \tag{2}$$

with $r = N_a/N_b$; N_a/N_b the total numbers of initially present functional groups, incl. those of the chain terminator

$$f_n = p^{n-1}(1\text{-}p) \tag{3}$$

with n = DP of individual chains

$$W_n = m_n \cdot p^{n-1}(1\text{-}p) \tag{4}$$

Definitions and Classifications

Step-growth Versus Chain-growth Polymerizations

The classical definition of polycondensation is based on two aspects: 1) elimination of a small byproduct (e.g. H_2O, HCl etc.) in every propagation step, 2) a step-growth kinetic resulting from an equal reactivity of all monomers, oligomers and polymers (with exceptions in the case of monomers). It is important to keep both aspects in mind, because a polymerization process obeying a chain-growth kinetic may involve elimination of small byproducts in every propagation step. In an ideal chain-growth polymerization involving elimination steps, an initiator starts the propagation, the polymers do not react with each other and the monomers exclusively react with the active chain end. However, in real experiments both kinds of kinetics may overlap, so that a clearcut classification is not always feasible. Chain-growth polymerizations involving condensation steps are not a curiosity, they are, in fact, the oldest polymerization process known on earth because all syntheses of biopolymers in living organism obey this reaction pattern.

Step-growth and chain-growth polymerizations involving elimination reactions in every propagation step have the same favorable thermodynamic situation, namely a reaction entropy close to zero. However, the different kinetic course has the consequence that the molecular weight distribution (MWDs) may be quite different. In a typical step-growth polymerization, the polydispersity is ≥ 2.0, whereas in a chain-growth polymerization, narrow molecular weight distributions with polydispersities ≤ 1.1 may be obtained. Several in vivo syntheses of biopolymers which all involve chain growth polymerizations produce in fact monodisperse biopolymers. "Polycondensations" showing a more or less pronounced tendency towards chain-growth polymerizations with condensation steps are, for instance, the oxidative coupling of ortho-disubstituted phenols[5,6] or syntheses of poly(phenylene sulfide).[7] These polymerizations involve radicals as reactive intermediates.

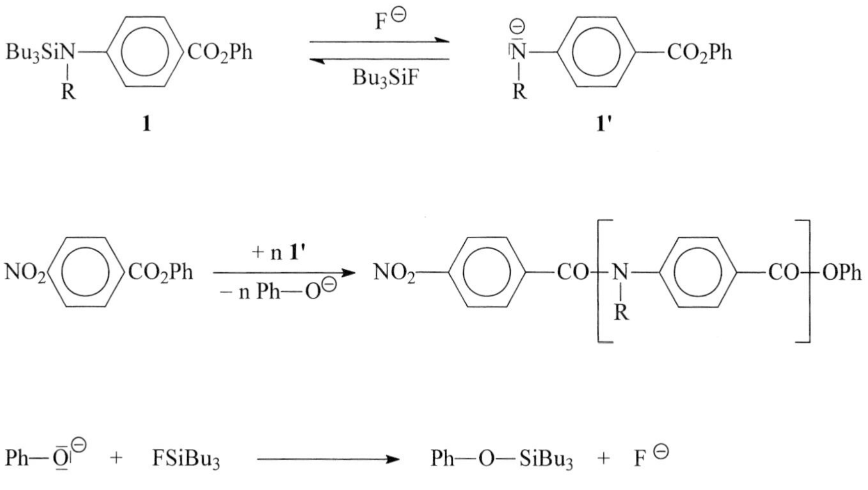

$Bu_3SiN(R)$–C_6H_4–CO_2Ph (**1**) $\underset{Bu_3SiF}{\overset{F^{\ominus}}{\rightleftharpoons}}$ $^{\ominus}N(R)$–C_6H_4–CO_2Ph (**1'**)

NO_2–C_6H_4–CO_2Ph $\xrightarrow[-\,n\ Ph-O^{\ominus}]{+\,n\ \mathbf{1'}}$ NO_2–C_6H_4–CO–[N(R)–C_6H_4–CO]–OPh

$Ph-O^{\ominus}$ + $FSiBu_3$ ⟶ $Ph-O-SiBu_3$ + $F^{\ominus}$

Scheme 1

More recent examples of chain-growth polymerizations involving condensation steps were published by Yokogawa and coworkers.[8-11] These polymerizations are based on nucleophilic substitutions of non-cyclic monomers, as exemplarily illustrated in Scheme 1. Those authors used the term "chain growth polycondensation" when the monomers exclusively reacted with an initiator and active endgroup of the growing chain. However, the long known[12] ring-opening polymerizations of cyclic amino acid anhydrides (Scheme 2) have the same kinetic and thermodynamic properties as the polymerizations of Yokosawa et al. and were defined as chain-growth (and ring-opening) polymerizations for many decades. Therefore, it is obvious that the term "polycondensation" is misleading, when applied to chain-growth polymerization, even when condensation steps are involved. The term "polycondensation" was and is defined for a polymerization process obeying a step-growth kinetic. Perhaps "chain growth condensation polymerization" is a terminology which is acceptable for polymerizations such as those outlined in Scheme 1.

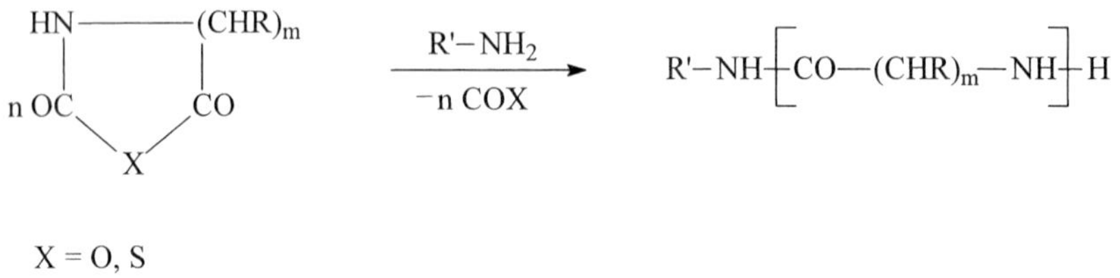

HN——(CHR)$_m$ | | n OC CO \ / X

$$\xrightarrow[-n\,COX]{R'-NH_2} R'-NH-\left[CO-(CHR)_m-NH\right]-H$$

X = O, S

Scheme 2

Kinetic Control Versus Thermodynamic Control

In all fields of chemistry the main product isolated from a reaction mixture may be the result of the most rapid reaction (kinetic control) or of an equilibration process (thermodynamic control). By coincidence the kinetically controlled reaction product may also be the thermodynamically most stable component of the reaction mixture, but in most cases it is not. In the chemistry of carbon compounds a general and systematic correlation between the kinetic and thermodynamic properties of a reaction mixture does not exist. A differentiation between a kinetically or thermodynamically controlled course can and should also be made for all polymerization processes. For instance, chain-growth polymerizations of α-olefins or vinyl monomers are kinetically controlled polymerizations resulting in thermodynamically instable polymers, which upon equilibration would collapse into (un)substituted cyclohexanes. However, for numerous ring-opening polymerizations and polycondensations both kinetically controlled and thermodynamically controlled reaction pathways exist. For a proper understanding of structure and properties of the resulting polymers it is important to know, if they were formed under kinetic or thermodynamic control. Characteristic for a kinetically controlled polycondensation (KCP) is the point that all products formed in early stages of the polymerization are stable during the later stages. Therefore, processing of such polymers at high temperature may change structure and properties of the polymer due to the influence of equilibration reactions. Such equilibrations typically involve the formation of cyclic oligomers by "back-biting degradation".[13,14] On the other hand, any kind of ordered structure, such as alternating sequences and block copolymers can only be obtained via a KCP. Furthermore, the molecular weight distributions, MWDs, may be different. Typical for thermodynamically controlled polycondensations (TCPs) are polydispersities in the range of 2.0-2.5,[4] whereas the products of KCPs may have broader distributions. For all these reasons it is important to classify polycondensations according to kinetic or thermodynamic control.

"a-b" Versus "a-a + b-b" Monomers

In most textbooks "a-b" and "a-a + b-b" monomers are discussed so as if their polycondensations and the structures of their resulting polymers are quite similar. However, both monomer systems differ largely in numerous aspects and these differences should be discussed in this section. The most obvious difference (usually mentioned in textbooks) is the fact that the "a-b" monomers have a built-in perfect stoichiometry of the functional groups. Furthermore, all oligomers and polymers possess the same combination of "a" and "b" endgroups. In contrast, polycondensations of "a-a" + "b-b" monomers automatically produce mixtures of three types of endgroup combinations. This difference is important when the polymers should be subjected to chemical modifications of endgroups, chain extension or crosslinking reactions.

Particularly important is the point that "a-b" monomers possess a considerably greater potential for syntheses of functional polymers and complex architectures.[15] For instance, addition of "a-a" or "b-b" monomers to a larger quantity of "a-b" monomers yields oligomers or polymers having either two "a" or two "b" endgroups.[16,17] Furthermore, the feed ratio of "a-b"/"a-a" (or "a-b"/"b-b") monomers allows a control of the $\overline{DP}$. The resulting telechelic polymers may be used as central blocks in syntheses of A-B-A triblock copolymers[18] or as components of multiblock copolymers. Secondly, copolycondensations of "a_n" monomers with an excess of "a-b" monomers yields star-shaped polymers. The number of the star arms depends on the functionality of "a_n" and the lengths of the star arms on the "a-b"/"a_n" ratio. In contrast the combination of "a_n" with "a-a" and "b-b" monomers yields networks. Thirdly, cocondensations of "a-b_n" + "a-b" monomers yield hyperbranched copolymers with variation of the branching density.[3,4,19,20,21] The presence of "b-b" monomers results again in crosslinking. Last but not least, all species in a sample of "a-b" type oligomers and polymers can, in principle, cyclize (see below), whereas in the "a-a" + "b-b" case 50% of the reaction products cannot cyclize. In summary, a classification of polycondensations based on "a-b" monomers and others based on "a-a" + "b-b" monomers is reasonable because of the quite different synthetic potential.

Stoichiometric Versus Non-stoichiometric Polycondensations

According to the classical theory of step-growth polymerizations, the highest molecular weights are obtained with a perfect 1:1 stoichiometry of the functional groups . However, in real polycondensations of "a-a" + "b-b" monomers it was found that an excess of one monomer may give higher molecular weights than the 1:1 feed ratio. For a proper understanding of so-called non-stoichiometric polycondensation, the first question which needs clarification is the structure (or composition) of the polymer. If the molar ratio of A and B units in the polymer backbone is 1:1, the chain growth was necessarily a stoichiometric process, even when the feed ratio was far from 1:1. Three reasons may account for the observation that an excess of one monomer in the feed may give optimum molecular weights of a perfectly stoichiometric polymer:

1) Side reactions of one monomer which do not significantly disturb the chain growth. A typical example is the hydrolysis of phosgene in the interfacial syntheses of polycarbonates, which is compensated by an excess of phosgene.

2) Physical reasons hindering one monomer to participate completely in the chain growth process. Such physical reasons are: distillation or sublimation from the reaction mixture, adsorption on a solid surface or complexation with solvents or other components of the reaction mixture.

3) A two-step propagation kinetic with a faster second step. This means that an intermediate is formed with a functional group which is more reactive than either the "a-a" or the "b-b" monomer. Therefore, an excess of the less reactive monomer will accelerate the chain growth and the degree of polymerization will be considerably higher than expected from the imbalance of the stoichiometry according to the classical theory. However, a 1:1 feed ratio will give the highest molecular weight, when the reaction time allows for high conversion . The polycondensation of 2,2-dichloro-4,5-benzodioxolane with diphenols (Scheme 3) is a typical example of such a non-stoichiometric polycondensation.[22] Another example is presented by M. Ueda and coworkers in this book below.

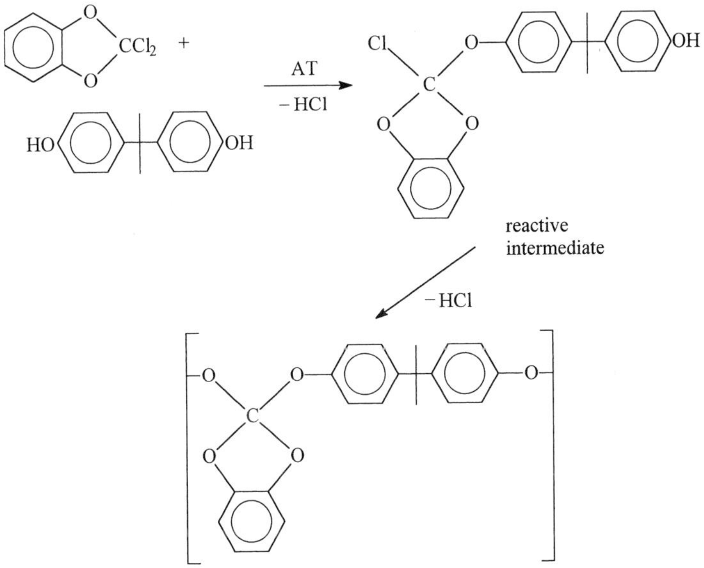

Scheme 3

Cyclic Monomers and Ring-opening Polycondensations

The experimental work of Carothers and Flory and the classical theory of polycondensation is based on linear or branched monomers. However, any kind of heterocycles containing at least two reactive bonds may, in principle, be used as monomers for "ring-opening polycondensation". None the less, in textbooks of polymer science cyclic monomers are exclusively discussed in connection with ring-opening polymerizations involving chain growth kinetics. Cyclic monomers can be used for polycondensations in two different combinations:

(I) Cyclic monomer + linear monomer

(II) Cyclic "a-a" monomer + cyclic "b-b" monomer.

The former combination (I) is far more versatile and allows for a much broader application of cyclic monomers than case (II) for which only one successful combination (eq. (5)) has recently been reported[23].

A versatile and commercial class of cyclic monomers useful for polycondensation are cyclic anhydrides. Polyester resins based on polycondensations of cyclic anhydrides and glycerol (or other polyols) were studied and commercialized more than seven decades ago (eq. (6)),[24] and quite recently resins based on hyperbranched poly(ester amide)s were developed.[25] Another group of versatile cyclic monomers containing Sn atoms has recently been explored by the author.[26,27] The Sn atom enhances the nucleophilicity of neighboring heteroatoms, and thus, favors polycondensations with various electrophilic "b-b" type monomers as exemplified in eq. (7). Numerous Sn containing heterocycles are easy to synthesize and they have the important advantage to allow for a combination of ring-opening polymerization and polycondensation in "one-pot procedures".[26] All ring-opening polycondensations have in common that their thermodynamic properties deviate from those of polycondensations exclusively involving linear monomers. In most cases (typical for combination (I)) the reaction entropy is negative.[26] In case (II) the reaction entropy depends very much on the ring size and ΔS may turn positive, but such a case has not been realized yet.[23]

$$\text{cyclic ethylene sulfite } (H_2C{-}CH_2,\ O{-}SO{-}O) + \text{cyclic anhydride } (OC{-}(A){-}CO,\ O) \xrightarrow[-\,SO_2]{\text{Cat.}} \left[O{-}(CH_2)_2O{-}CO{-}(A){-}CO \right] \quad (5)$$

$$2\ CH_2OH{-}CH(OH){-}CH_2OH + 3\ \text{cyclic anhydride } (OC{-}(A){-}CO,\ O) \xrightarrow[3\ H_2O]{} \text{Polyester Networks} \quad (6)$$

$$Bu_2Sn\langle X{-}CH_2{-}CH_2{-}X \rangle \xrightarrow[-\,Bu_2SnCl_2]{+\,ClCO{-}(A){-}COCl} \left[X{-}(CH_2)_2{-}X{-}CO{-}(A){-}CO \right] \quad (7)$$

$X = O, S$

Cyclic Versus Linear Polymers

For reasons discussed by Carothers[1] and Flory[3] those authors did not consider cyclization reactions to play an important role in their theory of step-growth polymerizations. In 1950 Jacobson and Stockmayer[13,14] proved that at least cyclic oligomers are formed in all TCPs due to "back-biting degradation" and they developed a mathematical treatment of equilibrium concentrations and molecular weight distribution of the cycles. More recently, theoretical calculations of KCPs (presented by Stepto[28,29] and Gordon et al.[30]) and experimental results obtained by MALDI-TOF mass spectroscopy (reported by Kricheldorf et al.[31-34]) proved that ring closure reactions play a decisive role in KCPs. Cyclization competes with propagation at any concentration and at any stage of a polycondensation. In an ideal KCP (free of side reactions and perfect stoichiometry) all reaction products will be cycles at 100% conversion.

Therefore eq. (1) needs to be replaced by eq. (8) and eq. (3) should be replaced by eq. (9), and the following consequences have to be considered.

$$\overline{DP} = \frac{1}{1-p\left(1-\frac{1}{X^{\alpha}}\right)} \tag{8}$$

with $\alpha = V_p/V_c$ ratio of propagation and cyclization rate

X = constant > 1.0 allowing for adaptation of the equation to individual experiments with variation of the concentration

$$\overline{DP} = \frac{1+r}{2r\left[1-p\left(1-\frac{1}{X^{\alpha}}\right)\right]+1-r} - 1 \tag{9}$$

Firstly, in real experiments involving a few side reactions and conversions below 100%, neither 100% cycles nor 100% linear chains will be obtained. Secondly, the rate of cyclization is decisive for the maximum molecular weight which can be obtained. The infinite molecular weight predicted by the classical "Carothers equation" (1) can never be achieved, even under ideal conditions. Thirdly, the MWDs are different from those of the classical theory formulated by Flory in eqs. (3) and (4). Fourthly, the "cascade theory"[3,4] describing the formation of hyperbranched polymers by polycondensation of "a-b_n" monomers also needs modification, because it does not include a consideration of cyclization reactions.

The polycondensation of an "a-b_n" monomer creates oligomers and polymers possessing one "a" and numerous "b" functional groups. Therefore, cyclization can compete with propagation at any stage of the polycondensation and eq. (8) is also valid for "a-b_n" monomers. Both chain growth and polydispersity are limited by the influence of cyclization in contrast to the calculations of Flory.[3,4] At 100% conversion all reaction products are cycles with hyperbranched side chains (Scheme 4).[35] The tree-shaped structures depicted in previous publications of numerous authors just present an intermediate state of an incomplete or imperfect polycondensation. The permanent competition of cyclization and propagation has also a strong influence on the structure of branched or crosslinked polymers resulting from KCPs of "a_n" + "b-b" monomers. However, because of the limited space of this article the rather complex course of "a_n" + "b-b" polycondensations should not be discussed here in more detail.

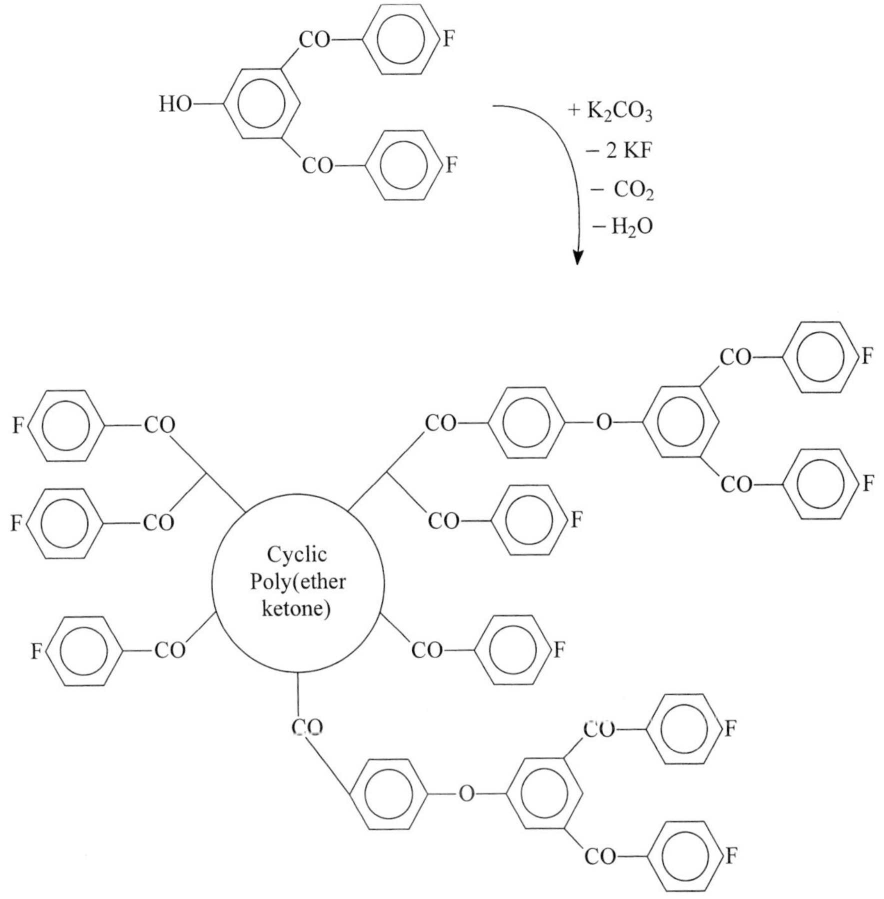

Scheme 4

Conclusion

The historical merits and the importance of the classical theory of step-growth polymerizations as elaborated by Carothers and Flory can never be overestimated. None the less, numerous experimental results and theoretical considerations which emerged over the past fifty years require revision, modification and expansion of the classical theory.

[1] W.H. Carothers, *J. Am. Chem. Soc.* **1929**, *31*, 2548.

[2] *"Collected Papers of W.H. Carothers on Polymerization"*, H. Mark and G.S. Whitby Eds., Wiley Interscience N.Y., 1940.

[3] P.J. Flory, *Chem. Rev.* **1946**, *39*, 137.

[4] P.J. Flory *"Principles of Polymer Chemistry"* Cornell University Press, Ithaca, N.Y. 1953, Chapters VIII and IX.

[5] W. Koch, W. Heitz, *Makromol. Chem.* **1983**, *184*, 779.

[6] H.R. Kricheldorf in *"Handbook of Polymer Syntheses"* (H.R. Kricheldorf, ed.), Marcel Dekker Publ., New York 1992, Chapter 9.

[7] W. Koch, W. Risse, W. Heitz, *Makromol. Chem. Suppl.* **1985**, *12*, 105.

[8] T. Yokozawa, S. Horio, *Polymer J.* **1966**, *28*, 633.

[9] T. Yokozawa, T. Asai, R. Sugi, S. Ishigooka. S. Hiraoka, *J. Am. Chem. Soc.* **2000**, *122*, 8313.

[10] T. Yokozawa, H. Suzuki, *J. Am. Chem. Soc.* **1999**, *121*, 11573.

[11] T. Yokozawa, Y. Suzuki, S. Hiraoka, *J. Am. Chem. Soc.* **2001**, *123*, 9902

[12] H.R. Kricheldorf *"α-Amino Acid N-Carboxyanhydrides and Related Heterocycles"*, Springer Publishers, Berlin, Heidelberg, N.Y. 1987.

[13] H. Jacobson , W.H. Stockmayer, *J. Chem. Phys.* **1950**, *18*, 1600.

[14] H. Jacobson, C.O. Beckmann, W.H. Stockmayer, *J. Chem. Phys.* **1950**, *18*, 1607.

[15] H.R. Kricheldorf, O. Stöber, G. Löhden, T. Stukenbrock, D. Lübbers in *"Step-Growth Polymers for High-Performance Materials"* J.L. Hedrick, J. L. Labbadie, eds.) ACS Symposium Series 624 (1996), Chapter 9.

[16] H.R. Kricheldorf, T. Adebahr, *Makromol. Chem.* **1993**, *194*, 2103.

[17] H.R. Kricheldorf, X. Chen, M. Al Masri, *Macromolecules* **1995**, *28*, 2112.

[18] H.R. Kricheldorf, T. Stukenbrock, C. Friedrich, J. Polym. *Sci. Part A Polym. Chem.* **1998**, *36*, 1387.

[19] H.R. Kricheldorf, Q.-Z. Zang, G. Schwarz, *POLYMER* **1982**, *23*, 1921.

[20] M. Jikei, K. Fuji, G. Yang, M. Kakimoto, *Macromolecules* **2000**, *33*, 6228.

[21] M. Jikei, K. Fuji, M. Kakimoto, *J. Polym. Sci., Part A, Polym. Chem.* **2001**, *39*, 3304

[22] N. Kihara, S. Komatsu, T. Takata, T. Endo, *Macromolecules* **1999**, *32*, 4776.

[23] H.R. Kricheldorf, O. Petermann, *Macromolecules* **2001**, *34*, 8841.

[24] R.H. Kienle, A.G. Hovey, *J. Am. Chem. Soc.* **1929**, *51*, 509.

[25] D. Muscat, R.A.T.M. van Benthem, *Topics Current Chem.* **2001**, *212*, 41.

[26] H.R. Kricheldorf, *Macromol. Rapid Commun.* **2000**, *21*, 528.

[27] H.R. Kricheldorf, B. Fechner, *Biomacromolecules* **2002**, *3*, 691.

[28] R.F.T. Stepto, D.R. Waywell, *Makromol. Chem.* **1972**, *152*, 263.

[29] J.L. Stanford, R.F.T. Stepto, D.R. Waywell, *J. Chem. Soc. Faraday Trans.* **1975**, *71*, 1308.

[30] M. Gordon, W. Temple, *Makromol. Chem.* **1972**, *152*, 277.
[31] H.R. Kricheldorf, M. Rabenstein, M. Maskos, M. Schmidt, *Macromolecules* **2001**, *34*, 713.
[32] H.R. Kricheldorf, S. Böhme, G. Schwarz, *Macromolecules* **2001**, *34*, 8879.
[33] H.R. Kricheldorf, S. Böhme, G. Schwarz, R.-P. Krüger, G. Schulz, *Macromolecules* **2001**, *34*, 8886.
[34] H.R. Kricheldorf, S. Böhme, G. Schwarz, C.-L. Schultz, *Macromol. Rapid Comm.* **2002**, *23*, 803.
[35] H.R. Kricheldorf, L. Vakhtangishvili, G. Schwarz, R.-P. Krüger, *Macromolecules*, submitted.

Macromol. Symp. **2003**, *199*, 15-22

The Role of Ring-Ring Equilibria in Thermodynamically Controlled Polycondensations

Hans R. Kricheldorf

Institut für Technische und Makromolekulare Chemie, Bundesstr. 45,
D-20146 Hamburg, Germany
Email: kricheld@chemie.uni-hamburg.de

Summary: Thermodynamically controlled polycondensations (TCPs) involve rapid equilibration reactions, such as transesterification, transamidation, etc. An important component of these equilibration reactions is the reversible formation of cyclic oligomers and polymers by "back-biting". Therefore, TCPs were described in the previous literature in terms of ring-chain equilibria. The present study presents a complementary theory saying that ring-chain equilibria automatically include ring-ring equilibria which gain in importance with higher conversions because the molar ratio of rings versus linear chains rapidly increases. At 100% conversion, all reaction products will be cycles and the ring-ring equilibria limit the chain growth. Several polycondensations cited from the literature are discussed in the light of the new theory.

Keywords: back-biting, cyclic polymers, equilibration, polycondensation, ring-chain equilibrium

Introduction

In the classical theory of step-growth polymerizations elaborated by Carothers[1] and Flory[2,3] no differentiation was made between kinetically controlled polycondensations (KCPs) and thermodynamically controlled ones (TCPs). Characteristic for TCPs are rapid equilibration reactions (e.g. transesterification or transamidation) and Stockmayer and coworkers proved the reversible formation of cyclic oligomers and polymers via "back-biting".[4,5] As a consequence of the Jacobson-Stockmayer (J.S.) theory TCPs were exclusively described in terms of ring-chain equilibria.[6,7] This J.S. theory was and is understood as a modification of the Carothers-Flory (C.F.) theory which predicts on the basis of equation (1) that an ideal polycondensation reaching 100% conversion will yield one giant chain containing all monomeric units. Modification with the J.S. theory means that the giant chain obtained at

 DOI: 10.1002/masy.200350902

100% conversion is at one or both ends in equilibrium with cyclic oligomers. The purpose of the present study is to propose and to discuss an alternative hypothesis saying that a proper understanding of TCPs requires a description in terms of ring-ring equilibria.

$$\overline{\mathrm{DP}} = \frac{1}{1-\mathrm{p}} \qquad (1)$$

$\overline{\mathrm{DP}}$ = average degree of polymerization

p = conversion

Results and Discussion

Experimental Results

On the basis of the J.-S. theory Flory has calculated[3] that a TCP conducted in bulk will contain around 2.5 weight% of cycles and one giant chain representing 97.5 weight% of the entire reaction product. In this article several previously published polycondensations should be cited which demonstrate that the reaction products do not contain a giant chain with 97-98 weight% of the entire mass, but mainly consist of cycles. In this connection, it should be mentioned that the final equilibrium of a TCP at 100% conversion does not depend on the reaction pathway, because no chemical equilibrium depends on the synthetic method and its kinetic implication. Therefore, an equilibration of monomeric or oligomeric cycles with a small amount of a catalyst (or without) is equivalent to a polycondensation with 100% conversion.

Recently, Brady et al.[8] studied the polycondensation of methyl esters of cholic acid derivatives (**1**) using solubilized potassium methoxide as transesterification catalyst. Those authors found an equilibrium of cyclic oligoesters free of high molecular weight linear polymers. Furthermore, a cyclic oligoester was isolated by HPLC and equilibrated with potassium methoxide. The same ring-ring equilibrium as that resulting from the polycondensation of **1** was obtained.

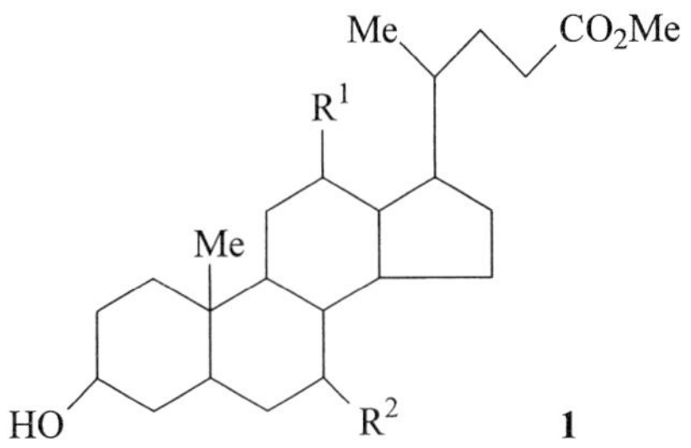

$$\text{Me, } CO_2Me,\ R^1,\ \text{Me},\ HO,\ R^2$$

1

$$Bu_2Sn(OMe)_2 \xrightarrow[-2\ MeOH]{+\ H\left(OCH_2CH_2\right)_nOH} Bu_2Sn\overbrace{\left[O-CH_2CH_2\right]_{n-1}O-CH_2CH_2-O}$$

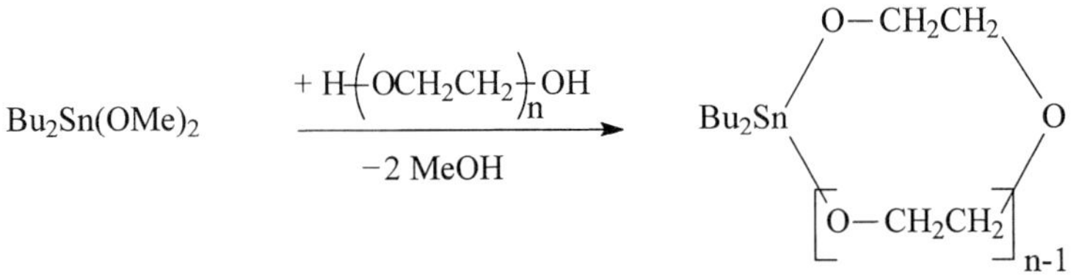

2

Polycondensations of dibutyltin dimethoxide with various monodisperse and polydisperse oligo- or poly(ethylene glycol)s were conducted by Kricheldorf and Langanke.[9] Regardless of the chain length of the diol mixtures, cyclic oligoethers (**2**) were isolated and no high molecular weight polyethers. Analogous results were obtained with poly(tetrahydrofuran)diols.[10] The equilibration of tin-alkoxide groups in these systems does not require a catalyst, because the exchange of alkoxide groups proceeds via O→Sn donor-acceptor interactions, even at room temperature.[11] The cyclic structure of tin-alkoxides, such as **2**, is proven by a variety of reactions and measurements,[12,13] but the sensitivity of the Sn-O bond to hydrolysis and alcoholysis prevents the direct characterization by "fast atom bombardment" or MALDI-TOF mass spectrometry.

Exclusively cyclic oligomers and no high molecular weight species (**3**) were isolated from polycondensations of dibutyltin bisacetate with various aliphatic dicarboxylic acids.[14]

Three different synthetic methods were compared which all yielded the same thermodynamically controlled reaction products.

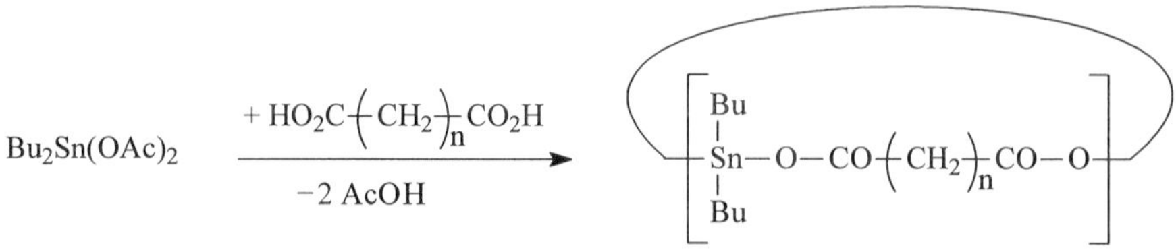

$$Bu_2Sn(OAc)_2 \xrightarrow[-2\,AcOH]{+\,HO_2C\left(CH_2\right)_nCO_2H} \left[-Sn(Bu)_2-O-CO\left(CH_2\right)_nCO-O- \right]_{cyclic}$$

3

$$\left[-O-Si(Me)_2- \right]_n^{cyclic} \xrightarrow{(R-O^{\ominus})} \left[-O-Si(Me)_2- \right]_x$$

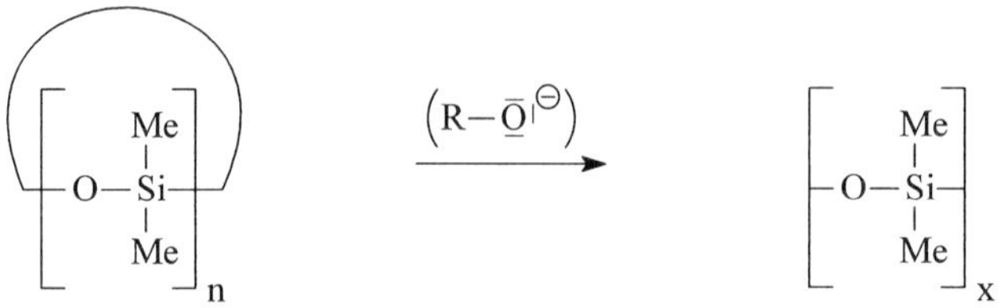

4

The equilibration of cyclosiloxanes in concentrated solution (23-24 weight%) or in bulk was studied by numerous authors.[6,15-17] All these studies were performed at a time when MALDI-TOF mass spectrometry was not available, so that the characterization of the small fraction of high molecular weight polysiloxanes was not feasible in those studies. However, the high fraction of clearly identified cyclosiloxanes (70-95 weight%) found in all studies is in contradiction to the small percentage expected from the J.S. theory.[3]

Finally, polycondensations of dibutyltin dimethoxide with penthaerythritol (**5a**) and ethoxylated derivatives (**5b,c**) should be mentioned.[18-20] According to the J.-S. and C.-F. theories, networks should be formed above 65% conversion. An infusible and insoluble gel was indeed obtained from **5a**, obviously because the 6-membered rings of the hypothetical spirocyclic oligomers **6a** possess ring strain. This ring chain shifts the ring-ring equilibrium to high molecular weight polymers which, in this case, are necessarily networks. However, the ethoxylated derivatives **5b** and **5c** yielded strain-free spirocycles (**6b** + **6c** and higher oligomers). At this point it should be mentioned that aliphatic ethers favor gauche conformations (in contrast to alkane chains) which, in turn, favor the formation of loops and larger cycles. The ring-ring equilibrium is then shifted to the side of the spirocycles, because the degradation of a gel to spirocycles includes a high gain in entropy.

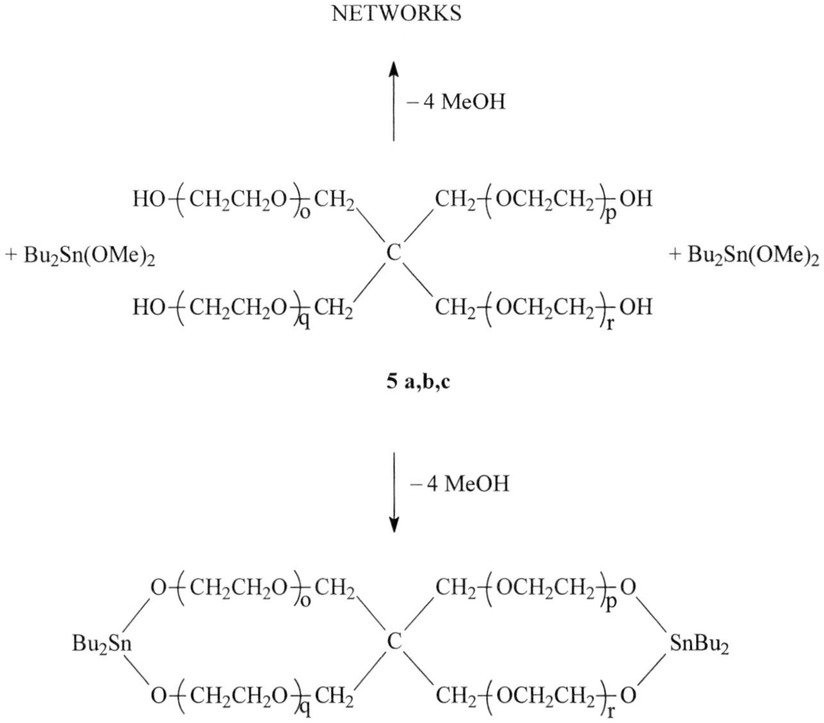

NETWORKS

↑ – 4 MeOH

$HO{-}(CH_2CH_2O)_o{-}CH_2$ $CH_2{-}(OCH_2CH_2)_p{-}OH$

$+ Bu_2Sn(OMe)_2$ C $+ Bu_2Sn(OMe)_2$

$HO{-}(CH_2CH_2O)_q{-}CH_2$ $CH_2{-}(OCH_2CH_2)_r{-}OH$

5 a,b,c

↓ – 4 MeOH

$O{-}(CH_2CH_2O)_o{-}CH_2$ $CH_2{-}(OCH_2CH_2)_p{-}O$

Bu_2Sn C $SnBu_2$

$O{-}(CH_2CH_2O)_q{-}CH_2$ $CH_2{-}(OCH_2CH_2)_r{-}O$

6 a,b,c
(+ higher oligomers)

a: $o + p + q + r = 0$

b: $o + p + q + r = 3$

c: $o + p + q + r = 15$

Discussion

The experiments presented above have in common that the reaction products were mixtures of cyclic oligomers and polymers and a significant fraction of high molecular weight linear polymers was not detectable. Hence, these results are in contradiction to the J.S. and C.F. theories, but agree with our theory of ring-ring equilibria.

Any TCP involves three kinds of equilibria: the chain-chain, the ring-chain and the ring-ring equilibrium because all components of a reaction mixture are engaged in the equilibration process. A description of TCPs exclusively in terms of a ring-chain equilibrium is a simplification which is helpful to simplify the mathematical treatment, but it is misleading when the consequences of high conversions are considered. The molar concentration of linear active species (including monomers) rapidly decreases with higher conversions as formulated in equation (2). Therefore, the molar ratio of cycles versus linear molecules permanently increases and all reaction products are necessarily cycles at 100% conversion. The assumption that 100% conversion yields one giant chain is a contradiction in itself, which was born at the time of Carothers when cyclization of long chains was considered to be almost impossible.[1,2] However, in connection with studies of KCPs, we have recently demonstrated[21] that polycarbonates with a DP of 200 easily cyclize and this DP corresponds to a DP of 10^3 for polyolefins. Further improvements of the mass spectrometry will certainly expand the limits for the identification of cyclic polymers to still higher masses.

$$[\mathrm{La}]_p = [\mathrm{La}]_o (1 - p) \tag{2}$$

La = linear active species including monomers

$$c(M)_x + c(M)_y \underset{}{\overset{K_c}{\rightleftharpoons}} c(M)_{x+y} \tag{3}$$

$$\overline{DP} = \frac{1}{1 - p(1 - X^{-K_c})} \tag{4}$$

When with increasing conversion the cycles begin to outnumber the linear species, a TCP may be described as a ring-ring equilibrium, because the chain-chain and ring-chain equilibria lose influence on the thermodynamical properties of the system. The final state of a TCP at 100% conversion is a neat ring-ring equilibrium which allows a schematic formulation by equation (3). The assumption of one single equilibrium constant (K_c) is, of course, a simplification, but it facilitates to illustrate the influence of ring-ring equilibria on the chain growth in a TCP. This influence is formulated in equation (4). The chain growth of an ideal TCP is limited by the ring-ring equilibrium, but in the case of rigid chains with high K_c values, equation (4) approaches the classical "Carothers equation" (1).

On the other hand, K_c values close to 0 indicate an almost complete formation of strain-free monomeric cycles as it is known from polycondensations of γ-hydroxy- or γ-aminobutyric acid which yield the thermodynamically stable γ-butyrolactone or γ-butyrolactam in nearly 100% yield. This short discussion demonstrates that the theory of ring-ring equilibria covers the full range of thermodynamically controlled cyclizations and polycondensations known from organic and macromolecular chemistry.

Finally, it should be discussed, why the fraction of cycles in technical TCPs are relatively low. Such technical TCPs conducted in bulk are, for instance, the syntheses of poly(ethylene terephthalate), poly(butylene terephthalate), nylon-6, nylon-6,6 and the syntheses of polycarbonates by transesterification of bisphenols with diphenyl carbonate. In all these cases, the content of cycles is seemingly only in the order of 2-5 weight% corresponding to 5-15 mol%. However, for technical polycondensates a high molecular weight is not desired, because a high melt viscosity is unfavorable for any processing from the melt. Therefore, the conversion is usually stopped around 98-99% or chain terminators are added. Furthermore, numerous side reactions may occur at high temperatures, such as decarboxylation, formation of vinyl and ether groups, formation of five-membered cycles (e.g. tetrahydrofuran from 1,4-butanediol) and Fries-rearrangement (in the case of aromatic ester or carbonate groups). Moreover, previous studies of cycles in technical TCPs are based on extraction of cyclic oligomers. These extractions are far from complete, so that these previous studies considerably underestimate the real content of cyclic oligomers and polymers. For all these reasons the rather low content of cyclic oligomers and polymers in technical polycondensates cannot serve as serious argument against the ring-ring equilibrium theory discussed above.

Conclusion

The total equilibration of all reaction partners and products in a TCP may be described by three kinds of equilibria: chain-chain, ring-chain and ring-ring equilibria. Whereas the J.S. theory[4] and its extension by Flory[3] is focussed on ring-chain equilibria, the present study emphasizes the role of ring-ring equilibria at high conversions. Increasing conversion means higher ring/chain ratios until all linear species disappear at 100% conversion. Considering our recent theory of KCPs[22,23] the following conclusion may be drawn: "It is the fundamental tendency of all step-growth polymerizations to yield cyclic oligomers and polymers as stable endproducts. The linear chains are nothing else but the reactive intermediates". A more detailed discussion of this topic has been published.[24]

[1] W.H. Carothers, *J. Am. Chem. Soc.* **1929**, *51*, 2548.
[2] P.J. Flory, *Chem. Rev.* **1946**, *39*, 137.
[3] P.J. Flory *"Principles of Polymer Chemistry"* Cornell University Press, Ithaca, N.Y., London 1953, Chapters III and VIII.
[4] H. Jacobson , W.H. Stockmayer, *J. Chem. Phys.* **1950**, *18*, 1600.
[5] H. Jacobson, C.O. Beckmann, W.H. Stockmayer, *J. Chem. Phys.* **1950**, *18*, 1607.
[6] J.A. Semlyen (ed.) *Cyclic Polymers*, Elsevier Applied Science, London 1986.
[7] G. Odian, *"Principles of Polymerization"* Wiley, Singapore 1991, 3rd. ed., Chapter 2.
[8] P.A. Brady, R.P. Bonar-Law, S.L. Rowan, C.J. Suckling, K.M. Sanders, *Chem. Commun.* **1996**, *319*.
[9] H.R. Kricheldorf, D. Langanke, *Macromol. Chem. Phys.* **1999**, *200*, 1174.
[10] H.R. Kricheldorf, D. Langanke, *Macromol. Chem. Phys.* **1999**, *200*, 1183.
[11] A.G. Davies *"Organotin Chemistry"*, VCH Publishers Weinheim, N.Y., 1997, Chapter 12.
[12] H.R. Kricheldorf, S.-R. Lee, N. Schittenhelm, *Macromol. Chem. Phys.* **1998**, *199*, 273.
[13] H.R. Kricheldorf, M. Al Masri, G. Schwarz, *Macromolecules*, in press.
[14] H.R. Kricheldorf, S. Böhme, R.-P. Krüger, *Macromol. Chem. Phys.* **2002**, *203*, 313.
[15] J.F. Brown, G.M.J. Slusarczuk, *J. Am. Chem. Soc.* **1965**, *87*, 931.
[16] S.J. Clarson, J.A. Semlyen, *POLYMER* **1986**, *27*, 91.
[17] J.A. Semlyen in *"Large Ring Molecules"* (J.A. Semlyen, ed.) J. Wiley & Sons, Chichester, N.Y., London 1996, Chapter 1.
[18] H.R. Kricheldorf, S.-R. Lee, *Macromolecules* **1996**, *29*, 8689.
[19] H.R. Kricheldorf, B. Fechner, *Biomacromolecules* **2002**, *3*, 691.
[20] A. Finne, A.-C. Albertsson, *Biomacromolecules* **2002**, *3*, 684.
[21] H.R. Kricheldorf, S. Böhme, G. Schwarz, C.-L. Schultz, *Macromol. Rapid Commun.* **2002**, *23*, 803.
[22] H.R. Kricheldorf, M. Rabenstein, M. Maskos, M. Schmidt, *Macromolecules* **2001**, *34*, 713.
[23] H.R. Kricheldorf, S. Böhme, G. Schwarz, *Macromolecules* **2001**, *34*, 8879.
[24] H.R. Kricheldorf, *Macromolecules* **2003**, in press (Macrocycles 21.).

Nonstoichiometric Polycondensation I. Synthesis of Polythioether from Dibromomethane and 4,4'-Thiobisbenzenethiol

*Hirokazu Iimori, Yuji Shibasaki, Shinji Ando, Mitsuru Ueda**

Department of Organic and Polymeric Materials, Graduate School of Science and Engineering, Tokyo Institute of Technology, 2-12-1 O-okayama, Meguro-ku, Tokyo 152-8552, Japan

Summary: High molecular weight poly(phenylene thioether) (**3**) was successfully obtained by the polycondensation of 4,4'-thiobisbenzenethiol (**1**) and dibromomethane (**2**) with a variety of feed ratios in the presence of 1,8-diazabicyclo[5,4,0]undec-7-ene (DBU) in 1-methyl-2-pyrrolidinone (NMP) at 75 °C. The resulting polymer showed the maximum inherent viscosity (η_{inh}) of 0.50 dL/g in 4 h when 1.5 equivalents excess of **2** was used. The model reaction using benzenethiol (**4**) and dichloromethane (**5**) in the presence of DBU in deuterated dimethylsulfoxide (DMSO-d_6) at 25 °C indicated that the rate of the second nucleophilic displacement reaction (k_2) is 61 times faster than that of the first one (k_1). The maximum of theoretical molecular weights calculated at various stoichiometric imbalance (S) under the condition of $k_2/k_1 = 61$ showed a good agreement with the experimental molecular weights at specific polymerization times.

Keywords: dibromomethane, kinetics, polycondensation, poly(phenylene thioether), stoichiometric imbalance

Introduction

In the synthesis of condensation polymers, the number-average degree of polymerization X_n is expressed in an extension of the Carothers equation[1-2] as

$$X_n = (1 + S) / (1 + S - 2p),$$

where S ($\geqq$ 1) is the stoichiometric ratio of functional groups and, p is the extent of reaction. Thus, a stoichiometric balance of monomers is a critical factor to obtain a high molecular weight polymer. This classical theory is based on the concept of equal reactivity of functional

 DOI: 10.1002/masy.200350903

groups. Thus, if the reactivity of a functional group is dependent on whether the other functional groups in the monomer have reacted, another theory should be applied for showing the relationship between X_n and S.

In the synthesis of poly(*p*-phenylene sulfide) by Edmons and Hill method,[3] significant deviations from classical theory was found, although the growth of the polymerization may be written as a series of conventional polycondensation; that is, polymer yields and molecular weights are adequate even at incomplete monomer conversion, and perfect stoichiometric monomer ratios are not required to achieve high molecular weight polymers. Odian et al. studied the kinetic of the polymerization with the change in reactivity of one functional group upon reaction of the other,[4] and showed the relationship between the degree of polymerization with time under various ratios (κ) of the reaction rate constants of the first and the second reacting functional groups. Kihara et al. investigated the polymerization of 2,2-dichloro-1,3-benzodioxole with 4,4'-isopropylidenediphenol and found that the highest molecular weight of polyorthocarbonate was obtained at 0.7 equivalent excess of 2,2-dichloro-1,3-benzodioxole.[5] Based on this finding, they concluded that the degree of polymerization is enhanced by stoichiometric imbalance if the first condensation of bifunctional monomer enhances the second condensation of the remaining functional group. Nomura et al. also reported the palladium catalyzed allylic substitution polymerization, where the unexpectedly high molecular weight polymer was obtained in non-stoichiometric conditions.[6] Quite recently, Hay et al. reported that high molecular weight aromatic polyformals were obtained by the polycondensation of potassium bisphenolate with excess amounts of **2** and pointed out the first intermediate, the bromomethyl ether, was much more reactive than **2**.[7] This polymerization would be another example to show the stoichiometric imbalance-enhanced polymerization. This system, however, is heterogeneous, and not suitable for conducting a kinetic study because too many kinetic parameters must be considered.

In the present study, we performed kinetic investigations on the polycondensation of **1** and **2** in the presence of DBU by changing the feed ratio of each monomer and found that this polycondensation is also the stoichiometric imbalance-enhanced polymerization.

Experimental Part

Materials. 4,4'-Thiobisbenzenthiol (**1**), dibromomethane (**2**), benzenethiol (**4**), dichloromethane (**5**), 1,8-diazabicyclo[5,4,0]undec-7-ene (DBU), and 1-methyl-2-pyrrolidinone (NMP) were distilled prior to use. Other reagents and solvents were used without further purification.

Measurements. FT-IR spectra were measured on a Horiba FT-720 spectrometer. Viscosity measurements were carried out by using an Ostwald viscometer in NMP at 30 °C. Thermal analyses were performed with a Seiko SSS5000 TG-DTA220 thermal analyzer at a heating rate of 10 °C min^{-1} for thermogravimetry (TGA) and a Seiko SSS5000 DSC220 at a heating rate of 10 °C min^{-1} for differential scanning calorimetry (DSC) under a nitrogen atmosphere, respectively. Solution state ^{1}H NMR spectra were recorded with a Bruker DPX300S spectrometer operating at 300.0MHz. Solid-state $^{13}C\{^{1}H\}$ cross-polarization magic angle spinning (CP/MAS) NMR spectra were recorded with a JEOL GSX-300 spectrometer operating at 75.45 MHz. The total suppression of spinning sideband (TOSS) pulse sequence was used to suppress spinning sidebands. The pulse sequence that combines TOSS and the dipolar dephasing (non-quaternary carbon suppression) (TOSS & DD) was also used for selective observation of non-protonated carbons. The contact time for CP was 2 ms, and the recycle was 5 s. The numbers of scans were 1500-1800. The chemical shifts were calibrated indirectly through the adamantane peak observed at lower frequency (29.5 ppm relative to tetramethylsilane (TMS)).

Model reaction using 4 and 5. Into a 1 mL of glass bottle were placed 0.917 g (8.32 mmol) of **4** and 1.270 g (7.30 mmol) of **5**. This was well mixed, and 40 μL of it was transferred to a new glass bottle. The mixture was diluted with 1 mL of dimethylsulfoxide-d_6 (DMSO-d_6) containing 1 μL of *tert*-butylanisole as an internal standard. From the bottle, 0.45 mL of the solution was transferred into a NMR test tube to determine the exact S value (2 $[\mathbf{5}]_0$ / $[\mathbf{4}]_0$) as 1.717 (function 2 arises from the mono functional group of **4**). The addition of 40 μL (2.7×10^{-4} mmol) of DBU initiated the nucleophilic reaction that was monitored by the change of integration ratio between signals derived from **5** and bis(phenylsulfanyl)methane (**7**) in ^{1}H NMR spectra.

Polycondensation of 4,4'-thiobisbenzenethiol (1) and dibromomethane (2). Into a two-necked flask equipped with a reflux condenser were placed 0.5 g (2 mmol) of **1**, 2 mL of NMP, and 0.7 mL (4.7 mmol) of DBU under nitrogen stream. The solution was warmed to 75 °C, and then, the set amounts of **2** were added. After 2 h or 4 h, the viscous solution was diluted with NMP and poured into methanol to give white fibrous polymer. This was dried in vacuo at 70 °C for 24 h. Yield: 78–96 %. η_{inh} = 0.35–0.50 dL/g in NMP at 30 °C. IR (KBr): 744, 713 (CSC). ^{13}C NMR (solid): δ = 29.7(methylene) 126.4 (phenyl), 127.2 (phenyl), 132.6(phenyl), 135.3 (s phenyl). $(C_{13}H_{10}S_3)_n(264.2)_n$: Calcd. C 59.5, H 3.84, S 36.6; Found C 59.8, H 4.29, S 36.8.

Results and Discussion

Estimation of κ by model reaction. To conduct the polycondensation of **2** with a nucleophilie in homogeneous state, DBU was selected as an organic base in place of potassium hydroxide. Furthermore, bisphenol (pKa = 10) as the nucleophilie was replaced to **1** due to the higher acidity (pKa = 7), which means that complete formation of thiolate anion is possible by using DBU (pKa = 12.5).

First, the model reaction of **4** with **2** was carried out in the presence of DBU in DMSO-d_6 for estimating the ratio (κ) of the rates of the first (k_1) and second nucleophilic reactions (k_2) between **1** and **2**. The reaction was monitored by successive measurements of NMR spectra of a drop of the reaction mixture at different times. The reaction, however, was too fast even at -20 °C. Thus, the model reaction using **5** in place of **2** was performed at 25°C (Scheme 1).

$$\underset{\mathbf{4}}{\text{Ph–SH}} + \underset{\mathbf{5}}{CH_2Cl_2} \xrightarrow[\text{DMSO-}d_6]{k_1\ \ \text{DBU}} \underset{\mathbf{6}}{\text{Ph–S–}CH_2Cl}$$

$$\underset{\mathbf{6}}{\text{Ph–S–}CH_2Cl} + \underset{\mathbf{4}}{\text{Ph–SH}} \xrightarrow[\text{DMSO-}d_6]{k_2\ \ \text{DBU}} \underset{\mathbf{7}}{\text{Ph–S–}CH_2\text{–S–Ph}}$$

Scheme 1

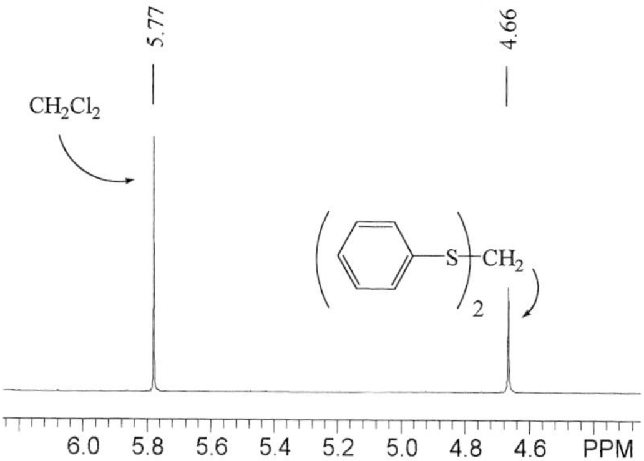

Fig. 1. Expanded ^{1}H NMR spectrum of the reaction mixture of **4** and **5** in DMSO-*d6* at 25 °C.

In the ^{1}H NMR spectra of the reaction mixture (Figure 1), product **7** was observed, but active intermediate **6** was not detected in any case. This suggests that k_2 is significantly larger than k_1. The ratio κ of the two rate constants (k_2/k_1) was determined from the time-dependent consumption of **5** according the theory in Appendix I.

Figure 2 shows the variations of the ratio (β)of the concentration of **5** at reaction time t ([**5**]) to the initial concentration of **5** ([**5**]$_0$).

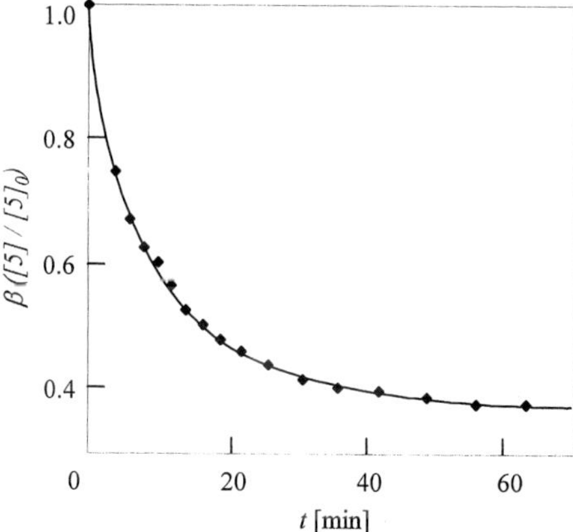

Fig. 2. Relationship between reaction time (t) and β ([**5**]/[**5**]$_0$) determined by ^{1}H NMR.

Using the array of β values obtained from the experiment, the corresponding τ values can be calculated for the array of t according to Eq.(14) (Appendix I) by assuming a certain κ value. As defined in Eq.(12), τ should be linearly proportional to t. Hence, the calculated τ values are plotted against t for a series of κ values, and the correlation coefficients (R^2) obtained from the fitting of linear equations using the least square method were evaluated as shown in Figure 3. Figure 4 shows the calculated R^2 values plotted against κ. Since the maximum R^2 was obtained at $\kappa = 61$, the second nucleophilic reaction was estimated to be 61 times faster than the first reaction.

Polymerization of 1 with 2. The polymerization of **1** with **2** was carried out with changing the molar ratio of each monomer. A solution of **1** and excess amounts of **2** in the presence of DBU in NMP was vigorously stirred at 75 °C for a specified time (Scheme 2).

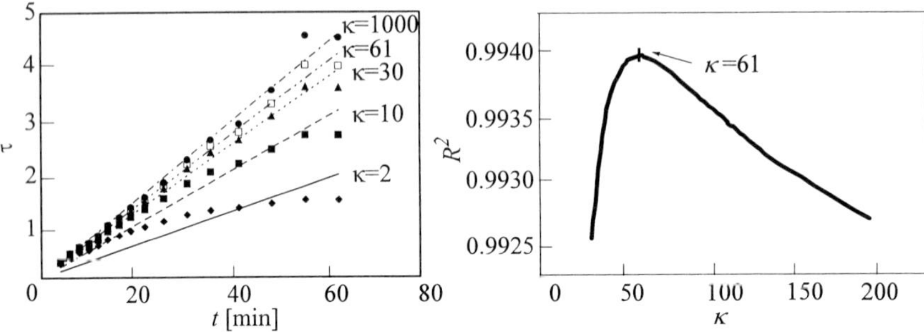

Fig. 3. Relationship between real time (t) and normalized time (τ) for various κ.

Fig. 4. Relationship between κ (k_2/k_1) and correlation coefficient R^2 (see text).

HS–C$_6$H$_4$–S–C$_6$H$_4$–SH (**1**) + CH_2Br_2 (**2**) $\xrightarrow[\text{NMP } 75^\circ\text{C}]{\text{DBU}}$ $\left(\text{S–C}_6\text{H}_4\text{–S–C}_6\text{H}_4\text{–S–CH}_2\right)_n$ (**3**)

Scheme 2

Figures 5(a) and 5(b) illustrate the dependence of η_{inh} of the polymers on the stoichiometric imbalance S for polymerization times 2 h and 4 h, respectively.

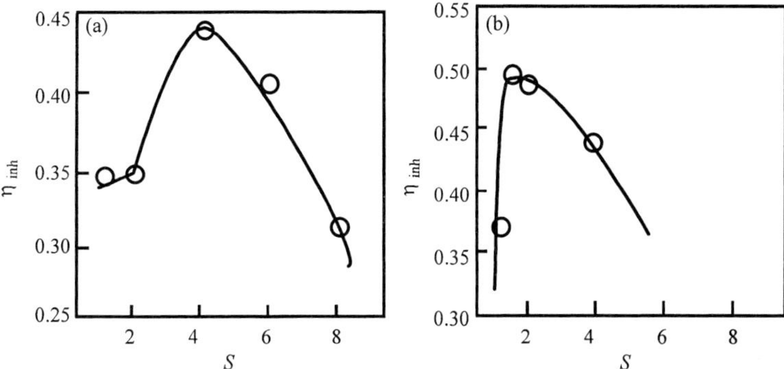

Fig. 5. Relationship between the monomer feed ratio S and the inherent viscosity η_{inh} of polymers obtained by the polycondensations for (a) 2 h and (b) 4 h.

The highest η_{inh} of the polymer was achieved for $S = 4$ at t =2 h and for $S = 1.5$ at t =4 h. These facts clearly indicate that the polymers having higher molecular weights an be obtained under non-stoichiometric conditions ($S > 1$), and the effect of non-stoichiometry is more significant at the shorter reaction time. In addition, the η_{inh} (0.44 dL/g) obtained for $S = 4$ at t = 2 h was not changed at $t = 4$ h, which indicates that the polymerization had already been terminated at 2 h under the highly non-stoichiometric condition. In contrast, the η_{inh}s (0.36 and 0.35 dL/g) obtained for $S = 1.5$ and 2 at 2 h are appreciably smaller than those at 4 h (η_{inh} = 0.50 and 0.48), indicating that the polymerization was still in progress at 2 h under the conditions with relatively smaller S values.

Time evolution of theoretical molecular weights. Theoretical number-averaged molecular weights (M_n) of polymers were calculated according to the theory in Appendix II by assuming that the second nucleophilic reaction is 61 times faster than that of the first reaction ($\kappa = 61$). Figures 6 shows the evolution of M_n with respect to the normalized polymerization time (τ) under different stoichiometric imbalance conditions. The highest M_n is realized in stoichiometric conditions ($S = 1$) at infinite time ($\tau > 110$), and it is obtained in larger stoichiometric imbalane at smaller τ. These phenomena support the experimental results; the η_{inh} values of the obtained polymers at $t = 4$ h decrease in the order of S =1.5 > 2.0 > 4.0 > 1.0, and those at $t = 2$ h decrease as S = 4.0 > 6.0 > 2.0 > 8.0 (Figure 5). These orders can be seen from 6.0 to 20 and 1.75 to 2.25 in τ, respectively (Figures 6 (b) and (c)). These indicate that at shorter polymerization time polymers obtained under larger stoichiometric imbalance show higher molecular weights when the second nucleophilic reaction is faster than the first one.

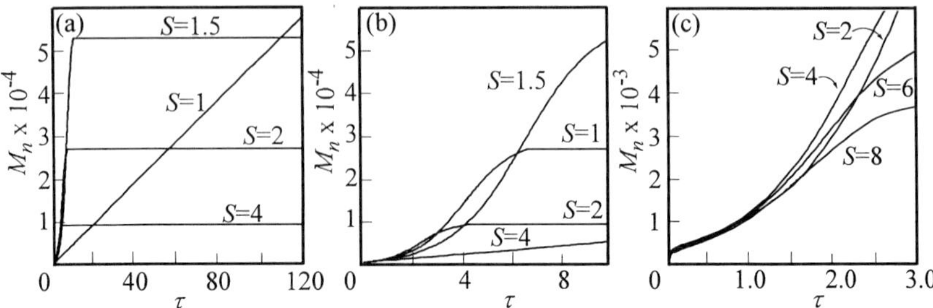

Fig. 6. Relationship between normalized polymerization time (τ) and theoretical M_n of **3** calculated for various S values using $\kappa = 61$. Expansion of the initial part is shown in (b) $\tau = 0 - 10$, and (c) $\tau = 0 - 3$.

Table 1. Optimized Geometries and Parameters Calculated from Density Functional Theory[a)]

		Cl, Cl, C, H H	PhS, Cl, C, H H	Br, Br, C, H H	PhS, Br, C, H H
Interatomic distance of C-X [angstrom]		1.7906	1.8290	1.9520	2.0022
Bond order of C-X		0.2086	0.1805	0.2237	0.1875
Net atomic charge on	carbone	-0.5312	-0.6183	-0.6041	-0.6495
	halogen	-0.0340	-0.0946	+0.0078	-0.0652

[a)] B3LYP/6-311G*.

Molecular orbital consideration on reactive intermediate. The density functional theory (DFT) calculations using the B3LYP hybrid functional with the 6-311++G(d,p) basis set were performed to elucidate the reactivity of the monomers and the intermediates (Table 1). The C–Cl bond length of the optimized geometries of intermediate **6** is longer than that of **5**, and the C–Cl bond-order of the former is smaller than that of the latter. In addition, the chlorine atom in **6** is more negative than those in **5**. All these parameters indicates that the C–Cl bond in **6** is more weaker than those in **5**, and intermediate **6** is more reactive than **5**. Similar phenomena are also observed for the bromide compounds as shown in Table 1, supporting the experimental results observed in the polymerization reactions.

Polymer characterization. Polymer **3** is a white fibrous solid and soluble only in NMP. The TGA curve (Figure 7(a)) shows that the 5 % weight loss temperature is 325 °C, and a large endothermic peak due to the melting was observed at 144 °C in the DSC curve (Figure 7(b)). This indicates that the resulting product is a semi-crystalline polymer. The chemical structure was characterized as the desired poly(phenylene thioether) by IR, NMR, and elemental analyses as described in the experimental part. The IR spectrum shows characteristic absorptions at 744 and 713 cm^{-1} due to C-S-C stretching. Elemental analysis

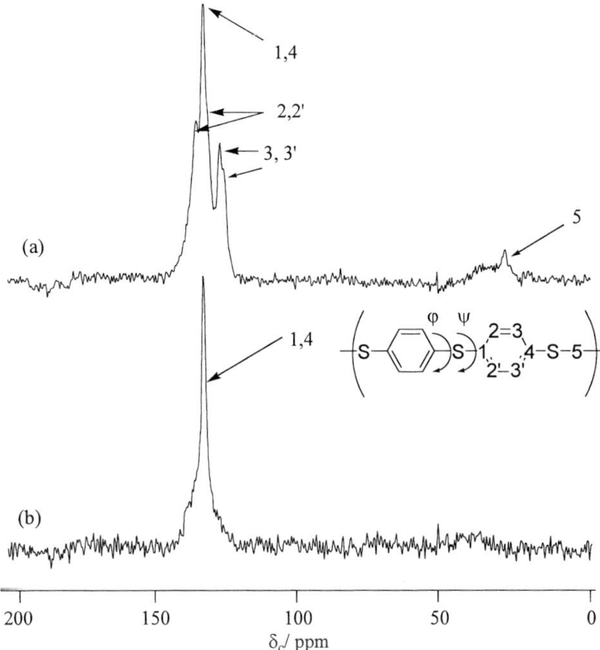

Fig. 7. Solid-state ^{13}C CP/MAS NMR spectra of **3** obtained using (a) TOSS and (b)TOSS & DD pulse sequences.

also supported the formation of expected polymer. The solid state ^{13}C CP/MAS spectrum (Figure 8(a)) shows well-resolved signals for the phenyl rings and a broad signal due to the methylene units. The NMR signals of the quaternary carbons observed in the TOSS&DD spectrum (Figure 8(b)) can be fitted by overlap of a sharp Lorentian and a broad Gaussian signals having the same chemical shifts (133.6 ppm). The former was assigned to carbon 1, and the latter was assigned to carbon 4. The broad signals of carbons 4 and 5 suggest the large distribution of the conformations at the phenyl-S-CH_2-S-phenyl moiety. In contrast, the sharp signals for carbons 1, 2, 2', 3, and 3' indicates the restricted conformation at the diphenyl thioether moiety. Furthermore, the single peak of carbon 1 and the split peaks of 2,2' and 3,3' indicate that the diphenyl thioether structure take the C_2 symmetry (propeller-like conformation), and the dihedral angles (ϕ, ψ) are estimated as (45, 45°)–(60°, 60°) according to the DFT calculations of the nuclear shieldings for dipenyl thioether.[8]

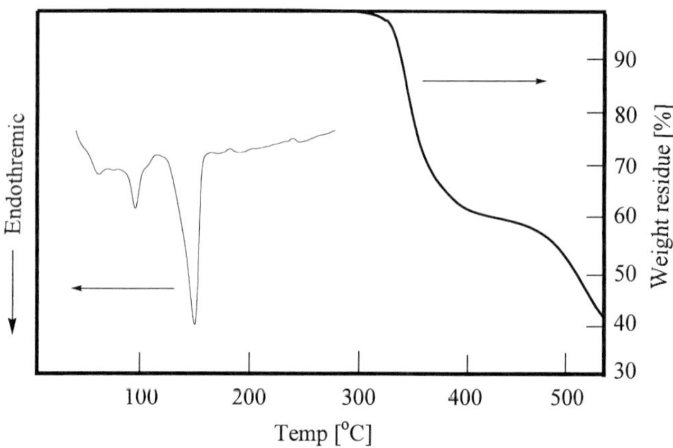

Fig. 8. TG and DSC traces of **3** measured under nitrogen.

Conclusions

The kinetics of polycondensation of **1** and **2** with a variety of monomer feed ratios in the presence of DBU in NMP at 75 °C was investigated. It was found that this polycondensation is a stoichiometric imbalance-enhanced polymerization, that is, the first condensation of **1** with the thiophenol group of **2** enhances the reactivity of the resulting bromomethyl thioether, and high molecular weight polymer **3** is obtained even in the presence of excess amount of **2**.

The ratio κ of two rate constants (k_2/k_1) was determined as 61 from the reaction of **4** and 5 in the presence of DBU in DMSO-d_6 at 25 °C. The time-evolution of the molecular weights of **3** calculated at various S values agrees well with inherent viscosities of polymers obtained by polycondensations for 2 h and 4 h. In addition, the higher reactivities of the intermediates compared with those of the monomers are rationalized by the density functional theory calculations.

[1] Flory, P. J. Principles of Polymer Chemistry; Cornell University Press: Ithaca, NY. 1953.
[2] Odian, G. Principle of Polymerization, 3rd ed; Wiley: New York, 1991; Chapter 2.
[3] Hill, H.; Wayne, Jr.; Edmonds, J. T., Jr. *Polym. Prepr., Amer. Chem. Soc., Div. Polym. Chem.* **1972**, *13(1)*, 603.
[4] Ozizmir, E.; Odian, G. *J. Polym. Sci., Polym. Chem. Ed.* **1980**, *18(3),* 1089.
[5] Kihara, N.; Komatsu, S.; Takata, T.; Endo, T. *Macromolecules*, **1999**, 32, 4776
[6] Nomura, N; Tsurugi, K; Okada, M. *Angew. Chem. Int. Ed.* **2001**, 40(10), 1932.
[7] Miyatake, K.; Hlil, A. R.; Hay, A. S. *Macromolecules*, **2001**, 34, 4288.
[8] Aimi, K.; Yamane A.; Ando, S. *J. Mol. Struc.*, **2001**, *602-603*, 417.

[Appendix 1]

Since the model reaction of **4** and **5** (Scheme 2) proceeds as a S_N2-type reaction, the kinetics can be expressed by the following equations.

$$-\frac{d}{dt}[\mathbf{5}] = k_1[\mathbf{4}][\mathbf{5}] \tag{1}$$

$$-\frac{d}{dt}[\mathbf{4}] = k_1[\mathbf{5}][\mathbf{4}] + k_2[\mathbf{6}][\mathbf{4}] \tag{2}$$

$$-\frac{d}{dt}[\mathbf{6}] = -k_1[\mathbf{4}][\mathbf{5}] + k_2[\mathbf{6}][\mathbf{4}] \tag{3}$$

The initial concentration of each compound is defined as $[\mathbf{4}]_0$, $[\mathbf{5}]_0$, $[\mathbf{6}]_0$, and $[7]_0$, respectively, where $[\mathbf{6}]_0=0$ and $[7]_0=0$. The variables α, β, γ and δ are introduced as

$$\alpha = \frac{[\mathbf{4}]}{[\mathbf{4}]_0},\quad \beta = \frac{[\mathbf{5}]}{[\mathbf{5}]_0},\quad \gamma = \frac{[\mathbf{6}]+[7]}{[\mathbf{5}]_0},\ \text{and}\ \delta = \frac{[\mathbf{6}]}{[\mathbf{5}]_0}, \tag{4)–(7}$$

and the parameters κ is defined as

$$\kappa = \frac{k_2}{k_1} \tag{8}$$

According to the definition of S, represented as follows in the model reaction

$$S = 2\frac{[\mathbf{5}]_0}{[\mathbf{4}]_0}. \tag{9}$$

Using the relationships, $[\mathbf{5}]_0 = [\mathbf{5}] + [\mathbf{6}] + [7]$ and $[\mathbf{4}]_0 = [\mathbf{4}] + [\mathbf{6}] + 2[7]$, γ and δ can be expressed as

$$\gamma = 1-\beta \text{ and } \delta = \alpha - 1 + S(1-\beta). \qquad (10),(11)$$

In addition, the normalized dimensionless time τ is defined as

$$\tau = k_1[\mathbf{5}]_0 t. \qquad (12)$$

By solving the coupled equations (1) and (3), α value at a specific β can be expressed as

$$\alpha = 1 - S + \frac{S\beta}{2(\kappa-1)}(2\kappa - 1 - \beta^{\kappa-1}). \qquad (13)$$

Furthermore, Eq.(1) can be solved as

$$\tau = \int_1^{1/\beta} \frac{\mathrm{d}w}{(\frac{2}{S} - 2)w + \frac{1}{\kappa - 1}(2\kappa - 1 - w^{1-\kappa})} \qquad (14)$$

The right side of Eq. (14) can be numerically calculated using Romberg's method. The computation program was written in Microsoft Basic language, and all non-integer variables were defined as double precisions in the program. The criteria for the convergence of iterative integrations was $1.0\text{x}10^{-12}$.

[Appendix 2]

In the polycondensation in Scheme 1, monomer a-a (**1**) reacts with monomer b-c (**2**), where only the reactivity of b-c changes after the first nucleophilic reaction. For convenience of using the equations deduced in Appendix I, we consider Scheme 3 as polycondensation reactions, where compounds **4** and **5** are treated as a-a and b-c type monomers, respectively. When the number of monomer molecules are denoted by N_{aa} and N_{bc}, the total number of ends in polymer ($N_{\text{E}}^{\text{Poly}}$) and monomer ($N_{\text{E}}^{\text{Mono}}$) can be expressed as

$$N_{\text{E}}^{\text{Poly}} = 2N_{aa}\alpha + 2N_{bc}(2 - \gamma - \delta) \qquad (15)$$

$$N_{\text{E}}^{\text{Mono}} = 2N_{aa}\alpha^2 + 2N_{bc}(1-\gamma). \qquad (16)$$

Since the total weight of polymer is $N_{\text{aa}}W_{\text{aa}}(1-\alpha^2)+N_{\text{bc}}W_{\text{bc}}\gamma$, the number-averaged molecular weight of polymer M_{n} is given by

$$M_{\text{n}} = \frac{N_{\text{aa}}W_{\text{aa}}(1-\alpha^2) + N_{\text{bc}}W_{\text{bc}}\gamma}{(N_{\text{E}}^{\text{Poly}} - N_{\text{E}}^{\text{Mono}})/2}, \qquad (17)$$

where the weights of reacted a-a and b-c are W_{aa} and W_{bc}, respectively.
According to the definition of S, $N_{\text{bc}}=SN_{\text{aa}}$ in polymerizarion and Using Eqs.(4)-(7), (8), (9), (15) and (16), M_{n} is rewritten as

$$M_n = \frac{W_{aa}(1-\alpha^2) + W_{bc}S(1-\beta)}{2\alpha - \alpha^2 + S - 1 - S\beta} \quad . \tag{18}$$

The procedure for estimating the τ dependence of M_n under specific values of S and κ is as follows.
Calculate W_{aa} and W_{bc} for the polymerization considered.
A certain value of β is taken between 1.0 and $(1-1/S)$.
The corresponding τ is calculated according to Eq.(14).
The value of α is calculated according to Eq.(13).
The M_n is calculated according to Eq.(18).
The M_n is plotted against the corresponding τ.
The value of β is gradually brought close to the limit of $(1-1/S)$, and repeat (2)-(6) for each β.

Macromol. Symp. **2003**, *199*, 37-46

Chain-Growth Polycondensation of Potassium 3-Cyano-4-fluorophenolate Derivatives for Well-Defined Poly(arylene ether)s

Yukimitsu Suzuki,[1,2] *Shuichi Hiraoka,*[1] *Akihiro Yokoyama,*[1] *Tsutomu Yokozawa**[1,2]

[1] Department of Applied Chemistry, Kanagawa University, Rokkakubashi, Kanagawa-ku, Yokohama 221-8686, Japan, and "Synthesis and Control", PRESTO, Japan
[2] Japan Science and Technology Corporation (JST), Japan
E-mail: yokozt01@kanagawa-u.ac.jp

Summary: Polycondensation normally proceeds in a step-growth reaction manner to give polymers with a wide range of molecular weights. However, the polycondensation of potassium 2-alkyl-5-cyano-4-fluorophenolate (**1**) proceeded at 150 °C in a chain polymerization manner from initiator, 4-fluoro-4'-trifluoromethyl benzophenone (**2**), to give aromatic polyethers having controlled molecular weights and low polydispersities ($M_w/M_n \leq 1.2$). The resulting polycondensation of **1** had all of the characteristics of living polymerization and displayed a linear correlation between molecular weight and monomer conversion, maintaining low polydispersities. Sulfolane was a better solvent for chain-growth polycondensation of **1** than other aprotic solvents. The polyether from **1** with a low polydispersity showed higher crystallinity than that with a broad molecular weight distribution, obtained by the conventional polycondensation of **1** without **2**.

Keyword: chain-growth polycondensation, crystallinity, living polymerization, MALDI-TOF mass, polyethers

Introduction

The nature of polycondensation that proceeds in a step-growth reaction manner results in polymers with broad molecular weight distributions, which generally contain not only linear polymers but also macrocyclic oligomers and polymers.[1] We have been recently successful in the production of only linear polycondensation polymers with defined molecular weights and low polydispersities by chain-growth polycondensation for polyamides,[2] where the monomer reacts with the polymer end group selectively, not with other monomers by virtue of different subsituent effects of the nucleophilic site on the electrophilic site between the monomer and polymer.[3] We now apply this chain-growth polycondensation to the synthesis of poly(arylene ether)s with low polydispersities and then have designed monomer **1**.[4]

 DOI: 10.1002/masy.200350904

We expect two approaches to the chain-growth polycondensation of **1**: electrophilic site propagation and nucleophilic site propagation (Scheme 1). In the electrophilic site propagation by using reactive initiator **2** bearing an electron-withdrawing group, **1** would react with **2** to yield ether **3** faster than with the aromatic fluorine of another **1** having the strong electron-donating phenoxide group. Monomer **1** would now react with **3** to yield a dimeric ether faster than with another **1**, because the ether linkage of **3** is a weaker electron-donating group than the phenoxide group of **1**, and the aromatic fluorine of **3** would be more reactive than that of the monomer. Growth would continue in a chain polymerization manner with the conversion of the strong electron-donating phenoxide group of **1** to the weak electron-donating ether linkage in polymer. In the nucleophilic site propagation, phenoxide is the propagating end group instead of fluorine. Thus, when potassium 4-methoxyphenoxide **4** having an electron-donating group is used as an initiator, **4** would react with **1**, and the ether linkage formed would make the polymer terminal phenoxide more reactive than the phenoxide of monomer, because ether linkage has a stronger electron-donating character than fluorine. Both approaches depend on whether there is enough difference of substituent effects between monomer and polymer.

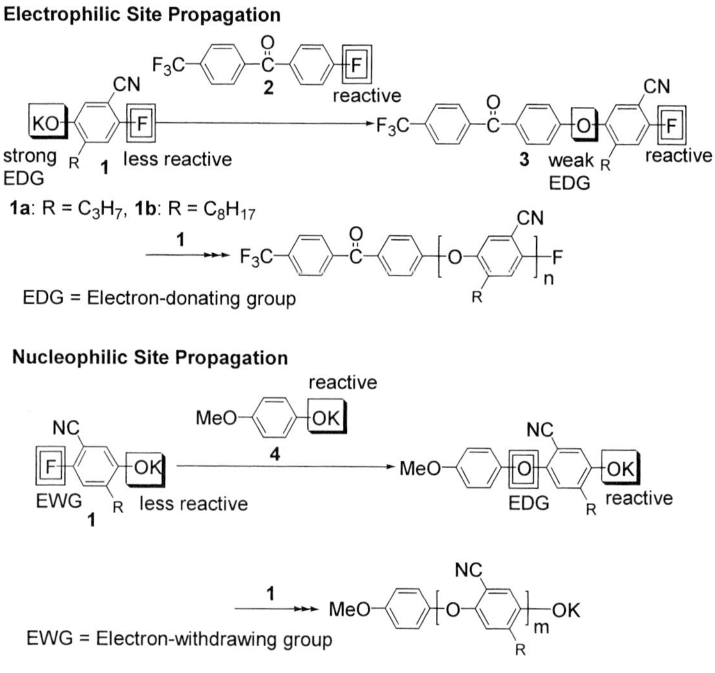

Electrophilic Site Propagation

1a: R = C_3H_7, **1b**: R = C_8H_{17}

EDG = Electron-donating group

Nucleophilic Site Propagation

EWG = Electron-withdrawing group

Scheme 1

In this paper, we report the chain-growth polycondensation of **1** for well-defined poly(arylene ether)s through the electrophilic site propagation. The polymerization behavior has all of the characteristics of living polymerization and displays a linear correlation between molecular weight and monomer conversion, maintaining low polydispersities. The MALDI-TOF mass spectrum of poly**1** also reveals that this polycondensation does not include conventional polycondensation which gives macrocycles and the polymer without **2** unit. Sulfolane is a suitable solvent for the chain-growth polycondensation of **1**. Furthermore, poly**1** with a low polydispersity from chain-growth polycondensation posesses higher crystallinity than that with a broad molecular weight distribution from conventional polycondensation.

Chain-growth Polycondensation of 1a

According to the two approaches to the chain-growth polycondendensation of **1** mentioned in Introduction, the polymerizations of **1a** with 7 mol% of initiator **2** for electrophilic site propagation and with 7 mol% of initiator **4** for nucleophilic site propagation were carried out at 150 °C in sulfolane, respectively. Surprisingly, the polymerization of **1a** with **4** did not proceed at all, whereas the polymerization with **2** took place to yield a polymer with a low polydispersity. This result implies that not only the reaction of **4** with **1a** but also the reaction of monomers **1a** with each other did not take place under this condition, and that the polymerization of **1a** in the presence of **2** did not involve step polymerization but was initiated with **2**. To elucidate whether chain-growth polymerization takes place from **2** in this polycondensation, the M_n values,[5] the M_w/M_n ratios, and the ratios of initiator unit to end group in polymer were plotted against monomer conversion in the polymerization of **1a** with 7 mol% of **2** (Figure 1(a)). In general polycondensations that proceed in a step polymerization manner, the molecular weight does not increase much in low conversion of monomer and is accelerated in high conversion, and the M_w/M_n ratios increase up to 2.0. The ratios of initiator unit to end group would be less than 1.0, because monomers react with not only an initiator but also other monomers. As shown in Figure 1(a), the M_n values increased in proportion to conversion, and the M_w/M_n ratios were less than 1.1 over the whole conversion range. The ratios of initiator unit to end group, which were easily determined by the ^{19}F NMR spectra of polymer, were constantly about 1.0 irrespective of conversion. Consequently, Figure 1(a) shows that the polycondensation of **1a** proceeds in a chain-growth polymerization manner like living polymerization. In another series of experiments, **1a** was polymerized with varying feed ratio ($[\mathbf{1a}]_0/[\mathbf{2}]_0$). As shown in Figure 1(b), the observed M_n values of polymers were in good agreement with those calculated with the assumption that one initiator molecule forms one polymer chain. The M_w/M_n ratios were less than 1.1 when the [**1a**]$_0$/[**2**]$_0$ ratios were more than 1.0.[6] This also agrees with the features of chain-growth polymerization.

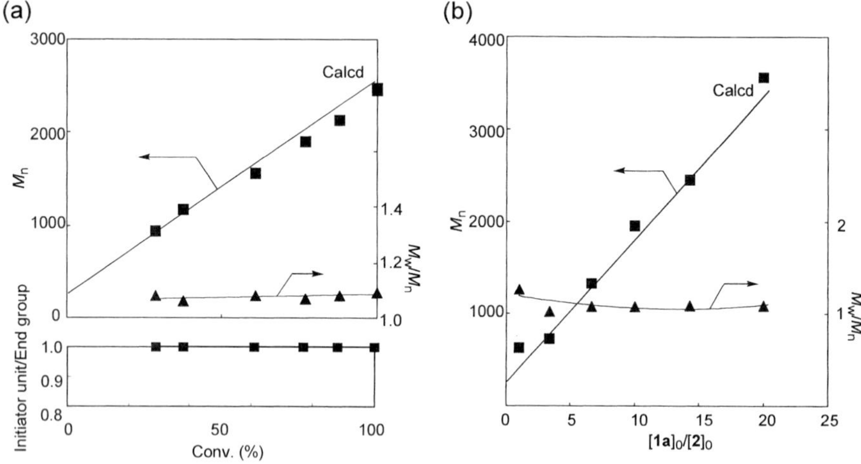

Fig. 1. (a) M_n and M_w/M_n values of poly**1a** and the ratios of initiator unit to end group in poly**1a**, obtained in the presence of **2** in sulfolane at 150 °C, as a function of monomer conversion: [**1a**]$_0$ = 0.17 M; [**2**]$_0$ = 11.7 mM. (b) M_n and M_w/M_n values of poly**1a**, obtained in the presence **2** in sulfolane at 150 °C, as a function of the feed ratio of **1a** to **2**: [**1a**]$_0$ = 0.17 M; [**2**]$_0$ = 8.3 –167 mM; conversion = 100%.

Solvent Effect on Chain-growth Polycondensation of 1b with 2

Solvent effect on the chain-growth polycondensation was studied by using potassium 5-cyano-4-fluoro-2-octylphenolate (**1b**), which was more soluble than **1a** in a variety of solvents. Thus, the polymerization of **1b** was carried out in the presence of **2** ([**1b**]$_0$/[**2**]$_0$ = 50) at 150 °C for 24 h in various solvents that dissolve **1b** and poly**1b** at the polycondensation temperature. When **1b** polymerized in THF, monomer conversion was low and polymer with a high polydispersity was obtained. In quinoline the polymerization proceeded very slowly but polymer with M_n of 6940 and a low polydispersity (M_w/M_n <1.3) was afforded in spite of low conversion, which was not a step polymerization nature but a chain polymerization nature. In *N,N*-dimethylimidazolidinone (DMI) and tetraglyme, **1b** was converted in more than 80%.

However, the ratios of initiator unit to fluorine end group were more than 1.0, indicating that termination took place by the substitution of the polymer terminal fluorine in the side reaction.[7] The ratios of initiator unit to fluorine end group of the polymer obtained in DMI were almost 1.0 for 1 h but less than 1.0 for 5 h. Since this ratio of less than 1.0 means that there are more end groups than the initiator units, both step-growth polymerization and chain-growth polymerization took place for 5 h. In sulfolane, the conversion of monomer was 100% and high moleclure weight polymer (M_n = 5900) was obtained, but the molecular weght distribution was rather broad ($M_w/M_n \geq 2.0$). This may be because poly**1b** was precipitated during polymerization. The polymerization of **1b** was then carried out at the feed ratios of $[\mathbf{1b}]_0/[\mathbf{2}]_0$ of less than 30. The polymerizaton proceeded homogeneously to yield polyethers with low polydispersities. The M_n values of polymers were in good agreement with those calculated M_n. The ratios of initiator unit to end group was constantly 1.0. Sulfolane was the best solvent for the chain-growth polycondensation of **1b** in all of the solvents we examined.

Table 1. Polymerization of **1b** with **2**.

Solv.	$[\mathbf{1b}]_0/[\mathbf{2}]_0$	Time (h)	Conv. (%)$^{a)}$	M_n (Calcd)	$M_n^{b)}$	$M_w/M_n^{b)}$	Initiator unit /End group
Quinoline	50	24	6	956	6940	1.22	c)
THF/18-crown-6	50	24	65	7716	3600	1.88	c)
Tetraglyme	50	24	82	10218	5160	2.11	1.18
DMI	50	1	29	3400	2150	2.15	0.98
	50	5	88	10318	3250	3.13	0.76
	50	24	87	10088	5090	1.92	2.44
Sulfolane	50	24	100	11725	5900	2.01	1.02
	10.2	24	95	2648	2240	1.13	1.01
	14.9	24	85	3560	2750	1.14	1.00
	26.8	24	85	5070	4550	1.17	1.02

The polymerization of **1b** was carried out in the presence of initiator **2** at 150 °C; $[\mathbf{1b}]_0$ = 0.17 M; $[\mathbf{2}]_0$ = 3.4 mM.

a) Determined by HPLC (eluent: $MeOH/H_2O$ = 70/30).

b) Determined by GPC based on polystyrene standards (eluent: THF).

c) Not determined.

MALDI-TOF Mass Spectra of Poly1b

MALDI-TOF mass spectrometry is an ideal technique for the detection of small quantities of side reaction products and could analyze chain-growth polycondensation polymer having the initiator unit and conventional polycondensation polymer bearing no initiator unit. Thus, the polymers obtained in DMI and in sulfolane, mentioned in the above section, were analyzed by the MALDI-TOF mass spectra, respectively.

Under the MALDI-TOF mass conditions used, poly**1b** is ionized to $[M + Ag]^+$ ion. The mass spectrum obtained from the product of the polymerization in DMI (conversion = 29%; the ratio of initiator unit to end group = 0.98) is shown in Figure 2. It contains two distributions. The distribution in the lower mass range corresponds to the Ag^+ adducts of poly**1b** without **2** unit from conventional polycondensation. For example, the 6-mer of this distribution is expected to produce a signal at *m/z* 6 × 229.14 (repeat unit) + 20.01 (HF) + 106.90 ($^{107}Ag^+$) = 1501.75, as indeed is observed at 1503.53. The distribution in the higer mass range is due to the Ag^+ adducts of poly**1b** with initiator **2** from chain-growth polycondensation. For example, *m/z* 3585.71 detected in Figure 2 is in good agreement with the expected signal of the 14-mer at *m/z* 14 × 229.14 (repeat unit) + 268.06 (**2**) + 106.90 ($^{107}Ag^+$) = 3582.92. Therefore, the polymerization of **1b** in DMI involves both chain-growth polycondensation and conventional polycondensation even in low conversion.

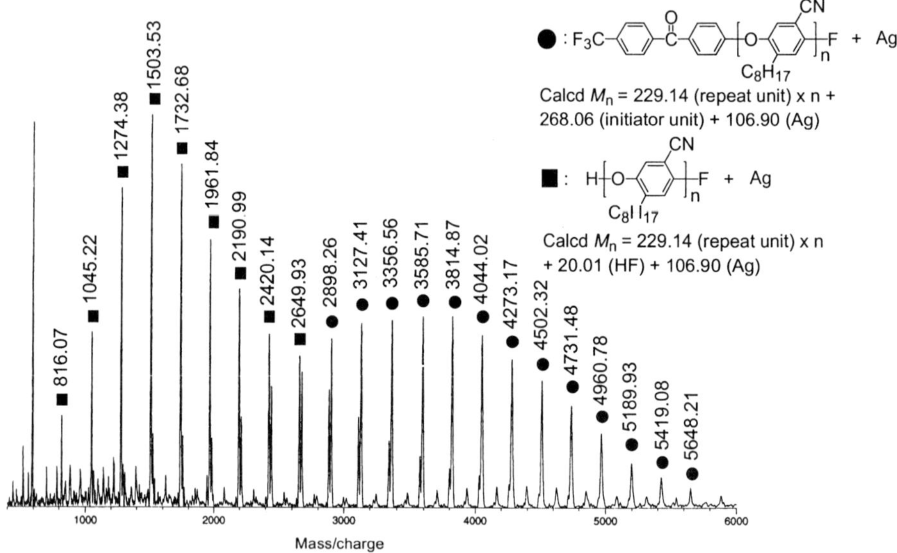

Fig. 2. MALDI-TOF mass spectrum of poly**1b** obtained in the presence of **2** ($[\mathbf{1b}]_0/[\mathbf{2}]_0$ = 50) in DMI at 150 °C (conversion = 29%; the ratio of initiator unit to end group = 0.98).

On the other hand, the mass spectrum obtained from the product of the polymerization in sulfolane (conversion = 85%; the ratio of initiator unit to end group = 1.00) contains only the one series of peaks, whose *m/z* values correspond to the Ag^+ adducts of the poly**1b** with **2** (Figure 3). For example, the 11-mer of this distribution is expected to produce a signal at *m/z* 11 × 229.14 (repeat unit) + 268.06 (**2**) + 106.90 ($^{107}Ag^+$) = 2895.50, as indeed is observed at 2896.15 as the highest signal. The linear polymer without **2** unit from conventional polycondensation (-268.06 Da + 20.01 Da (HF)) and macrocycles with no **2** unit (-268.06 Da) are absent (no *m/z* 2647 and 2627 detected in Figure 3). This mass spectrum indicates that the polymerization of **1b** in sulfolane proceeded in a chain polymerization manner from initiator **2** without accompanying conventional polycondensation.

Conseqauently, it turns out that the chain-growth polycondensation of **1b** for poly(arylene ether) is strongly depend on the reaction solvent; sulfolane is a specific solvent for the chain-growth polycondensation of **1b**.

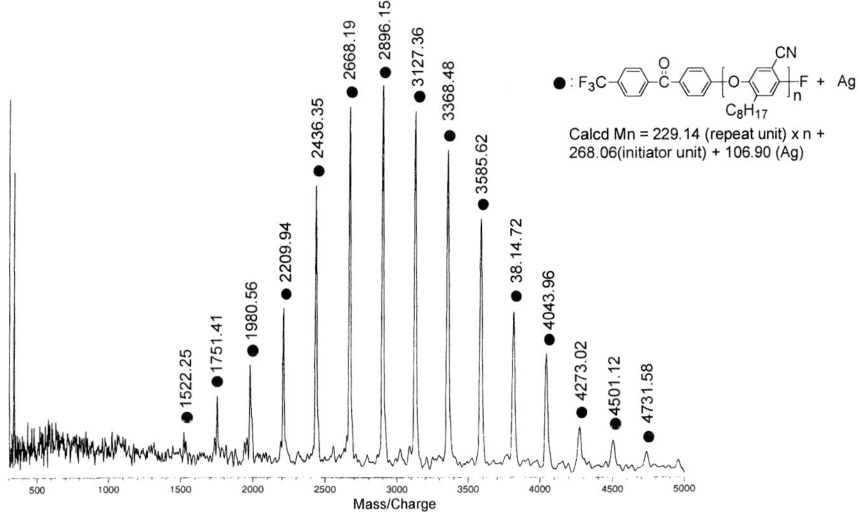

Fig. 3. MALDI-TOF Mass spectrum of poly**1b** obtained in the presence of **2** ($[\mathbf{1b}]_0/[\mathbf{2}]_0$ = 14.9) in sulfolane at 150 °C (conversion = 89%; the ratio of initiator unit to end group = 1.00).

Crystallinity of Polymers Effected by Polydispersity

We found that poly**1a** having initiator **2** unit with low polydispersity was less soluble in organic solvents than the polyether obtained by conventional polycondensation without **2**. The powder X-ray diffraction (XRD) pattern of both polymers with similar molecular weight.

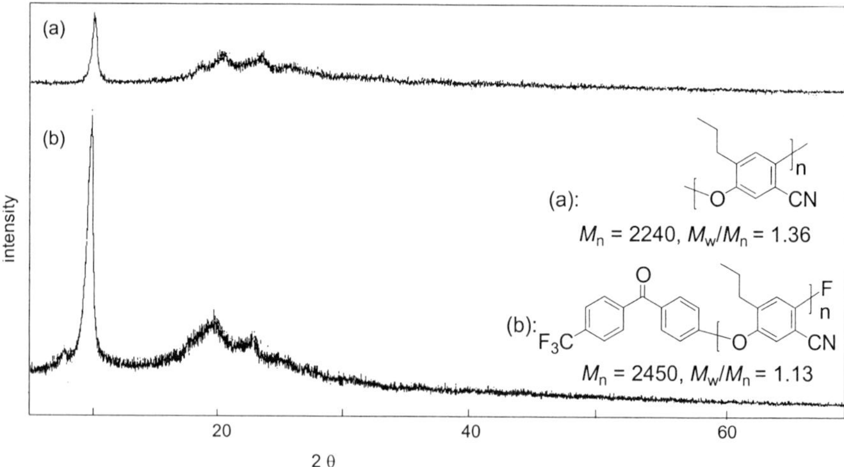

Fig. 4. XRD pattern: (a) poly**1a** obtained by conventional polycondensation without **2** in sulfolane at 150 °C: [**1a**]$_0$ = 0.17 M; (b) poly**1a** obtained in the presence of **2** in sulfolane at 150 °C: [**1a**]$_0$ = 0.17 M; [**2**]$_0$ = 11.7 mM.

showed that the crystallinity of the polymer obtained by chain-growth polycondensation (M_n = 2450, M_w/M_n = 1.13) was higher than that of the polymer obtained by conventional polycondensation (M_n = 2240, M_w/M_n = 1.36) (Figure 4). The differential scanning calorimetry (DSC) traces of the above conventional polycondensation polymer did not show any intense peak on both the first and second heatings (Figure 5(a)). On the other hand, the first heating of the polymer obtained by chain-growth polycondensation showed the melting point at 255 °C with a melting enthalpy of 20 J/g. The complete melting polymer was glassified by rapidly quenching to room temperature. On a second heating the glass transition temperture (T_g) appeared at 78 °C, and furthermore the exothermic peak attributed to cold crystallization at 172 °C with ΔH of 28 J/g and the melting point at 255 °C with ΔH of 21 J/g were observed (Figure 5(b)). The DSC analysis of the polymers distinctly also indicates that the polymer with low polydispersity obtained by chain-growth polycondensation possesses higher crystallinity. This result implies that the crystallinity of condensation polymers could be controlled by polydispersity of polymers.

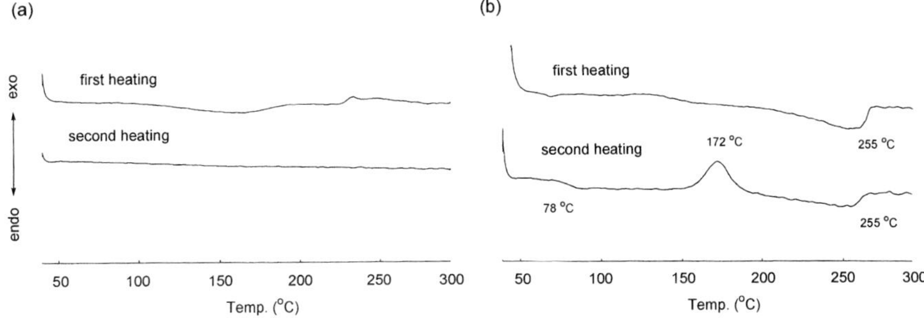

Fig. 5. DSC traces of poly**1a** (heating rate 10 °C/min): (a) poly**1a** obtained by conventional polycondensation in sulfolane ($[\mathbf{1a}]_0$ = 0.17 M) at 150 °C (M_n = 2240, M_w/M_n = 1.36); (b) poly**1a** obtained in the presence of **2** ($[\mathbf{1a}]_0$ = 0.17 M; $[\mathbf{2}]_0$ = 11.7 mM) at 150 °C in sulfolane (M_n = 2450, M_w/M_n = 1.13).

Conclusion

Our present results demonstrate that potassium fluorophenolate derivatives **1** undergo chain-growth polycondensation via electrophilic site propagation to yield aromatic polyethers having defined molecular weights and low polydispersities. The polycondensation of **1b** with **2** was carried out in a variety of solvents to find remarkable solvent effects on this polymerization. The MALDI-TOF mass spectrum of poly**1b** with **2** in sulfolane reveals that this polycondensation does not include conventional step-growth polycondensation which gives macrocycles and the polymer without initiator unit. Poly**1a** with a low polydispersity from chain-growth polycondensation posesses higher crystallinity than that with a broad molecular weight distribution from conventional polycondensation.

Acknowledgments

This work was supported in part by a Grant-in-Aid (12450377) for Scientific Research from the Ministry of Education, Science, and Culture, Japan.

[1] (a) H. Jacobson, W. H. Stockmayer, *J. Chem. Phys.* **1950**, *18*, 1600; (b) P. Maravigna, G. Montaudo,"*Comprehensive Polymer Science*", G. Allen, J. C. Bevington, Eds.; Pergamon: Oxford, 1989; Vol. 5, pp 63; (c) U. R. Süter, "*Comprehensive Polymer Science*", G. Allen, J. C. Bevington, Eds.; Pergamon: Oxford, 1989; Vol. 5, pp 91.
[2] T. Yokozawa, T. Asai, R. Sugi, S. Ishigooka, S. Hiraoka, *J. Am. Chem. Soc.* **2000**, *122*, 8313.
[3] For examples of polycondensation in which the reaction of monomer with polymer end group is faster than that of monomers with each other, although the polymerization behavior does not show the character of living polymerization, see: (a) R. W. Lenz, C. E. Handlovits, H. A. Smith, *J. Polym. Sci.* **1962**, *58*, 351. (b) A. B. Newton, J. B. Rose, *Polymer* **1972**, *13*, 465. (c) W. Risse, W. Heitz, *Makromol. Chem.* **1985**, *186*, 1835. (d) V. Percec, T. D. Shaffer, *J. Polym. Sci. Part C: Polym. Lett.* **1986**, *24*, 439. (e) V. Percec, J. H. Wang, *J. Polym. Sci. Part A: Polym. Chem.* **1991**, *29*, 63. (f) V. Percec, J. H. Wang, *Polym. Bull.* **1990**, *24*, 493. (g) J. H. Wang, V. Percec, *Polym. Bull.* **1991**, *25*, 33. (h) D. R. Robello, A. Ulman, E. J. Urankar, *Macromolecules* **1993**, *26*, 6718. (i) T. Yokozawa, H. Shimura, *J. Polym. Sci. Part A: Polym. Chem.*, **1999**, *37*, 2607.
[4] For preliminary report, see; T. Yokozawa, Y. Suzuki, S. Hiraoka, *J. Am. Chem. Soc.* **2001**, *123*, 9902.
[5] The M_n values of polymer were estimated by the ^{1}H NMR spectra based on the ratios of signal intensities of the repeating units to the initiator unit.
[6] When the $[\mathbf{1a}]_0/[\mathbf{2}]_0$ ratios were 25 or above, the polymer was precipitated during polymerization.
[7] H. R. Kricheldorf, S. Böhme, G. Schwarz, R. Krüger, G. Schulz, *Macromolecules* **2001**, *34*, 8886.

Polyaddition Reaction in a Dispersed Medium: Elaboration of New Core-Shell Polyurethane Particles

*Pierre Chambon, Bindushree Radhakrishnan, Eric Cloutet, Eric Papon, Henri Cramail**

Laboratoire de Chimie des Polymères Organiques UMR 5629
ENSCPB – Université Bordeaux-1 – CNRS
16, Avenue Pey-Berland, F-33607 Pessac Cedex, France
E-mail: cramail@enscpb.fr

Summary: The elaboration in a dispersed organic medium of calibrated polyurethane particles with a core-shell structure is presented in this paper. The objective could be achieved by using a series of reactive steric stabilizers of the type ω-$(OH)_x$-poly(*n*-butyl acrylate), -polystyrene, -polysiloxane or -polybutadiene (x=1 or 2) that play the role of surfmers during the polyaddition reaction between ethylene glycol and tolylene-2,4-diisocyanate, in cyclohexane as a dispersant medium. The final size of the polyurethane particles (0,5-10 μm) was found to be a function of the steric stabilizer characteristics (nature, molar mass and concentration) and of the addition procedure of the different reactants. These novel particles constituted of a polyurethane core and various shells depending on the stabilizer used exhibit specific and original properties.

Keywords: core-shell particles, dispersion, polyurethane

Introduction

It is with the objective to find novel applications for polymers obtained by step polymerization that we investigate the field of polyaddition reaction in a dispersed medium. The need for numerous applications (coatings, adhesives, etc.) to have materials easy to handle and process led us to evaluate the possibility to elaborate, by an heterogeneous polymerization process, "polycondensates" as a form of calibrated particles. Usually such a preparation is made possible by the presence, in the dispersant phase, of an ionic or a steric stabilizer that avoid the coagulation of the precipitating polymer and maintains it as a stable dispersion ("latex").

Among the steric stabilizers generally tested, amphipathic block copolymers (PS-b-PEO as an example) and reactive polymers (surfmers or macromonomers) can be discriminated. While the former enable the stabilization of the growing particle by physical adsorption, the latter react

 DOI: 10.1002/masy.200350905

with the growing polymer and remain covalently bonded to the final particle. If the literature is well documented in the field of free radical polymerization of vinyl monomers in heterogeneous conditions, very few data are given concerning the preparation of polymer materials issued from a step polymerization *via* a controlled heterogeneous process. The preparation of calibrated polyurethane particles in the size range 5-50 μm, by suspension technique in a non-aqueous medium, has been reported by Nippon.[1] In this work, oligomeric glycols were condensed with diisocyanates in the presence of poly(ethylene oxide)-b-poly(dimethylsiloxane) block copolymer as a steric stabilizer. The use of dihydroxy-terminated poly(dodecylmethacrylate)s as reactive steric macromonomers for the preparation of polyurethane particles in a dispersed organic medium has been recently reported by Sivaram *et al.*[2] We also explored the possibility to prepare polyurethane particles with a core-shell structure by a dispersion technique.[3],[4] A series of reactive steric stabilizers of the type ω-$(OH)_x$-poly(*n*-butyl acrylate) (P*n*BuA), -polystyrene (PS), -polydimethylsiloxane (PDMS) or -polybutadiene (PBut) (x=1 or 2) have been tested, leading to various core-shell polyurethane-based materials and exhibiting a large scope of properties depending on the chemical nature of the shell. Results are discussed in terms of particle characteristics (size and size distribution) along with some properties.

Results and Discussion

The preparation of polyurethane materials under the form of calibrated microspheres has been investigated in cyclohexane as the dispersant medium at 60°C in the presence of dibutyl tin dilaurate (DBTDL) as a catalyst. For this study, ethylene glycol (EG) and tolylene-2,4-di-isocyanate (TDI) were chosen as monomers at the ratio [NCO]/[OH] = 1,2. In order to prepare a series of novel core-shell polyurethane materials with specific properties, we designed a number of reactive stabilizers that differ by their chemical nature, molar mass, valence as well as solubility in the selected dispersant medium. The effect of each of the parameters indicated above were studied with respect to the ability to produce calibrated core-shell polyurethane particles with a tunable size and exhibiting specific properties.

Unless PBut-OH and PDMS-OH that have been purchased, all the reactive stabilizers were synthesized either by "living" anionic or "controlled" radical polymerizations to insure a very good control of their molar mass and valence.[3],[4] A complete list of these different functional polymers is given in Table 1 together with their solubility parameter as well as the solubility parameter of cyclohexane (dispersant medium) for comparison. The knowledge of the solubility parameter of each component is of importance in regards to the nucleation process as well as to the stabilization and shelf life of the latex.

Table 1. List of the reactive stabilizers used for the preparation of polyurethane particles in cyclohexane as a dispersant medium at 60°C.

Reactive stabilizer		Synthesis method	Solubility parameter δ (MPa)$^{1/2}$
PS-OH	$-(CH_2-CH(C_6H_5))_n-CH_2-CH_2OH$	Anionic [3]	18,6
PS-(OH)$_2$	$Br-(CH(C_6H_5)-CH_2)_n-CH(CH_3)-C(=O)-O-CH_2-C(CH_2OH)(CH_2CH_3)-CH_2OH$	Anionic[6]	18,6
PBut-OH	$-(CH_2-CH=CH-CH_2)_n(CH_2-CH(CH=CH_2))_{n'}-CH_2CH_2OH$	Anionic [6]	17,2
PBut-(OH)$_2$	$-(CH_2-CH=CH-CH_2)_n(CH_2-CH(CH=CH_2))_{n'}-CH_2CH_2O-C(=O)-CH_2-C(CH_2OH)(CH_3)-CH_2OH$	Anionic [6]	17,2
P*n*BuA-OH	$-(CH_2-CH(C(=O)O^nBu))_n-CH_2-CH(Br)-CH_2OH$	ATRP [3],[5]	20,4
P*n*BuA-(OH)$_2$	$-(CH(C(=O)O^nBu)-CH_2)_n-C(CH_3)_2-C(=O)-O-CH_2-C(CH_2OH)(CH_2CH_3)-CH_2OH$	ATRP [5]	20,4
PDMS-OH	$-(Si(CH_3)_2-O)_n(CH_2-CH_2-O)_2H$	commercial	7,4
cyclohexane			16,8

In addition to the chemical nature and valence of the steric and reactive stabilizer that govern the formation of a stable dispersion of the growing polyurethane chains, numerous other parameters such as the wt % concentration of the stabilizer, the kinetic and order of addition of the reactants, the concentration of the monomers (solid content), the stirring speed in the reactor, etc. may be tuned to enable the preparation of calibrated particles with a very narrow size distribution.

As shown in Scheme 1, the growing particles are constituted of homo-polyurethane chains together with either block- or graft- copolymers - depending on the valence (1 or 2) of the surfmer used - that play the role of steric stabilizers in the dispersion process.

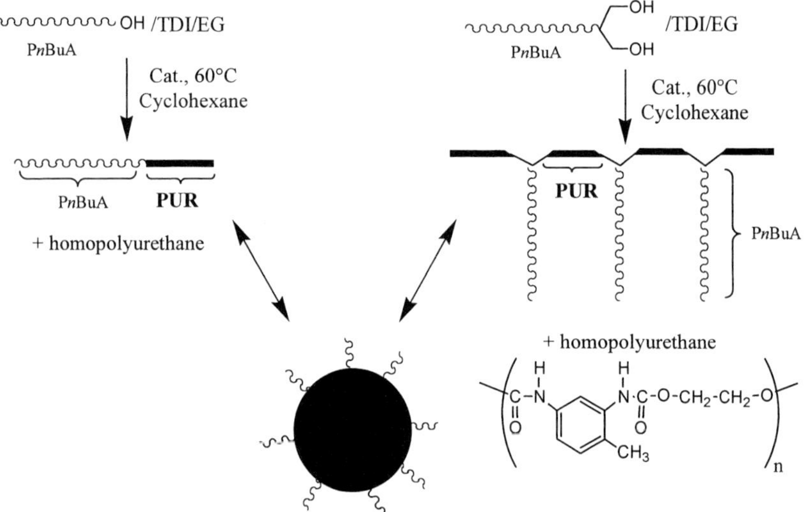

Scheme 1. synthetic route to core-shell polyurethane particles.

Influence of the Procedure

The order of addition of the reactants was found to be crucial in the procedure. Indeed, it was observed that ω-(di)hydroxyl steric stabilizers have to be pre-reacted first with a slight excess of tolylene-2,4-diisocyanate (TDI) leading to the formation of ω-(di)NCO-terminated polymers, considered as the true reactive stabilizers. EG is then added over a 30' period giving rise to a turbid medium due to the insolubility of this monomer and of growing PUR oligomers in cyclohexane. TDI, totally soluble in cyclohexane, was then added drop-wise over variable time periods. The time of TDI addition was also found to be important with respect to the size of the particles ; the longer the time of TDI addition, the lower the particle size. As an example in the specific case of PS-OH used as a stabilizer ($\overline{M}_n$=2100 g/mole; 10 wt%), the particle size goes from 2 μm to 0,5 μm while TDI is added over 1 or 6 hours.

The procedure of addition described above was followed in all the cases, whatever the reactive stabilizer used. Different parameters discussed in the next sections were then varied to compare the efficiency and behavior of each reactive stabilizer.

Influence of the wt% Concentration of the Reactive Stabilizer

The effect of the stabilizer concentration on the average size of the polyurethane particles has been evaluated for the various mono-hydroxy reactive stabilizers (see Figure 1).

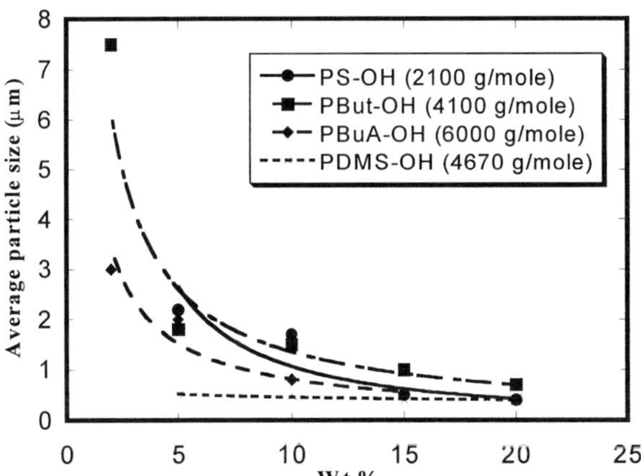

Fig. 1. Influence of the wt % concentration of the monovalent stabilizers on the PUR particle size.

As expected, the higher the stabilizer concentration, the lower the particle size. Indeed, the higher the concentration, the higher the surface coverage of the nuclei that are formed at the earliest stage of the polymerization, therefore preventing their coalescence. It is also interesting to note that the variation of the particle size with the stabilizer concentration is very much dependent to the chemical nature of the latter. While a pronounced effect on the particle size is found with PBut-OH concentration, no change was observed in the case of PDMS-OH used as a steric stabilizer. Whatever the wt % of PDMS-OH, PUR particles in the range 400-500 nm were systematically obtained.

Influence of the Molar Mass of the Reactive Stabilizer

The molar mass of the reactive stabilizer was found to have a dramatic effect on the formation of a stable latex. Indeed, a critical length of the stabilizer is required to control the nucleation step and thus to avoid coagulation of the growing particles. The domain of molar masses required to get a stable latex constituted of particles with a controlled size was observed to be very much dependent to the chemical nature (as well as to the valence, see next chapter) of the reactive stabilizer. As illustrated in Figure 2, the comparison of PS-OH and PBuA-OH allowed us to define $\overline{DP_n}$ intervals (20-40 and 40-200 respectively) where these polymers are efficient in the PUR particles stabilization process, in these experimental conditions. Outside these intervals, partial or complete coagulation of the latex was observed.

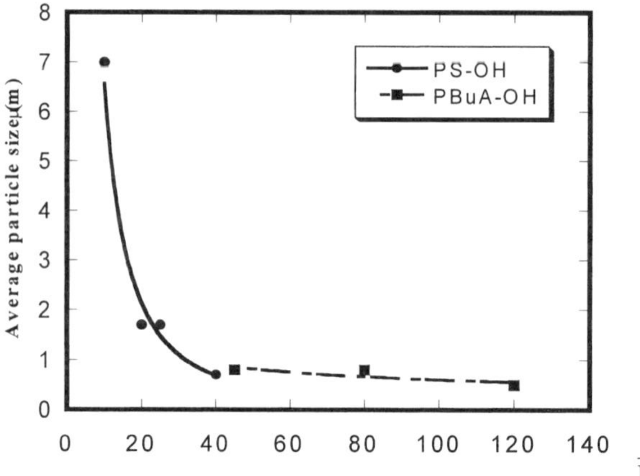

Fig. 2. Influence of the stabilizer molar mass on the PUR particle size.

Influence of the Valence of the Reactive Stabilizer

The valence of the stabilizer has an important effect on the formation of stable latex. Depending on the valence of the stabilizer (one or two), block or graft copolymers are formed during the polymerization (see Scheme 1). The ability of these block or graft copolymers to initiate the nucleation step and to stabilize the final latex may be different. Indeed, this behaviour is exemplified here (Figure 3) in the case of acrylate-based stabilizers (PBuA-OH and $PBuA(OH)_2$ respectively).

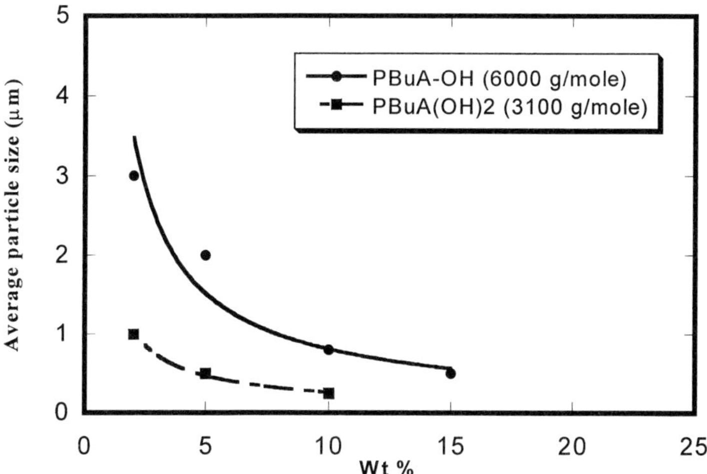

Fig. 3. Influence of the valence of the reactive stabilizer on the PUR particle size in the case of mono- and divalent poly(*n*-butyl acrylate)s (PBuA-OH & $PBuA(OH)_2$).

As shown in Figure 3, gemini-type $PBuA\text{-}(OH)_2$ macromonomer is much more efficient than its monovalent analog with respect to the formation of polyurethane particle with a controlled size. Low concentration of $PBuA\text{-}(OH)_2$ are sufficient to form a stable dispersion and the corresponding polyurethane particles are systematically smaller than the ones obtained with PBuA-OH. In addition it was observed that the required critical $\overline{DP_n}$ of $PBuA\text{-}(OH)_2$ to form a stable latex is much lower than the one needed with PBuA-OH.

These trends were also observed with the other reactive stabilizers used. These general observations show that the formation of first nuclei occurs more easily when divalent reactive stabilizers are used. A better repartition of the stabilizer chains at the surface of the PUR particles may explain this phenomenon.

Characterization of the PUR Particles

The characterizations of these new materials have been performed to check the true participation of the reactive stabilizer in the polyaddition reaction and thus to prove the core-shell structure of the particles. ^{1}H NMR, SEC and mechanical tests were systematically realized. The ^{1}H NMR characterization was implemented in DMSO at 50°C to permit the solubilization of all the chains packed in the particle. As shown in Figure 4 with poly(*n*-butyl acrylate) used as a stabilizer, the integration of the different peaks allowed us to estimate that more than 90% of the stabilizer initially introduced in the reactor vessel, do participate in the polyaddition reaction and is covalently bonded to urethane segments.

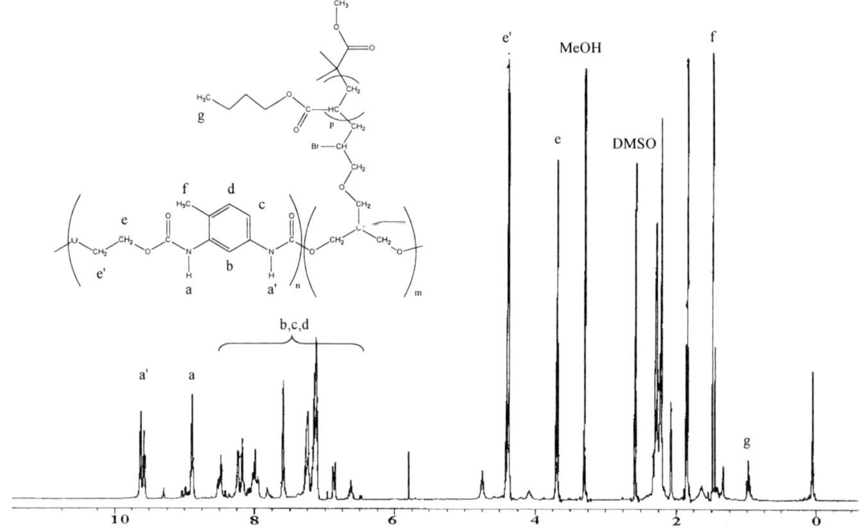

Fig. 4. ^{1}H NMR spectrum (DMSO-d_6, 200 MHz) of core-shell PUR-PBuA particles.

The SEC characterization of these materials was realized in DMF. The experimental molar masses based on polystyrene calibration were systematically found around 5000-8000 g/mole with a multimodal distribution. These values are rather low in agreement with the experimental conditions. No real correlation between the molar mass and the particle size was observed.

It is also worth noting that dynamic mechanical analyses of the PUR particles reveal two glass transition temperatures corresponding to the core and the shell moieties of the particles. An illustration of this is given in Figure 5, with polystyrene-OH used as a stabilizer.

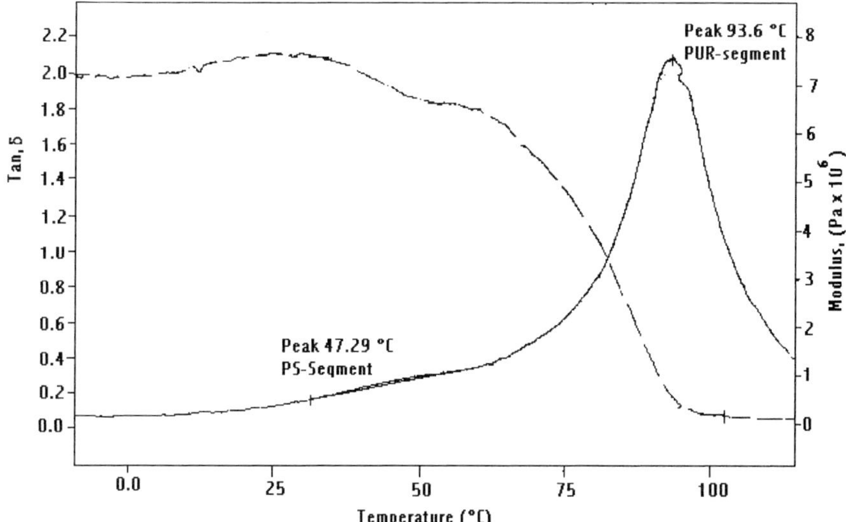

Fig. 5. DMA characterization of core-shell PUR-PS particles.

Adhesive Properties of the PUR Films

Films of PUR latexes prepared using poly(*n*-butyl acrylate) as the stabilizer were coated on different substrates (glass, aluminium, PET) to check their thermo-dynamical surface properties and their tacky behavior. After evaporation of the solvent, the particles aggregate to form hexagonal structures like a "bee-hive", as shown in Figure 6.

Contact angle measurements as well as tack tests clearly show that the coating conditions govern the surface properties. Interestingly, a PBuA-like behavior of the film is observed at room temperature due to a contribution on the surface properties of the PBuA chains. In addition, an increase of the poly(*n*-butyl acrylate) molar mass yields higher tack property. Conversely, a PUR-like behavior was found on heating above Tg at 120°C, due to a coalescence of the PUR core.

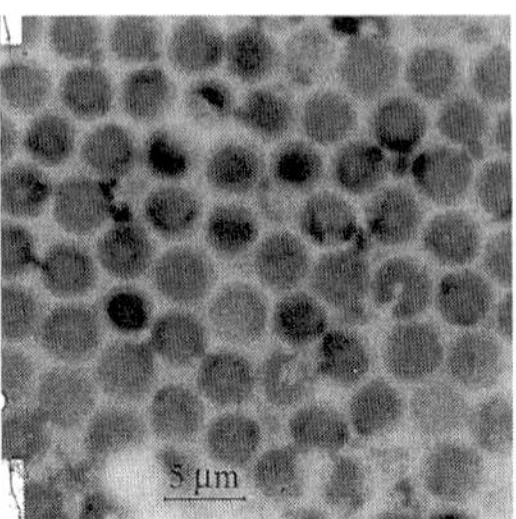

Fig. 6. TEM micrograph of a PUR latex film.

Derivatization of the PUR Particle Surface

The controlled character of the atom transfer radical polymerization of n-alkyl acrylates allowed us to prepare new stabilizers of the type P*t*BuA-b-P*n*BuA-OH with various compositions. Following the same procedure as described above, such block copolymers were then used as reactive polymers for the preparation of polyurethane particles in a dispersed medium. The derivatization of the *t*-butyl ester groups, located at the shell of the particle, into acrylic acid functions after treatment of the latex with trifluoro-acetic acid rendered the PUR particles re-dispersible in water. Typical TEM microscopy of PUR organic and aqueous latexes are shown in Figure 7.

(a) **(b)**

Fig. 7. TEM micrographs negatives of PUR particles obtained using (a) P*n*BuA$(OH)_2$, and (b) P*t*BuA-b-P*n*BuA-OH (particles redispersed in water after conversion of the P*t*BuA block in PAA block).

[1] Nippon Paint Co, *US Patent* 5,603,986, **1997**.
[2] L.S. Ramanathan, P.G. Shukla, S. Sivaram, *Pure & Appl. Chem.* **1998**, 70,1295.
[3] E. Cloutet, B. Radhakrishnan, H. Cramail, *Polym. Int.* **2002**, 51, 978-985 .
[4] B. Radhakrishnan, E. Cloutet, H. Cramail, *Colloids & Polym. Sci.* **2002,** 280, 1122-1130.
[5] B. Radhakrishnan, P. Chambon, E. Cloutet, H. Cramail, *Colloids & Polym. Sci.* accepted in **2002**.
[6] B. Radhakrishnan, E. Cloutet, H. Cramail, *Macromol. Chem. Phys.* submitted in **2002**.

Curing under Continuous Microwaves (2 450 MHz) of Thermosetting Epoxy Prepolymers: Final Statement

*Patrick Alazard, Michel Palumbo, Albert Gourdenne**

Ecole Nationale Supérieure des Ingénieurs en Arts Chimiques et Technologiques
Laboratoire Catalyse, Chimie Fine et Polymères, 118, route de Narbonne,
31077 Toulouse Cedex 4, France
Fax : (+33) 5 62 88 56 00; E-mail: Albert.Gourdenne@ensiacet.fr

Summary: After some recalls on the dielectric behavior of the organic materials, a parametrical study of the activation by continuous microwaves (2 450 MHz) of the curing reaction of an epoxy resin of DGEBA type in presence of diamino-biphenyl-methane used as crosslinking agent, is carefully described. The recording during the irradiation of the variations of the average temperature of the chemical medium and of the associated dielectric loss on the one hand, and the determination of the glassy transition temperature of the final networks on the other hand, allows optimization procedures of the electromagnetic treatment of the initial prepolymeric mixtures.

Keywords: activation, crosslinking, epoxy resin, kinetics, microwaves

Introduction

Microwave activation of polymerization reactions at 2 450 MHz first came into use in the early seventies. Experiments were carried out inside domestic ovens or multimode cavities. No significant results were obtained in such a way, because one resorted too much to broad empiricism. This situation was due to insufficient background of the implied researchers either in chemistry or in electromagnetism. The state of this research improved around later, when quantitative banks of microwave irradiation were built, allowing the possibility of recording the variations versus time of the average temperature of the organic materials under irradiation and of the associated dielectric loss due to activated dipolar relaxation.[1,2] The analysis of the data thus gained gave scientists access to the kinetics of the reactions and to a knowledge of the structural changes which occur inside the chemical medium during the polymerization. However, the geometrical dimensions of the wave guide elements compatible with the working frequency and used as applicators, disqualify this type of equipment for industrial transfer operations. Nevertheless, such experiments are necessary so as to identify the mechanisms of polymerization under microwave irradiation for future wide-scale

 DOI: 10.1002/masy.200350906

applications. More recently, the microwave frequency was replaced by a radiofrequency of 27.12 MHz which seems to be more efficient towards the curing of thermosetting prepolymers [3,4]. Both frequencies lead to similar conclusions as for the nature of the interactions between the oscillating electrical field and the organic materials to be polymerized and which work when curing reactions of epoxy resins (step polymerization) or of unsaturated polyester - styrene mixtures (chain polymerization) develop. Moreover, when microwaves are applied, two types of wave emission can be used, continuous or pulsed. When pulsed waves are applied, the kinetics of polymerization can be accelerated or delayed in comparison with those obtained in the case of the continuous modes of emission [4,5]. On the other hand, microwaves [6] or radiofrequencies [7] have been applied with success to activate the crosslinking reaction of composite materials with thermosetting polymeric matrix, which constitute model systems for the study of interaction between waves and bi-phased matter. But now, additional parameters should be taken into consideration, as the chemical nature of the fillers, transparent to the waves (silica), or electrically conductive (metals and carbon-black), and their shape and volume concentration.

The present paper opens with some comments on the dielectric loss in organic matter under electromagnetic irradiation; then, the microwave curing reaction, according to a step polymerization mechanism, of a DGEBA epoxy resin or diglycidylether of bisphenol A, in presence of 4,4'-diaminodiphenylmethane used as crosslinking agent, is carefully described. In the conclusion, the possible extension of the electromagnetic activation to various polymerization reactions, carried out in homogeneous (liquid state) or two-phase media.

Dielectric loss in molecular or polymeric organic materials

The dielectric loss, expressed as an electrical power or P_u, in organic molecular or polymeric matter submitted to electromagnetic waves of electrical power or P_0, is given by the general formula : $P_u = P_u(t) \propto 2\pi.f.\varepsilon''.v.E^2$, in which t, f, ε'', v and E are the time, the frequency of oscillation of the electrical field, the loss factor , with $\varepsilon^* = \varepsilon' - j.\varepsilon''$, the volume of the sample and E the modulus of the applied electrical field. At 2 450 MHz, P_u is due to dipolar relaxation, activated by the electromagnetic beam: the dipolar species tend to be in line with the oscillating field. So, the dielectric loss is of mechanical energy type. But the motion of relaxing entities is more or less hindered by their physical interactions with the surrounding chemical medium (viscosity effect). Consequently, this phenomenon of internal friction leads to a partial conversion of P_u into heat or Q. When the temperature is sufficiently high, the activation of polymerization reactions can operate. Thus, the conversion of electricity into heat includes two steps: from electrical energy (P_0), one reaches mechanical energy (P_u), with $P_u = \alpha.P_0$ and $\alpha < 1$, and heat (Q), with $Q = \beta.P_u$ and $\beta < 1$, or $Q = \alpha.\beta.\ P_0$, with $\alpha.\beta \ll 1$.

Electrical energy	⇨	Mechanical energy	⇨	Heat
(P_0)		(P_u)		(Q)

The residual dielectric loss, $(1 - \alpha).P_u$, or irrecoverable energy, is used to activate the Brownian motion inside the samples and other possible non-exothermic physical phenomena. Both average values of α and β, estimated from liquid epoxy resins, are close to 25%. Consequently, the energetic yield, or $\alpha.\beta$, is low (~ 6%).

Experimental

Microwave irradiation experiments at 2 450 MHz [1,2] are carried inside a wave guide element used as applicator. The waves are progressive - no stationary system is formed - and propagate according to the TE_{01} mode. Moreover, the electrical field is polarized and parallel to the small side of the section of the guide. The prepolymeric matter (20 ml) to be irradiated is poured into cylindrical pyrex pill boxes - transparent to the electromagnetic beam - used as reactors. The irradiation is carried out inside an element of wave guide. Watt-meters distributed along the wave guide line give access, for an applied electrical power P_0, to the values of the transmitted power or P_t, of the reflected power by the reactor or P_r, and of the power due to conduction loss in the metallic walls of the apparatus or P_a. The conduction loss at given P_0 is obtained from blanks. The powers P_t and P_r are absorbed through load charges, in order to avoid the establishment of systems of stationary waves which would create hot spots of energy. Moreover, each reactor constitutes an open thermodynamic system which exchanges energy with the external medium (air at 20 °C circulating inside the wave) through thermal convection. The dielectric loss, or P_u , due to dipolar relaxation, is given by the expression $P_u = P_0 - (P_t + P_r + P_a)$. The average temperature or T, is recorded through a non-interactive probe, immersed in silicon oil contained inside a pyrex tube plunged in the chemical medium. The variations versus time or t of T and P_u, at given P_0, are visualized on the screen of a computer and stored before mathematical treatment. Four types of curve are obtained : $T = T(t)$, $P_u = P_u(t)$, $(T)' = dT/dt$ and $(P_u)' = dP_u/dt$.

The organic system which is used as a model system for the study of the interactions between microwaves and reactive organic matter, consists of an epoxy prepolymer which associates an epoxy resin of DGEBA type (condensation index: 0.06) and 4,4'-diaminodiphenylmethane, or DDM, as curing agent, in stoechiometric ratio epoxide/amino-proton = 1).

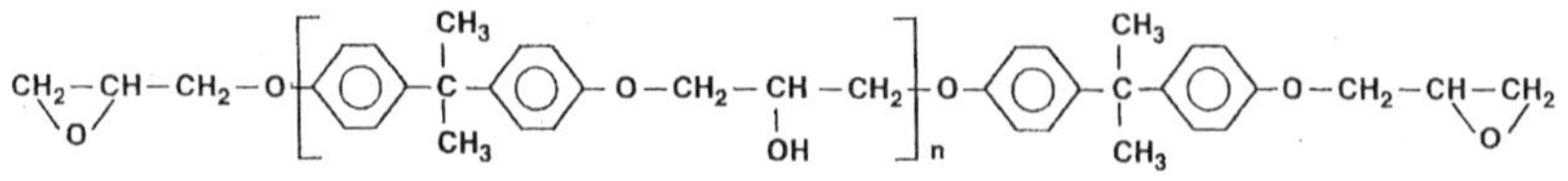

diglycidylether of bisphenol A

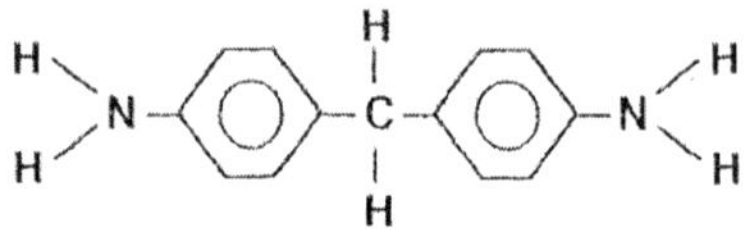

4,4'-diaminodiphenylmethane

The dielectric relaxation spectra of the initial prepolymeric mixture in fluid state shows that the absorption of the epoxides at room temperature appears as a broad peak centered around 3 GHz [8]. As for the absorption of the other polar groups (hydroxyls, primary and secondary amines), it is located at lower frequencies. So, the epoxides relax at the working microwave frequency or 2 450 MHz, and the electromagnetic activation of the polymerization reaction is possible.

The various steps of the thermally activated curing process (inside an oven) of the prepolymeric mixture - previously pre-cooled at - 18 °C during several hours - which leads to final crosslinked polymeric networks, are now well known. In the initial physical state, strong Van der Waals bonds have been developed between the various components of the thermosetting system which appears more or less turbid. The structure of the chemical medium is now that of a crystalline - like one. When the temperature is increased, the physical interactions are partially broken, the medium becomes fluid, transparent and, of course, less polar. Then, when the temperature is sufficient, the exothermic polymerization reaction starts working. The viscosity effect, associated with the progress of the reaction, hinders more and more the Brownian motion and lowers the kinetics of curing. Before long, the medium is completely gelled and consequently the relaxation of the polymeric chains is drastically restricted. The structural change which occurs in the epoxy system, when it passes from the fluid state to the gel, does not involve the formation of micro-gels, contrary to the case of the gelation process of the unsaturated polyesters - styrene resins [9]. The last step of the curing process which generates an exothermic peak, corresponds to the formation of a highly crosslinked network from the gelled state. All preceding variations will be encountered in the case of the microwave polymerization of the epoxy resins.

Results and discussion

Figures 1 and 2 display the curves $T = T(t)$, $P_u = P_u(t)$, $(T)' = dT/dt$ and $(P_u)' = dP_u/dt$, when a sample of epoxy prepolymer (20 ml), pre-cooled at -18°C, is irradiated under continuous microwaves, at P_0 = 40 watts. The data start being recorded from 0°C. The shape of the T curve underlines the usual steps of an exothermic curing reaction. T starts growing from 0°C.

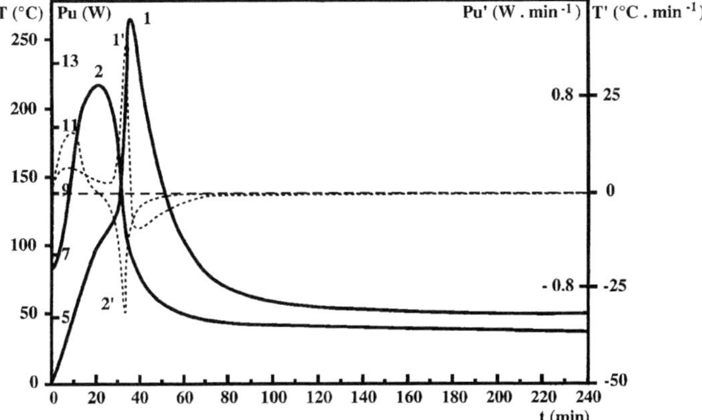

Figure 1. Curing under microwaves at P_0 = 40 W. (1): $T = T(t)$; (2): $P_u = P_u(t)$; (1'): $(T)' = dT/dt$; (2'): $(P_u)' = dP_u/dt$

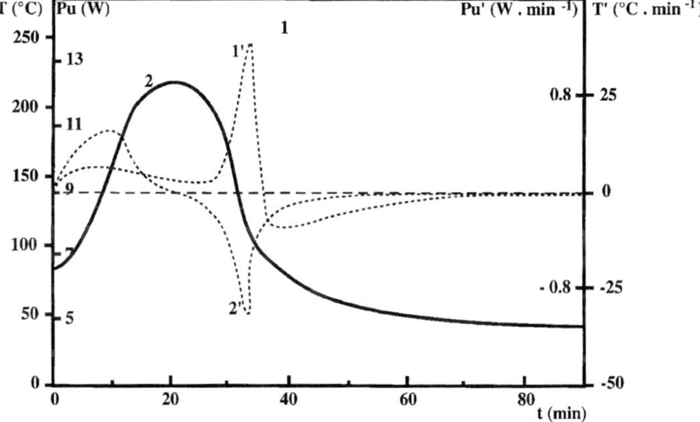

Figure 2. Curing under microwaves at P_0 = 40 W. (90 minutes). (1): $T = T(t)$; (2): $P_u = P_u(t)$; (1'): $(T)' = dT/dt$; (2'): $(P_u)' = dP_u/dt$

The fusion of the crystalline-like structure is quickly undergone; it is located in time at the first inflexion point, or at the first maximum of (T)', that is at 7.0 minutes. Then, the gelation process begins through the reaction of condensation. It is admitted that the complete gelled structure is located at the foot of the exothermic peak, when the corresponding experimental value of the extent of the reaction is assumed to be grossly around 70 %. The second inflexion point of the T curve, positioned in time at 24 minutes through the second maximum of (T)', is attributed to the sol-gel transition which separates the liquid state (easy dipolar relaxation) and the gelled state (restricted relaxation because of the high viscosity of the chemical medium).
The final crosslinking step which follows the complete gelation, is started by a fast exothermal conversion of epoxide functions within a short time (increase of 15 or 20 % of the extent of the condensation), and generates a great deal of energy within a few minutes: the temperature rapidly increases. Such a variation is only possible, because the thermal effect keeps the temperature of the medium at a level higher than the glassy transition of the epoxy network in formation, according to the Time - Temperature - Transformation (TTT) cure diagram [10], and consequently allows the condensation reaction. Now, let us consider the ascendant part of the peak of temperature which reaches its ceiling at 264.4 °C : it presents an inflexion point (33.3 min, 210.1 °C), which corresponds to the maximum of the heating rate or dT/dt, that is to say that the kinetics of conversion of the epoxides are the fastest at that time, since the rate of production of chemical heat varies as the rate of the reaction. Moreover, the fact that the temperature goes through a thermal maximum, means that the chemical heat which is still produced beyond the maximum, provided that the temperature is higher than the glassy transition temperature, can no more counterbalance the loss of energy through thermal convection from the sample towards the external medium (air at T_0 = 20 °C, circulating inside the wave guide applicator). So, beyond the maximum, the temperature does not cease going down, the viscosity increases and the polymerization finally stops. At the end, T takes a plateau value or T_p = 49.7 °C, much lower than the glassy transition temperature - determined through DSC, at a heating rate value of 10°C/min - of the formed network or T_g = 175.0 °C. This result constitutes a proof that the crosslinked epoxy resin still absorbs microwaves in the glassy state.

As for the P_u curve , it shows that, at the beginning of the electromagnetic treatment, the dielectric loss varies in the same way as the temperature. The electrical loss increases quickly because of the high polarity of the chemicals due to strong physical bonding in the crystalline-like medium. Then, an inflexion point is observed, beyond which the time dependence of P_u

changes : the fusion has become significant and the polarity of the prepolymeric matter drops, whereas the microwave activated dipolar relaxation inside the liquid rises. One assumes that the fusion transition is located at 8.5 minutes (abscissa of the inflexion point), whereas that determined from the T curve is 7.0 minutes. This time gradient, due to uncertainty in the measurements, is not significant : the interesting result is that both times are of the same magnitude. But, even when it is expected that, beyond the transition, the temperature reaches a plateau value, the dielectric loss goes down, whereas T still increases. Such a divergence finds its origin in the progress of the curing reaction which starts as soon as the preceding transition is undergone. Indeed, two phenomena with antagonistic effects enter into competition : i) the increase of temperature which helps the dipolar relaxation - P_u is expected to rise -, and ii) the increase of viscosity of the chemical medium in relation with the extent of the conversion of the epoxies, which progressively decreases the dielectric loss. Indeed, the increasing viscosity not only restricts the dipolar relaxation of the polar species working at the microwave frequency, but also induces a shift of the dielectric absorption towards low frequencies which initiates a progressive decoupling between microwaves and the organic prepolymer on curing. The conversion of epoxies which absorb in the 2 450 MHz region, has also to be taken into consideration. The presence of a maximum is a compromise between all preceding phenomena. By comparing the relative position in time of the T and P_u curves, the maximum of P_u can be attributed to the sol-gel transition, located now at 20.5 min, instead of the preceding value extrapolated from the T curve, that is 24.0 min. But, since the locating in time of a maximum is easier and more accurate than that of an inflexion point, the abscissa of the maximum of P_u is chosen for that of the sol-gel transition. Moreover, one can add that the variations of the dielectric loss are quasi immediate, whereas their thermal effect is consecutive and delayed in time. Consequently, the time gradient between both values is not surprising. The fact that P_u decreases, whereas T does not cease growing, means that the curing reaction does not need now any electromagnetic activation: the reaction is auto-activated through chemical heat on production. Therefore, microwave heating could be used in order to initiate the polymerization process until its auto-activation, and then removed. At the end, the dielectric loss tends towards a plateau value or P_{up} = 4.5 W.

An extended study of the P_0 dependence of polymerization kinetics has been carried out, when it varies from 10 to 50 W. The series of the corresponding T and P_u curves are respectively gathered in Figures 3 and 4, and the co-ordinates of the various maxima and the plateau values, T_p and P_{up}, collected in Table 1. All temperature curves present the same type of shape previously encountered for P_0 = 40 W, but some variations should be underlined as for their exothermic peak: when P_0 drops, it is shifted towards the high values of time - such a dependence means that that the kinetics of polymerization are more and more delayed -, and it broadens out. A similar tendency of P_0 dependence is observed in the case of P_u curves. At the highest values of time, the discrimination of the final segments of all T or P_u curves becomes impossible.

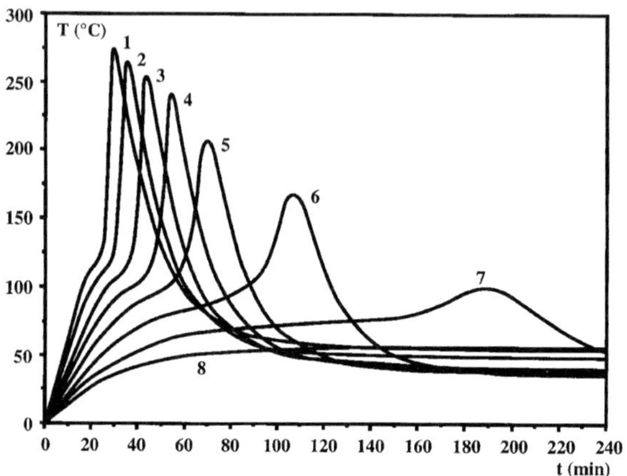

Figure 3. Curing under microwaves at variable P_0. Curves T = T(t). (1): 50W; (2): 40 W; (3): 35 W; (4): 30 W; (5): 25 W; (6): 20 W; (7): 15 W; (8): 10 W

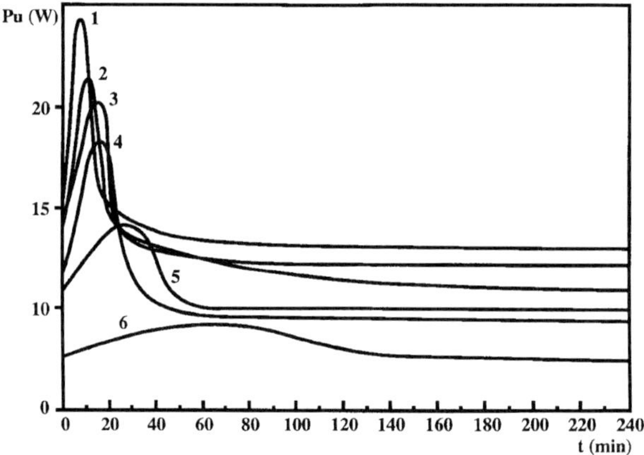

Figure 4. Curing under microwaves at variable P_0. Curves $P_u = P_u(t)$. (1) : 50 W; (2): 40 W; (3): 35 W; (4): 30 W; (5): 25 W; (6): 20 W; (7): 15 W; (8): 10 W

Table 1. P_0 dependence of the various kinetic parameters and of the glassy transition temperatures of the epoxy networks.

Curve N°	1	2	3	4	5	6	7	8
P_o (W)	50	40	35	30	25	20	15	10
P_u max								
Abscissa (min)	17.0	20.5	25.0	32.5	41.0	53.5	90.7	177.3
Ordinate (W)	15.0	12.3	9.8	8.9	7.6	5.7	3.9	2.7
T max								
Abscissa (min)	29.5	35.5	43.0	54.0	70.0	106.0	185.3	222.7
Ordinate (°C)	273.3	264.0	253.1	239.7	206.1	167.5	103.0	56.8
P_{up} (W)	6.6	4.5	4.0	3.0	2.9	2.4	1.9	2.1
T_p (°C)	55.2	49.7	42.4	41.1	38.2	36.4	53.1	56.2
T_g (°C)	175.9	175.0	174.6	172.6	172.0	/	/	/
$T_{g\infty}$ (°C)	183.5	183.8	183.8	185.0	187.8	/	/	/

The T dependence of P_u at given P_0 constitutes another approach of the kinetics of the step polymerization (Figure 5). All curves present a maximum which corresponds to the sol-gel transition. In addition, the final segments are discriminated for all P_0 values, except 10 W, because of lack of experimental precision. Curves of same type have previously encountered

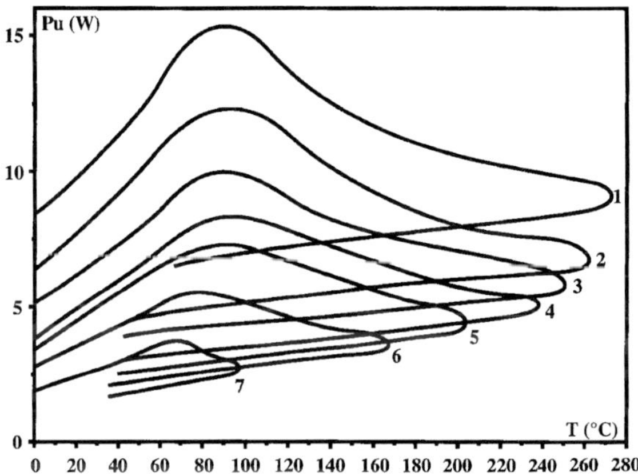

Figure 5. Curing under microwaves at variable P_0. Curves $P_u = P_u(T)$. (1): 50 W; (2): 40 W; (3): 35 W; (4): 30 W; (5): 25 W; (6): 20 W; (7): 15 W

when the temperature dependence of the loss factor or ε'' was recorded elsewhere, in the case of DGEBA/diaminodiphenylsulphone mixtures crosslinked under microwave irradiation [11]. But no information on the progress of the curing reaction and on the structural changes which occur during the polymerization, was provided and no attribution of the singular points of the various curves done. Nevertheless, such similarity between the T dependence of ε'' and P_u incites to think that the measure of the dielectric loss (P_u)gives an idea of the variations of the loss factor (ε'') during the polymerization reaction under microwaves ($P_u \propto K.\varepsilon''$).

The determination of the glassy transition temperature, or T_g, of the final crosslinked DGEBA/DDM networks, carried out through Differential Scanning Calorimetry or DSC, provides information upon the extent of the crosslinking reaction of the epoxy resin which varies as T_g. Figure 6 shows the P_0 dependence of T_g, determined at a heating rate value of 10 °C/min, when the power varies from 25 to 50 watts. All values are close to 174 °C. Lower levels of the applied power (20, 15 and 10 W) lead to fluid materials under-crosslinked: their thermal characterization can not be performed in good conditions. For P_0 = 10W, a polymerization reaction can not be excluded, but its kinetics are too slow to be appreciated with accuracy. The P_0 dependence of the glassy transition temperature which is assumed to vary as the progress of the curing, indicates that it slightly increases from 172.0 to 175.5 °C, when P_0 grows from 25 to 50 W, that is to say that the different values of the extent of conversion of the epoxies at given electrical power, are grossly of the same magnitude. Experiments carried out at higher P_0 values induce thermal degradation of the samples at core.

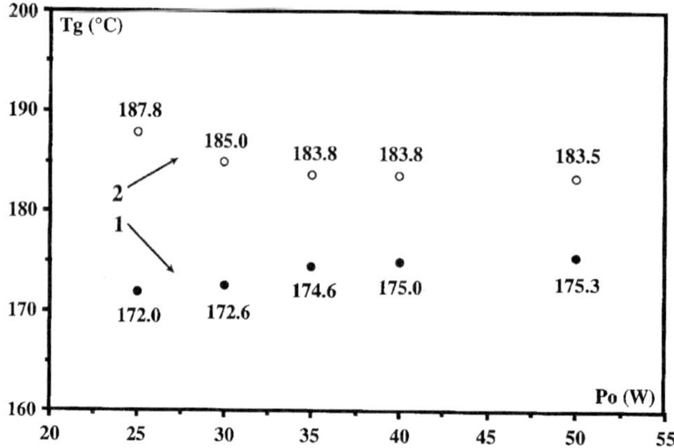

Figure 6. Po dependence of the glassy transition temperatures. (1): T_g; (2): $T_{g\infty}$

The P_0 dependence of the plateau values T_p and P_{up} (Figures 7 and 8) shows that it is of linear type from 25 W. Now, if one considers that the structure of all corresponding samples crosslinked in such conditions is quasi-identical, since their glassy transition temperature is comparable ($T_g \propto 174$ °C), the reported behavior is that of a given epoxy network. Such a linearity means that the same relaxation processes work, because no transition is detected. The expressions of T_p and P_{up} are respectively: $T_p = 19.73 + 0.71\ P_0$ and $P_{up} = -\ 2.30 + 0.18\ P_0$.

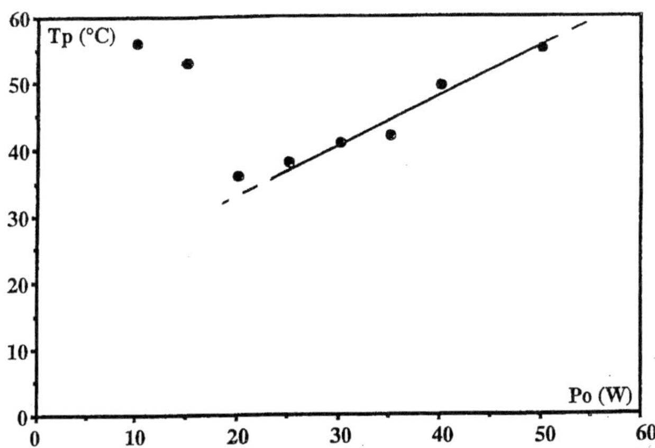

Figure 7. Curing under microwaves at variable P_0. Curves $T_p = T_p(P_0)$. (1): 50 W ; (2): 40 W; (3): 35 W; (4): 30 W; (5): 25 W; (6): 20 W; (7): 15 W; (8): 10 W

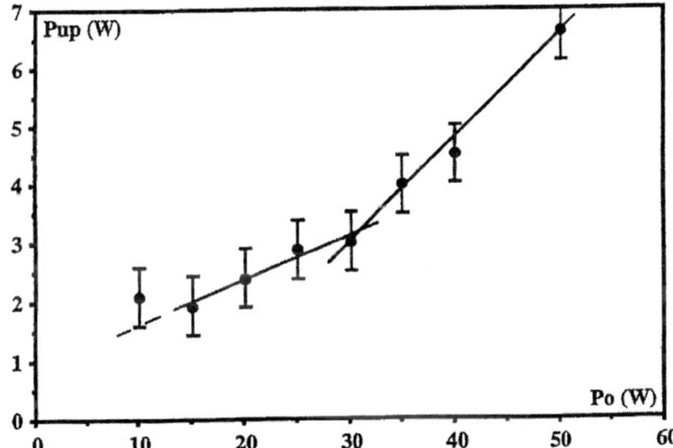

Figure 8. Curing under microwaves at variable P_0. Curves $P_{up} = P_{up}(P_0)$. (1): 50 W ; (2): 40 W; (3): 35 W; (4): 30 W; (5): 25 W ; (6): 20 W; (7): 15 W ; (8): 10 W

The constant term of the expression of T is 19.73, that is to say the temperature of the air circulating inside the wave guide used as applicator or 20 °C. Of course, the P_{up} dependence of T_p is also linear. The coefficient of P_0 in the expression of T_p can be deduced owing to the fact that the sample is in thermodynamic equilibrium. Indeed, the loss of energy of the polymeric sample by thermal convection towards the external medium (air) is given by the expression - $h.s.(T_p - T_0)$, in which h, s and T_o are respectively the convection coefficient, the external area of the sample and the temperature of the air or $T_0 = 20$ °C, whereas the thermal energy brought by microwaves is Q, with $Q = \beta.P_u = \alpha.\beta.P_0$. The thermodynamic equilibrium is reached when $Q - h.s.(T_p - T_0) = 0$. At the end, one obtains: $T_p = T_0 + (\alpha.\beta/h.s).\ P_0$, with $\alpha.\beta/h.s = 0.71$.

Now, when P_0 is lower than 25 W, T_p keeps values much higher than those extrapolated from the straight line, even if the applied power is smaller. On the other hand, the same tendency is observed for P_{up}, but with less spectacular deviations, since the values of the dielectric loss are small. All these anomalies are due to the fact that the crosslinked epoxy resins are under-polymerized and consequently more polar or less transparent to microwaves, that is to say abler to be heated through dielectric loss.

After their initial thermal characterization which provides T_g values, all samples crosslinked through microwaves at P_0 higher than 25 W, are thermally post-cured inside the oven of the DSC equipment through a series of successive analyses, in order to increase the extent of the curing reaction. A final glassy transition temperature, or $T_{g\infty}$, higher than T_g, is obtained (Figure 6). The P_0 dependence of $T_{g\infty}$ shows that it reaches its maximum value or 187.8 °C, for $P_0 = 25$ W, whereas that of the corresponding Tg is 172.0 C°, value which indicates that the used power leads to the lowest extent of the polycondensation reaction under microwaves, at least in the P_0 range studied. But, it is admitted that the structural homogeneity of the final post-cured samples varies as that of the networks just issued from polymerization reactions, and consequently, as that of the corresponding gels. If such a law is extended to microwave cured epoxy materials, one can obviously assume that the best structural homogeneity should be expected for $P_0 = 25$ W, since the rate of formation of the gel network is then the lowest, and that consequently its physical structure would have to be more homogeneous than that of the other gels obtained at upper values of P_0. Actually this is what is observed, because the corresponding $T_{g\infty}$ value is the highest of the series.

Conclusion

All results which have been described, and which are reproducible through a few hundred polymerization experiments, show that continuous microwaves are efficient to activate the curing reactions of epoxy resins of DGEBA type, when tetra-functional amino-compounds are used as crosslinking agents. The optimization of the electromagnetic treatment leads to final epoxy products with glassy transition temperatures comparable within two degrees to those of the products crosslinked through classical thermal activation carried out inside an oven. Moreover, the values of elasticity modulus of the epoxy networks, determined through uniaxial compression, are very close to 3 GPa, whatever the mode of activation of the polymerization reactions may be. So, one can assume that there is no specific effect of microwave energy on the three-dimensional epoxy structures; it is the mode of heating of the prepolymeric matter which is specific, since heat is created inside the organic medium, whereas it is brought through an external thermal source in the case of a classical thermal activation carried out inside an oven. On the other hand, direct comparisons between both kinetics of polymerization activated by classical or microwave heating can not obviously proposed, because, at given temperature, the corresponding physical states of the organic prepolymers are completely different. Indeed, when the curing is carried inside an oven, the motion of the dipolar species is random, whereas that working in the case of microwave heating is forced to be in line with the oscillating electrical field.

Now, when fillers are added to prepolymers in order to form composite materials, two cases should be considered. When the additives are inorganic and transparent to microwaves, such as glass fibers and silica powders, the electromagnetic beam is scattered when it goes through the fillers because of a gradient of optical index between the mineral and organic components. If they are electrically conductive, as metallic powders or carbon fibers, the waves are reflected by the fillers and an electrical loss or P_c, due to electronic conduction, has to be added to the dielectric loss. Moreover, in this last case, the propagation of the waves quickly becomes complicated, when the volume fraction of the additives increases, since various phenomena develop, such as multi-reflection of the beam, aggregation of the fillers and electric percolation.

Of course, the microwave activation works in all step polymerization reactions which can be thermally activated, and has been also applied to the initiation of chain polymerization reactions, as curing of unsaturated polyester - styrene mixtures. However, there are some limitation of the use of this electromagnetic process in polymer synthesis:

i) the experiments should be carried out in bulk, without any solvent - if this condition is not respected, the solvent absorbs microwaves and poor effects due to microwaves on the kinetics of polymerization are expected;
ii) micellar structures, usually encountered in radical polymerization (emulsion and suspension), can not be stabilized, when the electromagnetic beam is applied, they explode;
iii) in the case of heterogeneous polymerization using stereospecific catalysts in order to produce stereoregular polymeric chains, a change of the mechanisms of catalysis induced by electrical conduction at the level of the catalytic substrates is expected and would have to lead to a variation of configuration of the synthesized polymers, or even the non-formation of the monomer - catalytic site complex.

[1] A. Gourdenne, A. H. Maassarani, P. Monchaux, S. Aussudre, L. Thourel, *Polym. Prepr., Am. Chem. Soc., Div. Polym. Chem.* **1979**, *20(2),* 471.
[2] Q. Le Van, A. Gourdenne, *Eur. Polym. J.* **1987**, *23(10),* 777.
[3] P. Alazard, A. Gourdenne, *Material Research Society, Symposium Proceedings, Microwave Processing V* **1996**, *430*, 593.
[4] A. Gourdenne, *Ceram. Trans.* **1997**, *80*, 425.
[5] [5a] N. Beldjoudi, A. Bouazizi, D. Douibi, A. Gourdenne, *Eur. Polym. J.* **1988**, *24(1),* 49; [5b] N. Beldjoudi, A. Gourdenne, *Eur. Polym. J.* **1988**, *24(1),* 53; [5c] N. Beldjoudi, A. Gourdenne, *Eur. Polym. J.* **1988**, *24(3),* 265.
[6] [6a] Y. Baziard, S. Breton, S. Toutain, A. Gourdenne, *Eur. Polym. J.* **1988**, *24(6),* 521; [6b] Y. Baziard, S. Breton, S. Toutain, A. Gourdenne, *Eur. Polym. J.* **1988**, *24(7),* 633;
[6c] Y. Baziard, A. Gourdenne, *Eur. Polym. J.* **1988**, *24(9),* 873; [6d] Y. Baziard, A. Gourdenne, *Eur. Polym. J.* **1988**, *24(9),* 881.
[7] P. Alazard, "Activation des réactions de polymérisation par les hautes fréquences (27.12 MHz) - Application à la réticulation des matériaux composites à matrice polymère", *Thesis*, Institut National Polytechnique de Toulouse, 1997, N° 109.
[8] A. Gourdenne, *unpublished results.*
[9] K. Dušek, *Ang. Makromolek. Chem.* **1996**, *240*, 1.
[10] J. B. Enns, J. K. Gilham, *J. Appl. Polym. Sci.* **1983**, *28*, 2567.
[11] M. Ollivon, S. Quinquenet, M. Seras, M. Delmotte, C. More, *Thermochimica Acta* **1988**, *125*, 141.

Synthesis and Properties of Thermally Reversible Polyesters

*Toru Yamanaka, Akinori Kanomata, Toshihide Inoue**

Chemicals Research Laboratories, Toray Industries, Inc., 9-1 Oe-cho, Minato-ku, Nagoya 455-8502 Japan
E-mail: Toshihide_Inoue@nts.toray.co.jp

Summary: Thermally reversible polyesters were obtained by the ester formation reaction of thermoplastic polyesters with hydroxyl end groups and the diacid anhydride of tetra carboxylic acid as a thermally reversible chain extender. Typical example of the thermally reversible polyesters was obtained by the reaction of PBT (polybutylene terephthalate) and PMA (pyromellitic dianhydride). This material having twice as large molecular weight as the original PBT exhibited almost the same melt viscosity as the original. Also that thermally reversible chain extension reaction occurred without unfavorable side reaction such as cross-linking. This material shows both good processability and superior mechanical properties due to its thermally reversible characters.

Keywords: chain extension, polybutylene terephthalate, polyester, thermally reversible, viscosity

Introduction

Thermoplastic polyesters are of great importance due to their excellent balance of thermal and mechanical properties and processability in practical use. Same as other thermoplastics, properties of thermoplastic polyesters are much affected by their molecular weight. For instance, increasing the molecular weight of the polymer is one of the good solutions to realize superior toughness and impact strength. However, it is not so easy to increase the molecular weight of the polyesters polymerized by melt polycondensation process, and the processability of polyesters with high molecular weight is reduced due to high melt viscosity. In order to solve the conflict between high molecular weight and excellent processability, so called thermally reversible polymers have been proposed.[1] In the thermally reversible polymers, thermally reversible chain extension-fission manner is generally used, which should realize a polymeric material with low molecular weight at the elevated temperature during processing, and with high molecular weight at the ambient temperature for practical usage. Though many approaches to the polymers have been investigated by introducing thermally reversible

 DOI: 10.1002/masy.200350907

covalent linkage to polymers, there were few successful examples in practical sense. In most cases, chain extension or crosslinking reactions are possible, but thermally reversible chain dissociation reactions are difficult because of side reactions and thermally decomposition. One example of the trials of thermally reversible polyester[2] shows the difficulty of thermally reversible dissociation crosslinking due to the slow reaction and thermal decomposition of polyesters.

Strategy

Fig. 1 shows the strategy for the main chain type thermally reversible polymers, which is consisting of low molecular weight polymer and thermally reversible chain extender. Both components react with each other and form covalent chemical bond that dissociate at the elevated temperature. As the result of the thermally reversible covalent bond, it behaves like low molecular weight polymer at the elevated temperature and high molecular weight polymer at the ambient temperature. By the above-mentioned mechanism, it may exhibit both excellent mechanical properties due to high molecular weight and excellent processability due to low molecular weight. That character of thermally reversible polymers is suitable for the application of large structural parts and the matrices resins of fiber reinforced composite, and it also has good recyclability.

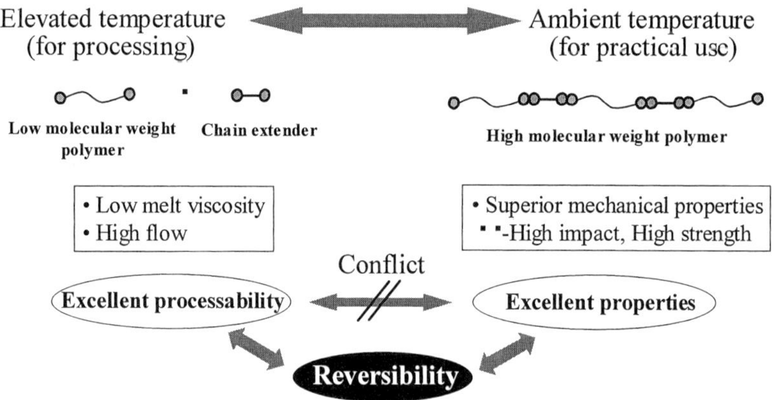

Fig. 1. Strategy of thermally reversible polymers.

We focused on the introduction of thermally reversible bond into the chain ends of polyester. We selected the reaction of diacid anhydride of tetracarboxylic acid and hydroxyl group as a thermally reversible reaction. It is very convenient to introduce hydroxyl groups to the chain ends of polyesters especially in case of polymers with low molecular weight.

Synthesis

First, we selected polybutylene terephthalate (PBT) as a basic polyester with hydroxyl end groups. And the thermally reversible PBT was obtained by melt mixing of the low molecular weight PBT with hydroxyl end groups and a diacid anhydride compound as a chain extender at 280 °C for 10 minutes followed by slow cooling. The molecular weight of the obtained thermally reversible PBT was determined by gel permeation chromatography (GPC).

The relation between the structure of chain extender and molecular weight increment after the reaction with low molecular weight PBT is shown in Fig. 2. All the chain extenders tried in this study exhibited a function of chain extender. Among these compounds, pyromellitic dianhydride was most effective.

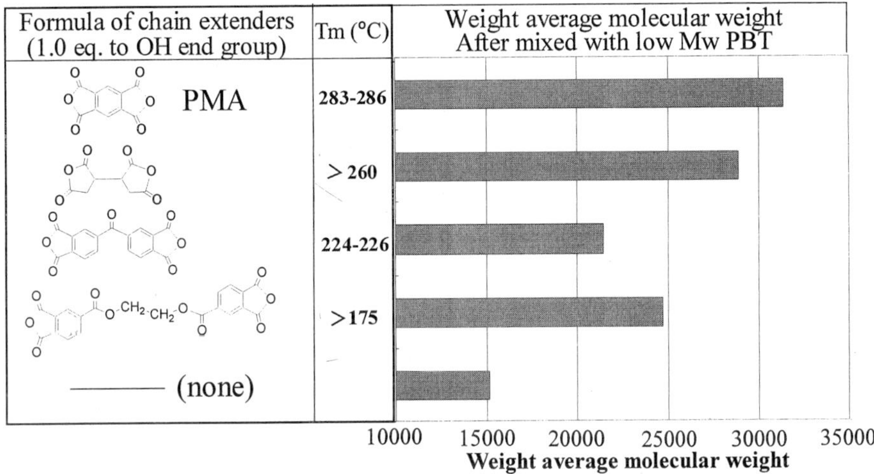

Fig. 2. Weight average molecular weight of polymers obtained by mixing of low Mw PBT and tetracarboxylic acid dianhydride compounds.

Thermally Reversible Phenomena

Fig. 3 shows the relation between melt viscosity measured at 280 °C and weight average molecular weight. This shows that the ordinary PBT has good correlation between melt viscosity and molecular weight. On the other hand, high molecular weight PBT obtained by the reaction of low molecular weight PBT having OH end groups and PMA exhibited as low melt viscosity as that of the low molecular weight PBT itself. This phenomenon explains that high molecular weight PBT changes its molecular weight at the elevated temperature. This phenomenon also means that this thermally reversible PBT may realize both high molecular weight and low viscosity.

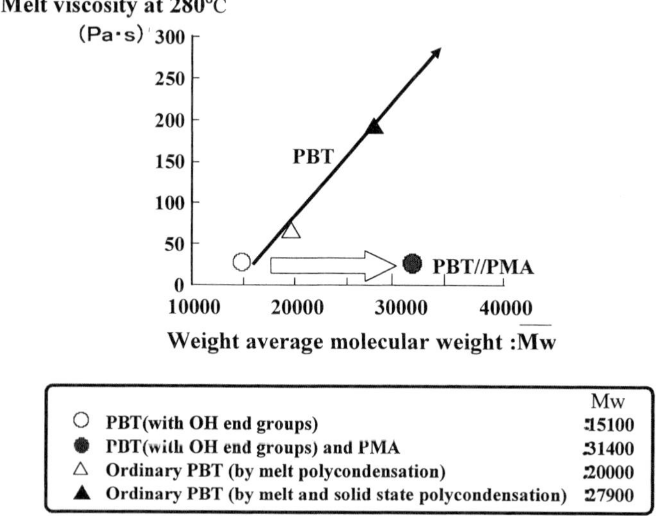

Fig. 3. Relation between melt viscosity and weight average molecular weight; Melt viscosity was determined by capillary-type rheometer.

Thermally Reversible Temperature

In order to determine the thermally reversible temperature, especially the temperature that the ester bonds dissociate, we proceeded following experiment. Low molecular weight PBT having OH end groups (Sample A) and PMA were charged in the test tube in the nitrogen atmosphere and heated up to 280 °C, then kept the temperature for 10 minutes followed by slow cooling. By this procedure, high molecular weight PBT with the weight average molecular weight of more than 30,000 (Sample B) was formed. The obtained Sample B was set on the hot plate and heated up to 280 °C followed by rapid quenching in liquid nitrogen to form Sample C. The molecular weight of Sample C was almost the same as Sample A. This means that the thermally reversible bond formed by the reaction OH end group and acid anhydride group was completely dissociated at 280 °C (Fig. 4). This method can be applied to determine the reversible temperature by changing the resident temperature on the hot plate.

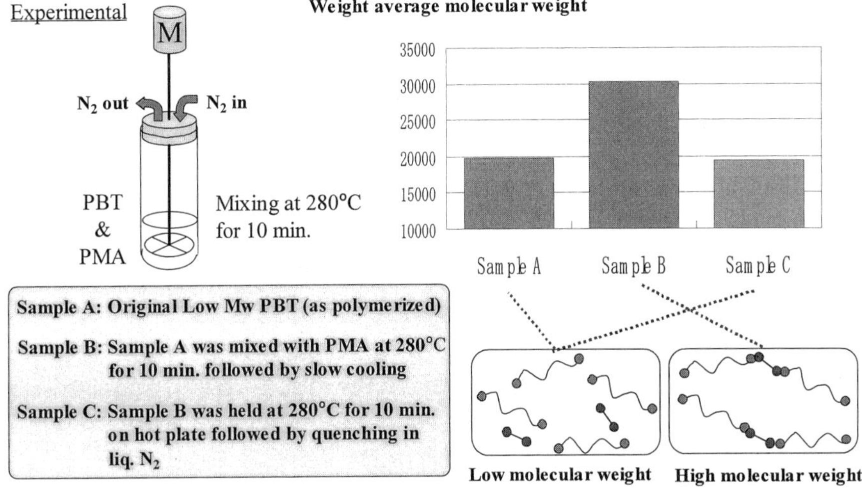

Fig. 4. Experiment for the determination of the thermally reversible temperature.

The result of the detection of reversible temperature is shown in Fig. 5. It was certain that the thermally reversible ester bond was completely dissociated above 240 °C. But a little amount of reversible ester bond was dissociated less than 230 °C.

The next step of the experiment is to determine the formation temperature of thermally reversible ester bond. We used the Sample C obtained by rapid quenching in liquid nitrogen from 280 °C as the starting material. Sample C was held at the varied resident temperature on the hot plate for 10 minutes followed by rapid quenching in liquid nitrogen, and then the molecular weight was measured by GPC. The result of the experiment was described in Fig. 6. The increment of the molecular weight was determined even in 180 °C, and the formation of thermally reversible ester bond was occurred less than 230 °C.

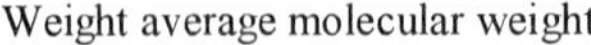

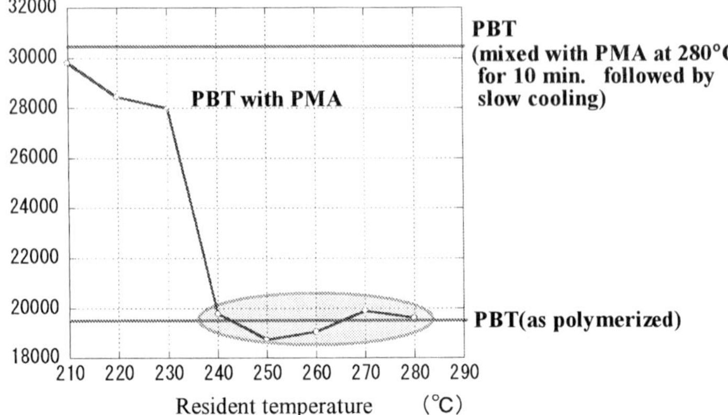

Fig. 5. Dissociation temperature of thermally reversible ester bond; Low Mw PBT and PMA were mixed at 280°C for 10 min followed by slow cooling, then held at the resident temperature on the hot plate for 10 minutes followed by rapid quenching in liquid nitrogen. Mw was measured by GPC.

Weight average molecular weight

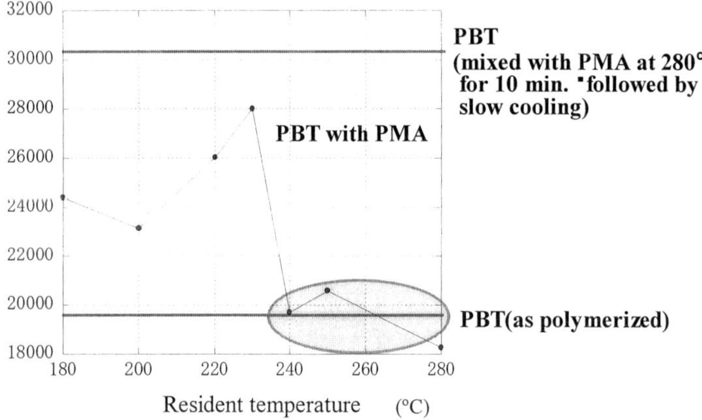

Fig. 6. Formation temperature of thermally reversible ester bond; Low Mw PBT and PMA were mixed at 280°C for 10 min followed by rapid quenching in liquid nitrogen, then held at the resident temperature on the hot plate for 10 minutes followed by rapid quenching in liquid nitrogen. Mw was measured by GPC.

Properties of Thermally Reversible PBT

We tried to clarify the structure of the thermally reversible PBT obtained by the reaction of low molecular weight PBT and PMA as a thermally reversible chain extender compared to a copolymer consisting of PBT and pyromellitic acid. We synthesized a copolymer of PBT and pyromellitic acid by ordinary method. We investigated the correlation between concentration and reduced viscosity. The result is shown in Fig. 7. It is clear that the ordinary PBT and the thermally reversible PBT using thermally reversible chain extender are the similar correlation, on the other hand, copolymer of PBT and pyromellitic acid exhibit steeper correlation compared to former two polymers.

This result seems to suggest that the thermally reversible PBT has linear structure and the copolymer of PBT and pyromellitic acid has branched structure. The pyromellitic acid acts as a branch generator when condensed with PBT. These two polymers have totally different structure.

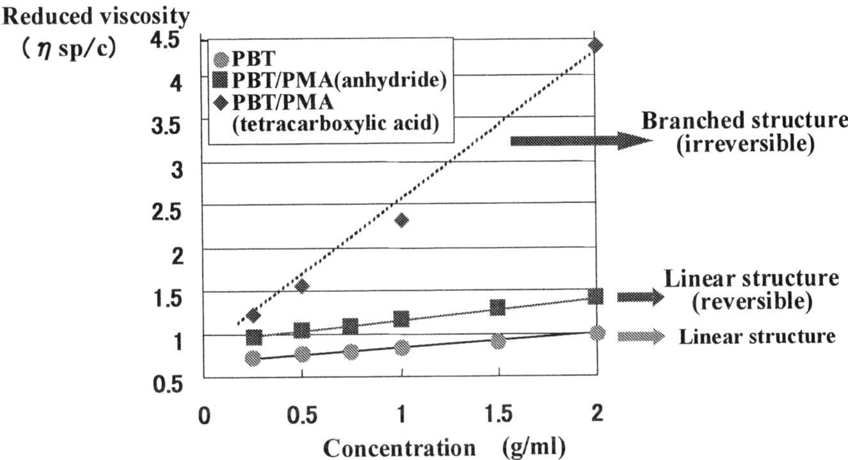

Fig. 7. Correlation between concentration and reduced viscosity (o-chlorophenol at 25 °C).

Thermal Properties

The DSC analysis of the thermally reversible PBT was shown in Table 1 compared to the ordinary PBT, which exhibits that the thermally reversible PBT has smaller heat of fusion and smaller heat of crystallization than those of the ordinary PBT without PMA, even it has almost the same melting point and crystallization point as the ordinary PBT. The results show the thermally reversible PBT has low crystallinity due to the introduction of PMA molecules in the main chain of PBT.

Table 1. Thermal properties of thermally reversible PBT.

sample	Thermal properties			
	Tm2(℃)	⊿Hm2(J/g)	Tc2(℃)	⊿Hc2(J/g)
PBT	224.6	48.5	183.8	58.4
PBT/PMA	223.5	33.2	182.0	48.0

Mechanical Properties

Mechanical properties of injection molded articles of the thermally reversible PBT are shown in Table 2 compared to the ordinary PBT and the low molecular weight PBT with OH end groups before mixing with PMA. The thermally reversible PBT has excellent toughness due to high molecular weight at the ambient temperature. Though the reason for the high modulus of it has not been clear, one possibility is the formation of pseudo-crosslinking structure by the interaction of between carboxyl groups formed by the reaction of OH end groups of PBT and PMA.

Table 2. Mechanical properties of thermally reversible PBT.

	Ordinary PBT (Mn:20000)	PBT with OH end group (Mn:15100)	PBT with OH end group and PMA (Mn:31400)
Notched Izod impact strength (J/m)	53.8	49.2	54.2
Flexural modulus (GPa)	2.40	2.40	2.53
Flexural strength (MPa)	77.6	77.8	82.7
Tensile strength (MPa)	56.3	54.1	57.8
Elongation at break (%)	15.1	8.2	52.4

Conclusion

Thermally reversible polyesters were obtained by the reaction of polyesters with OH end group and dianhydrides of tetracarboxylic acid. Obtained thermally reversible linear polyesters show unique properties such as,

-Low molecular weight and low melt viscosity at high temperature

-High molecular weight at ambient temperature

-Excellent mechanical properties due to high molecular weight

Thermally reversible reaction proceeded without annoying side reaction such as crosslinking or decomposition. A variety of molecular structure can be applied as the need arises.

[1] L. P. Engle and K. B. Wagener, *J. M. S.-Rev. Macromol. Chem.Phys.*, **1993,** *C33 (3)*, 239.
[2] J. R. Jones et al., *Macromolecules*, **1999**, *32*, 5786.

Synthesis of Fully Conjugated Aromatic Polyazomethine from AB-type Monomer Using Soluble Precursor Method

Toshihiko Matsumoto, Toshihiko Matsuoka, Yoshitaka Suzuki, Kenshi Miyazawa, Toshikazu Kurosaki, the late Takayuki Mizukami*

Center for Nanoscience and Technology, Tokyo Institute of Polytechnics, 1583 Iiyama, Atsugi, Kanagawa 243-0297 Japan
Email: matumoto@chem.t-kougei.ac.jp

Summary: A new AB-type monomer, *N,N*-bistrimethylsilylated *p*-aminobenzaldehyde diethyl acetal was prepared *via* three steps from *p*-bromoaniline as a starting material. The two-stage polymerization involving a soluble precursor polymer process gave a poly(*p*-phenylenevinylene)-type polyazomethine, poly(1,4-phenylenenitrilomethylidyne). The first stage of polymerization was carried out in tetrahydrofuran or hexamethylphosphoramide containing water at room temperature. In the second stage, the polymer was thermally converted into the final polyazomethine by heating over 300°C to form a free-standing film. The film was reddish brown and insoluble in common organic solvents. The investigation of the first-stage products by means of MALDI-TOF mass spectroscopy proved the oligomers with 4-11 repeating units per molecule. From the ^{1}H-NMR analysis of the model reaction, the polymerization mechanism was found to be a stepwise polycondensation of 4-diethoxymethylaniline which was formed by removal of two silyl groups of the monomer.

Keywords: conjugated polymers, poly(1,4-phenylenenitrilomethylidyne), polyazomethine, polycondensation, soluble precursor

Introduction

It has been known that fully conjugated polyazomethines formed from aromatic diamines and aromatic dialdehydes precipitate even at the initial stage of the polymerization due to their rigid-rod chain structures,[1] although they are expected to exhibit excellent properties such as electric conductivity, nonlinear optical property, light-emitting property, or many other interesting characteristics.[2-8] Several approaches have been undertaken to improve the processibility of conjugated polyazomethines, for example by introducing various substituted benzene ring in the main chain,[9] by using monomers containing certain heterocyclic units such as thiophene,[8,10] phenylquinoxaline ring,[11,12] or monomers with cardo-structure,[13] tetraphenylethene,[14] and others.[15] Processibility of insoluble polymers has generally approached *via* two methodologies

 DOI: 10.1002/masy.200350908

processing into thin film form, namely (a) the use of soluble precursor polymers which are thermally converted into insoluble materials after film formation,[16,17] and (b) chemical vapor deposition (CVD) involving co-sublimation of two or more reactive monomers which impinge on a substrate where they react, typically *via* a polycondensation mechanism, to form a polymer film.[18,19] The latter methodology, CVD, has been shown to be an effective and efficient means of producing polyazomethines with high conjugation lengths, and vacuum deposited aromatic polyazomethines were described for the preparation of organic light emitting devices.[20] Excellent examples of the former can be seen in the polyimide synthesis and poly(*p*-phenylenevinylene) (PPV) synthesis.[16] In 1987, we synthesized a full conjugated polyazomethine, poly(1,4-phenylenemethylidynenitrilo-1,3-phenylenenitrilomethylidyne) (*pm*PAM), using a soluble precursor method (see Scheme 1). The polyazomethine *pm*PAM was obtained by two different processes, namely (a) from terephthalaldehyde diethyl acetal and *N,N'*-bistrimethylsilylated 1,3-diaminobenzene,[21] and (b) from terephthalaldehyde and *N,N,N',N'*-tetrakistrimethylsilylated 1,3-diaminobenzene.[22] In both cases the resulting polyazomethine films were flexible but insoluble in common organic solvents.

(a) R: $SiMe_3$ or H — Soluble Precursor Polymer — polyazomethine poly(1,4-phenylenemethylidynenitrilo-1,3-phenylenenitrilomethylidyne) (*pm*PAM)

(b) Soluble Precursor Polymer — *pm*PAM

Scheme 1. A soluble precursor methodology for a polyazomethine synthesis from: (a) terephthalaldehyde diethyl acetal and *N,N'*-bistrimethylsilylated 1,3-diaminobenzene; (b) terephthalaldehyde and *N,N,N',N'*-tetrakistrimethylsilylated 1,3-diaminobenzene.

Optical properties such as a second harmonic generation (SHG) and a refractive phenomenon are known to depend strongly on the orientation degree of a net dipole moment of the monomer unit. Common polyazomethines prepared from two kinds of monomers, namely dialdehydes and diamines, have two azomethine linkages per a repeating unit. The dipole moments are counterbalanced each other (Figure 1(a)) when the discussion on the polymer structure is limited to one dimension. However, if an appropriate AB-type monomer is used, we can synthesize a polyazomethine having oriented dipole moments (Figure 1(b)), which is expected to show interesting properties, for example, a large intensity in SHG.

In the present article we report the synthesis of a new AB-type monomer, *N,N*-bistrimethylsilylated *p*-aminobenzaldehyde diethyl acetal, and the preparation of a fully conjugated polyazomethine, poly(1,4-phenylenenitrilomethylidyne), with oriented dipole moments using a soluble precursor route. The structure of the precursor polymer has been investigated using matrix-assisted laser desorption/ionization time-of-flight mass spectroscopy (MALDI-TOF MS). On the basis of ^{1}H-NMR analysis of the model reaction, a polymerization mechanism is proposed.

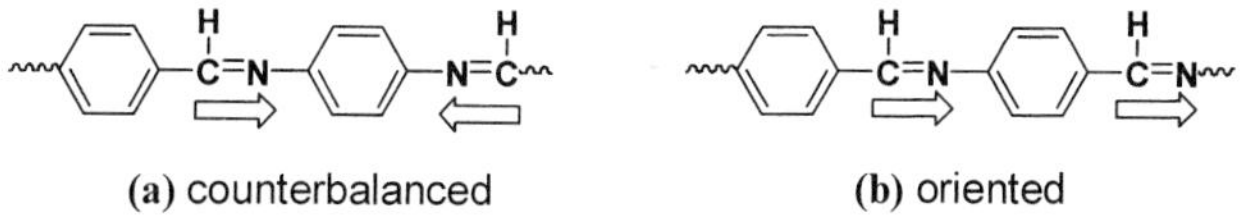

(a) counterbalanced (b) oriented

Fig. 1. The direction of dipole moments in the polyazomethines synthesized from: (a) a dialdehyde and a diamine (*counterbalanced*); and (b) an AB-type monomer (*oriented*).

Experimental Parts

Chemicals and Solvents

p-Bromoaniline, 1,1,1,3,3,3-hexamethyldisilazane (HMDS), trimethylchlorosilane (TMCS), bromoethane, magnesium, triethoxymethane, chloral hydrate, deuterated chloroform ($CDCl_3$), deuterated *N,N*-dimethylformamide (DMF-d_7) and hexamethylphosphoramide (HMPA) were used as received (Tokyo Kasei Organic Chemicals or Sigma Aldrich Chemical Co.). Tetrahydrofuran (THF) and hexane were refluxed with CaH_2 and then fractionally distilled. 1,8,9-Anthracenetriol (dithranol) as a matrix for MALDI-TOF MS was analytical-grade material (Sigma Aldrich Chemical Co.), used as supplied.

Characterization

Infrared spectra were obtained with a JASCO VALOR III Fourier transform spectrometer. ^{1}H and ^{13}C NMR spectra were run on solutions in $CDCl_3$ or DMF-d_7 and recorded using a JEOL JNM-LA 500 spectrometer. Dichloromethane was used as the internal standard for both monomer and polymers ($\delta(^1H)$=5.306 ppm; $\delta(^{13}C)$=53.37 ppm). MALDI-TOF MS experiments were performed on a Voyager-DE PRO-T system (Applied Biosystems). The desorption/ionization was induced by a pulsed N_2 laser emitting at 337 nm with a 3 ns pulse width. The mass spectra were obtained at 20 kV accelation voltage and recorded in linear and in reflection mode. 1,8,9-Anthracenetriol (dithranol) was used as the matrix. The samples for MALDI-TOF MS analysis were prepared by mixing adequate volumes of the matrix solution (dithranol, 1 mg/mL in THF) and the polymer solution (the polymer, 1mg/1mL in THF) to obtain a 1:1 v/v ratio. Then 1 μL of the mixture was spotted on the sample holder and slowly dried to allow matrix crystallization. The instrument was calibrated using a mixture of angiotensin (peak at 1297.51 g/mol), ACTHs (peaks at 2094.46, 2466.72, and 3660.19 g/mol), and insuline (peak at 5734.59 g/mol). Spectra were collected as the averages of 200 scans.

Monomer Synthesis

***N,N*-Bistrimethylsilylated *p*-bromoaniline (2):** The compound **2** was prepared *via* two steps from *p*-bromoaniline according to a modified method of the literature.[23] Yield: 30% based on *p*-bromoaniline. Bp: 81°C/1mmHg (lit., [24] 106°C/1.2mmHg). ^1H NMR ($CDCl_3$, 500MHz): δ= 0.09 (18H, s, Si($\underline{CH_3}$)$_3$), 6.79 (2H, d, H^2 and H^6, $J_{2,3}$=$J_{6,5}$=8Hz), 7.33 (2H, d, H^3 and H^5, $J_{3,2}$=$J_{5,6}$=8Hz). ^{13}C NMR ($CDCl_3$, 125MHz): δ= 2.0 (Si(CH_3)$_3$), 116.8 (C^4), 131.5 (C^3 and C^5), 131.8 (C^2 and C^6), 147.2 (C^1).

***N,N*-Bistrimethylsilylated *p*-aminobenzaldehyde diethyl acetal (3):** In a 2-L four-necked flask equipped with a Dimroth condenser, a mechanical stirrer, a dropping funnel, and a nitrogen inlet was placed 6.1 g (0.25 mol) of magnesium. The whole apparatus was dried as described above, then 300 mL of tetrahydrofuran and 1.1 g (0.01 mol)) of bromoethane to activate the magnesium were charged into the flask. The bistrimethylsilylated derivative **2** (79.1 g, 0.25 mol) was added to the solution through the dropping funnel and the mixture was heated at reflux temperature until the magnesium solid disappeared (ca. 12 h). To the resulting solution 37.1g (0.25 mol) of triethoxymethane was slowly added through the dropping funnel, and then the mixture was heated at reflux temperature for 13 h. The reaction mixture was treated as described above, and the compound **3** was isolated by fractional distillation; yield: 22%. Bp: 96°C (0.2 mmHg). ^1H NMR ($CDCl_3$, 500MHz): δ= 0.09 (18H, s, Si($\underline{CH_3}$)$_3$), 1.26 (6H, t, J=7Hz, $CH_2\underline{CH_3}$), 3.58 (2H, qd, $C\underline{H}_2CH_3$), 3.63 (2H, qd, $C\underline{H}_2CH_3$), 5.54 (1H, s, $C\underline{H}(OEt)_2$), 6.90 (2H, d, H^2 and H^6, $J_{2,3}$=$J_{6,5}$=8Hz), 7.34 (2H, d, H^3 and H^5, $J_{3,2}$=$J_{5,6}$=8Hz). ^{13}C NMR ($CDCl_3$, 125MHz): δ= 2.0 ($Si(CH_3)_3$), 15.2 ($OCH_2\underline{C}H_3$), 60.6 ($O\underline{C}H_2CH_3$), 101.3 ($\underline{C}H(OEt)_2$), 119.7 (C^4), 126.7 (C^3 and C^5), 129.7 (C^2 and C^6), 147.9 (C^1). $(C_{17}H_{33}NO_2Si_2)_n$: Calcd. C 60.12 H 9.80 N 4.12; Found C 60.29 H 9.61 N 4.51.

Polymer Synthesis for MALDI-TOF MS Analysis and Film Preparation

In a 10-mL glass bottle with a polyethylene cap a mixture of 0.41 g (1.2 mmol) of **3** and 0.40 g (22 mmol) of water in 2.0 g of THF or HMPA and a magnetic stirring bar were placed and stirred under a nitrogen atmosphere at room temperature for 2 weeks. The solution given by polymerization in HMPA was cast on a glass plate, and the glass plate was heated over 300 °C for 30 min.

Polymer Preparation for NMR Analysis

In an NMR sample tube (5 mm o.d. and 180 mm high) 0.0545 g (0.147 mmol) of **3**, 1.0 g of deuterated DMF-d_7), and a small amount of an internal standard (CH_2Cl_2) were charged, and 5.4 mg (0.294 mmol) of water was added into the tube using a micro-syringe just prior to measurement.

Results and Discussion

As shown in Scheme 2, a novel AB-type monomer, *N,N*-bistrimethylsilylated *p*-amino-benzaldehyde diethyl acetal (**3**) was prepared *via* three steps from *p*-bromoaniline as a starting material. The first step is an ordinary mono-trimethylsilylation of the amino group using excess HMDS with 10 mol-% TMCS present at a reflux temperature without solvent for 12 h. The second step involves reaction of the monosilylated derivative **1** with excess ethylmagnesium bromide in THF, followed by TMCS.[23] The final step is formylation reaction of triethoxymethane with Grignard reagent prepared from disilylated derivative **2**.[25] The usual precautions for protecting the reaction from moisture were taken through the whole process for preparation of the monomer. All moisture-sensitive substances (**1**, **2**, and **3**) were handled in a dry box, and all the glass apparatus was flame-dried thoroughly using a gas burner, nitrogen being flowed. The monomer **3** is stable when a deprotecting reagent such as water is absent, and it can be stored for a long time in a refrigerator. The stability is explained by the steric hindrance between the diethoxymethyl and the *N,N*-bistrimethylsilylated amino groups and/or by an electronic effect of the two groups, that is a push-pull effect of them through a benzene ring.

HMDS, TMCS, 77% — EtMgBr, TMCS, 39% — EtMgBr, $HC(OEt)_3$, 22%

1 bp: 78 · /1mmHg — **2** bp: 81 · /1mmHg — **3** bp: 96 · /0.2mmHg

Scheme 2. Synthetic route to a novel AB-type monomer, *N,N*-bistrimethylsilylated *p*-aminobenzaldehyde diethyl acetal (**3**).

We attempted to synthesize the polyazomethine, poly(1,4-phenylenenitrilomethylidyne) using two-stage method, that is, through a soluble precursor polymer. The expected synthetic route is illustrated in Scheme 3. The first stage polymerization is initiated with *N*-trimethylsilylated *p*-aminobenzaldehyde diethyl acetal **4** which is generated by removal of a trimethysilyl group with water. We have reported that the condensation of *N*-trimethylsilylated amino group with diethyl acetal group underwent to form polyazomethine with a high molecular weight *via* a precursor polymer, although the precursor has not been isolated and characterized yet.[21] In the second stage, the precursor polymer, if produced, is thermally converted into the corresponding polyazomethine. In the present study, the first stage polycondensation was carried out in THF or HMPA containing 15-16 wt-% of water. MALDI-TOF MS was a useful tool in the structural characterization of the precursor and the final form of these polyazomethines.[26] The result of products obtained in THF solvent using dithranol as the matrix is displayed in Figure 2. No cationization salt such as silver (I) trifluoroacetate was used, as it might oxidize the polymer. The correct repeat unit mass difference (103.12 amu) is observed between peaks, which is caused by the phenylene azomethine unit, $-C_6H_5-N=CH-$. The spectrum shows peaks belonging to more than four series, and that the products consist of polyazomethine oligomers terminated with different end groups. The most intense peak series, at m/z = 267.44 + 103.12(n-1) + 1, is due to the protonated ions of oligomer chains terminated with diethyl acetal group at one end and monosilylated amino group at the other end (species **Bn**), where n is equal to the number of monomer units. A second series of peaks at m/z = 195.26 + 103.12(n-1) + 1 can be assigned to the protonated ions of oligomer chains terminated with diethyl acetal group at one end and free amino group at the other end (species **An**). A third series of peaks appearing with low intensity at m/z = 399.62 + 103.12(n-1) + 1 corresponds to protonated ions of oligomer chains terminated with diethyl acetal group at one end and disilylated amino group at the other end (species **Cn**).

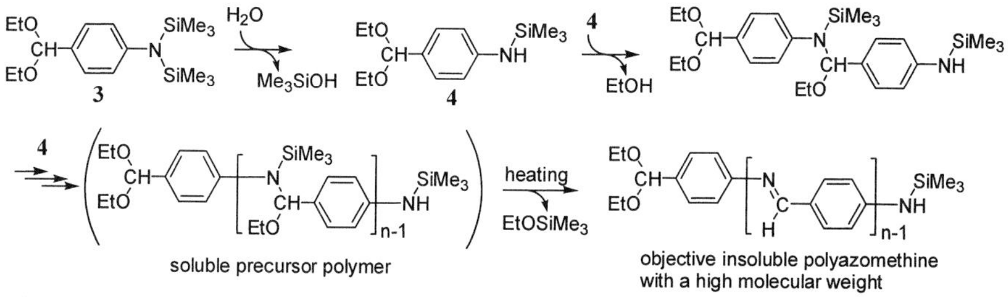

Scheme 3. Expected synthetic route to polyazomethine from *N,N*-bistrimethysilylated *p*-aminobenzaldehyde diethyl acetal (**3**) using soluble precursor method.

The large amount of oligomer chains containing free or monosilylated amino groups as end groups (series **An** or **Bn**) indicates that an extensive hydrolysis reaction of disilylated amino groups occurred, either due to the water added at initial stage of the polymerization or to the atomospheric air moisture when preparing the samples for MALDI-TOF MS analysis. The last series of peaks appearing with relatively high intensity at m/z = 432.70 + 103.12(n-2) + 1 is due to protonated ions of oligomer chains (species **Dn**). Although the structures corresponding to the mass peak series **Dn** are unknown, we propose some possible ones which may contain a azomethine precursor unit such as -C_6H_5-N($SiMe_3$)-CH(OEt)- (see Figure 3). The precursor polymers are thought to be very unstable, and once formed they may be quickly converted into polyazomethines. The azomethine oligomers having monomer units in a range from 4 to 11 were observed. At longer reaction time (after 1 week) the sample showed precipitate formation and was subjected to extraction with THF, to have a soluble portion suitable to MALDI-TOF MS analysis. As the precipitated polymers have higher molecular weights than those in the soluble portion, polymerization degrees of the polymers produced indeed may be more than 11.

% Intensity
mass (m/z)
An: 195.26 + 103.12(n-1) + 1
Bn: 267.44 + 103.12(n-1) + 1
Cn: 339.62 + 103.12(n-1) + 1

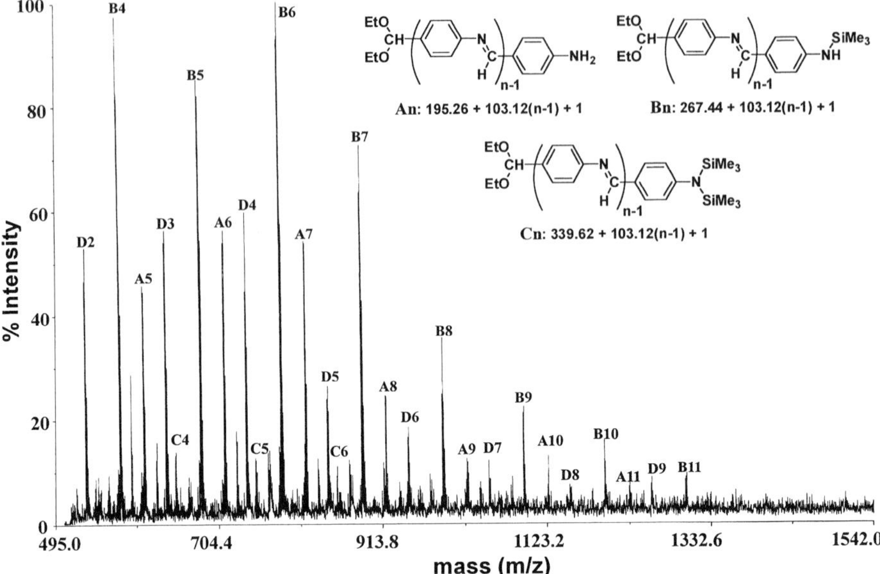

Fig. 2. MALDI-TOF mass spectrum, in the range 500-1500 g/mol, of the products obtained by reaction of **3** with water in THF at room temperature for 2 weeks. Species exist with an attached proton.

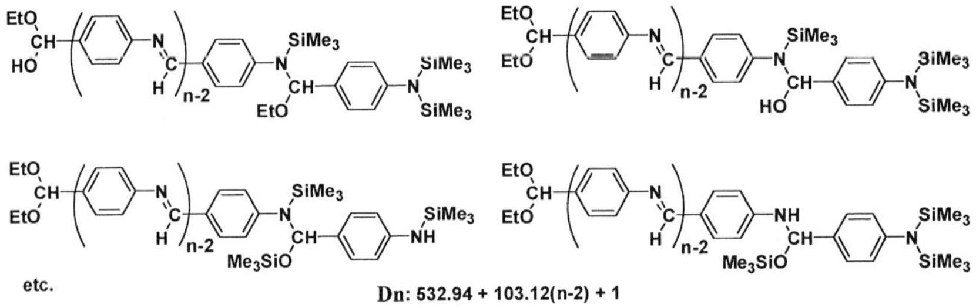

etc.
Dn: 532.94 + 103.12(n-2) + 1

Fig. 3. Possible structures corresponding to the mass peak series **Dn**.

A model reaction, which was carried out in an NMR tube containing a DMF-d_7 solution of **3** and two times molecular amount of water, was studied by ^{1}H NMR spectroscopy in order to estimate the polymerization mechanism. At each interval after adding water to the DMF-d_7 solution of **3**, ^{1}H NMR spectrum was determined. The aromatic proton region is expanded and displayed in Figure 4. The signal intensities of aromatic protons corresponding to **3** (signals α and β) were decreased gradually. Proton signals due to the desilylated compound (signals A and B), appeared just after adding water and were increased gradually as shown in Figure 4(a). The signals had the maximum intensity after several hours. It is noteworthy that no signals corresponding to the monosilylated derivative (compound **4** in Scheme 3) were observed at any time in DMF-d_7. The result indicates that the two trimethylsilyl groups were removed almost at the same time and that a life-time of the monosilylated derivative may be very short. It is also supported by the fact that a signal due to trimethylsilanol methyl proton ($(C\underline{H}_3)_3SiOH$) appeared at -0.43 ppm just after adding water and that no signal corresponding to monotrimehylsilylated compound ($(CH_3)_3Si$-NH-) was detected over the whole time. After approximately 30 min, aromatic proton signals assigned to the first formed azomethine (signals 1-4), which is a dimer terminated with an acetal group at one end and a free amino group at the other end, appeared with very low intensity. Judging from the spectral change, the dimerization rate is estimated to be slow, although the signal intensities became higher with an increase of time. After 14 days, the signals due to a trimer (signals 1'-6') are observed and those to the higher molecular weight products than a trimer appear with low intensity as shown in Figure 4(b). In ^{1}H-NMR spectrum, an azomethin proton signal, -C$\underline{H}$=N-, is observed in the range of 7.9-8.3 ppm. The azomethine proton region is expanded and displayed in Figure 5. At one hour after adding water, only a signal assigned to the first produced azomethine (a dimer) was observed at 7.92 ppm. The signal intensity became higher with an increase of time, and had a maximum intensity after several days. In the spectrum after 4 hours (Figure 5(a)), two azomethine proton signals, x and y, corresponding to a trimer can be also observed at 7.98 and 8.18 ppm, respectively. After 14 days, the signals assigned to high molecular weight azomethine oligomers such as a tetramer, a pentamer, etc., appear (signals z') other than those to a dimer and a trimer (Figure 5(b)). After several hours, a signal due to hexamethyldisiloxane ($[(CH_3)_3Si]_2O$) appeared at -0.39 ppm with very low intensity. The intensity of the signal increased with an increase in time, whereas that due to trimethylsilanol methyl proton at -0.43 ppm degreased. Signals corresponding to ethanol, which was produced by

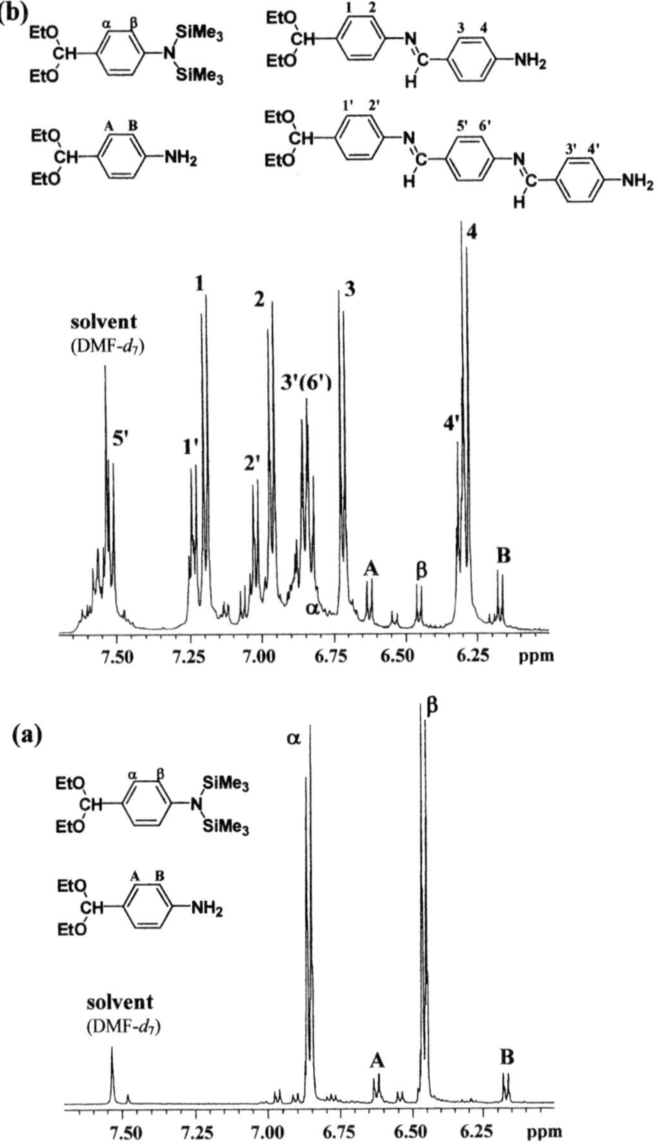

Fig. 4. Expanded ^{1}H NMR spectra (aromatic proton region, 6.0-7.7 ppm) of the mixture of **3** and water in DMF-d_7: (a) 0 h; (b) 14 days after adding water.

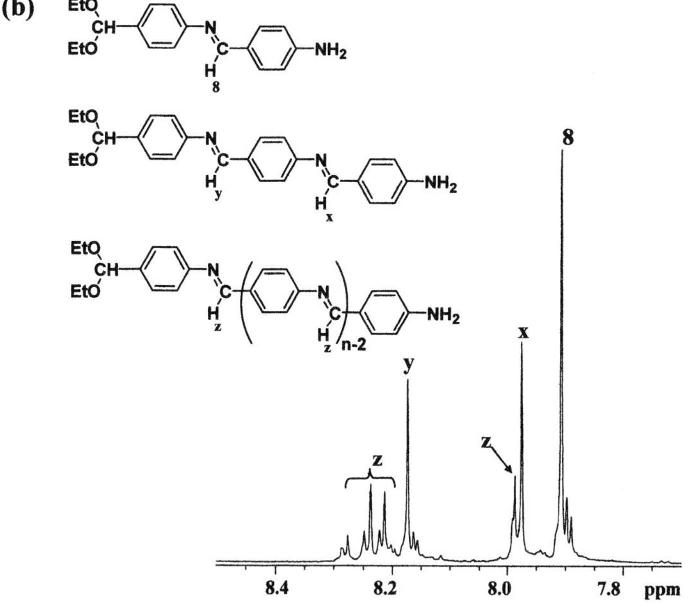

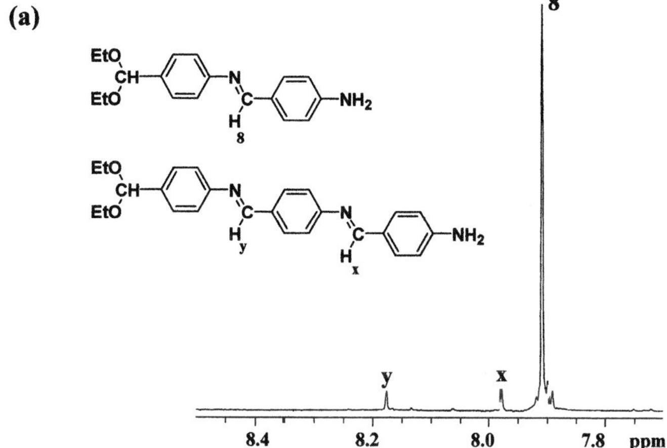

Fig. 5. Expanded ^{1}H NMR spectra (azomethine proton region, 7.7-8.5 ppm) of the mixture of **3** and water in DMF-d_7: (a) 4 h; (b) 14 days after adding water.

reaction of a free amino group with a diethoxymethyl (diethyl acetal) group, appeared after about 30 min, and the intensity increased gradually throughout the whole time. On the basis of these results, a possible mechanism for the model reaction in DMF-d_7 is assumed as shown Scheme 4. In the first step, the monomer **3** reacts with water and generates *p*-aminobenzaldehyde diethyl acetal **5** and trimethylsilanol. As trimethysilanol is very sensitive and unstable, it undergoes condensation to form hexamehyldisiloxane and water.[27] The next step involves condensation of two molecules of **5** to give the first azomethine **6** (a dimer) and ethanol. The compound **6** reacts with **5** to form a trimer and ethanol, and condensation between two molecules of **5** gives a tetramer.

In the NMR spectral analysis of the model reaction carried out in a NMR tube using DMF-d_7 as a solvent, unfortunately we cannot find the evidence for the existence of soluble precursor polymers. On the other hand, in the MALDI-TOF MS spectrum of the polymerization products obtained in THF solution, the peak series due to the oligomers terminated with an *N*-monotrimethylsilylated or an *N,N*-bistrimethylsilylated amino group at the one end were observed, and it would suggest that the polymerization proceeds *via* soluble precursor polymers with a very short life-time. The difference in the polymerization mechanism may be caused by the reaction condition, such as solvent, the ratio of monomer and water, etc. The further characterization of the polymer and the study of polymerization mechanism are currently under way.

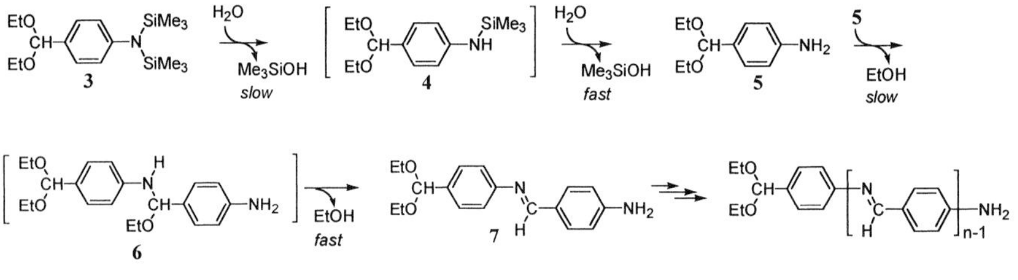

Scheme 4. A possible mechanism for the model reaction of **3** with water in DMF-d_7 to form polyazomethine oligomers.

The free-standing but somewhat brittle film was obtained by heating the polymerization solution (HMPA) on a glass plate. The film was insoluble in common organic solvents and its color was reddish brawn. Figure 6 shows the infrared spectrum of a converted polyazomethine film prepared from the precursor in HMPA with a final heat treatment of 0.5 h over 300°C. The peak at 1621 cm^{-1} is assign to a C=N streching band, and the other assignments are based on a correlation of the peaks to characteristic group frequencies which are listed in the reference.[16a]

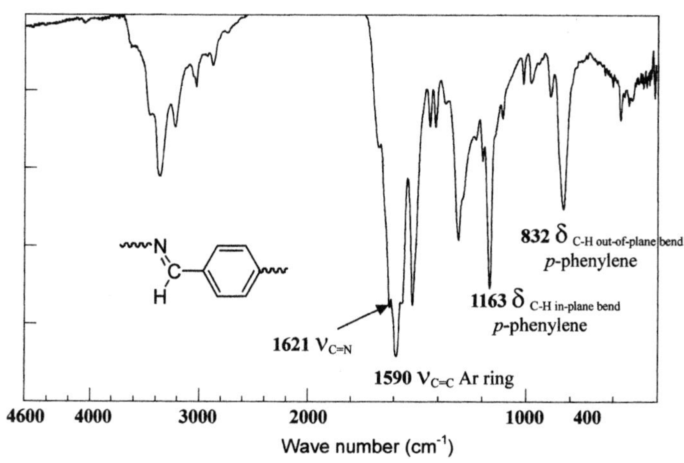

Fig. 6. FT-IR spectrum of a converted polyazomethine film prepared from the precursor in HMPA with a final heat treatment of 0.5 h over 300°C.

Conclusions

We synthesized a new AB monomer, *N,N*-bistrimethylsilylated *p*-aminobenzaldehyde diethyl acetal (**3**) via three steps from *p*-bromoaniline. A PPV-type polyazomethine, poly(1,4-phenylenenitrilomethylidyne), was prepared using a solution-processable precursor polymer. The first stage polymerization was carried out in THF or HMPA containing water at room temperature. In the second stage, the polymer was thermally converted into the final polyazomethine by heating over 300°C. The investigation of the first-stage products, which were THF-soluble fractions, by means of MALDI-TOF mass spectroscopy proved the oligomers with 4-11 repeating units per molecule. The ^{1}H-NMR spectral analysis of the products of the reaction in DMF-d_7 shows that the oligomers were produced by a stepwise polycondensation of 4-diethoxymethylaniline which was formed by desilylation of the monomer **3**. The final polyazometine was given as a free-standing film (somewhat brittle) with reddish brown color and insoluble in common organic solvents.

Acknowledgement

The authors acknowledge Mr. Masayuki Tanaka at Tokyo Institute of Polytechnics for his technical assistance. This work was partly supported by a Grant-in-Aid's for Academic Frontier from the Ministry of Education, Culture, Sports, Science and Technology, Japan.

[1] W. Steinkopf, W. Eger, *Justus Liebigs Ann. Chem.* **1938**, *533*, 270.
[2] (a) N. Ooba, S. Tomaru, T. Kurihara, Y. Mori, Y. Shuto, T. Kaino, *Chem. Phys. Lett.* **1993**, *207(4-6)*, 468. (b) S. Tatsuura, W. Sotoyama, K. Motoyoshi, A. Matsuura, T. Hayano, T. Yoshimura, *Appl. Phys. Lett.* **1993**, *62*, 2182.
[3] W. Li, M. Wan, *Solid State Commun.* **1994**, *92*, 629. d) S. Destri, W. Porzio, Y. Dubitsky, *Synth. Metals* **1995**, *75*, 25.
[4] R. Nomura, Y. So, A. Iizumi, Y. Nishihara, K. Yoshino, T. Masuda, *Chem. Lett.* **2001**, (9), 916.
[5] W. Fischer, F. Stelzer, F. Meghdadi, G. Leising, *Synth. Metals* **1996**, *76*, 201.
[6] L. S. Park, Y. S. Han, S. D. Kim, J. S. Hwang, *Mol. Cryst. Liq. Cryst.* **2001**, *371*, 309.
[7] K. Suruga, F. Takahashi, K. Ishikawa, H. Takezoe, T. Kanbara, T. Yamamoto, *Synth. Metals* **1996**, *79*, 149.
[8] C. Amari, C. Pelizzi, G. Prediere, S. Destri, W. Porzio, *Synth. Metals* **1995**, *72*, 7.
[9] S.-G. Kim., S.-J. Lee, M.-S. Gong, *Macromolecules* **1995**, *28*, 5638.
[10] C. Wang, S.Shieh, E LeGoff, M. G. Kanatzidis, *Macromolecules* **1996**, *29*, 3147.
[11] M. Bruma, B. Schulz, T. Köpnick, R. Dietel, B. Stiller, F. Mercer, V. N. Reddy, *High Perform. Polym.* **1998**, *10*, 207.
[12] M. Bruma, E. Hamciuc, F. Mercer, T. Köpnick, B. Schulz, *High Perform. Polym.* **2000**, *12*, 277.
[13] K. H. Park, T. Tani, M. Kakimoto, Y. Imai, *Makromol. Chem. Phys.* **1998**, 199, 1029.
[14] T. Matsumoto, F. Yamada, T. Kurosaki, *Macromolecules* **1997**, *30*, 3547.
[15] A. Farcas, M. Grigoras, *High Perform. Polym.* **2001**, *13*, 201.
[16] (a) D. D. C. Bradley, *J. Phys. D: Appl. Phys.* **1987**, *20*, 1389. (b) S. Antoun, F. E. Karasz, R. W. Lenz, *J. Polym. Sci. Polym. Chem. Ed.* **1988**, *26*, 1809.
[17] J. H. Burroughes, D. D. C. Bradley, A. R. Brown, R. N. Marks, K. Mackay, R. H. Friend, P. L. Burn, A. B. Holmes, *Nature* **1990**, *347*, 539.
[18] T. Yoshimura, T. Tatsuura, W. Sotoyama, A. Matsuura, *Appl. Phys. Lett.* **1992**, *60*, 268.
[19] J. McElvain, S. Tatsuura, F. Wudl, A. J. Heeger, *Synth. Metals* **1998**, 95, 101.
[20] M. S. Weaver, D. D. C. Bradley, *Synth. Metals* **1996**, *83*, 61.
[21] T. Muneto, K. Miyazawa, T. Matsumoto, T. Kurosaki, *Polymer Preprints Japan* **1987**, *36*, 325.
[22] K. Miyazawa, T Muneto, T. Matsumoto, T. Kurosaki, *Polymer Preprints Japan* **1987**, *36*, 324.
[23] K. Yamaguchi, A. Hirao, K. Suzuki, K. Takenaka, S. Nakahama, N. Yamazaki, *J. Polym. Sci., Polym. Lett. Ed.* **1983**, *21*, 395.
[24] D. R. M. Walton, *J. Chem. Soc. C* **1966**, 1706.
[25] L. I. Smith and J. Nichols, *J. Org. Chem.* **1941**, *6*, 489.
[26] B. Grimm, R.-P. Krüger, S. Schrader, D. Prescher, *J. Fluorine Chem.* **2002**, *113*, 85.
[27] L. Birkofer and O. Stuhl, "General synthetic pathways to organosilicon compounds", in: *The chemistry of organic silicon compounds, Part 1*, S. Patai, Z. Rappoport, Eds., John Wiley & Sons, Chichester 1989, chapter10, p723.

New Monomers and Polymers via Diels-Alder Cycloaddition

Alexander L. Rusanov,[1] *Zinaida B. Shifrina,*[1] *Elena G. Bulycheva,*[1] *Mukhamed L. Keshtov,*[1] *Marina S. Averina,*[1] *Yulia I. Fogel,*[1] *Klaus Muellen,*[2] *Frank W. Harris*[3]

[1] Nesmeyanov Institute of Organoelement Compounds, Russian Academy of Sciences, Vavilov 28, 119991, Moscow, Russia

E-mail: alrus@ineos.ac.ru

[2] Max-Planck-Institute for Polymer Research, Ackermannweg,10, Mainz, 55128, Germany

[3] The University of Akron, Akron,OH44325-3909, USA

Summary: The series of new bis(naphthalic anhydrides) was prepared through Diels-Alder cycloaddition. The Diels-Alder cycloaddition was used as a synthetic route to new phenylated monomers as well as to polymers. All polymers synthesized revealed to be soluble in a wide range of organic solvents such as toluene,THF, chloroform, and displayed high thermostability. Therefore, they can be processed easily and are promising candidates for advanced coating systems as well as for electrooptical applications.

Keywords: Diels-Alder polymers, gel-permeation chromatography, isomers, phenylated polyphenylenes, synthesis

Introduction

The Diels-Alder reaction[1,2,3] has proven of outmost importance in synthetic organic chemistry. In addition to its major role in the total synthesis of natural products,[4,5] the [2+4]cycloaddition has also achieved considerable impact on synthesis of polymers. Due to the increasing interest in practical applications of the obtained materials,[6,7,8] Diels-Alder polymerization is very active field of polymer chemistry. Here we introduce the Diels-Alder reaction as a simple route to new monomers for synthesis of polynaphthoylenebenzimidazoles and polynaphthylimides as well as directly to polyphenylenes. Polyphenylenes (PPs)[9-11] polynaphthylimides (PNIs)[14-16] and polynaphthoylenebenzimidazoles (PNBIs)[12,13] represent the important group of thermally stable polymers. However, the application of these polymers is limited by their high glass transition temperature and poor solubility in organic solvents. A successful approach to

 DOI: 10.1002/masy.200350909

overcome such difficulties consisted in the incorporation of phenyl substituents into the polymers chain and was applied earlier by some of us and other authors for different classes of polymers.[15-19] However, active demand for these polymers with valuable properties defines the necessity of diversification.

Experimental Section

^{1}H and ^{13}C NMR chemical shifts were obtained using a Bruker AMX 400 spectrometer (deuterated solvents as the internal standard) and expressed in parts per million (ppm). FTIR spectra were recorded on Perkin-Elmler 1720 X. Thermogravimetrical measurements were carried out on Perkin-Elmler TGA-4; the heating rate was 10°/min. GPC analyses were accomplished on a Waters SEC, equipped with a PPS gel column using polystyrene standards. Detection was achieved with a diffraction index detector.

1,3-diphenylacetone, 4-bromo-1,8-naphthalic anhydride, phenylacetylene, p-diethynylbenzene, benzidine, 3,3'-diaminobenzidine were commercially available. The synthesis of starting bis-α-diketones was performed according to literature.[20] The bis(cyclopentadienone)s 1 a–c were synthesized by methods analogous to those already published in the literature[21] and used without further purification.

1a ^{13}C NMR (400 MHz, chloroform-d, 25°C): δ=200.01 (C=O of cyclopentadienone), 154.69, 152.59, 137.11, 136.41, 132.33, 130.00, 128.84, 128.41,128.19, 128.11,127.95, 127.41, 127.10,126.91, 126.36,125.13

^{1}H NMR (400 MHz, chloroform-d, 25°C): δ=7.59 (m, 4H), 7.28-7.21 (br, 22H), 7.19 (t, 4H), 7.05 (d,4H), 6.94 (d, 4H)

Yield: 89%

1b ^{13}C NMR (400 MHz, chloroform-d, 25°C): δ=200.01 (C=O of cyclopentadienone), 154.73, 152.70, 137.37, 136.54, 132.41, 130.11, 128.97,128.49, 128.31, 128.13,128.09, 127.56, 127.21, 127.02, 126.44, 125.21

^{1}H NMR (400 MHz, chloroform-d, 25°C): δ=7.59 (m, 4H), 7.28-7.21 (br, 22H), 7.19 (t, 4H), 7.05 (d,4H), 6.94 (d, 4H)

Yield: 87%

1c ^{13}C NMR (400 MHz, chloroform-d, 25°C): δ=199.73 (C=O of cyclopentadienone), 195.21

(C=O of benzophenone), 154.03, 152.80, 137.57, 136.62, 132.66, 130.32, 130.01, 130.03, 129.99, 129.57,129.25, 129.05,128.66, 128.10, 127.98, 127.77,127.54, 126.32, 125.32

^{1}H NMR (400 MHz, chloroform-d, 25°C): δ=7.59 (m, 4H), 7.28-7.21 (br, 22H), 7.19 (t, 4H), 7.05 (d,4H), 6.94 (d, 4H). Yield: 89%

4-phenyl-ethynylenenaphthalic anhydride was synthesized according to [22].

General Procedure for the Synthesis of Bis(naphthalic Anhydride)s 2a-c.

A 50 cm^3 Schlenk tube was charged with a suspension of 1,4-bis(2,4,5-triphenylcyclopentadienone-3 –yl)arylenes (1a-c), 4-phenylethynylenenaphthalic anhydride and diphenylether and then degassed by a serried freeze-pump-thaw cycles to remove O_2. The reaction was carried out for 7 hours at 240°C, with stirring under argon atmosphere. After the reaction was complete (check by TLC and changing the color from deep purple to yellow) the solution was precipitated into methanol, filtered off, washed with an excess of methanol, and dried in vacuo at 80°C for 24 h.

2a ^{13}C NMR (400 MHz, chloroform-d, 25°C): δ= 160.36, 160.16 (C=O of anhydride), 147.72-116.14 (C_{arom});^{1}H NMR (400 MHz, chloroform-d, 25°C): δ=8.87 (d, 2H, J=8.1), 8.70 (d, 2H, J=8.1), 8.61 (d, 2H, J=8.1), 7.60-7.66 (t, 2H), 7.32 (d, 2H), 6.87-6.40 (br, 44H)

Elemental analysis $C_{90}H_{54}O_6$(found/calc.,%): C 87.77/87.78; H 4.96/4.42.

Yield: 83%

2b ^{13}C NMR (400 MHz, chloroform-d, 25°C): 160.42, 160.21 (C=O of anhydride),146.34-115.41(C_{arom}); ^{1}H NMR (400 MHz, chloroform-d, 25°C): δ=8.40 (d, 2H, J=8.1), 8.30 (d, 2H, J=8.1), 8.19 (d, 2H, J=8.1), 7.63-7.55 (t, 2H), 7.43 (d, 2H), 6.83-6.53 (br, 48H)

Elemental analysis $C_{96}H_{60}O_6$(found/calc.,%): C 86.42/88.19; H 4.77/4.47

Yield: 74%, mp =380-408°C (DSC)

2c ^{13}C NMR (400 MHz, chloroform-d, 25°C): δ=196.19, 196.14, 196.09 (C=O of benzophenone), 160.53, 160.33 (C=O of anhydride),147.54-116.59 (C_{arom})

^{1}H NMR (400 MHz, chloroform-d, 25°C): δ=8.42 (d, 2H, J=8.1), 8.32 (d, 2H, J=8.1), 8.22 (d, 2H, J=8.1), 7.60-7.66 (t, 2H), 7.43 (d, 2H), 7.14-6.50 (br, 48H)

Elemental analysis $C_{97}H_{60}O_7$ (found/calc.,%): C 87.37/87.24; H 4.60/4.37

Yield: 91%

General Procedure for the Synthesis of Polyphenylenes 3b-c

A 50 cm^3 Schlenk tube was charged with a suspension of 1,4-bis(2,4,5-triphenylcyclopentadienone-3 –yl)arylenes (1b-c) (1 eq.), freshly sublimated p-diethynylbenzene (1 eq.) and diphenylether and then degassed by a serried freeze-pump-thaw cycles to remove O_2. The common concentration of the starting compounds was 0.7 mol/l. The reaction was carried out for 72 hours at 240°C, with stirring under argon atmosphere. After the reaction was complete the clear viscous yellow solution was precipitated into methanol, filtered off, washed with an excess of methanol, and dried under vacuum at 80°C for 24 h.

3b η_{red}=0.93 dl/g (N-MP, 25°C)

Yield: 93%, M_w =192.300 g • mol^{-1} (GPC)

3c η_{red}=0.82 dl/g (N-MP, 25°C) Yield: 91%, M_w=207600 g • mol^{-1} (GPC)

General Procedure for Synthesis of Polynaphthylimides 4a-c and Polynaphthoylenebenzimidazoles 5a-c

The three-necked flask equipped with argon inlet and a stirrer was charged with a corresponding di- or tetraamine (0.01mol), phenylated bis(naphthalic anhydride) (0.01 mol), benzoic acid (0.01 mol), benzimidazole (0.01 mol) and molten phenol. Temperature was raised stepwise from 40°C to 170°C. Reaction was carried out for 20 hours. Reaction mixture was precipitated into ethanol, filtered off, extracted by means of Soxhlet extractor and dried under vacuum at 120°C for 24 h.

Results and Discussions

Monomers

For the preparation of new processable PNI's and PNBI's highly phenylated bis(naphthalic anhydrides) (BNA's) were used as starting materials. The synthetic route leading to BNAs involved the Knoevenagel reaction of different bis(α-diketones) with a two-fold molar amount of 1,3-diphenylacetone to result phenylated bis(cyclopentadienones). At the next stage these compounds have been reacted with a two-fold molar amount of 4-(phenylethynylene)naphthalic anhydride under Diels-Alder conditions.

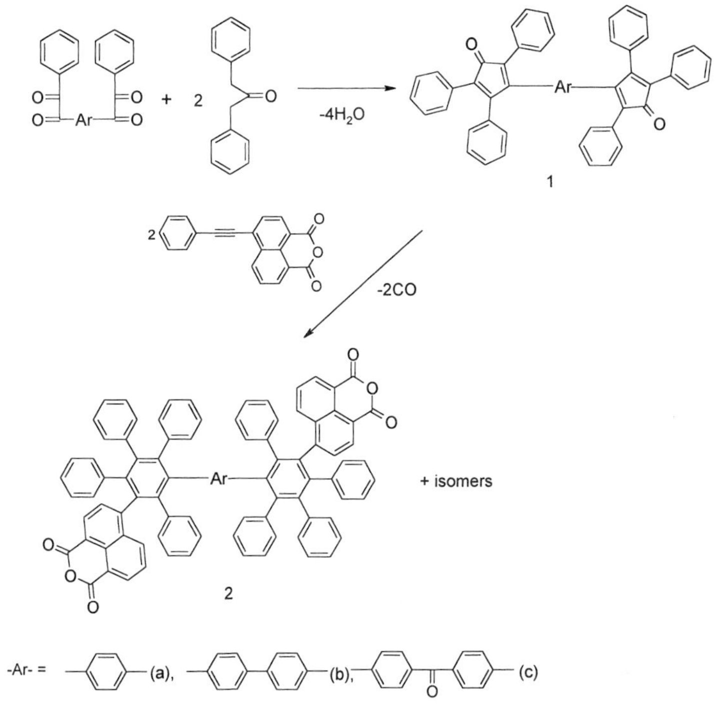

Scheme 1

Phenylated BNA's obtained by this route are not individual compounds, but mixtures of isomers. In some cases, it is necessary to separate the isomers to prepare ordered polymers, but in our research we escaped that because polyheteroarylenes based on the mixture of isomeric monomers are less ordered and exhibit improved solubility and enhanced free volume, useful for low dielectric constant materials applications. All reactions were running smoothly and resulted in desired products with high enough yield. Even the use of less reactive 4-(phenylethynylene)naphthalic anhydride containing non-terminal ethynylene groups (instead of very active terminal ethynylic compounds) in the Diels-Alder cycloaddition did not lead to dramatic reduction of the product yield. The conversion of the reactions was checked by TLC and by reaction mixture color changing from deep purple to yellow-brown, indicating the depletion of bis(cyclopentadienone) in the Diels-Alder reactions. The compositions of the

BNA's as well as intermediate bis-cyclopentadienones were confirmed by elemental analysis, H^1 and C^{13}NMR and FTIR spectroscopy. In the FTIR spectra of the bis-cyclopentadienones there are very intense absorption bands of the C=O vibration in the range of 1700-1720 cm^{-1} . Also in the FTIR spectrum of bis-cyclopentadienone **1(c)** together with absorption band at 1710cm^{-1} there is a carbonyl mode at 1660 cm^{-1}. In the ^{13}C NMR spectra of all cyclopentadienones there is characteristic resonance of the carbonyl carbon of the cyclopentadienone at about δ=200 ppm. Besides in ^{13}C NMR spectrum of **1c** there is a band at 196 ppm of carbonyl carbon of benzophenone moiety. ^{1}H NMR spectra of the bis(cyclopentadienone)s comprised the proton signals just in aromatic region. ^{13}C NMR spectra of bis(naphthalic anhydrides) **2 a-c** displayed characteristic signals of unequal carbons of anhydride moiety at 160.3 and 160.5 ppm whereas resonance of the carbonyl carbon of cyclopentadienones at 200 ppm disappeared. As well as for 1c the ^{13}C NMR spectrum of 2c included the signals at about 196 ppm are attributed to carbonyl carbon of benzophenone fragment. However, it is worth to note that band at 196 ppm represented three signals (196.19, 196.14, 196.09 ppm) of carbonyl carbon instead of single one as in case of **1c**. This phenomenon should be treated in term of confirmation of the isomer mixture formation in course of Diels-Alder reaction between bis-(cyclopentadienone)s and 4-phenylethynylenenaphthalic anhydride. In general the NMR spectra of dianhydrides were more complex in aromatic regions than bis(cyclopentadionone)'s ones, caused by the increasing number of benzene rings. FTIR spectra of BNA's displayed the absorption bands at about 1730 and 1780 cm^{-1} attributed to C=O vibration of dianhydrides whereas the modes of C=O of cyclopentadienone moiety in the range of 1700 to 1720 cm^{-1} disappeared.

Polymers

A. Polyphenylenes

Synthesis of phenylated polyphenylenes was performed from bis(cyclopentadienone)s **1b,c** and p-diethynylbenzene via Diels-Alder reaction. Since earlier the optimal conditions for such synthesis have been found by some of us[23] for bis(cyclopentadienone) **1a** we suggested the using it in the reported experiments to be rather expedient.

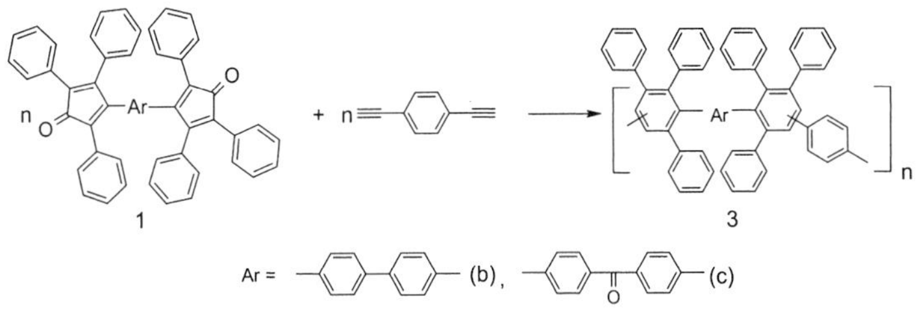

Scheme 2

Thus, the polycycloaddition was carried out in Shlenk tube under argon atmosphere in diphenylether at 240°C. The change of the color from magenta to yellow indicating the accomplishment of the reaction was observed after 7-9 hours depending on bis(cyclopentadienone)s used. However the time of reaction was expanded to 72 hours to achieve the maximum of molecular weight. The polymers remained to be soluble during the reaction time and were isolated by precipitation into methanol, intensively washed with hot ethanol and dried. Polyphenylenes were obtained with yields above 90 %, demonstrating a very high conversion of monomers. The ^{1}H NMR, ^{13}C NMR, FT-IR spectroscopy where used for characterization of the new polymers. In the FT-IR spectra of new polymers there are no ν bands at 2100 cm^{-1} of ethynyl moiety and at 1710cm^{-1} of carbonyl of cyclopentadienone. Also the signals of cyclopentadienone carbonyl carbon ($\approx$200 ppm) and ethynyl carbon ($\approx$90 ppm) could not be detected in the ^{13}C NMR spectra. The ^{13}C NMR and ^{1}H NMR spectra of the polymers were rather complex in the aromatic region and did not significantly depend on bis(cyclopentadienone)s used. The molecular weights of the polymers were determined by gel-permeation chromatography calibrated with respect to polystyrene standards. The weight-average molecular weight M_w was determined to be 192.300 g • mol^{-1} for **3b** with a polydispersity index M_w/M_n of about 2.6 and of 207600 g • mol^{-1} for **3c** with a polydispersity index M_w/M_n=3.5. The polymers were found to be readily soluble in toluene, THF, 1,4-dioxane, chloroform, N,N-dimethylformamid, N-methylpyrrolidone. The unique solubility of polymers should be attribute both to formation of meta- and para- coupling in course of Diels-Alder polycycloaddition and to bulky phenyl substituents. High molecular weights of polymers obtained combined with solubility in a wide range of solvents makes their use in coatings systems very promising.

Thermogravimetric analysis (TGA) reflects the high thermal stability of polymers **3b** and **3c** in air. The thermograms displayed starting weight loss of 10% at 450-470°C (ΔT=4.5°/min) and at 600-610 °C (ΔT=20°/min) for polyphenylenes **3b** and **3c**. Thermal degradation started even at high temperature in argon atmosphere.

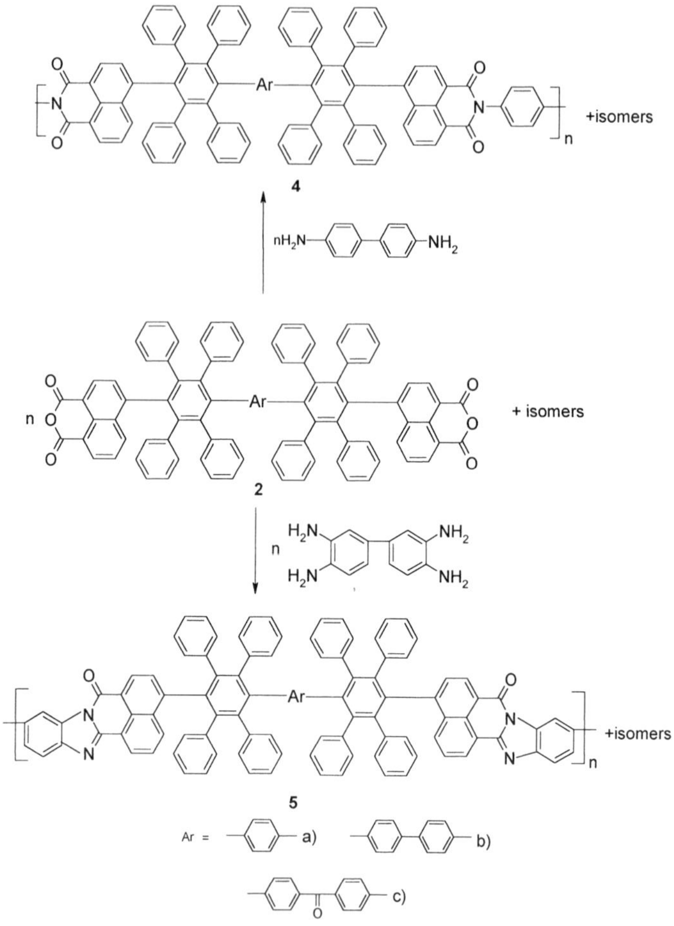

Scheme 3

B. Polynaphthylimides and Polynaphthoylenebenzimidazoles

There was revealed that disorder in polymer chains facilitates the solubility of polymers as well as a presence of phenyl substituents. Because one of the research goal was the synthesis of soluble high-molecular weight polymers we did not attempt to separate the isomers of bis(naphthalic anhydrides) **2a-c** and used the isomer mixture in synthesis of polynaphthoylenebenzimidazoles and polynaphthylimides.

The reactions were performed under high-temperature polycondensation conditions in phenolic solvents using the benzoic acid and benzimidazole as catalysts. Within reaction time all reactions proceeded homogenously and resulted in phenylated polymers demonstrating very high conversion of monomers and extraordinary solubility.

B1. Polynaphthylimides

The polynaphthylimides **4a-c** were characterized by FTIR and NMR spectroscopy. FTIR spectra display ν bands at around 1713 cm^{-1} of the C=O vibration of the naphthylimide cycle. The ^{13}C NMR spectra of polymers show characteristic carbonyl resonances of the naphthylimide moiety at δ=164.0 ppm and 164.4 ppm. Besides, in spectrum of **4c** there is a broad signal at δ=196 ppm attributed to carbonyl carbon of benzophenone fragment. ^{13}C and ^{1}H NMR spectra of **4a-c** in aromatic region were rather complex to reliably assign polymer structure.

Synthesized polymers **4a-c** were predictably soluble in a wide range of organic solvents including N-MP, chloroform, THF. Their molecular weights (M_w) were in the range of 60000 to 121000 g mol^{-1}. Polymers appeared to be extremely thermostable. Thermograms show the starting weight loss of 10% in air at 480-500 **°C** (ΔT=4.5°/min) and at 580-600°C (ΔT=20°/min) for **4a, 4b** and **4c**. T_g of polynaphthylimide **4c** was 398°C whereas for **4a,b** T_g's were not observed before 500°C. Such a phenomenon could be rationalized in terms of new condensed structures formation due to intramolecular oxidative cyclodehydrogenation under the heating of the sample (see Scheme 4).

The presence of the bridge carbonyl fragment in polynaphthylimides **4c** decreased the T_g significantly what is very useful for the processing.

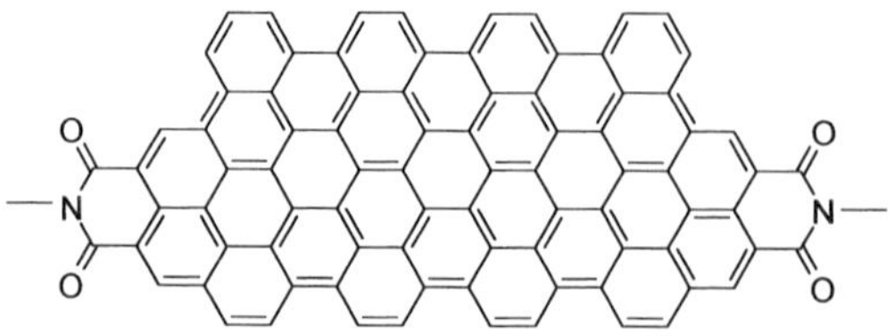

Sheme 4

B2. Polynaphthoylenebenzimidazoles

The polynaphthoylenebenzimidazoles **5a-c** were subjected to FTIR and Raman scattering analysis. FTIR spectra of **5a-c** differed from ones of **4a-c** not significantly and displayed ν bands at around 1702 cm^{-1} of the C=O vibration and weak signals at 1660 and 1551 cm^{-1} attributed to =C=N- group of the naphthoylenebezimidazole cycle. The Raman spectra of polynaphthoylenebenzimidazoles strongly proved the conversion of o-aminonaphthylimide cycles into naphthoylenebenzimidazole ones. In general, the polymers **5a-c** demonstrated the properties very closed to polymers **4a-c.** They proved to be soluble in standard organic solvents, including chloroform and THF and possessed high thermal stability in air in a temperature range of 510-620°C. As well as for **4a,b** we did not observed the T_g's for **5a,b** before decomposition started. For **5c** the T_g was 425°C allowing the processing of this polymer. In all the polymers **5a-c** were more thermostable than **4a-c** to be concerned with more stability of naphthoylenebenzimidazole cycle when compared with naphthylimide one.[24] The molecular weights of polymers (M_w) were determined by gel permeation chromatography and vary from 182000 g mol^{-1} for 5a to 378100 g mol^{-1} for **5c** with polydispersity index 2.1 and 2.55, accordingly.

The mechanical, electrooptical properties of films elaborated from such polymers are in progress.

Conclusion

It appears that the Diels-Alder cycloaddition is a very simple synthetic route to new phenylated monomers and polymers. The target phenylated polymers obtained by using of this method directly (polyphenylenes) or for synthesis of phenylated monomers (polynaphthylimides and polynaphthoylenebenzimidazoles) revealed to be soluble in a wide range of organic solvents such as toluene,THF, chloroform, and displayed high thermostability. Therefore, they can be processed easily and are promising candidates for advanced coating systems as well as for electrooptical applications.

Acknowledges

Financial support by the Russian Foundation for Basic Research (Grant #02-03-32426) is gratefully acknowledged

[1] O. Diels, K. Alder, *Liebigs Ann.Chem.* **1928**, *460*, 98.
[2] J. Suer, R. Sustman, *Angew.Chem.Int.Ed.Engl.* **1980**, *19*, 779.
[3] P .Wetzel, *Nachr.Chem.Techn.Lab.* **1983**, *31*, 979.
[4] R.B. Woodward, F. Sondheimer, D. Taub, K. Heusler, W.M. McLamore, *J.Am.Chem.Soc.* **1951**, *73*, 2403.
[5] G. Stork, E.F. van Tamelen, L.J. Freidman, A.W. Burgstahler, *J.Am.Chem.Soc.* **1951**, *73*, 4501.
[6] T. Horn, S. Wegener, K. Muellen, *Macromol.Chem.Phys.* **1995**, *196*, 2463; A. Mueller, R. Stadler, *Macromol.Chem.Phys.* **1996**, *197*, 1373.
[7] A.D. Schlueter, *Polym.Prepr.* **1995**, *36*, 592.
[8] R.A. Kirchhoff, K.J. Bruza , *Adv.Polym.Sci.* **1994**, *117*, 66.
[9] P. Kovacic, M.B. Jones, *Chem.Rev.* **1987**, *87*, 357.
[10] J. Economy, in: "*Contemporary Topics in Polymer Science*" ed., E.J. Vandenberg, New-York: Plenum 1984.
[11] R.H. Baughman, J.L. Bredas, R.R. Chance, R.L. Elsenhaumer, L.W. Shacklette, *Chem. Rev.* **1982**, *82*, 209.
[12] A.L. Rusanov, *Russian Chem. Rev.* **1992**, *61*, 815.
[13] A.L. Rusanov, *Adv. Polym. Sci.* **1994**, *111*, 116.
[14] A.L. Rusanov, L.B. Elshina, E.G. Bulycheva, K. Muellen, *Polym Sci.* **1999**, *41*, 2.
[15] V.V. Korshak, A.L. Rusanov, *Russian Chem. Rev.* **1983**, *52*, 812.
[16] G.K. Noren, J.K. Stille, *Macromol. Rev.* **1971**, *5*, 385.
[17] J.K. Stille, F.W. Harris, H. Mukamal, R.O. Rakutis, C.L. Shilling, G.K. Noren, J.A. Reed, *Advanced Chemistry* **1969**, *91*, 628.
[18] H. Mukamal, F.W. Harris F W, J.K. Stille, *J Polymer Sci.* **1967**, *5*, 272.
[19] M.L. Keshtov, A.L. Rusanov, A.A. Askadskii, V.V. Kireev, A.A. Kirillov, S.V. Keshtova, F.W. Harris, *Polymer Sci.* **2001**, *43*, 399.
[20] M.A. Ogliaruso, L.A. Shadoff, E.I. Becker, *J Org. Chem.* **1963**, *28*, 2725; N.M. Kofman, Ph D Thesis, Moscow, Russia, 1977; M.L. Keshtov, *Habilitation Thesis,* Moscow, Russia, 2002.
[21] M.A. Ogliaruso, M.G. Romanelli, E.I. Becker, *Chem.Rev.* **1965**, *65*, 261; W. Broser, J. Reusch, H. Kurreck, P. Siegle, *Chem.Ber.* **1969**, *102*, 1715.
[22] I.A. Khotina, A.L. Rusanov, *Russ. Chem. Bull.* **1995**, *44*, 514.
[23]Z.B. Shifrina, M.S. Averina, A.L. Rusanov, M. Wagner, K. Muellen, *Macromolecules* **2000**, *33*, 3525.
[24] V.V. Korshak, S.A. Pavlova, P.N. Gribkova, L.A. Mikadze, A.L. Rusanov, L.Kh. Plieva, T.V. Lekae, *Russian Chem Bull.* **1977**, *6*, 1381.

Macromol. Symp. **2003**, *199*, 109-124

Control of Macromolecular Architecture of Polyamides by Poly-functional Agents. 2. Use of Oligomerization in Polycondensation Study

Cuiming Yuan,[1] *Giuseppe Di Silvestro,**[1] *Franco Speroni,*[2] *Cesare Guaita,*[2] *Haichun Zhang*[2]

[1] Università di Milano, Dipartimento di Chimica Organica e Industriale, INSTM Unità di ricerca di Milano, Via Venezian 21, I- 20133 Milano, Italy
[2] Rhodia Engineering Plastics, Via 1 Maggio, 80, Ceriano Laghetto (MI), Italy

Summary: In the first paper of the series, a statistical model for star-branched polycondenzation of AB type monomers in the presence of a polyfunctional agent RA_f was completely developed. The analytical expressions obtained for the number-average ($\overline{DP_n}$) and weight-average ($\overline{DP_w}$) degree of polymerization, and the dispersion index (D) for whole polymer species, linear and star macromolecular chains, are now derived as function of the feed and of end-group analysis. Also the important molecular parameter, mole fraction of star-branched polymer, can be evaluated. Some numerical examples are presented. It is illustrated that the molecular weight properties of the linear and star-branched polymers in the mixture of the products, very important factors for the application of this kind of polymeric materials, can be determined starting from the feed and terminal group analysis.
Polymerization and oligomerization of 6-aminocaproic acid were carried out in the presence of trimesic (T3) acid and 2,2,6,6-tetra(β-carboxyethyl)cyclohexanone (T4) and EDTA as tri- and tetra-functional agents. The molecular weights calculated are in good agreement with those obtained by Size Exclusion Chromatography (SEC), end group analysis and NMR spectra.

Keywords: molecular weight distribution, polycondensation, size exclusion chromatography (SEC), star polymers

Introduction

Star-branched polymers are characterised by a branch point from which emanate a number of polymer chains or "arms". Star-branched polymers have been actively studied since the first star-shaped polystyrene was prepared by linking living anionic polystyrene with silicon tetrachloride.[1] The interest in star-branched polymers increased rapidly over past decades due

 DOI: 10.1002/masy.200350910

to their unique physical and chemical properties as compared to their linear analogues. It is known that star branching chain decreases the molecular dimensions of polymers and results in a decrease in viscosity. This reduction of melt viscosity allows for processing at lower temperatures and pressures as well as the ability to process higher molecular weight polymers. Additionally, reduced viscosity allows also for improved extrudability as well as injection moulding.[2-4] Both the physical properties of polymeric materials and their processing behaviour at elevated temperatures depend essentially on the average molecular weight and the molecular weight distribution (MWD). It is therefore essential to give a precise description of the molecular weight distribution.

The synthesis of macromolecules with star architecture is produced with almost all polymerisation methods by two general approaches: 1) by terminating reaction of linear polymers with a multisited reagent; 2) by initiating polymerisation from a polyfunctional agent. For star–branched polyamides very few paper are present in the literature; Wilkes presented a paper devoted to a crystallisation behaviour[5] of star nylon-6 prepared by using as a comonomer the second-generation starburst polyethyleneimine described by Tomalia.[6] Anionic polymerisation of caprolactame to give six-arms nylon-6 was studied by Mathias and Sikes.[7,8] In our experience both these method cannot be used in hydrolytic polymerisation conditions due to the low thermal stability of comonomer used or to the not well determined molecular mass distribution in caprolactame anionic polymerisation.

In the previous paper we discussed some experimental results which confirm our theoretical model.[9] In the paper we illustrated that the molecular weight properties of the linear and star-branched polymers in the mixture of the products are very important factors for the application of this kind of polymeric materials.

Theoretical Aspects

Theoretical studies and some experimental approach to produce star-branched polyamides are present in the literature from the beginning of polyamides industrial production. More than 50 years ago, the molecular weight distribution of the resulting polymer by the polycondensation of AB type monomers with a polyfunctional agent RA_f (f is the functionality of active group A), at completion of the reaction, were studied by Flory[10,11] and Schulz.[12] The molecular weight distribution becomes narrower with increasing functionality of the agent RA_f. Dispersion index D depends only on the functionality f and equals to 1 + 1/f. However, Farina[13] proposed that a) the final product of AB monomer polymerisation in the presence of a polyfunctional agent RA_f is a mixture composed solely of linear chains and of star-branched macromolecules with f-arm; b) the length of the linear chains is equal to that of a branch of the star polymer. According to these hypotheses, Farina observed that during the polymerization for f>2 the dispersion index first increases, reaches a maximum and decreases until to the limiting value equal to (1 + 1/f).

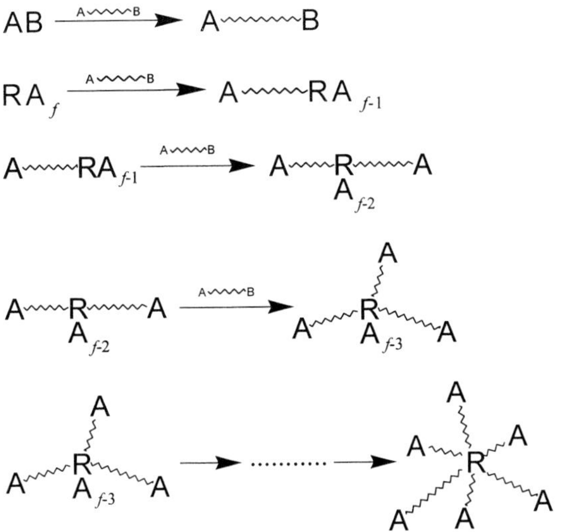

Scheme 1. Reaction scheme between the RA_f and AB monomer.

In our recent paper,[9] we assume (Scheme 1) that all the kinetic constants, in the polycondensation of AB type monomer with a polyfunctional agent RA_f, are independent of chain length and the same reactivity for all the functional groups in both polyfunctional compounds and macromolecular ones; a complete statistical and kinetic model was developed for the star-branched polycondensation of AB type monomers in the presence of a polyfunctional agent RA_f. The analytical expressions, as function of conversion of the functional group of RA_f, for the number-average ($\overline{DP_w}$) and weight-average ($\overline{DP_w}$) degree of polymerization, and the dispersion index for various polymer species were derived through independent methods: both kinetic and statistical ones. In order to easily use the theoretical results, in this work, we have derived the analytical expressions for the molecular parameters as function of the concentrations of terminal groups (for example: A=-COOH and B= $-NH_2$). In the next, we will refer to a multifuntional RA_f where A is a -COOH function.

The concentrations of the linear primary chain having r monomeric units (P_r) and the branched polymer having r monomeric units on i branches (P_r^i) are given by

$$P_r = [NH_2]\frac{[COOH]}{N_0 + fC_0}\left\{\frac{N_0 - [NH_2]}{N_0 + fC_0}\right\}^{r-1} \tag{1}$$

$$P_r^i = C_0\binom{f}{i}\binom{r-1}{i-1}\left\{\frac{[COOH]}{N_0 + fC_0}\right\}^f\left\{\frac{N_0 - [NH_2]}{N_0 + fC_0}\right\}^r \tag{2}$$

All the molecular weight properties can be derived from Equation (1) and (2).

The mole fraction of the total star polymer (f_S) and linear polymer in the (f_L)is given by

$$f_S = 1 - f_L \tag{3}$$

$$f_L = \left\{[NH_2] + \frac{fC_0(N_0 - [NH_2])}{[COOH]}\left(\frac{[COOH]}{N_0 + fC_0}\right)^f\right\} \Big/ \left\{[NH_2] + C_0\left(1 - \left(\frac{[COOH]}{N_0 + fC_0}\right)^f\right)\right\} \tag{4}$$

where C_0 and N_0 are the initial concentrations of polyfunctional agent RA_f (with f functionality) and monomer AB, $[NH_2]$ and [COOH] are the mole concentrations of terminal groups.

The number- and weight-average degree of polymerization and dispersion index are as follows:

$$\overline{DP_n} = \frac{N_0}{[NH_2] + C_0\left(1 - \left(\frac{[COOH]}{N_0 + fC_0}\right)^f\right)} \tag{5}$$

$$\overline{DP_w} = \frac{2N_0 + fC_0 - [NH_2]}{[COOH]} + \frac{C_0}{N_0} f(f-1)\left(\frac{N_0 - [NH_2]}{[COOH]}\right)^2 \tag{6}$$

$$D = \frac{\overline{DP_w}}{\overline{DP_n}} \tag{7}$$

When the reaction goes to completion, i.e. $[NH_2] = 0$, the limiting values of number- and weight-average degree of polymerization and dispersion index were be obtained.

$$\overline{DP_n}(\lim) = \frac{1}{a(1 - (\frac{fa}{1 + fa})^f)} \tag{8}$$

$$\overline{DP_w}(\lim) = 1 + \frac{1}{a}\left(1 + \frac{1}{f}\right) \tag{9}$$

$$D(\lim) = \left(a + 1 + \frac{1}{f}\right)\left(1 - \left(\frac{fa}{1 + fa}\right)^f\right) \tag{10}$$

where a is the initial ratio of C_0 to N_0.

Experimental

2,2,6,6-tetra(β-carboxyethyl)cyclohexanone (T4) was synthesized according to literature.[14] 6-aminocaproic acid (AB), trimesic (T3) acids, EDTA and octadecylamine (n-C18-NH2) (Fluka) were used without further purification.

The mixture of monomer and agent (total 200 g) was poured into a glass reactor beforehand dried by heating above 100°C under 1 Torr pressure. Reactions were carried out in a glass reactor at 275°C with mechanical stirring under the flow of N_2 (5 l/h) for 6 hours.

The SEC analyses were performed in anhydrous CH_2Cl_2 at 20°C by using a six Ultrastyragel column set (10^5, $2*10^4$, $2*10^3$ and 500 Å). The molecular mass of relative polyamides were determined through N-trifluroacetylation.[15] Column calibration was performed by polystryrene standards; the linear polyamide calibration curve was calculated according to the universal calibration principle.[16,17]

The titration of functional groups of $-NH_2$ and –COOH in the polyamides was carried out by potentiometer Metrohm 716 DMS in a solution of 0.7 – 2% polyamide dissolved in 85% trifluoroethanol and H_2O. The solution of polyamide was titrated by HCl N/50 for the determination of the concentration of $-NH_2$. The concentration of –COOH was titrated by NaOH N/50. NMR spectra were carried out in CF3COOD solution with a Bruker 400 MHz spectrometer.

Results and Discussion

In order to visualise the relationship between the molecular parameters and reaction conditions, some numerical examples are given. For a tri- and tetra-agent and initial ratio of RA_f to monomer 6-aminocaproic acid of 0.003, the molecular weight distributions of linear and star species are reported vs. degree of polymerization (r) in Figs. 1-4.

Figures 1,2 and 3,4 report in the case of a tri- and tetra-functional agent respectively, the molecular weight distribution of linear and star polymers at different state of polymerization passing from 0.99 to 0.999 of the complete reaction ($[NH_2]= 0$). It is evident from the figures that the molecular weight of the relative polymer and the star-branched fraction increase with increasing the conversion and f ($f \geq 3$). Final properties of the polymeric material will depend on the balance between linear and star macromolecules as on their molecular mass.

Equation (5) gives us the general relationship between the dispersion index D and the polymerization conditions, i.e. f value and initial ratio ($a=C_0/N_0$) of RA_f vs end group titration which is shown in Figure 5. Dispersion index increases with decreasing concentration of $[NH_2]$, reaches a maximum and then goes to its limiting value (Eq. 8). In the Figure 5 only the D value at high conversion is reported.

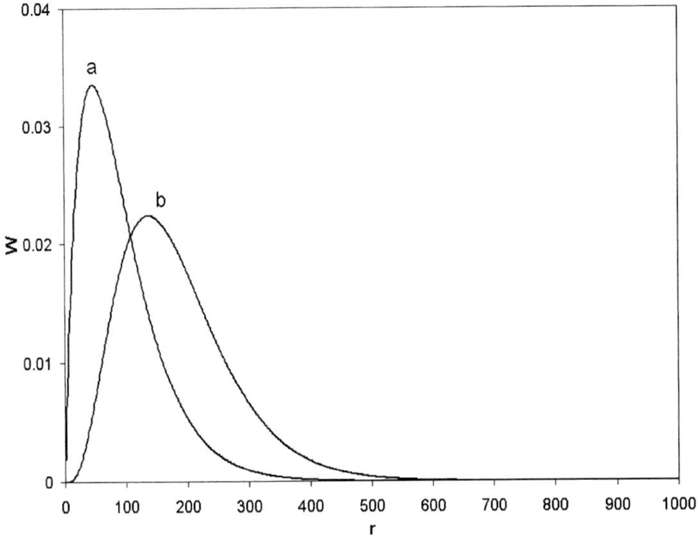

Fig. 1. Molecular weight distribution of linear and star-branched polymer. T3/No = 0.004 at 99% of the limiting conversion. a: Linear; b: 3 arm branched polymer.

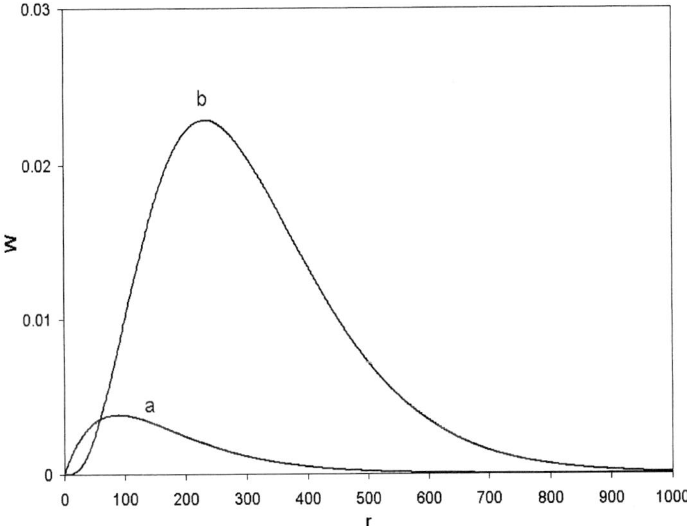

Fig. 2. Molecular weight distribution of linear and star-branched polymer. T3/No = 0.004 at 99.9 % of the limiting conversion. a: linear; b: 3 arm branched polymer.

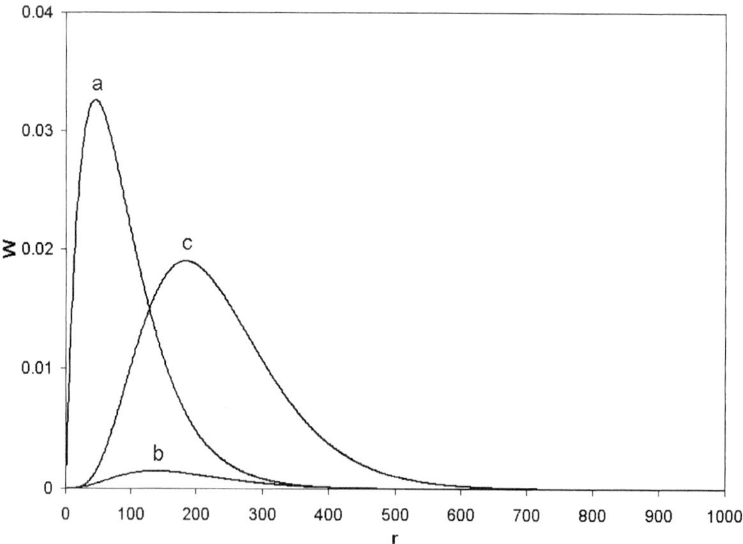

Fig. 3. Molecular weight distribution of linear and star-branched polymer: T4/No = 0.003; at 99 % of the limiting conversion. a: linear; b: 3 arm star polymer; c: 4 arm star polymer.

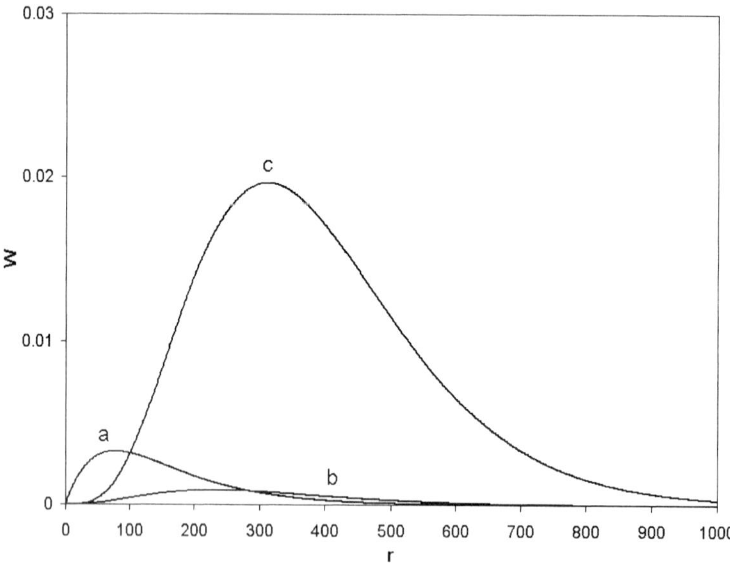

Fig. 4. Molecular weight distribution of linear and star-branched polymer. T4/No = 0.003, at 99.9 % of the limiting conversion; a: linear; b: 3 arm star polymer ; c: 4 arm star polymer.

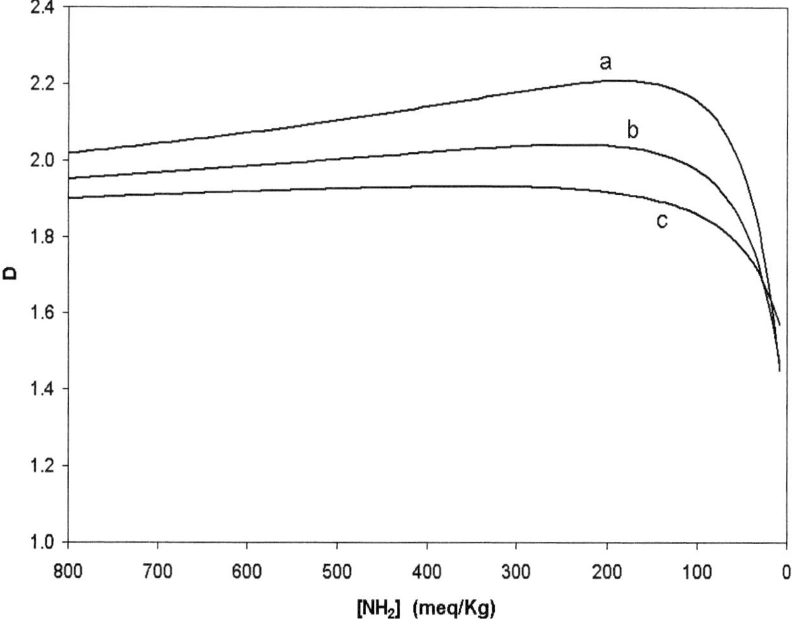

Fig. 5. Evaluated D values vs end group analysis [NH_2]: a: T4/No = 0.003; b: T3/No = 0.004; c: T2 /No = 0.006.

Therefore, some important conclusions can be obtained from the proposed model.

The molecular weight of the reaction system depends essentially on the ratio $a = C_0 / N_0$: the smaller value of a, the larger molecular weight can be obtained when the reaction goes to completion.

The molecular weight distribution becomes narrower with increasing functionality of the agent RA_f.

Dispersion index D depends not only on the functionality f but also on the mole ratio a.

When $f \geq 2$ and $a \leq 0.01$, the limiting values of number-average degree of polymerization and dispersion index are about equal to $1/a$ and $1/a + 1 + 1/f$ which is close to the value obtained by Flory.[2]

At the beginning of polymerization most of the polymers are linear, and the star-branched polymer will form only at very high conversion.

Polymerization Results

Polymerizations of 6-aminocaproic acid were carried out in the presence of tri- and tetra-functional acids. The theoretical and experimental data, as obtained by Size Exclusion Chromatography (Figure 6), for linear, tri- and tetra-branched polyamide are reported in table 1.

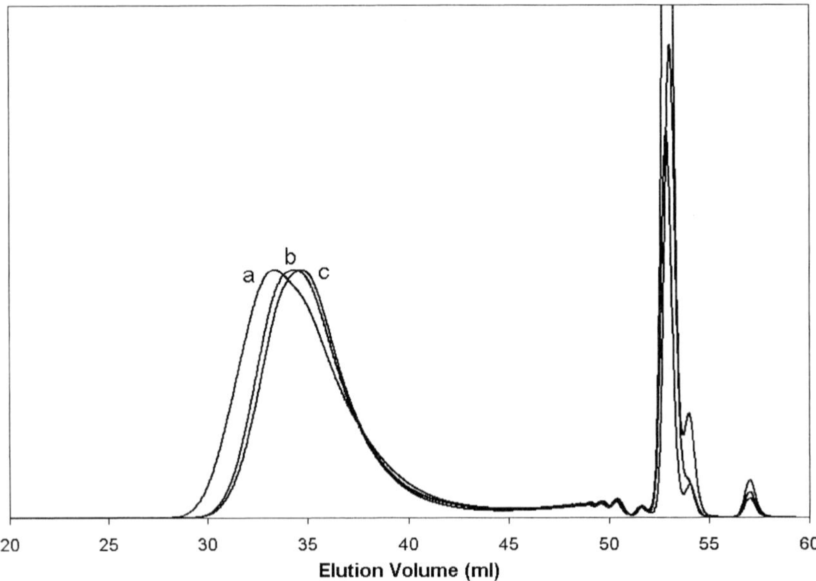

Fig. 6. Examples of SEC curves of PA6. a: linear; b: star polymer (T3 = 0.3% in mole); c: star polymer (T4 = 0.3% in mole).

Table 1. Feed, end group analyses and a comparison between evaluated and experimental molecular mass data (SEC).

Sample	NH2	COOH	Mn		Mw		D	
	(meq/Kg)	(meq/Kg)	model	SEC	model	SEC	Model	SEC
Nylon 6	45.50	41.97	21978	20832	43956	41576	2	1.996
T3 0,3%	20.07	102.17	21066	20534	35011	33733	1.662	1.643
T4 0,3%	14.11	133.25	22211	19118	34852	31840	1.569	1.665

We note that the SEC calibration was performed with linear PA6 and so the agreement between experimental data and the evaluated ones confirms the validity of the proposed reaction scheme.

We note also that all the SEC curves in figure 6 present the plateau due to cyclic oligomeric species; this is a further proof of the presence of linear chains in the obtained polymer.

Therefore, we can conclude that the optimisation of mechanical properties of material can, really, be obtained by a proper choice of the functionality and initial concentration of RA_f and of final conversion of amino group; the fraction of star-branched and linear polymer, and their molecular weights are very important for application of this kind of polymeric materials, because the first one offers reduction in melt viscosity and the linear ones gives good mechanical behaviours.

The key point in the synthesis of star-branched polycondensates is the choice of the multifunctional agent; the selected molecules must be stable in the polymerization and final processing conditions. In the case of polyamides polymerization and processing temperatures are very high (up to 300 °C). We discuss now an approach to solve this crucial problem.

Oligomerization Results

The first method generally used for a test of thermal stability is a DSC or a TG analysis; it should be noted that polymerization conditions differ from DSC analysis. In the thermal analysis the tested molecule is pure, solid or liquid according its phase; as a consequence, parasite reactions occurring in the thermal treatment cannot be the same as in the polymerization process. In the last case, the added reagent is in a dilute and reactive condition. At contrary, during oligomerization condition the substance can react with the monomer and parasite reactions can be better observed by the usual analytical methods.

By oligomerization, we tested several possible compounds as multifunctional molecules both polyamino or polyacid. Figure 7 reports SEC curves of 6–aminocaproic acid oligomerization in the presence of T4 and EDTA. The effect of parasite reactions in polymerization conditions is evident and agrees with the observed thermal decomposition of EDTA.

Oligomerization at different feeds can confirm the thermal stability of a multifunctional reagent; figure 8 reports SEC curves of PA6 oligomers obtained at T4 different feeds.

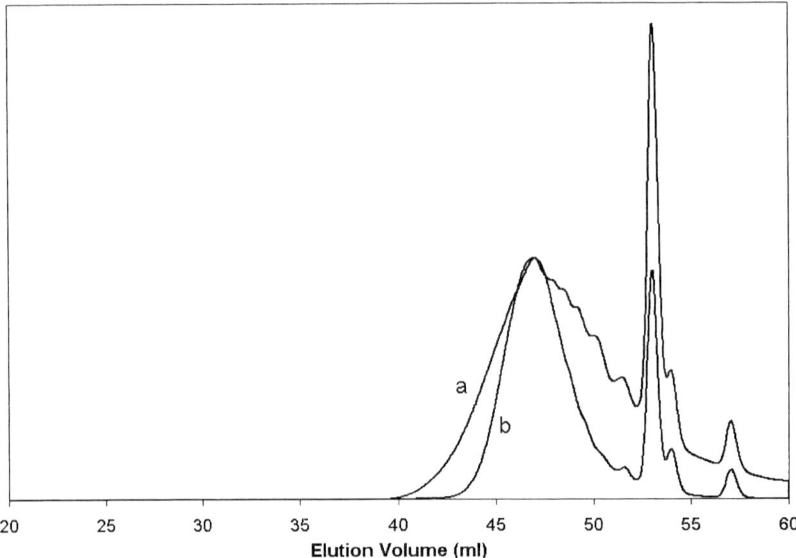

Fig. 7. SEC curves: a: AB /EDTA = 6; b: AB/T4 =6.

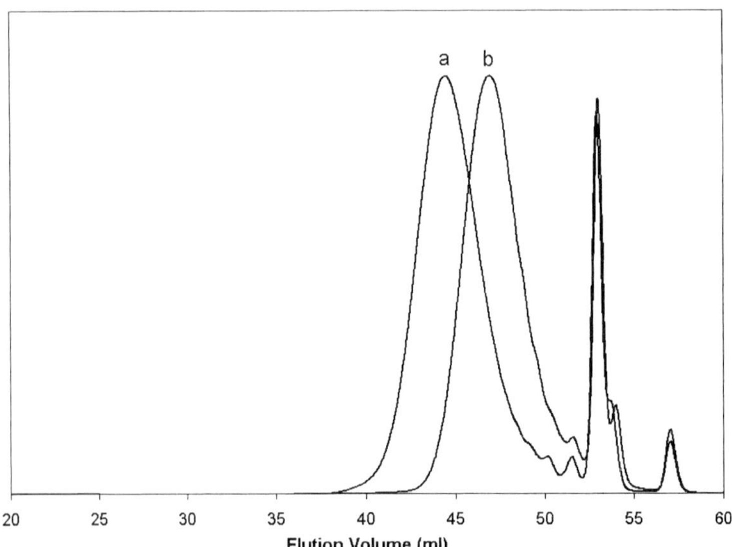

Fig. 8. SEC curves of 6–aminocaproic acid in the presence of T4: a: AB/T4 = 12; b: AB/T4 = 6.

In principle, the presence of the multifunctional agent can be observed in a NMR spectrum of oligomers. Figure 9 reports an example of a ^{1}H NMR spectrum of oligomer mixture obtained in the presence of T4. In this case, no clear signals are present for a correct determination of degree of reaction and or of anomalous reaction pathway. In particular, we searched for the presence of imidation reaction between amino group of the monomer and two different acid function of two different arms. However, degree of polymerization agrees with SEC results. Work is in progress on these mixtures by MALDI-TOF.

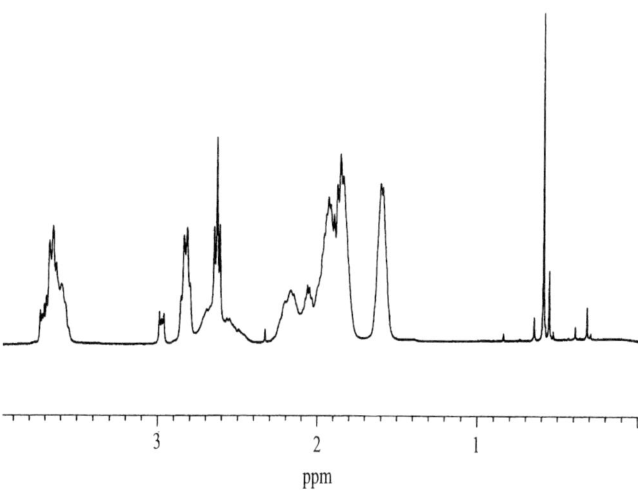

Fig. 9. ^{1}H NMR spectrum of oligomers. AB/T4 = 6.

The so far discussed difficulties in T4 oligomerization can be attributed to the absence of NMR signals far from those of the macromolecular chain. Assuredly to this hypothesis, we select trimesic acid as multifunctional reagent.

In order to have qualitative and quantitative results from NMR spectra we reacted T3 with n-C18-NH2 in different ratios. MS-, proton and carbon NMR spectra confirm the structure of products.

Figure 10 reports formulas and assignment of the aromatic region of 1H NMR spectrum of a mixture; from this spectrum quantitative results about product distribution vs feed can be obtained.

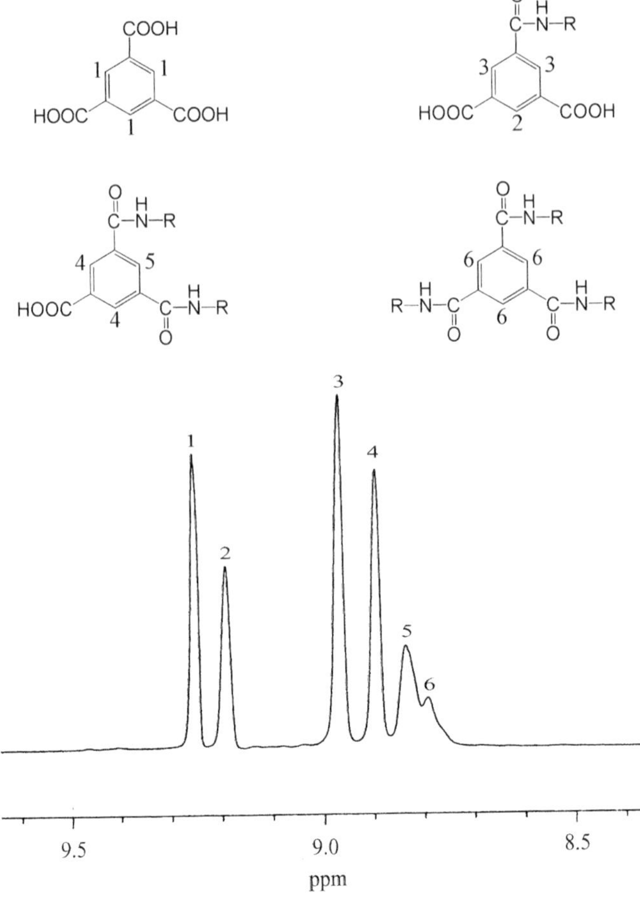

Fig. 10. Formulas and aromatic region of ^{1}H NMR spectrum of T3/6–aminocaproic oligomers.

Several oligomerization reaction were performed between ε–aminocaproic and trimesic acid in different ratios; as in T4 case, SEC data confirm thermal stability of the comonomer (two example are in Figure 11).

NMR spectra confirm the absence of parasite reaction and allow a quantitative determination of oligomeric species. Also in this case, no evidences of imide group is present in NMR and MS spectra. A more detailed study of these mixture by ^{1}H, ^{13}C NMR and MALDI-TOF is in progress.

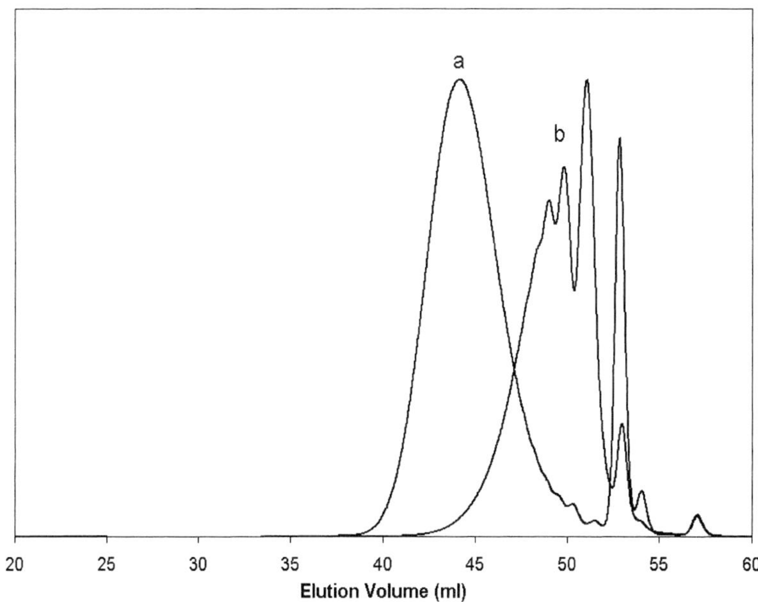

Fig. 11. SEC curve of T3/6–aminocaproic oligomers. a: AB/T3 = 18; b: AB/T3 = 4,5.

Two examples of oligomer distribution as obtained from end group analysis and ^{1}H NMR spectra are presented in table 2 for two different feed.

Table 2. Oligomer distribution from NMR data and end group analysis according to the model.

Sample	COOH (meq/Kg)	*T3*		*T3(NHR)*		*T3(NHR)$_2$*		*T3(NHR)$_3$*	
		NMR	model	NMR	model	NMR	model	NMR	model
AB/T3=4.5	5931.12	0.1869	0.1804	0.4094	0.4166	0.3140	0.3207	0.0897	0.823
AB/T3=18	1617.28	0	0.0051	0.0736	0.0738	0.3515	0.3542	0.5749	0.5669

The agreement between these data is excellent and confirm the assumption of equal reactivity of functional groups since the beginning of macromolecular chain growth independently of the chain length.

[1] M. Morton, T. E. Helminiak, F. Bueche, *J. Polym. Sci.* **1962**, *57*, 471
[2] S. Bywater, *Adv. Polym. Sci.* **1979**, *30*, 89.
[3] J. Rooves, *Encycl. Polym. Sci. Technol.* **1985**, *2*, 478.
[4] PCT Int. Appl. WO 97 24,388 (1997). Nylthec Italia and SNIA-Ricerche, invs.: A. Cucinella, G. Di Silvestro, C. Guaita, F. Speroni, H. Zhang; Chem. Abstr. **1997**, 149536c.
[5] B.G. Risch, G.L. Wilkes and J.M. Warakomski, *Polymer*, **1993**, *34*, 2330 .
[6] D.A. Tomalia, V. Berry, M. Hall and D.M. Hedstrand, *Macromolecules*, **1987**, *20,* 1167. 1167
[7] L.J. Mathias and A.M. Sikes, *Polym. Mat. Sci. Eng.* **1987**, *56*, 87.
[8] L.J. Mathias and A.M. Sikes, *ACS Symp. Ser.*, **1988**, *367*, 66.
[9] C. M. Yuan, G. Di Silvestro, F. Speroni, C. Guiata, H. Zhang, *Macromol. Chem. Phys.* **2001**, *202*, 2086.
[10] P. J. Flory, *J. Am. Chem. Soc.* **1936**, *58*, 1877.
[11] R. Schaefgen and P.J. Flory, , *J. Am. Chem. Soc.* **1948**, *70*, 2709.
[12] G.V. Schulz, Z. Phyys. Chem. **1939,** *B43*, 25
[13] R. Filippini, M. Fornaroli, M. Farina, *VI Convegno Italiano di Scienza delle Macromolecole dell'AIM*, 1979.
[14] A. H. Buison, T. W. Riener, *JACS*, **1942**, *64*, 2850.
[15] H. Schuttenburg, R. C. Schulz, *Angew. Chem.* **1976**, *88*, 848.
[16] G. Di Silvestro, C. Guaita, P. Sozzani, S. Andoni, *VIII Convegno Italiano di Scienza delle Macromolecole dell'AIM, Milano,* 1987 p. 363.
[17] G. Di Silvestro, Y. Cuiming, C. Guaita, *2nd China-Italy Joint polymer Symposium*, Hangzhou 17-20 Oct. 1995 p. 69.

Macromol. Symp. **2003**, *199*, 125-133

Synthesis and Characterization of Multiblock Copolyesters Containing Poly(dimethylsiloxane) in the Soft Segments

Miroslawa El Fray, Volker Altstaedt*

Department of Polymer Engineering, University of Bayreuth, Universitaetsstr. 30, D- 95447 Bayreuth, Germany
E-mail: mirfray@ps.pl

Summary: Synthesis and thermal properties of poly(aliphatic/aromatic-ester) copolymers containing additionally poly(dimethylsiloxane) (PDMS) chains in the soft segments are discussed. A two step method of transesterification and polycondensation from the melt was carried out in a presence of magnesium-titanate catalyst. An aliphatic dimer fatty acid was used as a component of the soft segments while poly(butylene terephthalate) (PBT) constituted the hard blocks. Effectiveness of the incorporation of PDMS into polymer chain was confirmed by the Soxhlet extraction and infrared spectroscopy of an excess of 1,4-butane diol destilled off from the polycondensation reaction. Multiblock copolymers showed microphase separation as determined by differential scanning calorimetry. Incorporation of a small amount of PDMS (up to 14.5 wt.-%) into polymer chain containg low concentration of hard segments of PBT lead to decrease in crystallinity of such copolymers. This may indicate that semicrystalline PBT are dissolved in the amorphous matrix of the soft segments.

Keywords: dimer fatty acid, multiblock copolymers, poly(butylene terephthalate), poly(dimethylsiloxane), polycondensation

Introduction

Thermoplastic elastomers (TPE) are very interesting class of polymers which possess both plastic and rubbery properties. They are composed from different fragments (segments) of macromolecule which are alternated along the main chain. These are tighted through labile connections being a consequence of glass transition of segments, crystallization, they might be ionic associations or groups connected by the hydrogen bonds. These groups have features of physical cross-links and they are called the "hard segments". In contrast to chemically cross-linked materials, the physical cross-links are reversible and can be disrupted at elevated temperatures or in solvents, which give the material its good processability. Other segments which are characterized by "small" elastic modulus, low glass transition temperature and low bond cohesion energy are called the "soft segments".[1,2]

 DOI: 10.1002/masy.200350911

By varying the hard/soft segments composition and concentration and additionally using thermal and mechanical conditioning, the physical, mechanical and chemical properties as well as morphology of the multiblock copolymers can be changed. The resulting polymers can behave as classical thermoplasts or polymeric materials of high elasticity or change from crystalline to amorphous polymers.

Widely known examples of multiblock thermoplastic elastomers (TPE) include poly(ester-urethane)s, poly(ether-amide)s, poly(ester-ether)s.[3,4] Polymers containing aliphatic units (derived from aliphatic dicarboxylic acids) and polyester hard segments are also known.[5] Poly(dimethylsiloxane) (PDMS) has also been used in multiblock terpoly(ester-ether-siloxanes),[6,7] poly(siloxane-ester-imides)[8] and thermoplastic poly(ester-siloxane)s[9,10] as a component of the soft segments. Incorporation of highly hydrophobic PDMS into polymer chain impart improved antistick/antiblock and mould release properties as well as improved film and fiber properties.[11,12]

Taking into consideration that the phase behaviour of segmented block copolymers in general, is controlled by the differences in the Hildebrand solubility parameters (δ) it is known that PDMS has one of the smaller solubility parameters (7.3 - 7.6 $(cal/cm^3)^{1/2}$)[13] in comparision with other organic polymers, however aliphatic dimer fatty acid is characterized also by the small solubility parameter, namely 7.9 $(cal/cm^3)^{1/2}$ (calculated based on the Small method).[14] Thus, the formation of one homogenous soft phase is expected in such copolymers.

In this study, the multiblock poly(aliphatic/aromatic-ester) copolymers have been prepared and poly(dimethylsiloxane) has been introduced into polymer melt during the synthesis in order to produce material of highly hydrophobic properties.

Experimental

Synthesis of Polymers

Synthesis of hydrophobized poly(aliphatic/aromatic-ester) copolymers (PEDMS) was carried out by transesterification and polycondensation from the melt. Following materials were used: dimetylterephthalate (DMT) – "Elana S.A." Torun, Poland; 1,4 - butanediol (1,4 -BD) – BASF, Gemany; dimer fatty acid (DFA) – commercial name Pripol 1009, molecular weight 570 g/mol, Uniqema Chemie, BV , Gouda, The Netherlands; poly(dimethylsiloxane) (PDMS) of M_n = 1000 g/mol, Wacker, Germany; magnesium-titanate catalyst, prepared as described in.[21]

The apparatus for polymer synthesis was constructed as a two separated reactors: the upper reactor for the transesterification and the buttom one for the polycondensation. The reactors were connected by a launching tube where the valve was located.

The reaction mixture consisting of DMT, 1,4-BD and magnesium-titanate catalyst (2 g/kg of polymer) was heated to 200 °C with a heating rate of 1.5 °C/min and constant mixing rate (60 rpm) in the reactor for transesterification. The molar ratio of BD and DMT was as 1.8:1. The reaction was stopped when more than 95% of the stoichiometric amount of methanol was evaporated. Under these conditions and under reduced pressure oligomer of butylene terephthalate was obtained.

During the second stage of the reaction a dimer fatty acid was added along with a catalyst and α,ω-disilanol PDMS to the first reactor and the reaction mixture was transferred to the bottom reactor. The polycondensation was carried out at 245-255°C and 0.5-0.6 mm Hg of vacuum. The process was considered complete on the basis of the observed power consumption of the stirrer motor when the product of highest melt viscosity was obtained (up to a constant value of a power consumption by the reactor stirrer).

The reaction mass was extruded into water by compressed nitrogen and extrudate was granulated. The optimum yield from the apparatus was 1 kg of polymer.

Polymers Characterization

The limiting viscosity number (inherent viscosity) [η] of the polymers was measured at 30°C using polymer solutions in $PhOH/C_2Cl_3H$/toluene (1:1:2 by vol.).

Granules of the polymer were placed into the extraction thimble and were extracted with refluxing methanol for 6 hours. After extraction polymers were dried in a vacuum oven at 50 °C for 3 hours to constant mass.

IR spectra of 1,4-butanediol destilled off from the polycondensation reaction were recorded on a Specord M80 Carl Zeiss Jena Spectrometer using KBr pellets. Solid samples were characterized by Fourier transform attenuated total reflection infrared spectroscopy (ATR-FTIR). A Nexus FTIR spectrometer (Nicolet Instrument Corporation, U.S.A.) equipped with the Golden Gate Single Reflection Diamond ATR head, (Specac INC, U.S.A.) scanning between 600 and 4000 cm^{-1}.

Water contact angle was measured using an optical bench-type contact angle goniometer (OCA Contact Angle System, Data Physics, USA). Drops of destilled water (3 μl) were deposited onto the compression moulded polymer film surfaces and the direct microscopic measurement of the contact angles was done with the goniometer.

Differential scanning calorimetry (DSC) was performed with a TA Instruments (DSC-2920) apparatus. The samples were dried in a vacuum at 70 °C. The process was carried out in three cycles: heating, cooling, and heating, in the temperature ranges -100 to 220 °C. The rate of heating and cooling was 10 °C/min. Temperature and enthalpy calibration were carried out using an indium standard. The glass transition temperature (T_g) was determined from the temperature diagrams as the temperature corresponding to mid point.

Results and Discussion

Three series of hydrophobized poly(aliphatic/aromatic-ester) (PEDMS) copolymers, containing the ester hard segments of poly(butylene terephthalate)(PBT) and an aliphatic soft segments composed of dimer fatty acid extended with poly(dimethylsiloxane) were synthesiszed. Polymers differed in chemical composition and concentration (mass ratio) of used monomers as presented in Table 1.

Table 1. Composition of PEDMS copolymers containing poly(dimethylsiloxane) (PDMS) of M_n = 1000 g/mol and their inherent viscosities [η].

No	Mass ratio (at feed) (wt.-%)			[η] (dL/g)
	PBT	DFA	$PDMS_{1000}$	
I a	26	74	-	0.9
I b	26	70	4	0.83
I c	26	66.5	7.5	0.86
I d	26	63	11	0.81
I e	26	59.5	14.5	0.82
	PBT	DFA	$PDMS_{4000}$	
II b	26	70	4	0.75
II c	26	66.5	7.5	0.76
II d	26	63	11	0.72
	PBT	DFA	$PDMS_{1000}$	
IIIa	50	50	-	0.94
IIIb	50	47.5	2.5	0.93
IIIc	50	45	5	0.97
IIId	50	42.5	7.5	0.92
IIId	50	40	10	0.94

Extraction of the polymers revealed that polymers contained small amount (about 1%) of extractables which indicates on effectiveness of the PDMS incorporation into the polyester. Infrared spectroscopy has been also used to monitor the incorporation of poly(dimethylsiloxane) into the polymer chain by taking spectra of a low-molecular weight diol (1,4-butanediol) distilled off from the polycondensation reaction. Figure 1 presents spectra of a pure 1,4-butanediol.

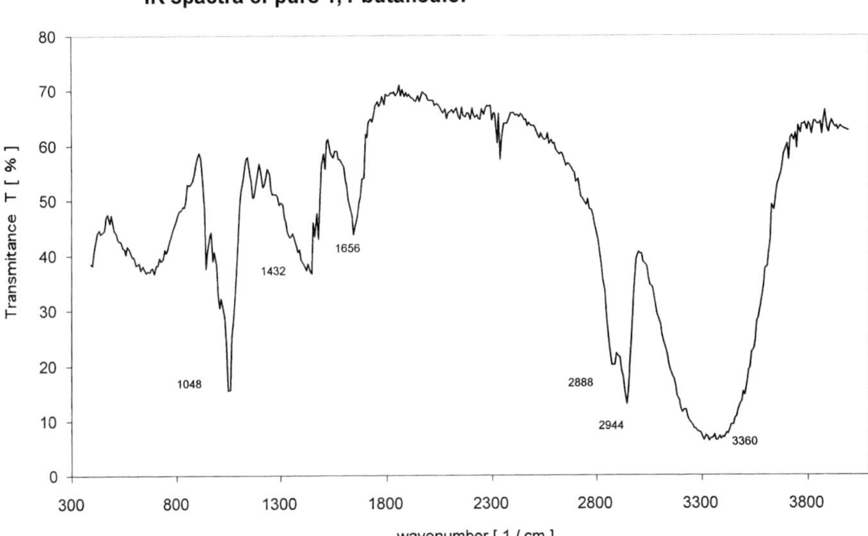

Fig. 1. IR spectra of a pure 1,4-butanediol.

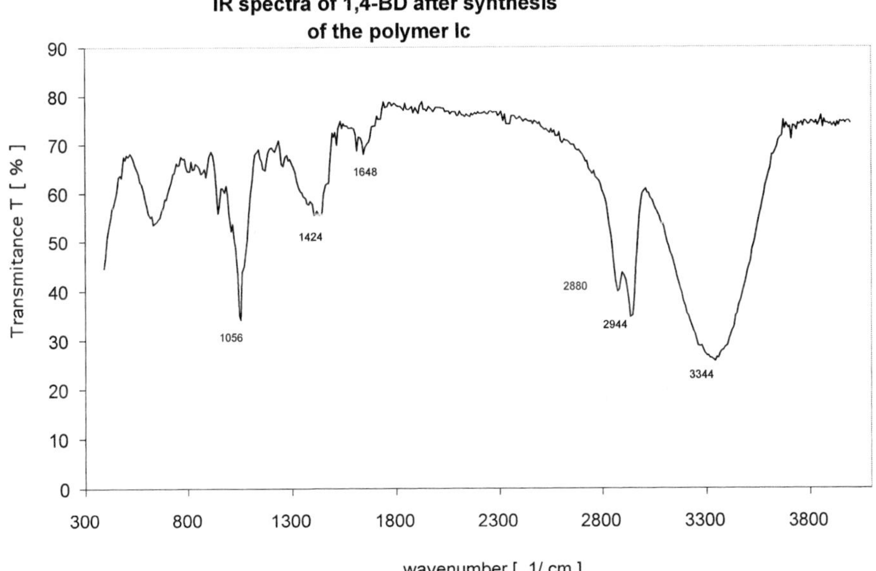

Fig. 2. IR spectra of 1,4-BD destilled from the polycondensation reaction (polymer I c).

Three characteristic bands are present: the bands at 2888 and 2944cm^{-1} correspond to ν(C-H) stretching; the bands at 3360, 1048 and1432 cm^{-1} correspond to O-H bending. This spectra was compared with the spectra (Figure 2) taken for 1,4-butanediol after polycondensation step. Good agreement between chemical structure of a pure 1,4-butanediol and 1,4-BD destilled from the reaction indicate the effectiveness of the synthesis.

The chemical structure of solid PEDMS polymer films was examined by ATRFT-IR spectroscopy. Infrared absorption spectrum shown as example for sample Ic on Figure 3 reveal presence of characteristic PDMS bands: Si-O-Si stretching vibrations at 1015.5 and 1098.8 cm^{-1}; Si-C stretching and CH_3 rocking at 794.1 cm^{-1}; CH_3 symmetric deformation of Si-CH_3 at 1266.4 cm^{-1}; and CH_3 asymetric deformation of Si-CH_3 at 1409.05 cm^{-1}.

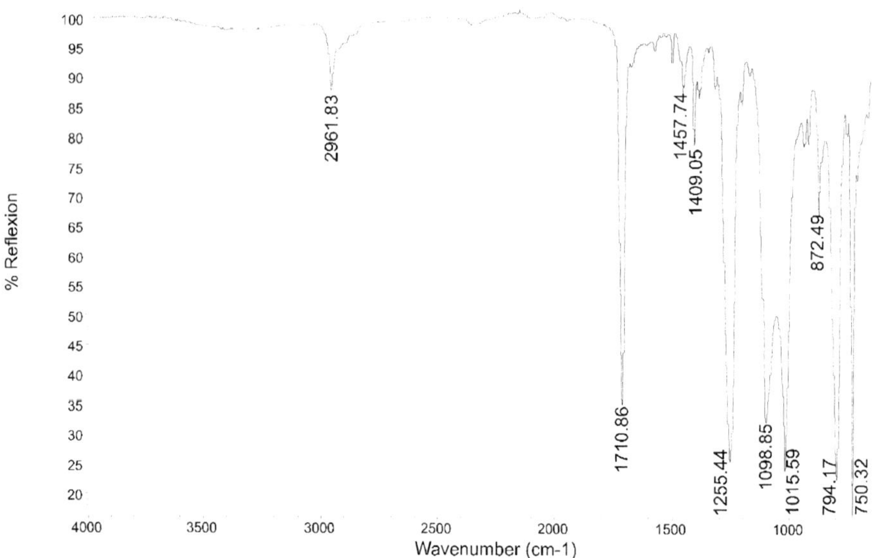

Fig. 3. ATR-FTIR spectra of a PEDMS sample (I c).

As it was discussed in the introduction part, a highly hydrophobic PDMS can be introduced into existing polymer substrates to improve surface properties without changing the bulk properties[15] (a silicon rubber has a water contact angle of 111°).[16] PEDMS copolymers (Figure 4) containing PDMS chains exhibit higher water contact angles compared to non modified. The effect of PDMS molecular weight on hydrophobicity of polymers is not much significant but copolymers containing $PDMS_{4000}$ (M_n = 4000 g/mol) shows the highest values of the contact angle.

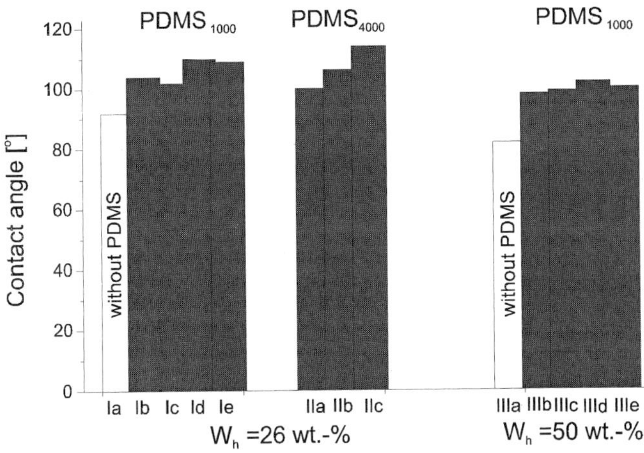

Fig. 4. Water contact angle of PEDMS polymer surfaces.

The PEDMS synthesised are semi-crystalline polymers. Results from differential scanning calorimetry show an appearance of two characteristic temperature transitions: a single low glass transition temperature and a single melting temperature confirming their multiblock structure. However, analysis of the low temperature transition for the „pure" DFA soft segment in series I and II (P_k = 1) and DFA-PDMS soft segments material indicate a slight decrease in the T_g of the soft segments (Table 2) what is associated with a higher degree of phase separation.[17] Opposite tendency is observed for the polymers of series III (P_k = 2.8) where the shift to higher temperatures is observed (DFA segments are better mixed with the PDMS material). The detected single glass transition, which shifts to higher temperature with increasing PDMS content in series III, indicates that the siloxane and aliphatic units form a homogenous amorphous phase at a PBT weight content of 50 % (P_k = 2.8).

Based on the experimental heat of fusion ΔH_m (J/g of polymer) from DSC measurements, the crystalline phase content in the polymer ($W_{c,h}$) was calculated from the following equation:

$$W_{c,h} = \Delta H_m / \Delta H_f \, 100 \qquad (1)$$

where ΔH_m is the heat of melting of the hard segments determined from DSC (Table 2) and ΔH_f is the heat of melting of crystalline PBT phase (ΔH_f = 144.5 J/g).[18]

Table 2. DSC transitions of PEDMS copolymers.

Symbol	Soft segments		Hard segments				
	T_{g1} (°C)	ΔC_p (J/g/°C)	T_{g2} (°C)	T_{m2} (°C)	ΔH_{m2} (J/g)	T_{c2} (°C)	$W_{c,h}$ (%)
I a	-41.7	0.331	53.3	117.2	10.6	24.3	7.3
I b	-41.9	0.374	49.1	102.2	8.8	26.1	6.1
I c	-39.7	0.471	48.4	99.8	9.4	32.5	6.5
I d	-42.1	0.326	48.5	121.9	8.5	43.3	5.8
I e	-42.8	0.361	47.4	126.1	9.3	75.4	6.4
II b	-41.3	0.317	55.1	100.7	9.8	32.3	6.8
II c	-41.7	0.394	52.7	104.7	10.1	26.7	7.4
II d	-42.4	0.436	53.6	96.9	12.5	30.1	8.6
III a	-33.0	0.249	55.7	170.8	16.8	122.3	11.6
III b	-30.8	0.261	51.8	171.1	18.8	114.3	13.0
III c	-28.0	0.268	52.8	173.8	20.7	115.2	14.3
III d	-27.6	0.225	53.8	175.6	21.0	120.1	14.5
III e	-29.3	o.225	48.8	182.1	23.4	132.0	16.1

T_{g1} - glass transition of the soft segments,
ΔC_p - change of the heat capacity at glass transition;
T_{g2}, T_{m2}, T_{c2} - glass transition, melting, and crystallization temperatures, respectively, of the hard segments;
ΔH_{m2} - melting enthalpy of the hard segments; $W_{c,h}$ - mass content of PBT crystallites in the polymer.

According to the DSC measurements, the crystalline fraction is higher for polymers containing higher amount of PBT hard segments (series III, Table 2). For series I, the crystalline phase content decreases with increasing $PDMS_{1000}$ concentration. These polymers contain a very low amount of crystalline phase (26 wt.-%). This indicate that crystallization behaviour is disturbed by the addition of poly(dimethylsiloxane) (PBT starts to dissolve in the soft amorphous phase). Increase of the molecular weight of PDMS to 4000 g/mol leads to opposite effect: an increase of the crystallinity was observed for series II when the hard segment charge was kept at the some weight level). Addition of the $PDMS_{1000}$ to polymers containing 50 wt.-% of the hard segments increases their crystalinity with increasing $PDMS_{1000}$ content.

Conclusions

The two step method of transesterification to produce an oligomer of poly(butylene terephthalate) and then polycondensation of this product with dimer fatty acid and silanol-terminated poly(dimethylsiloxane)s of different molecular weight has beed succesfully applied for preparation of multiblock copolymers. Their multiblock structure has been confirmed by differential scanning calorimetry. Incorporation of small amount of $PDMS_{1000}$ (up to 14.5 wt.-%) into polymers containing low concentration of crystalline hard segments of PBT has effected polymers crystallinity. This may indicate that semicrystalline PBT dissolves in the amorphous matrix of the soft segments. The influence of the PDMS molecular weight and concentration on mechanical properties, especially on the fatigue behaviour will be discussed in a further publication.

Acknowledgements

The authors would like to thank the Deutsche Forschungsgemeinschaft (DFG) for the support of this research.

[1] G. Holden, N.R. Legge, R. Quirk, H.E. Schroeder, editors "*Thermoplastic Elastomers*" Hanser Publishers, Munich Vienna New York 1996, p. 222.
[2] N.R Legge, G. Holden, H.E. Schroeder, editors. „*Thermoplastic Elastomers*", Carl Hanser Verlag, New York 1987.
[3] R. Walker , Ch.P. Rader. editors, „*Handbook of Thermoplastic Elastomers*", van Nostrand Reinhold Comp., New York, 1988, p. 207.
[4] R.J. Cella. *J. Polym. Sci., Polym. Symp.* **1973**, *42*, 727.
[5] M. El Fray , J. Slonecki, *Angew. Makromol. Chem.* **1999**, *266*, 30.
[6] Z. Roslaniec, *Polymer* **1992**, *33*, 1717.
[7] Z. Roslaniec, *Polimery (Warsaw)* **1997**, *42*, 376.
[8] M. Cazacu, C. Racles, A. Vlad , M. Marcu, *Europ. Polym. J.* **2001**, *37*, 2465.
[9] V.V. Antic, M.R. Balaban, J. Djonlagic, *Polym. Int.* **2001**, *50(11),* 1201.
[10] D.A. Schiraldi, *Polymer Preprints* **2001**, *42(1),* 221.
[11] P.R. Ginnings, US Patent 4 496 704; 1985.
[12] N. Yamamoto, H. Mori, A. Nakata, M. Suehiro, US Patent 4 894 427, 1990.
[13] O. Olabisi, L.M. Robertson, M.T. Shaw, "*Polymer-Polymer Miscibility*", Academic Press, New York 1979, p. 54.
[14] J.L. Gardon, in: *Encyclopedia of Polym. Sci. & Technol.,* John Willey & Sons, New York, 1965; p. 851
[15] Y. Ikada, *Biomaterials* **1994**, *15*, 725
[16] D.W. Grainger, S.W. Kim, J. Feijen, *J. Biomed. Mater. Res.* **1998**, *22*, 231.
[17] M. El Fray, J. Slonecki, *J. Macromol. Sci. – Phys.* **1998**, *B37(2)*, 143.
[18] Y. Camberlin, J.P. Pascault, *J. Polym. Sci. – Polym. Chem.* **1983**, *21*, 415.

polyamide-6 are known,[5-8] however the crystallization of the polyamide is slow and incomplete and thus the materials are hardly crystalline. The rate of crystallization of the polyamide is decreased by the high viscosity of PPE.[9,10]

A new polymer system based on PPE segments and fast crystallizing uniform amide units that is a segmented copolymer is described in this article. Uniform di-amide units were found to crystallize very fast and complete upon cooling from the melt in a copolymer with poly(tetramethylene oxide) that has a T_g of -65°C.[11-13] However the melting temperature is not so high. A higher melting temperature can be obtained with tetra-amide units.[14,15] For fast crystallization and good phase separation it is important that the crystallizable units have uniform length.[16-19]

The structure of segmented copolymers of PPE-2T segments[20] and tetra-amide units (T6T6T-dimethyl)[15] that are linked via dodecanediol is shown in Figure 1. These polymers are made via polycondensation. PPE-2T is bifunctional PPE with terephthalic methyl ester endgroups and a molecular weight of ~3100 g/mol [20]. T6T6T-dimethyl is a tetra-amide segment with terephthalic methyl ester endgroups that is based on two-and-a-half repeating unit of nylon-6,T.[15] The length of this segment (without the endgroups) is ~4 nm.

$$\text{-T-}[\text{-PPE}_{\sim3100}\text{ - T - C}_{12}\text{ - T6T6T - C}_{12}\text{ -}]_n$$

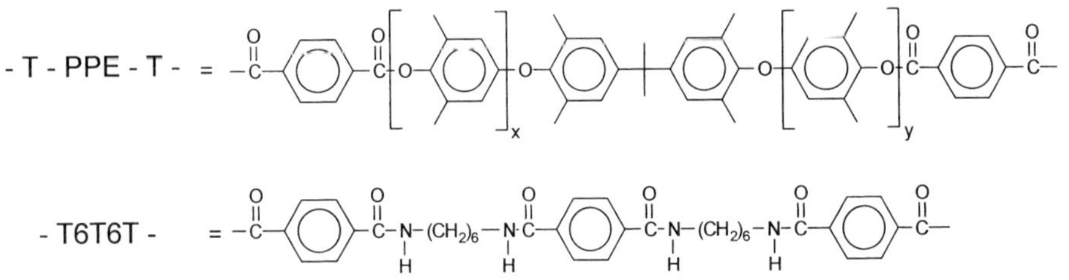

- T - PPE - T - =

- T6T6T - =

Fig. 1. Schematic structure of a PPE-2T/C12/T6T6T copolymer.

It is expected that the morphology of segmented copolymers based on PPE and uniform tetra-amide units is comparable with that of PTMO(/DMT) with uniform di-amide units.[21] A model for the crystalline structure of segmented copolymers that contain a low content (<20 wt%) of fast crystallizing short uniform amide segments is given in Figure 2.[21] The uniform amide crystalline segments crystallize into threads or ribbon-like structures (C) of ~4 nm thickness with high aspect ratio and some will be amorphous (B) and mixed with the amorphous PPE phase (A). [21-23] The length of the crystalline ribbons can be 100-500 nm, as was measured by AFM experiments.[21] When the crystalline content is low the copolymer is transparent. At high amide contents (>30 wt%) spherulitic structures can be formed and transparency will be lost.[23] Possibly the amide units are still ordered in the melt.[24-26] It is thought that crystallization of T6T6T takes place from these ordered T6T6T in the melt upon cooling. However, due to the high T_g of PPE, crystallization is incomplete and some T6T6T will be amorphous.

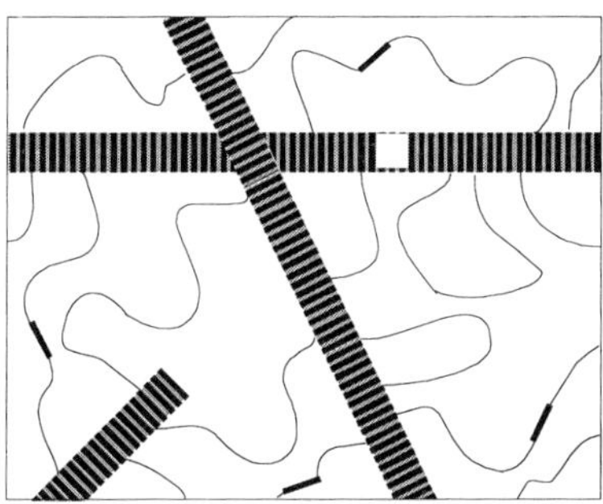

Fig. 2. Schematic representation of the morphology of crystallised segmented copolymers with uniform crystallizable segments: A = amorphous phase; B = amorphous crystallizable segments; C = crystalline ribbons.

In this review the synthesis and structure-property relationships of a segmented copolymer based on PPE-2T, dodecanediol and 13 wt% of T6T6T-dimethyl are described. The goal of this work is to obtain a semi-crystalline material with a high glass transition temperature (>150°C) and a high T_g/T_m ratio above 0.8 (T_m < 300°C). The synthesis and results of thermal-mechanical analysis, DSC, WAXD, water absorption and rheology experiments will be discussed successively.

Experimental

Materials. 1,12-Dodecanediol (C12), tetraisopropyl orthotitanate ($Ti(i\text{-}OC_3H_7)_4$) and N-methyl-2-pyrrolidone (NMP) were purchased from Merck. PPO-803® (11.000 g/mol) and Noryl-GTX® (GTX914) were obtained from GE Plastics (The Netherlands). All chemicals were used as received. T6T6T-dimethyl[15] and PPE-2T of 3100 g/mol (573 μmol OCH_3/gram)[20] were synthesized as described before.

Polymer Synthesis. The PPE-2T/C12/T6T6T copolymer was synthesized via a polycondensation reaction. The reaction was carried out in a 50 ml glass reactor with a nitrogen inlet and mechanical stirrer. The vessel was loaded with PPE-2T (10.0 g, 5.73 mmol OCH_3), dodecanediol (1.15 g, 5.73 mmol), T6T6T-dimethyl (1.97 g, 2.87 mmol), 20 ml NMP and catalyst solution (0.6 ml of 0.05M $Ti(i\text{-}OC_3H_7)_4$ in m-xylene). This mixture was first heated in an oil bath to 180°C under nitrogen flow. Then the temperature was raised in steps: 30 min 180°C, 30 min 220°C, 60 min 250°C and 60 min 280°C. The pressure was then carefully reduced ($P<20$ mbar) to distil off the remaining NMP and then further reduced ($P<1$ mbar) for 60 minutes. Finally, the vessel was allowed to slowly cool to room temperature whilst maintaining the low pressure. Then the polymer was cut out of the reactor and crushed.

Viscometry. The inherent viscosity of the polymers was determined with a capillary Ubbelohde type 1B at 25°C, using a polymer solution with a concentration of 0.1 g/dl in phenol/1,1,2,2-tetrachloroethane (50/50, mol/mol).

Dynamical Mechanical Analysis (DMA). Samples for the DMA test (70x9x2 mm) were prepared on an Arburg-H manual injection molding machine and dried in a vacuum oven at 80°C overnight. The torsion behavior was studied at a frequency of 1 Hz, a strain of 0.1% and a heating rate of 1°C/min using a Myrenne ATM3 torsion pendulum. The storage modulus G' and loss modulus G" were measured as a function of temperature starting at −100°C. The glass transition temperature (T_g) was expressed as the temperature at which the loss modulus G" has a maximum. The modulus of the rubbery plateau was determined at 40°C above the T_g. The flow temperature (T_{flow}) was defined as the temperature where the storage modulus G' reaches 0.5 MPa. The flow temperature is close to the melting temperature (T_m).

Differential Scanning Calorimetry (DSC). DSC spectra were recorded on a Perkin Elmer DSC7 apparatus, equipped with a PE7700 computer and TAS-7 software. Dried samples of

5-10 mg polymer in aluminum pans were measured with a heating and cooling rate of 20°C/min. The samples were heated to 300°C, kept at that temperature for 2 minutes, cooled to 100°C and reheated to 300°C. The (peak) melting temperature and enthalpy were obtained from the second heating scan. The crystallization temperature was defined as the maximum of the peak in the cooling scan.

Wide Angle X-ray Diffraction (WAXD) as Function of Temperature. Diffraction patterns at different temperatures were obtained using a Philips X'Pert-MPD diffractometer (curved graphite monochromator $CuK_{\alpha 1}$, radiation of 1.54056 Å). Melt-pressed samples of approximately 1 mm thickness were mounted in a sample holder in an Anton Paar HTK-16 temperature chamber. The measurements were carried out in nitrogen atmosphere and the heating and cooling rate were 2°C/min. The data were collected in a range of $2\Theta = 4\text{-}60°$.

Water Absorption. The absorption of water was measured as the weight gain after conditioning. DMA test bars were dried at 70°C in a vacuum oven for several days and weighed (w_0). Then the samples were conditioned in a dessicator over water at room temperature for 28 days (100% RH). The samples were reweighed after 7 and 28 days (w). The water absorption (in %) was calculated as $(w-w_0)/w_0 \times 100\%$.

Melt Viscosity. The melt viscosity was measured using a Kayeness capillary flow rheometer at 300°C. The length and diameter of the capillary were 20.32 and 1.016 mm respectively. The diameter of the barrel was 9.525 mm. The pressure was measured at different flow rates by applying varying piston speeds of 10, 20, 50, 100, 130, 200 and 500 mm/min (shear rates of 115, 230, 576, 1154, 1499, 2304 and 5760 sec^{-1} respectively).

Results and Discussion

Synthesis. Copolymers of PPE-2T, dodecanediol (C12) and T6T6T-dimethyl were made via a polycondensation reaction with a maximum temperature of 280°C. During the first part of the reaction NMP was used as a solvent because of the high melting temperature of the tetra-amide segment. The melting temperature of T6T6T-dimethyl is 303°C.[15] After one hour at 250°C, the reaction had progressed enough to allow the final part of the reaction to be performed in the melt. Most of the NMP was stripped off at 280°C. During the last hour a vacuum of <1 mbar was applied to strip off any methanol formed and to obtain polymers with high molecular weights. For copolymers with T6T6T-dimethyl as crystallizable segment the reaction mixture was gradually transferred from a clear solution into a clear melt. The clear melt indicates that melt phasing or liquid-liquid demixing between both segments is absent or only present at a nano-scale level.

The copolymer has an inherent viscosity of 0.41 dl/g (M_n~15.000 g/mol) which is higher than that of PPO-803® (M_n = 11.000 g/mol). A molecular weight above 10.000 g/mol is preferred for good toughness of the material.

Thermal-mechanical properties. The thermal-mechanical properties of segmented copolymers based on PPE-2T, dodecanediol and T6T6T-dimethyl were measured by DMA. The polymers were injection molded (with an unheated mold) into bars and dried in a vacuum oven at 80°C. The test bars are slightly transparent. When the material is transparent this indicates that no spherulites are present.[23] Light scattering also occurs when phase separated domains with a size above 100 nm are present. Apparently such domains are not present in PPE-2T/C12/T6T6T with 13 wt% T6T6T.

Table 1. Properties of the PPE-2T/C12/T6T6T copolymers.

	T6T6T content [wt%]	η_{inh} [dl/g]	T_g [°C]	G' (at T_g + 40°C) [MPa]	T_{flow} [°C]	T_g/T_{flow} [-]
PPE-2T[a]	-	0.18[b]	168[c]	-	-	-
PPO-803®	-	0.37[b]	200	-	222	-
PPE-2T/C12	0	0.31[b]	169	-	193	-
PPE-2T/C12/T6T6T	13	0.41	169	10	269	0.82

(a), bimodal product made by one-pot synthesis (3100 g/mol); [20]
(b), chloroform was used as a solvent instead of phenol/tetrachloroethane;
(c), measured by DSC instead of DMA;
(d), properties of test bar after heat treatment in a press (10 bar) for 5 minutes at 240°C and then slowly (5°C/min) cooled down to room temperature.

In Figure 3 the storage and loss moduli as measured by DMA are given for the PPE-2T/C12/T6T6T and amorphous PPE-2T/C12 copolymer compared to pure PPE (PPO-803®). The results are summarized in Table 1.

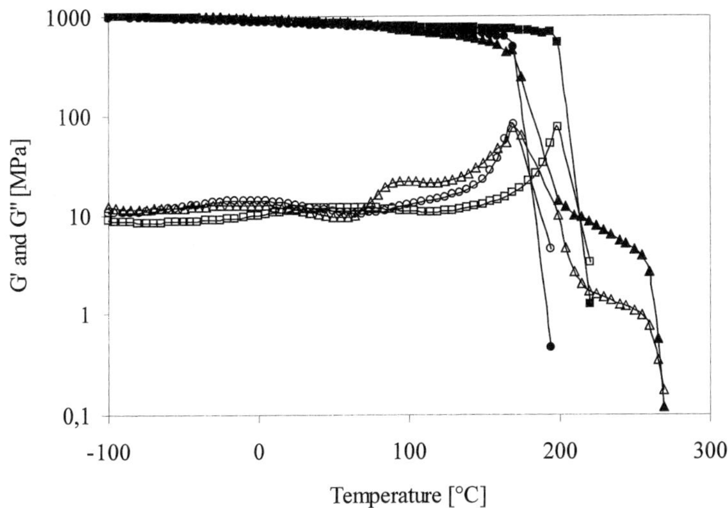

Fig. 3. Storage (solid symbols) and loss (open symbols) modulus of: (▲, △), PPE-2T/C12/T6T6T (13 wt%); (●, ○), PPE-2T/C12; (■, □), PPO-803®.

PPO-803® has a high and constant modulus up to the T_g at 200°C. The amorphous PPE-2T/C12 copolymer has a high and constant modulus up to the T_g at 169°C. The T_g is decreased compared to pure PPE because of incorporation of the flexible dodecanediol.[27]

The copolymer of PPE-2T, dodecanediol and T6T6T-dimethyl is a semi-crystalline material. It shows a rubbery plateau above the glass transition temperature of the amorphous PPE phase. This result is very surprising, because the T_g/T_{flow} ratio is very high (~0.82). The modulus is high and constant up to the T_g of 169°C. However the modulus drops somewhat above 70°C and the loss modulus shows a peak around 100°C. This small effect can possibly be attributed to a glass transition in the (semi-crystalline) T6T6T phase. The T_g of nylon-6,T is 125°C.[28] So the peak in the loss modulus at 100°C in the copolymer with T6T6T is probably originating from a T6T6T/C12 phase that has a little lower T_g than pure nylon-6,T.

The rubber modulus of the copolymer with 13 wt% T6T6T is 10 MPa and the flow temperature is 269°C. The flow temperature is sharp, which indicates that crystalline ribbons with a constant thickness are present. Compared to the commercial blend Noryl-GTX®, PPE-2T/C12/T6T6T has a higher modulus between 50 and 200°C. The T_g and T_m are comparable.

DSC. With DSC the melting and crystallization behavior of PPE-2T/C12/T6T6T with 13 wt% uniform T6T6T was studied. The T_m of T6T6T in the copolymer is 268°C and the enthalpy of melting is 14 J/g, which corresponds to 109 J/g T6T6T. From the enthalpy of melting of T6T6T-dimethyl (152 J/g)[15] it can roughly be calculated that the crystallinity of T6T6T is around 70%. The crystallization temperature is 250°C and thus the undercooling is only 18°C, which indicates that the crystallization of T6T6T is very fast.

WAXD. The crystalline structure of PPE-2T/C12/T6T6T with 13 wt% T6T6T was studied with WAXD. A melt-pressed sample of 1 mm thickness was prepared in a press at 300°C during 5 minutes with a pressure of 10 bar (cooling rate 5°C/min). The WAXD spectrum was measured at room temperature and at 300°C (Figure 4).

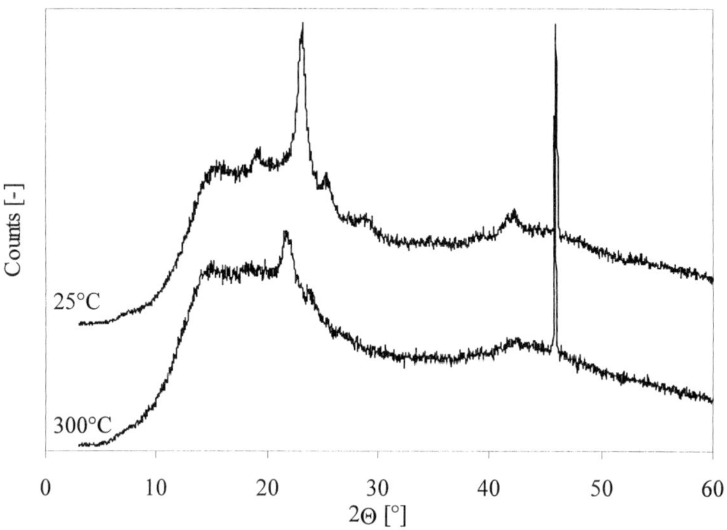

Fig. 4. WAXD data for PPE-2T/C12/T6T6T (13 wt%) at 25 and 300°C.

The main peak at 2Θ = 22-23° shifts to lower position and decreases in height when the temperature is increased. This indicates an increase in crystal dimensions and a decrease in crystalline order with increasing temperature. At 300°C, which is well above the melting temperature of ~268°C, the peak at 2Θ = 22-23° has still 30% of its original height. So in the melt there is still an ordered T6T6T phase present. In this phase the length of the hydrogen bonds is increased and the distance between the T6T6T units is larger than in crystalline T6T6T. For copolymers of PTMO(/DMT) and di-amide units[24-26] it was found by IR experiments that the di-amide units remain hydrogen bonded in the melt and that the average hydrogen bond strength decreases with increasing temperature.

Water Absorption. The water absorption of DMA test bars of PPE-2T/C12/T6T6T with 13 wt% uniform T6T6T was measured. The water absorption is compared with that of Noryl-GTX® and PPO-803®. The results are given in Table 2.

PPO-803® has a very low water absorption of 0.24% after one week. The water absorption of the PPE-2T/C12/T6T6T copolymer is 0.48% after one week. The water absorption does not increase further after 7 days. The water absorption of Noryl-GTX® is 2.1% after 7 days and 3.3% after 28 days. PPE-2T/C12/T6T6T copolymer with 13 wt% T6T6T has a lower water absorption than Noryl-GTX®. Polyamides are known to have high water absorption. The new copolymer has a lower polyamide content (about 3 times) and the crystallinity of the polyamide phase is higher than in Noryl-GTX®. Crystalline polyamide adsorbs less water than amorphous polyamide.

Table 2. Water absorption of PPE-2T/C12/T6T6T with 13 wt% uniform T6T6T compared to Noryl-GTX® and PPO-803® (20°C, 100% RH).

	7 days [%]	28 days [%]
PPE-2T/C12/T6T6T	0.48	0.48
Noryl-GTX®	2.1	3.3
PPO-803®	0.24	0.24

Rheology. The melt viscosity of PPE-2T/C12/T6T6T with 13 wt% uniform T6T6T (~15.000 g/mol) was studied using a capillary flow rheometer at 300°C. The data of the copolymer are compared with that of Noryl-GTX® (GTX914) and PPO-803® in Figure 5.

The melt viscosity of PPO-803® is very high. By blending with polyamide as in Noryl-GTX® the processability is improved a lot. At 300°C the PPE-2T/C12/T6T6T copolymers have a much lower melt viscosity than Noryl-GTX®. The melt viscosity increases slightly with increasing molecular weight. It can be concluded that the processability of a copolymer of PPE and T6T6T with only 13 wt% of T6T6T is much better than that of PPO-803® and Noryl-GTX®

with 40 wt% of polyamide.

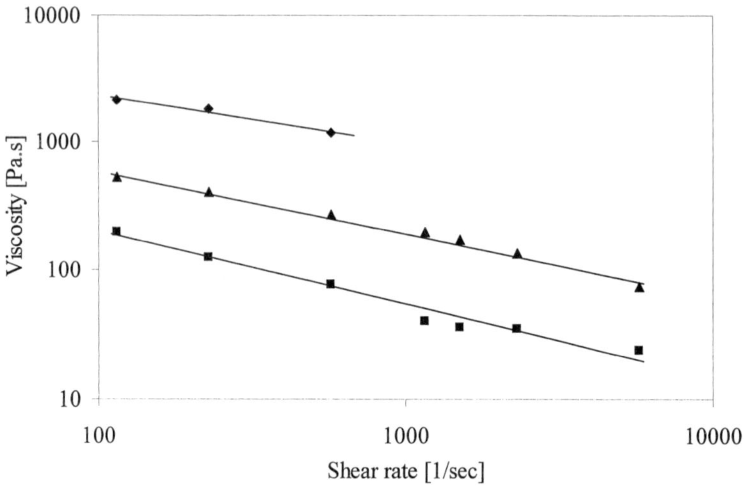

Figure 5: Melt viscosity at 300°C as a function of shear rate: (■), PPE-2T/C12/T6T6T (13 wt%, ~15000 g/mol); (▲),, Noryl-GTX®; (◆), PPO-803® (11.000 g/mol).

Conclusions

Copolymerisation of PPE with uniform crystallizable T6T6T units and dodecanediol as an extender is a good method to obtain a semi-crystalline material with a very high T_g/T_m ratio, good solvent resistance and good processability. The copolymer that contains only 13 wt% of T6T6T has a T_g of 169°C and T_{flow} of 269°C (T_m = 268°C). The polymer has a high and constant modulus up to the T_g. The modulus in the rubbery plateau is 10 MPa. It is very particular that the T6T6T units can actually crystallize in these copolymers despite the very high T_g/T_{flow} ratio of 0.82 and low hard segment concentration. The crystallinity of T6T6T is around 70%. The undercooling as measured by DSC is 18°C, which indicates that the crystallization is very fast. The crystallization is facilitated by ordering of T6T6T in the melt.

The PPE-2T/C12/T6T6T copolymers are interesting for applications where a high modulus up to the T_g of PPE is essential in combination with good solvent resistance and good processability. The segmented copolymers can also be useful as compatibilizer for blends of

PPE with polyamide. This method can also be useful to modify other high T_g amorphous polymers such at polycarbonate.

[1] D.M. White, in: *"Comprehensive polymer science"*, S.G. Allen, J.C. Bevington, Pergamon Press, New York 1989, p.473.
[2] D. Aycock, V. Abolins, D.M. White, in: *"Encyclopedia Of Polymer Science and Engineering"*, H.F. Mark, N.M. Bikales, C.G. Overberger, G. Menges, John Wiley & Sons, New York 1988, p.1.
[3] H.R. Kricheldorf, in: *"Handbook of Polymer Synthesis"*, Part A, Dekker, New York 1992, p.545.
[4] F.E. Karasz, J.M. O'Reilly, *J. Polym. Sci. Polym. Lett. Ed.* **1965**, 3, 561.
[5] J. Stehlicek, J. Kovarova, F. Lednicky, *Collect. Czech. Chem. Commun.* **1993**, 58, 2437.
[6] J. Stehlicek, R. Puffr, *Collect. Czech. Chem. Commun.* **1993**, 58, 2574.
[7] C.S. Han, S.C. Kim, *Polym. Bull.* **1995**, 35, 407.
[8] M. Sato, S. Ujiie, Y. Tada, *Eur. Polym. J.* **1998**, 34, 405.
[9] D.W. van Krevelen, *"Properties of Polymers"*, Elsevier, Amsterdam 1990, Chapter 19, p.585.
[10] J. Bicerano, *J. Macromol. Sci.* **1998**, C38, 391.
[11] M.C.E.J. Niesten, J. Feijen, R.J. Gaymans, *Polymer* **2000**, 41, 8487.
[12] R.J. Gaymans, J. de Haan, *Polymer* **1993**, 34, 4360.
[13] M.C.E.J. Niesten, J.W. ten Brinke, R.J. Gaymans, *Polymer* **2001**, 42, 1461.
[14] J.D. Hoffman, J.J. Weeks, *J. Res. Nat. Bur. Stand.* **1962**, Sect. A, 66, 13.
[15] J. Krijgsman, J.Feijen, R.J. Gaymans, *Polymer*, submitted.
[16] L.L. Harrell, *Macromolecules* **1969**, 2, 607.
[17] H.N. Ng, A.E. Allegrazza, R.W. Seymour, S.L. Cooper, *Polymer* **1973**, 14, 255.
[18] C.D. Eisenbach, M. Baumgartner, G. Gunter, in: *"Advances in Elastomer and Rubber Elasticity, Proc. Symposium"*, J. Lal and J. E. Mark Eds., Plenum Press, New York 1985, p.51.
[19] J.A. Miller, B.L. Shaow, K.K.S. Hwang, K.S. Wu, P.E. Gibson, S.L. Cooper, *Macromolecules* **1985**, 18, 32.
[20] J. Krijgsman, J.Feijen, R.J. Gaymans, *Polymer*, submitted.
[21] M.C.E.J. Niesten, R.J. Gaymans, *Polymer* **2001**, 42, 6199.
[22] L. Zhu, G. Wegner, *Makromol. Chem.* **1981**, 182, 3625.
[23] B.B. Sauer, R.S. McLean, R.R. Thomas, *Polym. Int.* **2000**, 49, 449.
[24] M.C.E.J. Niesten, S. Harkema, E. van der Heide, R.J. Gaymans, *Polymer* **2001**, 42, 1131.
[25] A.C.M. van Bennekom, R.J. Gaymans, *Polymer* **1997**, 38, 657.
[26] P.F. van Hutten,, R.M. Mangnus, R.J. Gaymans, *Polymer* **1993**, 35, 4193.
[27] J. Krijgsman, J.Feijen, R.J. Gaymans, *Polymer*, submitted.
[28] P.W. Morgan, S.L. Kwolek, *Macromolecules* **1975**, 8, 104.

Macromol. Symp. **2003**, *199*, 147-162

Synthesis and Properties of Poly(butylene terephthalate)-Poly(ethylene oxide)-Poly(dimethylsiloxane) Block Copolymers

*M. Dahrouch,**[1] *A. Schmidt,*[2] *L. Leemans,*[2] *H. Linssen,*[2] *H. Götz*[2]

[1] Dpto. Química Orgánica, Facultad de Ciencias Químicas, Universidad de Concepción, Casilla 160-C, Concepción, Chile
Fax: 56-41 245974, E-mail: mdahrouch@udec.cl
[2] DSM Research, P.O. Box 18, 6160 MD Geleen, The Netherlands
Fax: (31) 47 4763949, E-mail: heide.goetz@dsm.com

Summary:

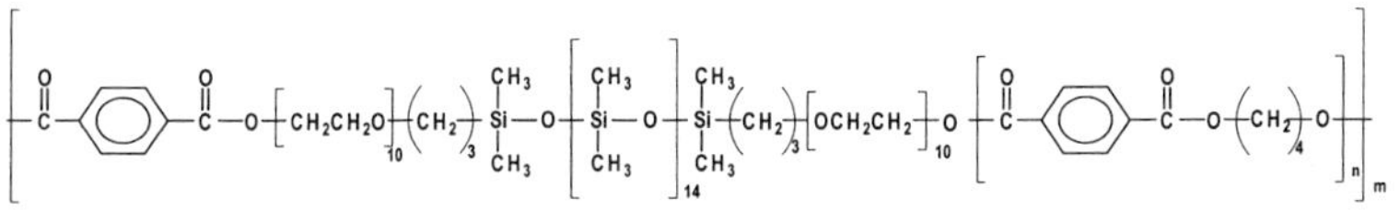

$$\left[\left[-\overset{O}{\overset{\|}{C}}-C_6H_4-\overset{O}{\overset{\|}{C}}-O-\left[CH_2CH_2O\right]_{10}-\left(CH_2\right)_3-Si(CH_3)_2-O-\left[Si(CH_3)_2-O\right]_{14}-Si(CH_3)_2-\left(CH_2\right)_3-\left[OCH_2CH_2\right]_{10}-O\right] - \left[\overset{O}{\overset{\|}{C}}-C_6H_4-\overset{O}{\overset{\|}{C}}-O-\left(CH_2\right)_4-O\right]_n \right]_m$$

Poly(butylene terephthalate)-poly(ethylene oxide)-poly(dimethyl siloxane)-poly(ethylene oxide) block copolymers, $(PBT\text{-}PEO\text{-}PDMS\text{-}PEO)_m$, are synthesized by polycondensation (PC) of dimethylterephtalate (DMT), 1,4-butanediol (BDO) and PEO-PDMS-PEO. The soft block has been incorporated from 10 to 70 wt-%; the total molecular weight (MW) of the block-copolymers amounts to 16000 – 20000 g/mol.
One major problem of polyether-PBT thermoplastic elastomers is their poor thermo-oxidative stability. Due to the excellent heat stability of PDMS, the resistance of this new thermoplastic elastomer against thermo-oxidative degradation has been increased 80 %!
From differential scanning calorimetry (DSC) and dynamic mechanical thermal analysis (DMTA) in the PEO-PDMS-PEO based COPEs, three phases can be distinguished. Besides the crystalline PBT phase, an amorphous mixed phase of PBT and PEO and an almost pure PDMS phase have been found. Due to the high concentration of the mixed PBT-PEO phase, the low temperature modulus and the glass transition temperature, T_g, are not dominated by the pure PDMS phase (T_g = -114 °C). Depending on the amount of PBT and PEO present, the main glass transition lies in the range of –50 °C to 50 °C.

Keywords: morphology, polycondensation, polysiloxanes, thermo-oxidative stability

 DOI: 10.1002/masy.200350913

1 Introduction

COPEs are segmented co-polymers with an $[\text{-A-B-}]_m$ structure with A being a polyether and B an aromatic polyester segment. COPEs are semicrystalline polymers with a crystalline polyester hard phase and a mixed amorphous phase.[1] These thermoplastic elastomers combine the elasticity and flexibility of rubbers and the easy processability of thermoplastics. Thus, injection molding, extrusion and blow molding can be applied for processing the material. In contrast to rubbers, COPEs can be recycled. Due to their good mechanical properties at a broad range of temperatures, high abrasion and solvent resistance and tear fatigue, they are used in a broad spectrum of applications. The automotive industry is the main end user, COPE being applied in boots and bellows, air ducts, and airbag covers. COPE films are used in roofing and textile lamination. Furthermore, COPEs are used in electric and electronic applications, for example for cable insulation and connectors. Moreover, mechanical goods as belting, tubes and hoses and keypads are made from COPEs.[1]

Silicones possess an excellent high- and low temperature stability, UV resistance, very good low temperature flexibility due to their low $T_g \sim -123$ °C, high gas permeability and a low surface tension. Incorporating them into a polyester backbone would combine these advantages with the melt processability of a plastic. Ten years ago, first works about incorporation of PDMS (instead of polyether) chains into polyester backbones have been published.[2-8] R. Mikami and coworkers encountered difficulties to co-polymerize DMT, BDO and hydroxy-terminated polyalkylsiloxanes. A significant fraction of siloxane was not incorporated into the polyester backbone. They explained this by the lack of compatibility between the polar PBT and the apolar siloxane.[7] D. A. Schiraldi studied the synthesis and properties of PDMS-PBT block copolymers [8]. The author reported that for concentrations higher than 15 wt-%, a loss of mechanical properties has been observed, and at higher contents, the polymers showed a lack of cohesiveness due to the incompatibility between the PBT and PDMS phases. D. J. Young and co-workers claim polyether-polysiloxane-polyether triblock copolymers with alkyl and polyether substituents (Figure 1) to increase the hydrophilicity of the surface of fibers and films.[5] The hydrophilic ether substituents of the siloxane allow complete miscibility with the aromatic polyester and due to the OH termination also covalent incorporation into the polyester backbone.

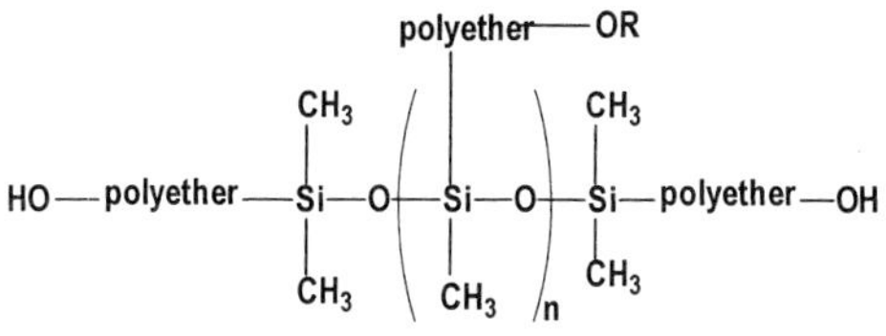

$$HO{-}\text{polyether}{-}\overset{CH_3}{\underset{CH_3}{Si}}{-}O{-}\left(\overset{\text{polyether}-OR}{\underset{CH_3}{Si}}{-}O\right)_n{-}\overset{CH_3}{\underset{CH_3}{Si}}{-}\text{polyether}{-}OH$$

Fig. 1. Polyether-polysiloxane-polyether triblock copolymers with alkyl and polyether substituents.[5]

In this article, we describe the synthesis of PBT-PEO-PDMS-PEO block copolymers. The PEO serves as a compatiblizer between the hydrophobic PDMS and the hydrophilic PBT. Thus, the PEO-PDMS-PEO soft block is not completely miscible as the copolymer shown in Figure 1, but also not as immiscible as pure hydroxyl-terminated PDMS.

2 Results and Discussion

First, the preparation of $(\text{PBT-PEO-PDMS-PEO})_m$ block copolymers is discussed. Then, the thermo-oxidative properties and morphology of these new COPEs are presented.

2.1 Synthesis of $(\text{PBT-PEO-PDMS-PEO})_m$ Block Copolymers

The copolymers are synthesized in two reaction steps: First, the transesterification (TE) takes place between DMT and BDO to form PBT oligomers (Scheme 1) by gradually heating the mixture to 220°C. BDO is present with 40 mol-% excess. The DMT can also react with the OH end groups of the soft block and form a prepolymer. Then, during polycondensation (PC) the prepolymers react under evaporation of the excess BDO and the molecular weight (MW) increases (Scheme 2). The PC is carried out by gradually raising the temperature from 230 – 245 °C under high vacuum (< 1mbar). The TE and PC are catalyzed by tetrabutyl titanate (TBT) and magnesium acetate tetrahydrate, added at equimolar quantities. The catalyst and co-catalyst concentration amount to 1.128 mmol for 800 g polymer, assuming 100% conversion. 0.5 wt-% of the phenolic anti-oxidant Irganox 1330 are added to avoid oxidative degradation. Silicon oil is used as anti-foaming agent.

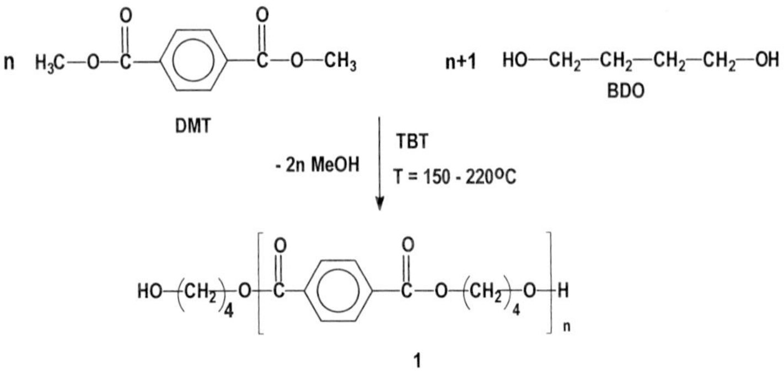

n $H_3C-O-C(=O)-C_6H_4-C(=O)-O-CH_3$ (DMT) + n+1 $HO-CH_2-CH_2-CH_2-CH_2-OH$ (BDO)

$\xrightarrow[\text{- 2n MeOH}]{\text{TBT, } T = 150 - 220^{o}C}$

$HO-(CH_2)_4-O-[C(=O)-C_6H_4-C(=O)-O-(CH_2)_4-O]_n-H$ **1**

Scheme 1. Transesterification and formation of PBT-oligomers.

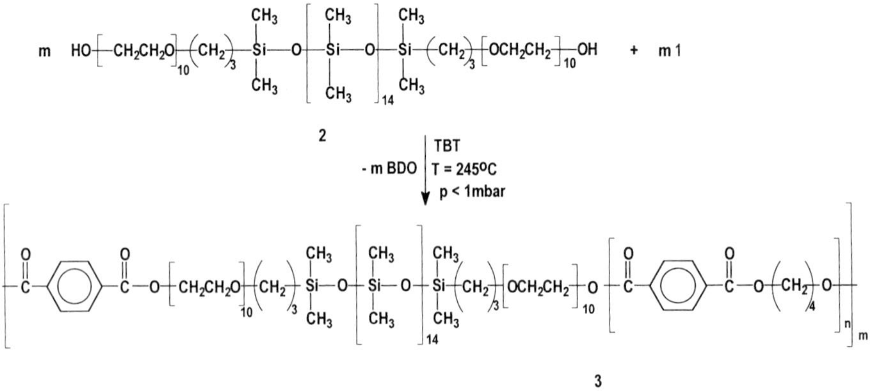

m $HO-[CH_2CH_2O]_{10}-(CH_2)_3-Si(CH_3)_2-O-[Si(CH_3)_2-O]_{14}-Si(CH_3)_2-(CH_2)_3-[OCH_2CH_2]_{10}-OH$ (**2**) + m **1**

$\xrightarrow[\text{- m BDO}]{\text{TBT, } T = 245^{o}C, \; p < 1 mbar}$

$[-C(=O)-C_6H_4-C(=O)-O-[CH_2CH_2O]_{10}-(CH_2)_3-Si(CH_3)_2-O-[Si(CH_3)_2-O]_{14}-Si(CH_3)_2-(CH_2)_3-[OCH_2CH_2]_{10}-O-[C(=O)-C_6H_4-C(=O)-O-(CH_2)_4-O]_n-]_m$ **3**

Scheme 2. Polycondensation.

The polymerization reaction is stopped, when the target torque is reached. The concentration of the soft block **2** has been varied from 10 – 70 wt-% (Table 1).

Up to 60 wt-% of **2** (MW = 2200 g/mol), melt phasing occurred at T = 245 °C; for higher concentrations the melt was clear. A milky appearance of the melt means that the soft and hard segments are no longer compatible under the conditions of the PC. Generally, at the transition from the one phase regime to the two phase regime, the effects of melt phasing are small and the copolymers exhibit one T_g and one melting point (T_m). Deep in the two-phase regime, two liquid phases exit, one rich in soft block and the other rich in hard block. These polymer blends have then also two T_g and two T_m [9]. According to [10], the resulting polymers have low tensile and tear strength and have a short elongation at break.

Table 1. Content of **2**, appearance of the melt at 245 °C, PC-time and torque at the end of PC at 20 rpm of PBT-PEO-PDMS-PEO block copolymers.

Sample	**2** [wt-%]	Melt	PC-Time [min]	Torque at 20 rpm [Nm]
S1	10	Milky	36	2.1
S2	30	Milky	40	2.4
S3	45	Milky	45	2.3
S4	50	Milky	75	2.30
S5	55	Turbid	100	2.2
S6			164	4.0
S7	65	Clear	225	1.8
S8	70	Clear	195	0.6

The purity of **2** has been investigated with ^1H and ^{13}C NMR. In the ^1H spectrum, the main resonances can be located at 3.55 ppm (m) $OC\underline{H}_2$ PEO, 3.3 ppm (t) $OC\underline{H}_2$ PEO/PDMS transition, 1.5 ppm (m) $C\underline{H}_2$ PDMS, 0.5 ppm (m) CH_2 PDMS, 0 ppm (s) Si-$(C\underline{H}_3)_2$ PDMS. The R-O-CH_2-CH_2-OH end groups could not be assigned unambiguously in the ^1H spectrum, in the ^{13}C spectrum these resonances were easily identified by the –$C\underline{H}_2$-OH resonance at 61.0 ppm. These signals are of higher intensity then those of the PEO-PDMS transitions at 23.7 and 14.5 ppm, indicating the presence of free PEO. Assuming the same average chain length, the free PEO amounts up to 25 wt-% (30 mol-%). The calculated MW of the PEO block is 830 g/mol and of the PDMS block 1160 g/mol.

The ^1H spectrum also reveals several end group resonances at 6.1 ppm (dd), 5.9 ppm (dd), 4.7 ppm (dq) and 4.3 ppm (dq) that have been assigned to both cis and trans propenyl (vinylene) end groups CH_3-$C\underline{H}$=$C\underline{H}$-OR. Their amount is 15 and 12 % of the total number of RO-CH_2-CH_2-OH end groups. Mostly, PDMS is linked to PEO by hydrosilylation of PEO containing an OH and a vinyl end group [5]. As PEO is added in excess, PEO-PDMS-PEO block copolymers contain free PEO. Each free PEO chain is terminated on one end by a vinyl group. Therefore, the percentage of unsaturated end groups (27 %) and the mol-% free PEO, determined from the excess –$C\underline{H}_2$-OH resonance (30 mol %), correspond very well.

As the free PEO chains have only one OH end group, they act as chain stoppers during PC; thus, the PC-time increases with the soft block content.

The number average molecular weight (M_n) of all samples has been determined from the concentration of hydroxyl (E_{OH}), carboxyl (E_C) and unsaturated end groups (E_{UNS}), which are listed in Table 2.

Table 2. E_C, E_{OH}, E_{UN} and M_n of all $(PBT\text{-}PEO\text{-}PDMS\text{-}PEO)_m$ block-copolymers.

Sample	E_C [meq/kg]	E_{OH} [meq/kg]	E_{UNS} [a] [meq/kg]	M_n [g/mol]
S1	11	79	10	20000
S2	15	73	30	17000
S3	13	67	45	16000
S4	20	50	50	17000
S5	20	49	55	16000
S6	14	48	55	17000
S7	18	34	65	17000
S8	15	37	70	16000

[a] The unsaturations of **2** are known from ^1H NMR; E_{UNS} has been calculated assuming that no further unsaturations are formed during PC.

The most striking observation is that independent of the final torque at 20 rpm, M_n amounts to about 17000g/mol. Even S5 and S6, where there is 100 % difference in torque, the M_n are nearly identical. The torque is determined by the melt viscosity and thus the weight average molecular weight (M_w) for non entangled chains or the viscosity average molecular weight for entangled chains.[11] We hypothesize that due to the high E_{UNS} crosslinking occurs, effecting more M_W than M_n. This might explain, why for S5 and S6 a 100 % increase in torque leads only to 6 % increase in M_n.

The incorporation of **2** has also been verified with ^1H NMR. In these spectra, the main resonances can be found at 8.0 ppm (s) aromatic H PBT, 4.38 ppm (t) $OC\underline{H}_2$ PBT, 3.57 ppm (t) $OC\underline{H}_2$ PEO, 1.91 ppm (t) $C\underline{H}_2$ PBT and 0.05 ppm (s) $Si\text{-}(C\underline{H}_3)_2$ PDMS. At a lower level, the signals from PEO-PDMS transitions can be found at 4.44 ppm (t) $C(O)OC\underline{H}_2$, 3.78 ppm (t) $C(O)OCH_2\text{-}C\underline{H}_2\text{-}O$, 3.37 ppm (t) $OC\underline{H}_2$, 1.68 ppm $C\underline{H}_2$ PDMS and 1.50 ppm $\text{-}C\underline{H}_2Si$ PDMS. Based on these signals, the overall composition of the copolymers has been calculated (Table 3).

Table 3. Overall composition of PBT-PEO-PDMS-PEO block copolymers determined with ^{1}H NMR.

Sample	PBT	PBT	PEO	PEO	PDMS	PDMS
	wt-%	MW [g/mol]	wt-%	MW [g/mol]	wt-%	MW [g/mol]
S2	72	3500	13	780	15	1100
S6	46	1000	25	810	29	1000

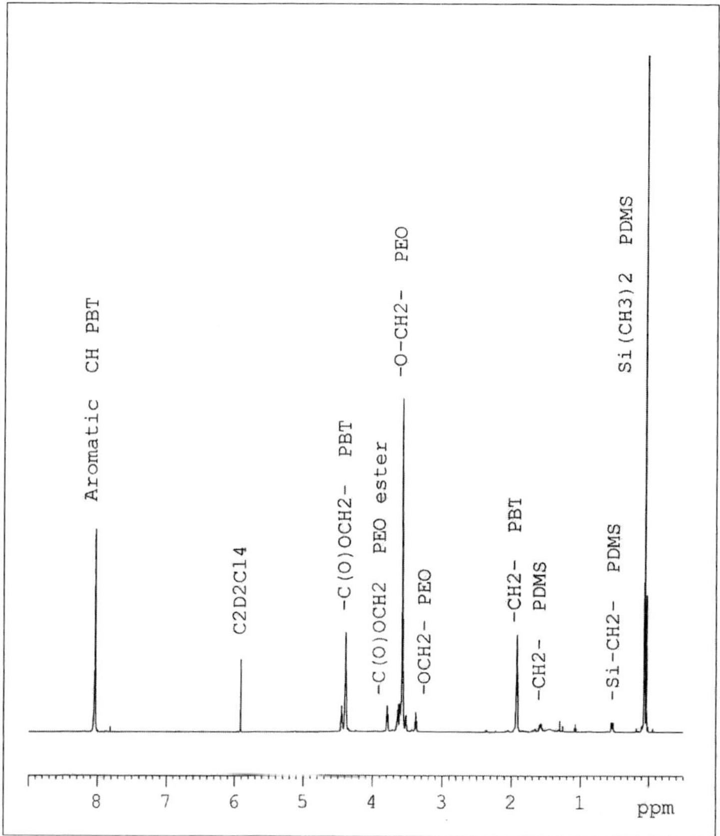

Fig. 2. ^{1}H NMR of S6.

With increasing PBT content, the MW of the PBT blocks increases. As expected, the MWs of the PEO and PDMS blocks do not change during polymerization. For PBT, the MW is calculated from the signals at 8.0 and 4.38 ppm. The MW of the PDMS block is determined from the

$Si(CH_3)_2$ signals at 0.05 ppm and the PEO-PDMS transitions at 0.50 and 1.5 ppm. In Figure 2, a typical ^1H NMR spectrum is shown.

2.2 Morphology

In Figure 3a, the atomic force microcopic (AFM) picture of sample S6 with 55 wt-% soft block is depicted. As a comparison, the picture of a PBT-polyether block-copolymer with 55 wt-% of a poly(ethylene oxide)/poly(propylene oxide) (PEO/PPO) copolyether is shown in Figure 3b. The AFM was operated in tapping mode, thus the contrast comes from the difference in hardness of the different phases. The hard crystalline PBT phase is represented by a light color, while the soft amorphous phase is dark. (The long light thin lines are arte facts of the preparation of the sample, which result from scratches on the teflon.) From Figure 3 it becomes obvious that the silicon based COPE shows less crystalline PBT than the corresponding COPE based on PEO/PPO soft blocks. Moreover, the crystalline PBT is co-continuous in case of the PBT-polyether block copolymer, while it is disperse in the silicon based COPE. Such a disperse morphology was also observed for PBT-polyether block-copolymer after stretching beyond their yield point.

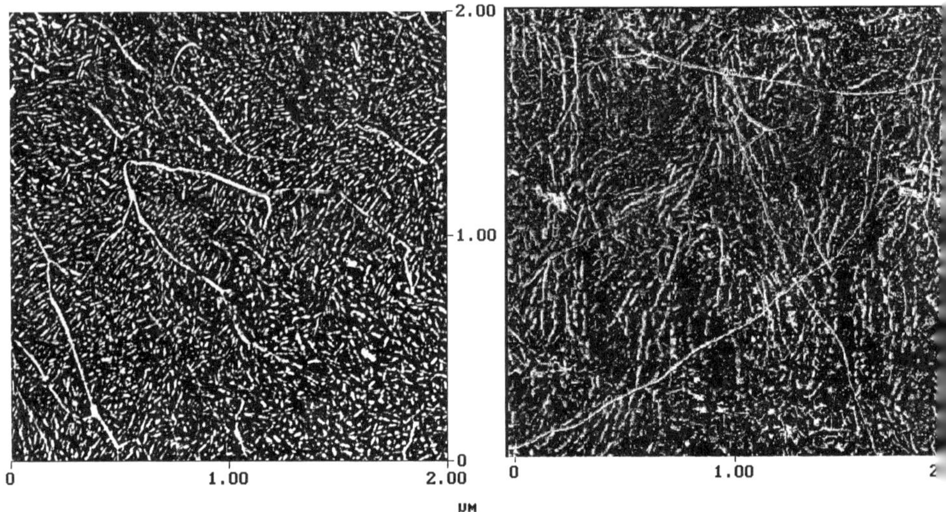

Fig. 3a/b. Tapping mode AFM on sample S6 with 55 wt-% PEO-PDMS-PEO soft block (left) and on a PBT-polyether blockcopolymer based on 55 wt-% PEO/PPO copolyether.

2.3 Thermal Analysis

In Table 4, the results of the DSC are listed. For all samples, a T_g at –114 °C has been observed. This T_g is close to that of pure PDMS (-123 °C), indicating that the PDMS is strongly phase separated. Although the appearance of the melt for the samples up to 55 wt-% soft block has been milky, all samples only show one melting peak. This is typical for not macro-phase separated COPEs. With increasing amount of soft block, T_m decreases from 217 °C to ca. 153 °C. This decrease can be explained by the decrease of the average degree of polymerization of the PBT blocks with increasing amount of soft block. For polyether based COPE, the T_ms of the PBT are strongly coupled to the average degree of polymerization of the PBT blocks and are independent of the type of ether soft block, as can be seen from Figure 4. However, the PDMS based materials seem to behave differently. Comparing polyether based COPEs and PEO-PDMS-PEO based COPEs with the same average degree of polymerization of the PBT, the silicon based materials show up to 40 °C lower melting points for the PBT. This indicates that the PBT crystals in silicon based COPEs are smaller and less perfect than in polyether (PTMO and PEO/PPO) based systems. The ability for the PBT to crystallize seems to be reduced in the PEO-PDMS-PEO based COPEs, which is also reflected in the relatively low crystallization temperatures (T_c) and the lower degree of crystallinity of PBT. While polyether-PBT block copolymers exhibit a PBT crystallinity of ca 40 %, for $(PEO\text{-}PDMS\text{-}PEO\text{-}PBT)_m$ the crystallinity was only about 25 % as determined from DSC. So far, this phenomenon cannot be explained.

Table 4. DSC data of $(PEO\text{-}PDMS\text{-}PEO\text{-}PBT)_m$.

Sample	SB [wt-%]	X_n (PBT)	T_g [°C]	T_c [°C]	T_m [°C]	ΔH [J/g]	Cryst. PBT [%]
S2	30	23	-114	160	217	24.5	24
S5	55	8	-114	131	191	16.4	25
S6	55	8	-114	122	187	17.8	27
S7	65	6	-114	79	153	8.5	17

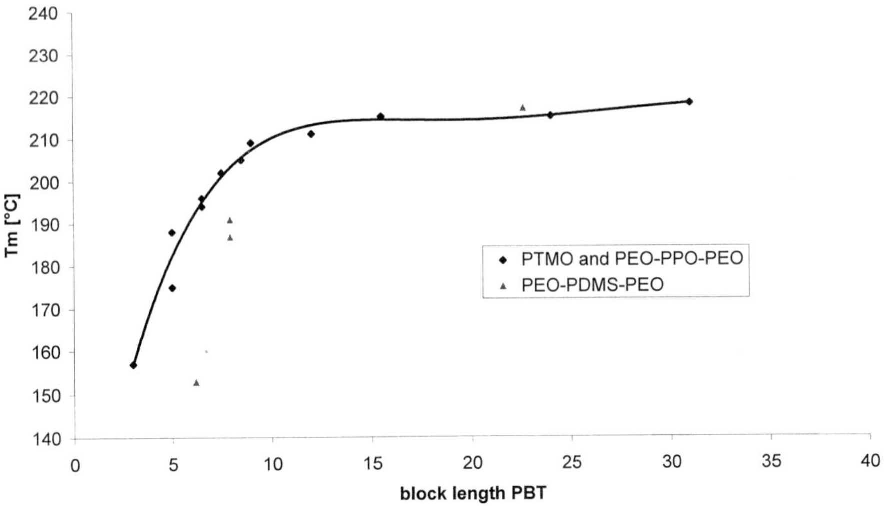

Fig. 4. T_m of the crystalline PBT with polyether and PEO-PDMS-PEO soft blocks as a function of the X_n of PBT.

2.4 Dynamic Mechanical Analysis

Figure 5 shows the results of the dynamic mechanical analysis of sample S2 with 30 wt-% of the PEO-PDMS-PEO soft block. At a temperature of –120 °C, a T_g is observed in the tan δ plot, which can be attributed to the glass transition of the almost pure PDMS phase. At 5 °C, a second transition occurs, which is linked to the T_g of a mixed amorphous phase of PBT and PEO. If the amount of soft block is increased to 55 wt-% (Figure 6), the T_g of the PDMS phase hardly shifts (-115 °C). From this, we conclude, that the PDMS phase is almost pure. In contrast, the T_g of the mixed PBT/PEO phase shifts to –48 °C due to the increase in PEO/PBT ratio.

If we compare the storage modulus E' of the PEO-PDMS-PEO based systems with those of PEO/PPO based COPES with similar MW and same amount of soft block (Figure 5/6), we observe a lower modulus up to a temperature of –60 °C for the PEO-PDMS-PEO based COPEs, in the temperature range of –60 to +50 °C, the PEO-PDMS-PEO based COPEs show a higher modulus. Since the PDMS almost completely demixes from the PBT, it has no plasticising effect on the amorphous PBT, which could further reduce the T_g. The T_g of the system is mainly determined by the amount of PEO present.

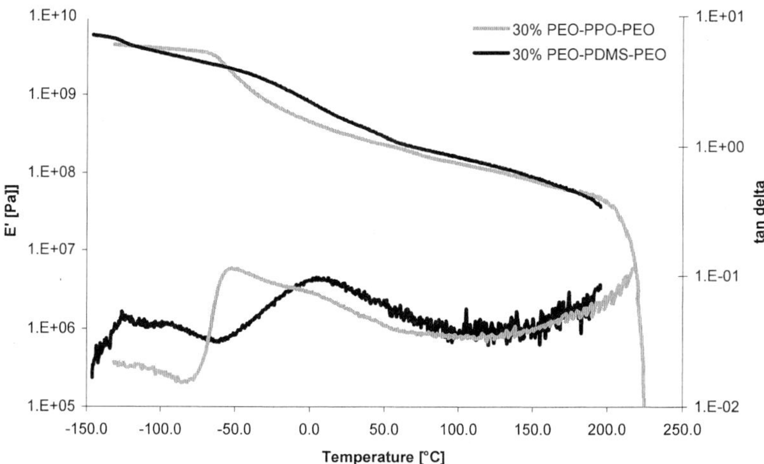

Fig. 5. DMTA of sample S2 with 30 wt-% soft block and a comparable COPE with 30 wt-% of PEO/PPO soft block.

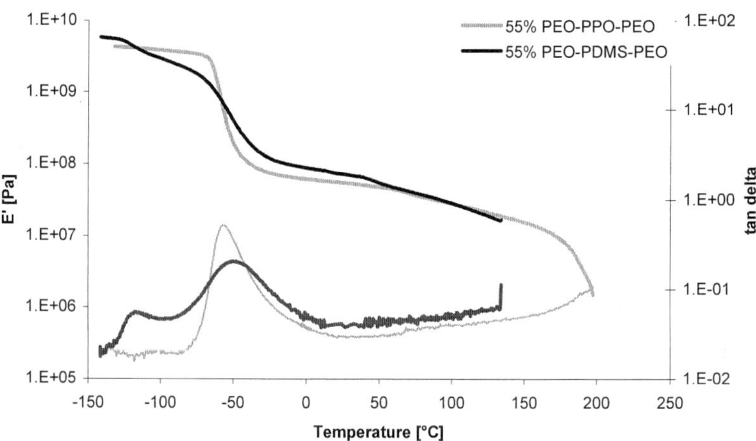

Fig. 6. DMTA of sample S6 with 55 wt-% soft block and a comparable COPE with 55 wt-% of PEO/PPO soft block.

2.5 Heat Aging

The thermo-oxidative stability in air at 135 °C has been investigated by measuring the relative solution viscosity (η_{rel}) of S1 and S2 at different times. The samples are stabilised against thermo-oxidative degradation with 0.5 wt-% of the phenolic anti-oxidant Irganox 1330. In Figure 7, η_{rel} and E_C of S1 and S2 are plotted versus the aging time.

The H-atoms in α-position next to the ether bond are labile and can react with oxygen from air. Due to chain cleavage of the ether-carbon and carbon-carbon bond formaldehyde is formed from PEO. The aldehyde will be oxidized further to formic acid, as shown in Scheme 3.[12]

$$\textbf{-O-CH}_2\textbf{-\textbackslash CH}_2\textbf{-O\textbackslash -CH}_2\textbf{- + O}_2 \rightarrow \textbf{HCOOH}$$

Scheme 3. Formation formic acid from PEO.

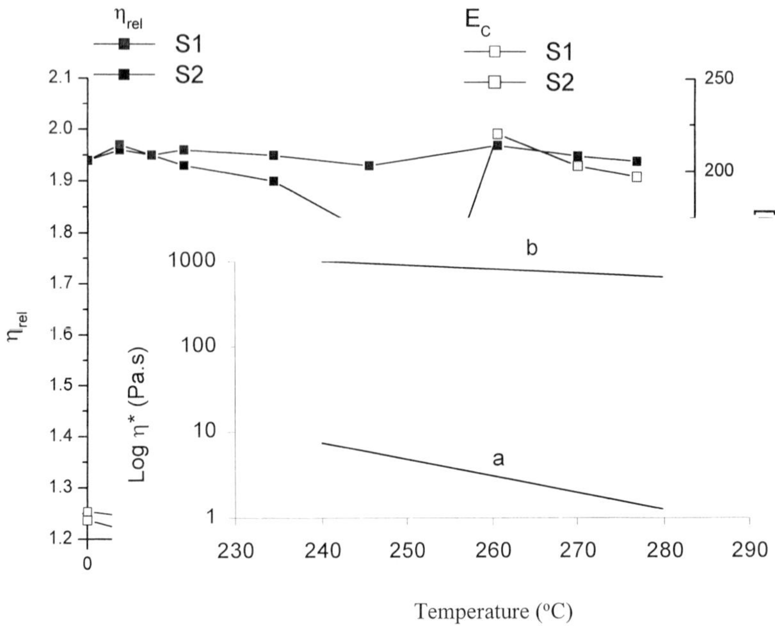

Fig. 7. η_{rel} and E_C of S1 and S2 during heat aging in air at 135 °C.

The chain cleavage causes a decrease in MW and thus η_{rel}; the formation of carboxylic acid leads to an increase in COOH. Due to the lower PEO content of S1 compared to S2, η_{rel} of S1 has not decreased after 400 h at 135 °C in air.

In Figure 8, η_{rel} of S2 and a PBT-PEO/PPO block copolymer with 30 wt-% polyether, stabilized with 0.5 wt-% Irganox 1330 during heat aging at 135 °C are shown. Due to the increased stability of PDMS compared to PPO, the thermo-oxidative stability of S1 is about 80 % higher than of PL580. Therefore, we hypothesize that an excellent heat stability should be obtained, if pure siloxane would be applied as soft block.

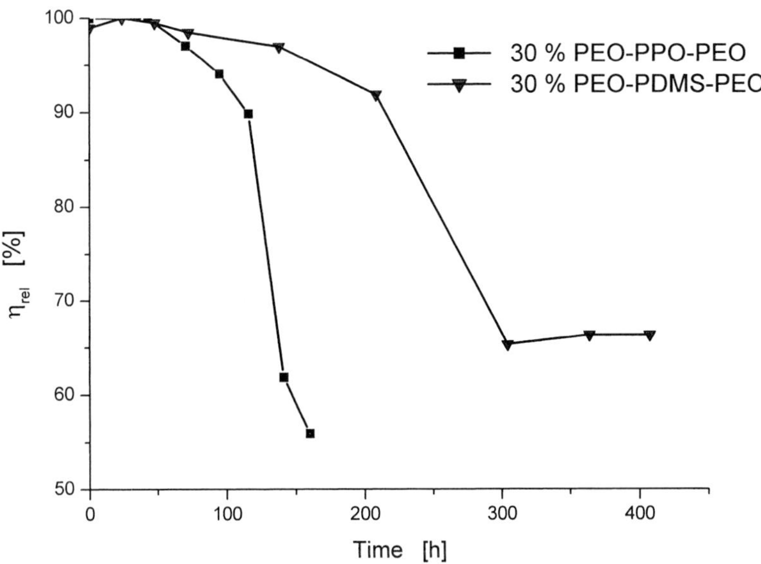

Fig. 8. η_{rel} of S2 and PBT-PEO/PPO block copolymer during heat aging in air at 135 °C.

3 Experimental Section

Materials. DMT from BP/Amoco, BDO from BASF, polysiloxane-polyether **1** from Wacker, TBT from Flucka and $Mg(OAc)_2.4H_2O$ from Aldrich were used as received.

General procedure for the preparation of (PBT-PEO-PDMS-PEO)$_m$ block copolymers, 3. DMT (330.4 g, 1.703 mol), BDO (214.6 g, 2.384 mol), **2** (440 g, 0.203 mol), TBT (383.5 mg, 1.128 mmol) $Mg(OAc)_2.4H_2O$ (241.4 mg, 1.128 mmol) and 15 droplets of silicone oil are placed in the reactor under nitrogen. The reaction mixture is heated slowly from room temperature to 220 °C under stirring and a slight nitrogen flow. The TE-reaction begins around 160 °C and methanol is distilled off. After this first step, the melt temperature is increased until 245 °C and under vacuum (< 1mbar) the distillation of BDO takes place. The PC-reaction is stopped, when the desired melt viscosity is reached. The torque is proportional to the melt viscosity and is measured by monitoring the electrical current, required to maintain an agitation rate of 20 rpm. At the end of the reaction, the melt is released from the reactor under nitrogen pressure into a water bath and rolled up to form a thread. This thread is later chopped into pellets, ready for analysis and processing.

^{1}H NMR. BRUKER Avance 500 MHz NMR spectrometer, concentration: 2% (wt/vol) in $C_2D_2Cl_4$, temperature: 80 °C, 90 ° pulse angle, 20 s relaxation delay, 128 scans, C_2HDCl_4 present in small amount in the solvent was used as an internal reference at 5.91 ppm.

^{13}C NMR. BRUKER ARX 400 NMR spectrometer under quantitative measuring conditions, concentration: 10 % (wt/vol) in $C_2D_2Cl_4$, temperature: 80 °C, 300 scans, 90 ° pulse angle, 10 s relaxation delay, inverse gated decoupling used to suppress any NOE; the internal reference: $\underline{C}_2D_2Cl_4$ at 74.2 ppm.

Viscometry. Ubbelohde capillary viscosimeter (Schott), determination of the flow time of a 10 wt-% solution with m-cresol as solvent, correction according to Hagenbach, temperature: 25 °C.

Titration of carboxylic acid end groups. Titration with potassium hydroxide, equlibrium determination with Methrohm photometer E662.

OH end groups. Esterification of the OH groups with anthracenoylchloride, high performance liquid chromatography (HPLC) of pure COPE and COPE treated with anthracenoylchloride, stainless steel column: 10 cm length, 2.1 mm ∅, 50 °C, Hypersil ODS of 5 μm particle size, UV detector, internal standard 1-hexadecanol in 1,1,1,3,3,3-hexafluoro-2-propanol; from the differences in these two chromatograms, E_{OH} is calculated.

AFM. AFM measurements were performed on a Nanoscope IIIa scanning probe microscope equipped with a J-scanner. The granulate was pressed into films using glass fiber reinforced Teflon foil (same procedure as used for DMTA measurements). Images were obtained under ambient conditions while operating the instrument in tapping mode using commercial Si cantilevers. Height and phase images were recorded simultaneously at the fundamental resonance frequency of the cantilever (typically 299-364 kHz) and at a scan speed of 0.5 lines/s (only phase images are depicted here.).

DSC. Perkin Elmer DSC7, heating and cooling rate: 20 °C min^{-1}, thermal analysis procedure: (i) heating from room temperature to 250 °C, (ii) 1 min at 250 °C, (iii) cooling from 250 °C to (-160 °C), (iv) 1 min at –160 °C, (v) heating from (– 160 °C) to 250 °C.

DMTA. Rheometrics Solids Analyser II, frequency: 1 Hz, compression molded samples with dimensions: 20 x 2 x 0.06 mm, heating rate: 5 °C/min, temperature range: -130 °C to 240 °C.

Thermo-oxidative aging. Air circulating oven (Heraeus Instruments LUT 6050E), temperature: 135 °C, granules: 2 mm ∅, 3 mm length.

4 Conclusions

$(PBT\text{-}PEO\text{-}PDMS\text{-}PEO)_m$ block copolymers have been synthesized by polycondensation of DMT, BDO and PEO-PDMS-PEO with TBT and $Mg(OAc)_2.4H_2O$ as catalysts. The soft block has been incorporated from 10 – 70 wt-%.

One major problem of polyether-PBT thermoplastic elastomers is their poor thermo-oxidative stability. Due to the excellent heat stability of PDMS, the resistance of this new thermoplastic elastomer against thermo-oxidative degradation has been increased 80 %!

The morphology of the PEO-PDMS-PEO based COPEs with a high soft block content is characterized by a crystalline PBT phase, a pure amorphous PDMS phase and a PEO/PBT mixed phase. In contrast to PTMO and PEO-PPO-PEO based COPEs, showing a co-continuous crystalline PBT phase, the crystalline PBT of the silicon Arnitels is disperse.

The melting points of the PBT in the $(PBT\text{-}PEO\text{-}PDMS\text{-}PEO)_m$ block copolymers are up to 40 °C lower than those of PTMO and PEO/PPO based COPEs. In addition, the degree of PBT crystallinity and the crystallization temperature of the PBT are also reduced.

[1] R. W. M. van Berkel, R. J. M. Borgreve, C. L. van der Sluis, G. H. Werumeus Buning, "Polyester-Based Thermoplastic Elastomers", in: *"Handbook of Thermoplastics"*, O. Olabisi, Ed., Marcel Dekker Inc., New York, Basel, Hong Kong 1997, p. 397 ff.
[2] J. K. Haken, N. Harahap, R. P. Burdford, *J. Chromatography* **1990**, *500*, 367.
[3] U.S. 4, 496, 704 (1995), inv: P. R. Ginnings.
[4] Vovelle Patent, 9318086 (1993), invs.: E. Fleury, P. Michaud, L. Tabus.
[5] U.S. 5, 132, 392 (1992), invs.: D. J. Young, G. J. Murphy, J. J. Deyoung.
[6] U.S. 4, 927, 895 (1990), invs.: T. Nakane, K. Hijikata, Y. Kagayama, K. Takahashi.
[7] Mikami, R.; Yoshitake, M.; Okawa, T. *U.S. Patent,* 4, 894, 427 (1990).
[8] D. A. Schiraldi, *Polymer Preprints* **2001**, *42(1)*, 221.
[9] C. Shih, J. M. McKenna, IUPAC Meeting Amherst, Mass. (1982).
[10] R. K. Adams, G. K. Hoeschele, "Thermoplastic Polyester Elastomers" in: *Thermoplastic Elastomers*, N. R. Legge, G. Holden, H. E. Schroeder, (Eds.) Hanser, Munich, Vienna, New York, 1987, p. 163 ff.
[11] P. C. Hiemenz, *"Polymer Chemistry"*, Marcel Dekker Inc. New York, Basel 1984.
[12] H. Zweifel, *"Stabilisation of Polymeric Materials"*, Springer-Verlag, Berlin 1998.

Synthesis and Characterization of Poly(ethylene glycol) Methyl Ether Endcapped Poly(ethylene terephthalate)

Qin Lin,[1] *Serkan Unal,*[1] *Ann R. Fornof,*[1] *Yuping Wei,*[1] *Huimin Li,*[1]
R. Scott Armentrout,[2] *Timothy E. Long**[1]

[1] Department of Chemistry, Virginia Polytechnic Institute and State University, Blacksburg, VA 24061-0212, USA
[2] Eastman Chemical Company, Kingsport, TN 37622, USA

Summary: Linear and branched poly(ethylene terephthalate) (PET) copolymers with poly(ethylene glycol) (PEG) methyl ether (700 or 2000 g/mol) end groups were synthesized using conventional melt polymerization. DSC analysis demonstrated that low levels of PEG end groups accelerated PET crystallization. The incorporated PEG end groups also decreased the crystallization temperature of PET dramatically, and copolymers with a high content of PEG (>17.6 wt%) were able to crystallize at room temperature. Rheological analysis demonstrated that the presence of PEG end groups effectively decreased the melt viscosities and facilitated melt processing. XPS and ATR-FTIR revealed that the PEG end groups tended to aggregate on the surface, and the surface of compression molded films containing 34.0 wt% PEG were PEG rich (85 wt% PEG). PEG end-capped PET (34.0 wt% PEG) and PET films were immersed into a fibrinogen solution (0.7 mg/mL BSA) for 72 h to investigate the propensity for protein adhesion. XPS demonstrated that the concentration of nitrogen (1.05 %) on the surface of PEG endcapped PET film was statistically lower than PET (7.67%). SEM analysis was consistent with XPS results, and revealed the presence of adsorbed protein on the surface of PET films.

Keywords: biocompatibility, crystallization rate, melt polymerization, poly(ethylene glycol) (PEG), poly(ethylene terephthalate) (PET)

Introduction

Poly(ethylene terephthalate) (PET) is an important polyester composition for a myriad of applications such as textile fibers, packaging materials, films, and container products.[1] However, PET has not emerged as a preferred engineering material or adhesive in many applications due to a low crystallization rate and poor compatibility with other substrates,

 DOI: 10.1002/masy.200350914

fillers and polymers.[2-3] To improve the performance of PET, several methodologies were developed.[2-8] For example, inorganic nucleating agents such as talc were used to accelerate the crystallization rate.[2] However, heterogeneous particles often act as stress concentrators that decrease impact strength, and glass fibers are required to reinforce the nucleated PET.[2] Thus, the development of novel compositions that exhibit a high crystallization rate will promote the utility of PET as an engineering material.

Recently, Xue and coworkers demonstrated that the crystallization rate of PET was improved via precipitating PET from a PEG solution at a low temperature.[9-11] In addition, PET segmented block copolymers prepared from difunctional aliphatic polyethers, such as PEG and PTMO, have been extensively studied.[12] However, the random incorporation of aliphatic polyether segments decreased the chain regularity and resulted in soft multiphase elastomers with a low level of crystallinity.[12-15] Earlier research has demonstrated that the incorporation of functional groups as end groups did not disrupt the regularity of the polymer backbone.[3] In order to eliminate the disruptive effect of random incorporation of PEG, mono-functional PEG was used to synthesize semicrystalline PET with PEG end groups. In addition, PET has been widely used in biomedical applications such as non-resorbable structures, tendons, ligaments, and facial implants.[16] PEG has also been used as an effective surface modifier to prepare biomaterials with low levels of protein adsorption and cell adhesion.[16-19] The incorporation of PEG end groups provides a novel methodology for the surface modification of semicrystalline PET.

In this manuscript, the synthesis of PEG end-capped linear and branched PET copolymers is reported. Low levels of incorporated PEG effectively accelerated the crystallization rate. Moreover, a high level of incorporated PEG resulted in a PEG rich layer on the film surface. This hydrophilic surface improved the biocompatibility and decreased protein adsorption on film surfaces.

Experimental

Synthesis of linear PEG endcapped PETs. Linear PEG endcapped PET copolymers were prepared via the melt polymerization of a PET oligomer with various levels of PEG endcapping reagents. Antimony oxide (150 ppm) was added to facilitate polycondensation.

The reactor consisted of a 250 mL round-bottomed flask equipped with an overhead mechanical stirrer, nitrogen inlet, and condenser. The reactor containing the monomers and catalysts was degassed using vacuum and nitrogen three times, and subsequently heated to 275 °C. The reaction was maintained at 275 °C for 1 h, and vacuum was gradually applied to 0.5 mm Hg and polycondensation continued for 2 h at 275 °C. The copolymers were termed PET-x-y, for example PET-700-1.5, wherein x denotes the molecular weight of PEG endcapper, and y denotes the molar percentage of endcapping reagent versus PET repeating units.

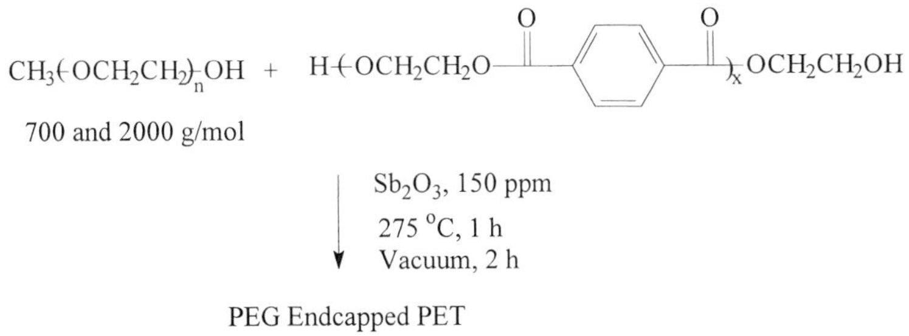

$CH_3(OCH_2CH_2)_nOH$ + $H(OCH_2CH_2O-C(=O)-C_6H_4-C(=O))_xOCH_2CH_2OH$

700 and 2000 g/mol

Sb_2O_3, 150 ppm
275 °C, 1 h
Vacuum, 2 h

PEG Endcapped PET

Scheme 1. Synthesis of linear PEG endcapped PETs via melt polycondensation.

Synthesis of PEG endcapped and branched PET. PEG end-capped branched PET was prepared via the melt condensation of dimethylterephthalate (DMT), ethylene glycol (EG) and 3 mol% branching agent (trimethyl 1, 3, 5-benzenetricarboxylate) (Scheme 2). Both titanium tetra(isopropoxide) (20 ppm) and antimony oxide (150 ppm) were added to facilitate ester exchange and subsequent polycondensation. The reactor consisted of a 250 mL round-bottomed flask equipped with an overhead mechanical stirrer, nitrogen inlet, and condenser. The reaction was maintained at 190 °C for 2 h, and the temperature was increased to 275 °C over 2 h. The reaction was allowed to proceed for 30 min at 275 °C. Vacuum was gradually applied to 0.5 mm Hg and polycondensation continued for 2 h at 275 °C. The product is termed BPET-2000-5, where B signifies a branched PET composition.

Protein Adhesion. Fibrinogen (50.0 mg) was dissolved in 100 mL PBS solution (pH = 7.4), and was filtered using a syringe filter. The relative concentration of protein was measured using a "Lowry protein assay". BSA standard tubes containing 0, 10, 20, 40, 70, 100 μg BSA

(Sigma) in a total volume of 100 μL were prepared. Standard agent was prepared as follows: 0.20 mL of 4.00% K Na tartrate and 0.20 ml of 1.28% $CuSO_4$ were mixed together, and then 10.0 mL of 3.00% Na_2CO_3 dissolved in 0.10 N NaOH was added. A standard agent (0.10 mL) and 0.10 mL phenol agent (Sigma) were added into each BSA tube. The absorbance of solutions at 750 nm was measured to generate a standard curve. The concentration of fibrinogen was measured using similar procedures, which was approximately 0.72 mg/mL BSA. PET and BPET-2000-5 films were initially immersed in a PBS solution for 2 h at 37 °C, and then a fibrinogen solution for different times (24, 48 and 72 h) at 37 °C. The films were rinsed using PBS solution for 20 seconds to remove weakly adsorbed fibrinogen, and dried at room temperature for 72 h for further analysis.

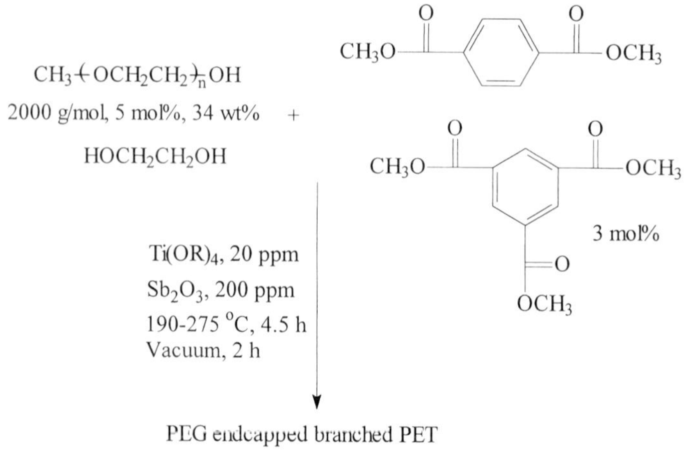

$CH_3(OCH_2CH_2)_nOH$
2000 g/mol, 5 mol%, 34 wt% +
$HOCH_2CH_2OH$

CH_3O–CO–C_6H_4–CO–OCH_3

CH_3O–CO–C_6H_3(–CO–OCH_3)–CO–OCH_3
3 mol%

Ti(OR)$_4$, 20 ppm
Sb_2O_3, 200 ppm
190-275 °C, 4.5 h
Vacuum, 2 h

PEG endcapped branched PET

Scheme 2. Synthesis of PEG endcapped and branched PET, BPET-2000-5

Characterization. The inherent viscosities of the samples were measured at 25 °C in a capillary viscometer using a 0.5 g/dL solution in a 60/40 w/w mixture of phenol and tetrachloroethane. ^{1}H NMR spectra were recorded on a Varian 400 MHz spectrometer. Trifluoroactetic acid–d was suitable as a NMR solvent. Thermal transitions were determined on a Perkin-Elmer DSC Pyris 1 under N_2 purge. Thermogravimetric analysis (TGA) was performed on a Perkin-Elmer TGA 7 under a nitrogen atmosphere at a heating rate of 10 °C/min. Contact angle measurements were conducted using the static drop method with a

Rame-Hart NRL contact angle goniometer. Angular dependent X-ray photoelectronic spectroscopy (XPS) was performed on a Perkin-Elmer physical electronic model 5400 with a hemisphere analyzer and a position sensitive detector. The rheological analyses were performed using a TA Instruments AR 1000 melt rheometer. SEM was performed on a (JEOL-JSM 5800), and samples were coated with Au prior to the analysis.

Results and Discussion

Synthesis. Our previous research demonstrated that residual catalysts exerted a pronounced effect on crystallization rate.[3] To minimize the effect of residual catalysts, the linear PEG endcapped PETs were synthesized using PET oligomers in the absence of transesterification catalysts (Scheme 1). In addition, a low level of antimony oxide (150 ppm) was used to facilitate polycondensation. Elemental analysis demonstrated that the actual level of residual catalyst agreed well with the charged amounts (Ti: ±1.0 ppm; Sb: ± 10 ppm).

A major limitation of endcapping methodologies is decreased molecular weight with an increase in the level of endcapping reagent; however, the simultaneous incorporation of a low level of branching agent will increase the number of end groups and permit higher weight average molecular weights. Thus, trimethyl 1, 3, 5-benzenetricarboxylate was used as a branching agent for the preparation of high molecular weigh PET with a high concentration of PEG end groups. This telechelic polyester was synthesized using a one step polymerization of DMT, excess EG, PEG endcapper (5 mol%), and branching reagent (3 mol%) (Scheme 2).

Table 1. ^{1}H NMR analysis and inherent viscosity of PEG end-capped PETs.

Sample	Charged PEG (mol%)	Charged PEG (wt%)	Residual PEG (wt%)[b]	$\eta_{inherent}$ (dl/g)[a]
PET	0	0	0	0.47
PET-700-1.5	1.5	5.2	5.0	0.42
PET-2000-1	1.0	9.4	8.9	0.47
PET-2000-1.5	1.5	13.5	12.7	0.46
PET-2000-2	2.0	17.6	16.5	0.44
BPET-2000-5	5.0	34.0	31.2	0.81

[a]: Determined at 25 °C using a capillary viscometer in a 0.5 g/dL solution of 60/40 w/w mixture of phenol and tetrachloroethane. [b]: Determined using ^{1}H NMR spectroscopy.

The composition of poly(ethylene terephthalates) with PEG endcappers (700 and 2000 g/mol) were verified using ^{1}H NMR spectroscopy. To ensure the absence of unreacted PEG, the films were immersed in water for 2 h prior to the NMR experiment. The results of the ^{1}H NMR analysis (Table 1) demonstrated that the PEG endcapper was quantitatively incorporated into PET as end groups. Linear copolymers had solution viscosities ranging from 0.47 to 0.42 dL/g. As expected, the presence of PEG end groups slightly decreased the molecular weight of the polyester products, i.e. the molecular weight decreased as level of end capping reagent increased. If a perfect telechelic copolymer, PEG-PET-PEG, was achieved, the molecular weight can be estimated using the equation reported in a previous report.[3] In this manuscript, only a low level (< 3 mol%) of PEG endcapper was used, and the estimated number average molecular weight for a perfect PEG-PET-PEG copolymers was 22000 g/mol (0.80 dL/g). Without further solid state polymerization, high molecular weight products were not obtained. Thus, the products were not perfect triblock structures, but a mixture of copolymers with one or two PEG end groups. Furthermore, it was presumed that the perfect telechelic structure would not offer advantageous properties, and perfect triblock polymers with high molecular weights are not described in this manuscript.

Thermal transitions and rheological analysis. Thermogravimetric analysis (TGA) demonstrated that the presence of PEG end groups did not exert a pronounced effect on the thermal stability of the final products. The telechelic polyesters exhibited a similar weight loss profile versus temperature as PET homopolymers, and the onset of degradation was approximately 360 °C. In order to study the crystallization behavior of PEG endcapped PET, amorphous films were compression molded. The polymer film was quenched using ice water, and transparent amorphous films were obtained immediately after quenching. However, BPET-2000-5 film developed haze within several minutes after quenching. PET-2000-2 formed a soft film immediately upon quenching. This film became opaque and the modulus gradually increased in two weeks presumably due to slow crystallization at room temperature. DSC analysis of the opaque films (two weeks after quenching) demonstrated that only a melt transition associated with PET (240 °C) was observed, which indicated that these two copolymers were able to crystallize at room temperature, and the crystallization rate increased with an increase in the level of PEG.

The thermal transitions of the transparent films were examined using DSC at a heating rate of 10 °C/min (Table 2). DSC results demonstrated that PEG endcapped PETs exhibited that the onset of crystallization was very close to the glass transition. As a result, it was difficult to measure the glass transition temperature accurately; however, the onset of the glass transition decreased with an increase in the PEG level (Table 1). Moreover, crystallization also occurred at lower temperatures compared to PET homopolymers at equivalent molecular weight, and both the onset of crystallization and the peak maximum of crystallization decreased with an increase in the PEG level. The heats of fusion for the copolymers were used to characterize the approximate levels of crystallinity (Table 2). PET copolymers and homo-PET had heats of fusion at 45 - 48 J/g (~35 – 40 % crystallinity based on 120.0 J/g for 100 % crystallinity)[1] (Table 2), and an increase in the level of crystallinity in blends of PET/PEG was not observed.[9-10] Isothermal crystallization of the transparent films was also performed at 85 °C. PET was obviously not able to crystallize at this temperature because it was just slightly higher than its glass transition temperature (78 °C). However, the PEG endcapped PET copolymers were able to crystallize at this temperature since the presence of PEG end groups improved polyester diffusion. The onset time (t_{onset}) and half time of crystallization ($t_{1/2}$) decreased with an increase in the PEG level, and isothermal crystallization data were analyzed using the Avrami equation,

$$\ln[-\ln(1-x_t)] = \ln K + n \ln t \qquad \text{(Eqn. 1)}$$

where x_t is the weight fraction of materials crystallized at time t, K is the kinetic growth constant, and n is the Avrami exponent. The values of n were determined to be in the range 2.0 -2.5 (the linear regression correlation coefficients were $r^2 = 0.98$), which were indicative of two dimensional crystal growth.

DSC analysis demonstrated that the incorporated PEG end groups effectively increased the PET crystallization rate. Melt rheological analysis was also performed to investigate PEG endgroup effects on viscosity. The results of the temperature sweep demonstrated that the copolymers exhibited lower melt viscosities than PET homopolymers with nearly equivalent inherent viscosities, and the difference in melt viscosity increased with an increase in the PEG content (Figure 1).

Table 2. Thermal transitions and water contact angle of quenched films.

Sample	T_g[a] (°C)	T_{hc} (°C)	ΔH_{hc} (J/g)	T_m (°C)	ΔH_m (J/g)	ΔH_m[b] (J/g)*	Contact Angle (°)
PET	62	138.5	16.2	237.6	45.2	45.2	82
PET-700-1.5	50	111.1	28.11	240.0	44.1	46.2	73
PET-2000-1	45	106.4	27.1	240.1	43.0	47.5	46
PET-2000-1.5	35	100.5	25.7	239.7	42.2	48.8	35
PET-2000-2	20	90.6	18.4	236.5	38.3	46.4	23
BPET-2000-5	-10	-----	-----	239.4	28.1	42.6	----

[a]: T_g (onset); [b]: after weight fraction calibration

Table 3. Isothermal analysis of quenched films.

Sample	t_{onset} (minute)	$t_{1/2}$ (minute)	ΔH (J/g)	ΔH[a] (J/g)	n
PET-700-1.5	1.61	6.78	25.1	26.5	2.50
PET-2000-1	0.68	3.21	21.9	24.2	2.33
PET-2000-1.5	0.38	1.10	19.4	22.4	2.22
PET-2000-2	0.21	0.42	19.0	23.1	2.29

[a]: after weight fraction calibration

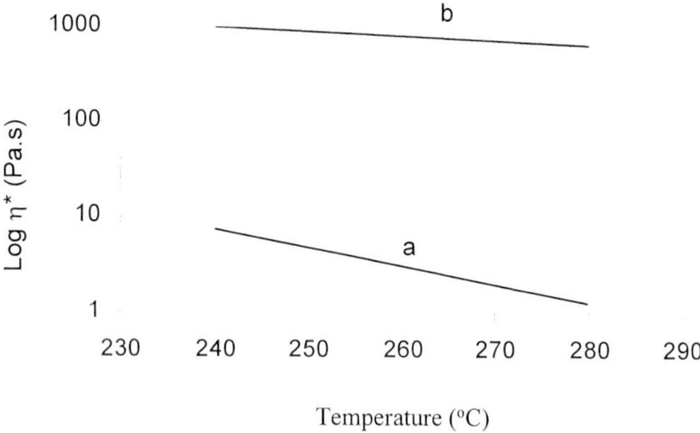

Fig. 1. Rheological analysis of BPET-2000-5 (a) and PET (b) with identical inherent solution viscosities (0.81 dL/g).

Surface analysis and biocompatibility. Water contact angles of PEG endcapped PET copolymers films are listed in Table 2, which showed a gradual change from a hydrophobic surface to a hydrophilic surface with an increase in PEG concentration. XPS revealed a clean PEG rich (85 wt%) layer on the surface of BPET-2000-5 (34 wt% charged PEG). To investigate potential biocompatibility, the PET and BPET-2000-5 films were immersed into a fibrinogen solution. The control PET and BPET-2000-5 films exhibited no signal related to the presence of nitrogen prior to immersion. However, the N_{1s} peak appeared in the XPS spectra of immersed films due to the presence of adsorbed protein on the surface (Table 4), which increased with an increase in time. Furthermore, the concentration of nitrogen on the surface of BPET-2000-5 films was much lower than PET homopolymer films with an identical time of immersion (Table 4). SEM micrographs revealed adsorbed protein on the surface of PET films immersed into fibrinogen solution for 72 h (Figure 2). However, adsorbed protein was not observed on the surface of BPET-2000-5 film under identical immersion conditions. These results confirmed that a PEG layer on the PET surface was able to improve the biocompatibility and decrease protein adsorption.

Table 4. Concentration of nitrogen (mol%) on the surface of films dipped into fibrinogen solutions for different times.

Sample	0 h	24 h	48 h	72 h
PET	0	5.48 %	6.89 %	7.37 %
BPET-2000-5	0	0.56 %	1.05 %	1.31 %

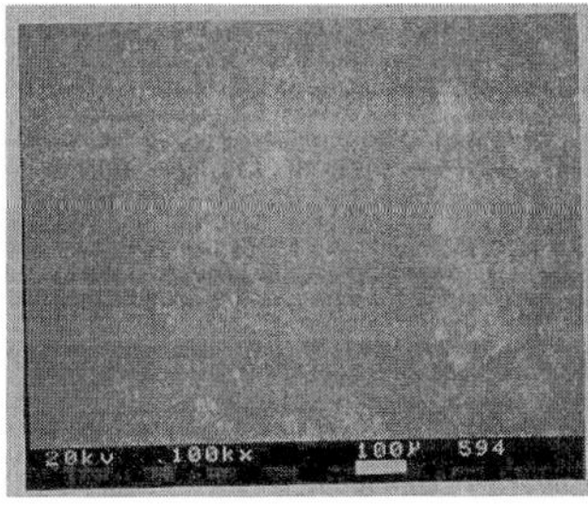

a

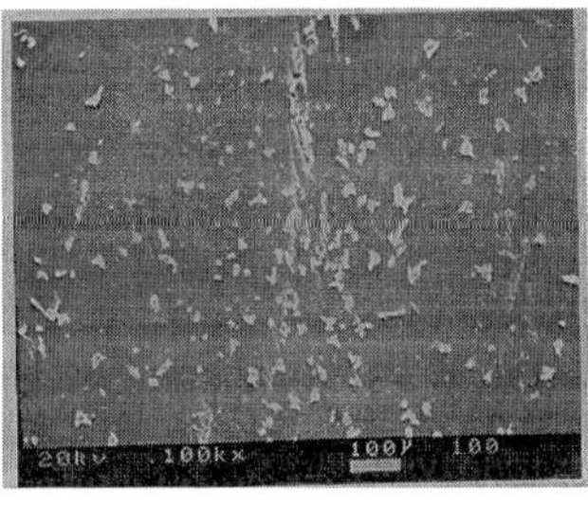

b

Fig. 2. SEM analysis: (a): BPET-2000-5 immersed into a fibrinogen solution for 72 h; (b): PET immersed into a fibrinogen solution for 72 h.

Conclusions

PEG endcapped linear PETs were synthesized using PEG (700 and 2000 g/mol) and PET oligomer. High molecular weight PEG endcapped PET with a high level of PEG was prepared using a branching agent. DSC analysis demonstrated that low levels of incorporated PEG accelerated the crystallization of PET, and decreased the crystallization temperature dramatically. A PEG rich layer (85 wt%) on the surface of PEG endcapped branched PET (34 wt%) was revealed using XPS and ATR-FTIR. This hydrophilic surface improved the biocompatibility and decreased protein adsorption. XPS analysis indicated that the PEG endgroups effectively retarded the adsorption of fibrinogen on compressed films compared to conventional PET homopolymers. SEM analysis agreed well with the XPS results, and revealed adsorbed protein on the surface of PET films without PEG end groups, and significantly less protein adhesion for PEG modified PET.

Acknowledgements

The authors thank Eastman Chemical Co. for financial and analytical support, Dr. David L. Phopham (Department of Biology, Virginia Tech) for his assistance in the investigation of protein adhesion, and Frank Cromer for his assistance with XPS and SEM.

[1] I. Goodman, R. J. Sheanan, *Eur. Polym. J.* **1990**, 26, 1081.
[2] I. Goodman, M. T. Rodriguez, *Macrol. Chem. Phys.* **1994**, 195, 1705.
[3] H. Y. Kang, Q. Lin, R. S. Armentrout, T. E. Long, *Macromolecules* **2002**, 35, 8738.
[4] J. E. Flanigan, G. A. Mortimer, *J. Polym. Sci.: Polym. Chem. Edn.* **1978**, 16, 1221.
[5] Q. Lin, J. Pasta, Z. H. Wang, R. Varian, G. L. Wilkes, T. E. Long, *Polymer International* **2002**, 51, 540.
[6] B. Gordon III, A. E. Mera, *Polym. Bull.* **1989**, 22, 273.
[7] N. Nagata, T. Kiyotsukuri, S. Minami, N. Tsutsumi, W. Sakai, *Polym. Int.* **1996**, 39, 83.
[8] D. P. Kint, A. Martinez de Llarduya, S. Munoz-Guuerra, *J. Polym. Sci.: Part A* **2000**, 38, 3761.
[9] G. Xue, G. D. Ji, H. Yan, M. M. Guo, *Macromolecules* **1998**, 31, 7706.
[10] G. Xue, G. D. Ji, Y. Q. Li, *J. Polym. Sci.: Part B* **1998**, 36, 1219.
[11] G. D. Ji, G. Xue, X. N. Zhang, B. Liu, D. S. Zhou, X. H. Gu, *Macromol. Chem. Physic.* **1996**, 197, 2149.
[12] J. A. Miller, J. M. McKenna, G. Pruckmayer, J. E. Epperson, S. L. Cooper, *Macromolecules* **1985**, 18, 1727.
[13] H. Veenstra, R. M. Hoovgvliet, B. Norder, A. P. de Boer, *J. Polym. Sci.: Part B* **1998**, 36, 1795.
[14] L. Zhu, G. Wenger, U. Bandara, *Macromol. Chem.* **1981**, 182, 3639.
[15] W. Gabriellse, M. Soliman, K. Dijlstra, *Macromolecules* **2001**, 34, 1685.
[16] P. Mougenot, J. Marchand-Brynaert, *Macromolecules* **1996**, 29, 3552.
[17] Q. Zhao, A. K. McNally, A. P. Urbranski, K. J. Stokes, ect, *Biomed. Mater. Res.* **1993**, 27, 379.
[18] W. Chen, J. McCarthy, *Macromolecules* **1998**, 31, 3648.
[19] D. Cohn, T. Stern, *Macromolecules* **2000**, 33, 137.

Macromol. Symp. **2003**, *199,* 173-186

Synthesis and Characterization of Semifluorinated Polymers and Block Copolymers

Doris Pospiech, Hartmut Komber, Dieter Voigt, Dieter Jehnichen, Liane Häußler, Antje Gottwald, Wolfram Kollig, Kathrin Eckstein, Angela Baier, Karina Grundke*

Institute of Polymer Research Dresden, Hohe Str. 6, 01069 Dresden, Germany
E-mail: pospiech@ipfdd.de

Summary: The goal of the investigation presented here was to evaluate the influence of semifluorinated side chains on the bulk structure and the surface properties of polysulfones with different chain structure. Thus, segmented block copolymers consisting of polysulfone and semifluorinated aromatic polyester segments as well as polysulfones having semifluorinated side chains randomly distributed over the polymer backbone were synthesized and characterized. Oxydecylperfluorodecyl side chains were used because of their strong tendency for self-organization. The influence of the chain architecture on the self-organization as well as on the surface properties, particularly the wetting behavior, was examined. It could be shown that despite of the higher self-organizing tendency of block copolymers the surface properties of both polymer types are comparable and depend only on the concentration of side chains.

Keywords: polysulfone, self-organization, semifluorinated polymers, ultrahydrophobicity, wetting

Introduction

Structural segments consisting of alkyl and perfluoroalkyl groups covalently linked by a C-C bond, called semifluorinated (SF) compounds, are known for their microphase separation resulting in highly ordered bulk structures.[1] The microphase separation is caused by the thermodynamic immiscibility between alkyl and perfluorinated alkyl segments. Additionally, the strong tendency of the perfluorinated parts in such compounds to surface segregation results in surface structures having $-CF_3$ groups regularly ordered at the surface, yielding in low surface free energy. The same behavior has also been reported for polymers with semifluorinated side chains, for example poly(styrene),[2] poly(methylmethacrylate),[3,4,5] poly(acrylates),[6,7] poly(styrene-*b*-isoprene) diblock copolymers,[8,9] poly(siloxanes),[10] and poly(vinyl ether) analogues.[11] One of the first examples of polycondensation polymers with SF-side chains were semifluorinated aromatic polyesters based on poly(p-phenylene isophthalate).[12] The rigid backbone requires longer ($-O-(CH_2)_{10}-(CF_2)_9-CF_3$, H10F10) semifluorinated side chains with ten C-atoms in the alkyl and ten C-atoms in the

 DOI: 10.1002/masy.200350915

perfluoroalkyl part to generate a self-ordered structure in the bulk and also at the surface,[13,14] resulting in an extremely low surface free energy of 9 mN/m. The aim of the work presented here was to use the same semifluorinated side chains to modify the surface properties of another polycondensation polymer, polysulfone (PSU). Therefore, two synthetic concepts have been explored, the first being the synthesis of segmented block copolymers (BCP) having the semifluorinated side chains in the polyester segments, connected by PSU segments, and the second way being the synthesis of polysulfones with semifluorinated H10F10 side chains randomly distributed along the PSU backbone. The schematic structures of both types of polymers are given in Figure 1. Thus, the influence of the chain architecture on both, the self-organization in the bulk as well as the surface properties could be explored.

Fig. 1. Schematic representation of the chain architectures of the semifluorinated polymers under investigation (left: segmented block copolymers with PSU and SF-PES segments; right: PSU with randomly distributed SF-side chains).

Experimental

The synthesis of segmented BCP with PSU and SF-PES was performed as described in refs.[16,17] The polysulfones with randomly distributed SF-side chains were prepared by nucleophilic aromatic substitution using bisphenol A, 4,4'-bis(chlorophenyl) sulfone, and 4,4'-bis(hydroxy-phenyl valeric acid), according to Koch et al.[18] The semifluorinated side chains were attached by Mitsunobu conditions[19] using HO-H10F10 (1mol), diethyl azodicarboxylate (1.5 mol/COOH-side group), triphenyl phosphine (1.2 mol/COOH-side group) in dry THF (50°C, 24 hrs).

SEC was carried out using a modular built Knauer SEC with RI detection. Solvents and separation columns as given in Tables 2 and 3. The molecular weights are relative to narrowly distributed polystyrene standards. NMR was performed on a Bruker DRX-500 spectrometer using $CDCl_3$/TFA-d (1:1 v/v) as solvent. The spectra were referenced on the $CHCl_3$ signal ($\delta(^1H)$ = 7.26 ppm, $\delta(^{13}C)$ = 77.0 ppm). DSC measurements were carried out on a Perkin Elmer DSC 7 (heating and cooling rate: 20 K/min). Temperature-dependent SAXS/WAXS measurements were performed at beamline A2 (DESY Hamburg, HASYLAB), heating and cooling rate 3 K/min. Thin films for surface characterization were prepared by spin coating (Headway Inc., USA, 2000 rpm/60 s) of polymer solutions (c_{pol}: 1 wt%, solvents: pentafluorophenol/ chloroform for BCP and chloroform for random copolymers) on precleaned silicon wafers. The films were dried at 50 °C for 4 h and then annealed at 200 °C in vac. for 4 h. Dynamic contact angles were measured using ADSA-P[15] and water as probe.

Results and Discussion

Synthesis of Segmented Block Copolymers Having PSU and SF-PES Segments

The SF polyester originally reported[12] was synthesized by Higashi polycondensation to prevent the cleavage of the ether linkage to the substituents. The low molecular weights obtained thus suggested the use of melt polycondensation of the semi-fluorinated isophthalic acid and hydroquinone diacetate (HQDA, in excess to compensate sublimation), resulting in higher molecular weights in the range of 50,000 g/mol for M_w.[16] Comparable conditions were applied to the synthesis of the segmented BCP copolymers with PSU and SF-PES blocks, recently reported by us in ref.[17] Two ways were explored: the synthesis of BCP with preformed oligomers (acetoxy group-terminated PSU and COOH-group-terminated SF-PES), and *in situ* preparation of the semifluorinated blocks during block copolymer synthesis with acetoxy-terminated PSU, as outlined in Figure 2. Different molecular weights of starting PSU and SF-PES blocks were used.

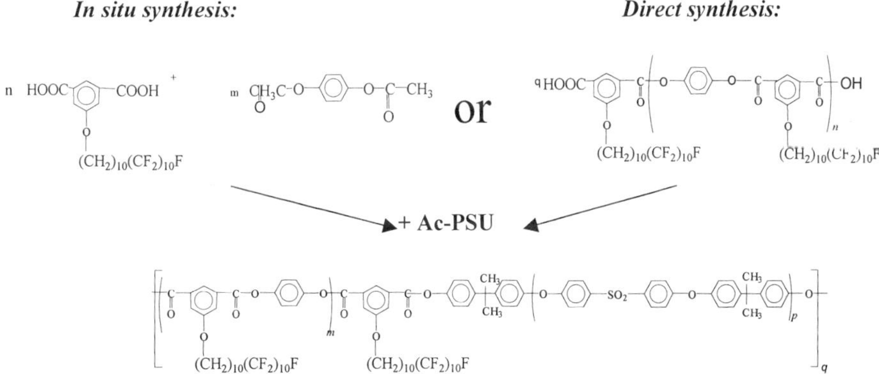

Fig. 2. Synthesis scheme of segmented block copolymers having polysulfone and semifluorinated polyester blocks.

The linking groups can be detected by their carbonyl signals at 166.9 – 167.6 ppm in the ^{13}C NMR spectrum. The ^{1}H NMR spectra gave additional information about the linkage between SF-PES and PSU. The signal of the 2-proton in the SF-IPA-ring at about 8.7 ppm, as shown in Figure 3, is split into three signals, where the first one at 8.71 ppm is caused by the triad HQ-SF-IPA-HQ, the second one at 8.68 ppm is caused by the triad PSU-SF-IPA-HQ and the third one at 8.67 ppm by the triad PSU-IPA-PSU. A comparable splitting is observed for the 2/6-protons (Figure 3).

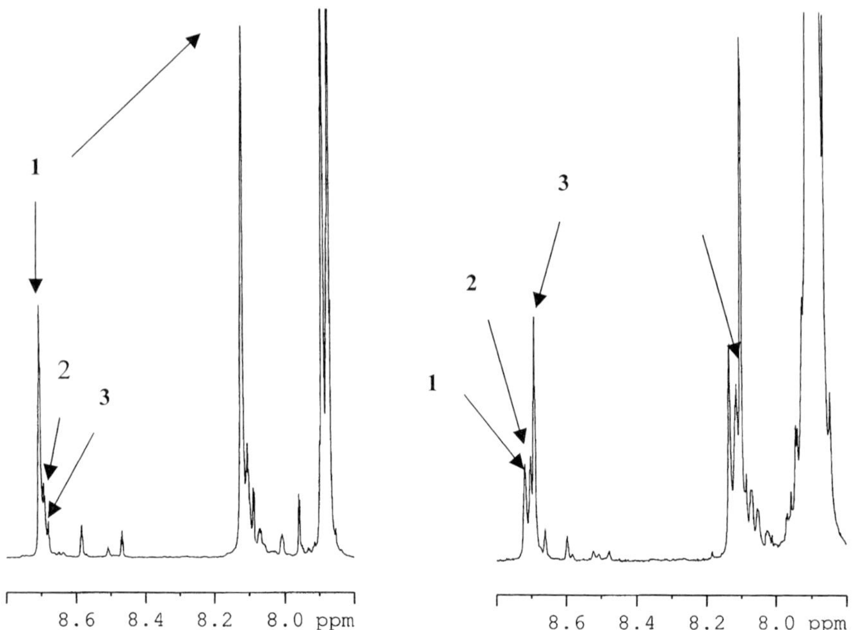

Fig. 3. ^{1}H NMR spectra of segmented block copolymers consisting of polysulfone and semifluorinated polyester segments (triads: 1-HQ-IPF-HQ; 2-PSU-IPA-HQ; 3-PSU-IPA-PSU).

Table 1 contains the concentration of triads obtained from the intensities of these protons in selected samples. The triad HQ-IPA-HQ can only occur within the semifluorinated polyester block. Triad PSU-IPA-HQ is the linkage desired for block copolymer formation and the triad PSU-IPA-PSU refers to the direct coupling of two PSU segments by a single semifluorinated IPA unit which can be generated by transesterification reactions. The content of PSU-IPA-PSU triads shows that transesterification within the SF-PES blocks occurs at times during direct synthesis. The direct coupling of SF-IPA to PSU is preferred during *in situ* synthesis of the SF-PES blocks, indicating that the reactivity of acetoxy-PSU to SF-IPA is obviously higher than the reactivity of HQDA. The concentration of this triad shows that the molecular weights of SF-blocks formed during *in situ* synthesis are lower than calculated from the initial ratio of IPA/HQDA (Table 2).

Table 1. Intensities of triads detected in ^{1}H NMR spectra of segmented PSU-SF-PES BCP, synthesized by the *in situ* method and by direct polycondensation.

PSU[a)] (g/mol)	COOH-SF[b)] (g/mol)	Ratio HQDA/SF-PES	HQ-IPA-HQ/ (%)	PSU-IPA-HQ/ (%)	PSU-IPA-PSU (%)
2,300	5,400	-	84	12	4
2,300	2,500	-	56	22	22
4,900	2,300	-	82	7	11
9,000	8,330	-	90	10	0
2,300	-	5/4	26	21	53
4,900	-	5/4	3	25	72

a) Molecular weight M_n of the acetoxy-terminated PSU, obtained by titration of OH-groups.
b) Molecular weight of COOH-terminated SF-PES blocks, obtained by ^{1}H NMR.

Table 2. Characterization of PSU-SF-PES block copolymers.

Ac-PSU	**SF-Oligomer**		**Block copolymer**		**Phase**
M_n[a)] (g/mol)	M_n (g/mol)	ratio SF/HQDA[c)]	M_w[e)] (g/mol)	M_w/M_n[e)]	**separation?**
2,400	2,500[b)]	-	19,300	3.33	yes
2,400	-	4/3	20,000	2.63	yes
2,400	5,400[b)]	-	33,200	2.89	yes
2,400	1,850[d)]	5/4	32,200	2.84	yes
2,400	-	10/9	21,000	2.13	yes
4,900	1,800[d)]	5/4	42,100	2.63	yes
4,900	2,300[b)]	-	13,000	2.71	yes
9,000	840	-	44,000	2.16	no
9,000	-	2/1	30,400	2.64	no
9,000	8,330[b)]	-	18,800	3.36	yes

a) Determined by titration of OH-groups of the OH-terminated PSU and from ^{1}H NMR spectra.
b) Determined by ^{1}H NMR of COOH-terminated SF-oligomers.
c) Ratio of the monomers used in the *in situ* synthesis of SF-PES blocks during block copolymer formation.
d) Molecular weight of SF-blocks calculated from ^{1}H NMR spectra.
e) SEC in pentafluorphenol/chloroform mixture, PL Mini MIX P separation columns.

Synthesis of PSU with Randomly Distributed Semifluorinated Side Chains

A two-step synthesis to obtain these polymers have been exploited. In the first step, copolymers of bisphenol A based polysulfone with 4,4'-bis(hydroxyphenyl) valeric acid with different concentration of valeric units have been synthesized by nucleophilic aromatic substitution using K_2CO_3 in NMP/toluene as solvent, according to Koch et al.[18] In the second step, the COOH groups were esterified by perfluorodecyldecanol according to Mitsunobu et al.[19] as shown in Figure 4.

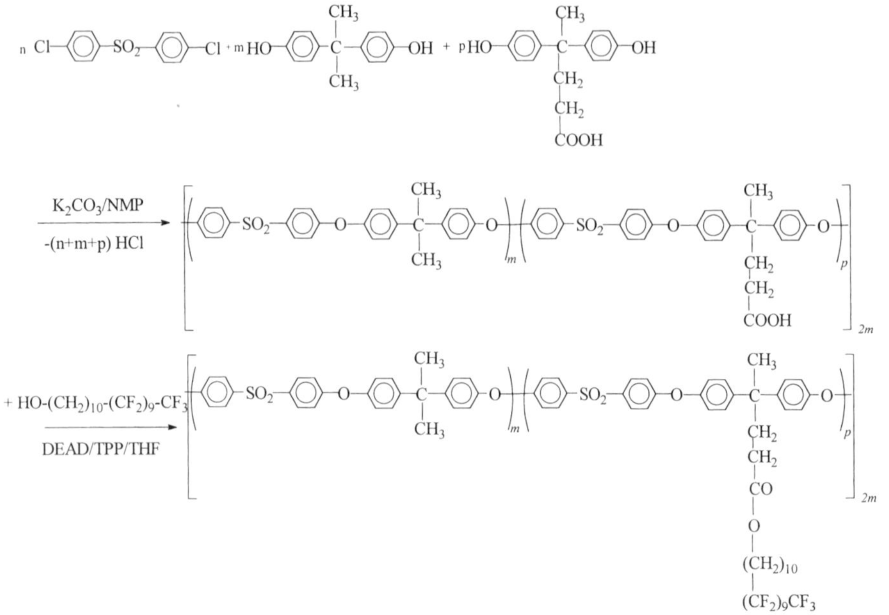

Fig. 4. Synthesis scheme of semifluorinated polysulfones with random chain structure.

The PSU's with COOH-side groups are very polar and only soluble in solvents as THF and DMSO. Esterification with the semifluorinated alcohol OH-H10F10 gives unpolar polymers, soluble in chloroform which could be characterized by SEC (Table 3).

The valeric acid comonomer is incorporated almost according to the stoichiometry used. At low side chain concentration, the polymer analogous reaction proceeds completely. High concentration of COOH-side chains obviously shields the reactive centers due to the polyelectrolyte character of the copolymer[20] and results, thus, in incomplete conversions of the esterification. The molecular weights of the copolymers are high, with broad molecular weight distributions that can be assumed from the synthesis method. The ^1H NMR spectra of a COOH-containing PSU and a H10F10-containing PSU are illustrated in Figure 5. The protons of CH_2-groups on the valeric acid (1 and 2) are slightly shifted after esterification with OH-H10F10 (1' and 2' in Figure 5b) in addition to the protons of the CH_2-groups of the SF side chain next to the ester bond (4).

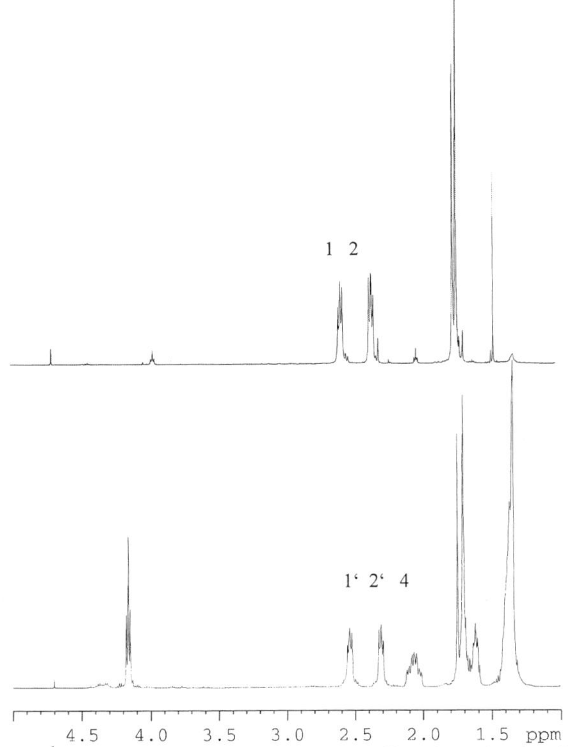

Fig. 5. ^{1}H NMR spectra of random a semifluorinated polysulfone (above: COOH-side chains, below: semifluorinated side chains).

Table 3. Characterization of polysulfone copolymers with randomly distributed COOH- and H10F10 side chains.

Concentration of COOH comonomer during polycondensation (mol%)	Concentration of COOH comonomer in the product [a)] (mol%)	Concentration of H10F10 side chains [b)] (mol%)	M_w [c)] (g/mol)	M_w/M_n [c)]
0	0	0	62,300	5.93
10	13		13,300	5.58
25	24	24	46,000	4.13
50	54	33	36,200	3.48
75	67	42	40,800	4.27
100	100	85	41,900	3.31

a) Calculated from 1H NMR spectra from the intensity ratio of aromatic protons (6.5-8.0 ppm) to CH_2 protons (2.06 ppm).

b) Calculated from 1H NMR spectra from the intensity ratio of CH_2 protons of valeric acid (2.06 ppm, 2H) to CH_2 protons of semifluorinated side chain (4.02 ppm, 2H).

Bulk Structure of Block and Random Copolymers

The structure of PSU-SF-PES block copolymers is determined by microphase separation. As indicated in Table 2, phase separation between the amorphous PSU phase and semifluorinated polyester domains occurs at very low molecular weights of SF segments having only a few monomeric units. The reasons therefore are the high interaction parameter between SF-PES and PSU (13.6)[16] as well as the high tendency for self-organization of the semifluorinated polyester itself, driving to a highly ordered bulk structure. The structure of the SF poly(p-phenylene isophthalate) was reported earlier.[21] By a combination of methods, the structural changes upon heating and cooling could be understood.[14,22] Phase separated block copolymers of PSU and the SF-PES should have both, the glass transition of polysulfone as indication of an amorphous matrix, as well as the thermal transitions of the SF-PES.[22] Based on the results obtained by DSC, SAXS, and TEM, a structural model was proposed recently[16,17] showing a random distribution of well-ordered SF domains within an amorphous matrix (Figure 6).

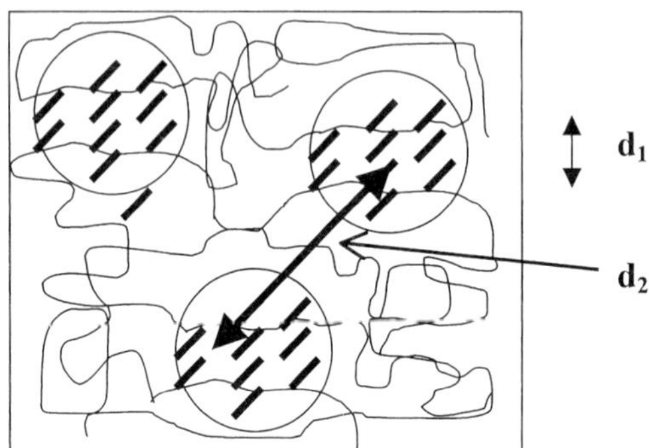

Fig. 6. Structural model of segmented PSU-SF-PES block copolymers, as published in ref.[16] indicating the periodic distances obtained in the SAXS measurements.

The SAXS curves show two scattering maxima at d_1 being exactly the layer distance obtained in the pure SF-PES. The larger scattering maximum d_2 (Figure 6 and 7) refers to the periodic distance between the SF-PES domains in the PSU matrix. A correlation between this distance and the domain size was postulated recently.[17] The random semifluorinated polysulfones show a somewhat different structure compared to the BCP. In DSC investigations, only a glass transition can be found, without indications of side chains melting and isotropization.

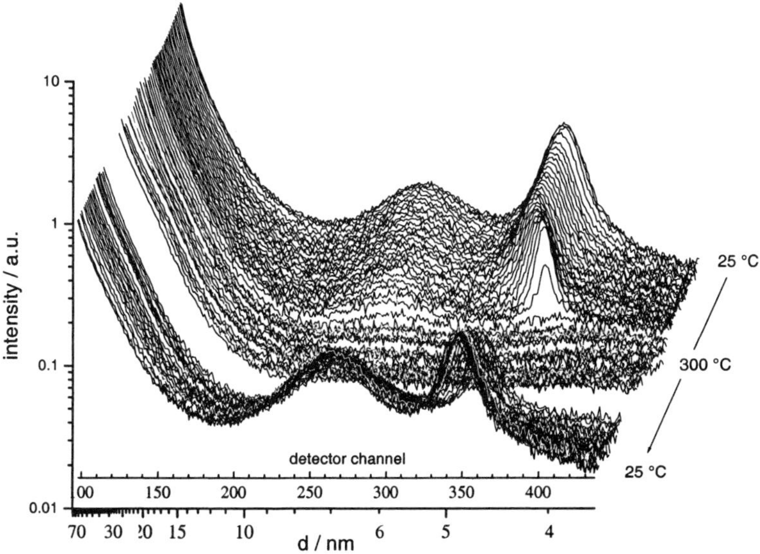

Fig. 7. Temperature-dependent SAXS curves of a BCP with PSU 2300 g/mol and SF-PES 2300 g/mol segments.

The T_g's strongly depend on the structure of the side chains (Figure 8). The T_g of PSU with COOH-side chains increases slightly with the side chain concentration referring to higher intermolecular interactions. In contrast, the T_g's of PSU with SF-side chains drop down significantly with increasing side chain concentration, showing that the SF-side chains alter the amorphous rigid structure of unmodified PSU.

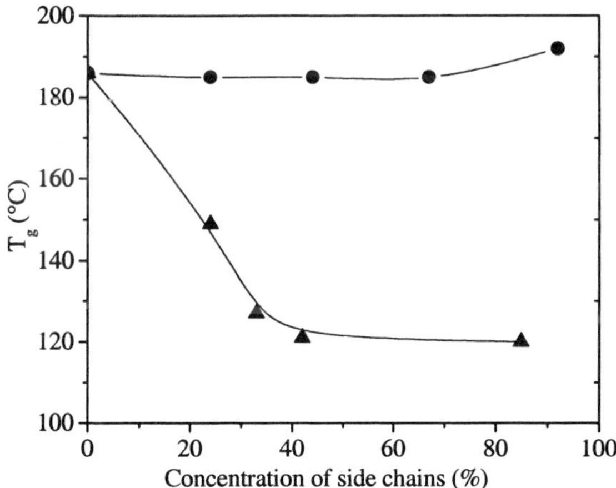

Fig. 8. Glass transitions temperatures of COOH-terminated polysulfones (circles) in comparison to polysulfones with semifluorinated side chains (triangles).

In contrast to the DSC results, temperature-dependent SAXS measurements clearly indicate the existence of an ordered phase having semifluorinated layers (Figure 9). At high side chain concentrations, this phase is already detected in the as-synthesized samples and at lower side chain concentrations, the phase develops upon heating. Even in the polymer with only 25 % SF-side chain, this phase can be detected, as illustrated in Figure 9.

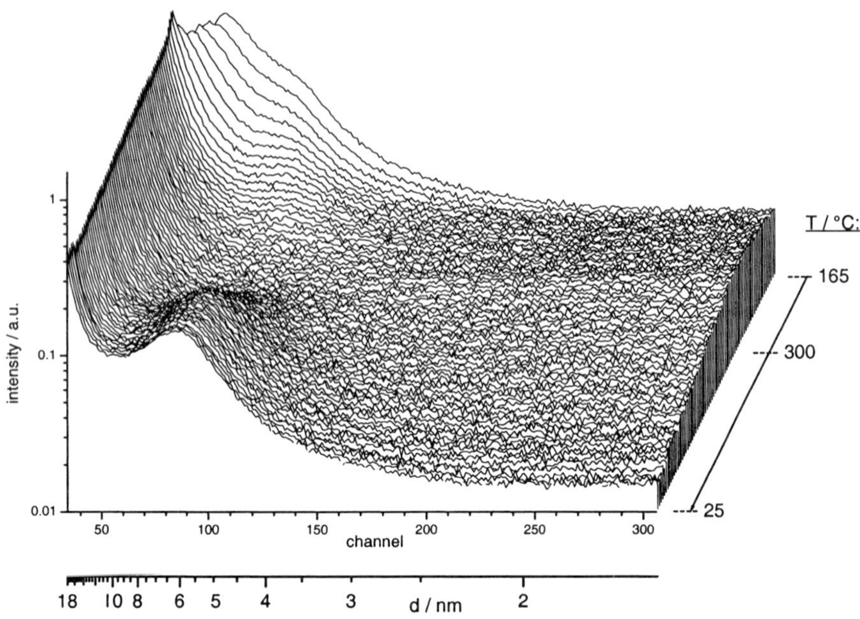

Fig. 9. Temperature-dependent SAXS curves of a random semifluorinated polysulfone (concentration of side chains: 25 mol%, 1st heating to 300 °C with 3 K/min).

The influence of side chain concentration is particularly reflected by the slight shifts in layer distance (Table 4). Higher concentration of semifluorinated side chains forces the side chains to arrange stronger into the smectic layers, thus giving lower layer distances. The copolymer with the highest concentration of SF side chains (85 mol%) shows about the same layer distance than the semifluorinated polyester synthesized by melt polycondensation.[22]

Table 4. Layer distances in the semifluorinated domains obtained for random semifluorinated polysulfones with different molar concentration of SF-side chains (after the first heating to 300 °C and cooling to 25 °C).

Molar concentration of SF-side chain (mol%)	Structure at room temperature?	Layer distance d after cooling (nm)
85	yes	5.2
42	yes	5.3
33	no	5.5
25	no	6.6
0	no	-

Surface Properties of the Block and Random Copolymers

Thin films were prepared by spin coating to investigate the wetting behavior of the polymers. Using similar concentrations of the polymers for spin coating should confirm comparable thickness of the produced copolymer films. The wetting behavior was examined by contact angle measurements using water as probe.

Figure 10 shows the results obtained for PSU-SF-PES BCP and Figure 11 the contact angles for PSU's with randomly distributed COOH- and SF-side chains with different concentration. Regarding the difference in the wetting behavior of both polymer types, it has to be stated that the segmented BCP's show a stronger dependence on the polymer composition than the random copolymers with SF-side chains. Whereas the BCP's show a clear increase of water advancing contact angle with the molecular weight of the SF-PES block (but without dependence on the molecular weight of PSU blocks), a clear dependence on the concentration of SF-side chains in the random copolymers cannot be found. All polymers with SF-side chains have high contact angles between 115 and 120°, that means, a clearly more hydrophobic behavior than the unmodified PSU, but a dependence on SF-side chain concentration does not appear. In contrast, PSU's with COOH-side chains show a more hydrophilic behavior as compared to PSU, as it was expected, but again without dependence on side chain concentration. The reason therefore might be discussed in terms of the much lower bulk order of the PSU copolymers.

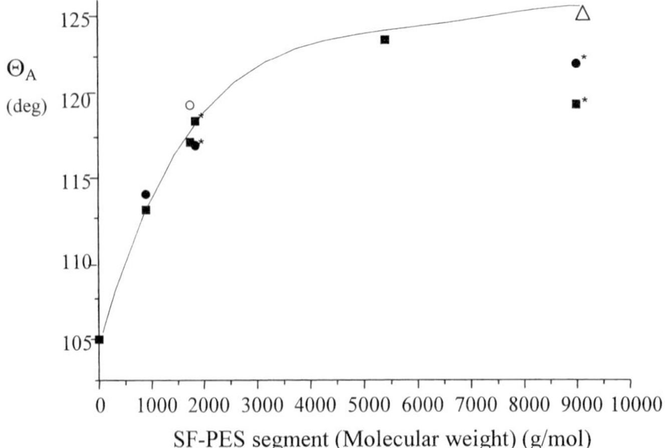

Fig. 10. Advancing contact angles of water on thin films of PSU-SF-PES block copolymers (*: molecular weights calculated by ^{1}H NMR, see Table 2; △: SF-PES; ■: PSU block molecular weight – 2,300 g/mol; ○: PSU block molecular weight – 4,900 g/mol; ●: PSU block molecular weight – 9,000 g/mol) plotted versus the molecular weight of the semifluorinated polyester blocks.

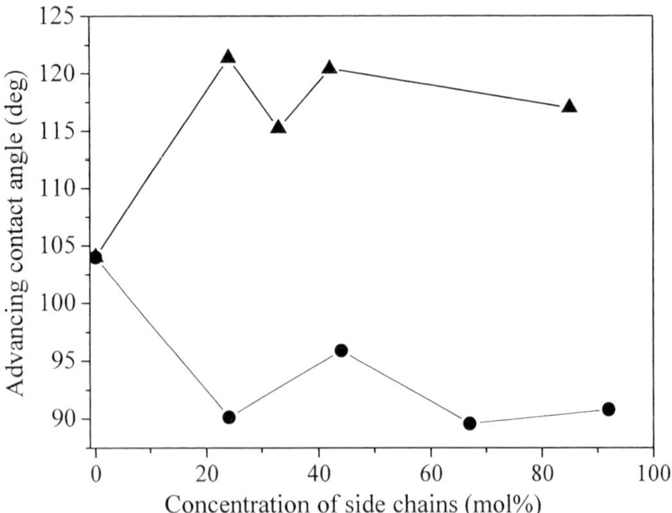

Fig. 11. Advancing contact angles of water on thin films of polysulfones with COOH- and SF-side chains (●: COOH-side chains; ▲: SF-side chains) plotted versus the molar concentration of semifluorinated side chains.

Conclusions

BCP with PSU and SF-PES segments show a stronger tendency of self-organization of semifluorinated side chains than PSU's with randomly distributed SF-side chains, which can be understood in terms of a lower distance and therefore higher interaction of the fluorinated parts in the BCP, leading to a higher ordered structure in the bulk. The surface properties of both polymer types are almost comparable, showing extremely high water contact angles, i.e., low surface free energies, at high concentration of SF-side chains. In both cases, the contact angle can be controlled by the chemical composition (i.e., the concentration of semifluorinated side chains) of the polymer. Both types can be used in blends, as reported recently.[23]

Acknowledgements

Financial support by German Science Foundation (DFG Po 575/2-1, 2-2) and NATO is gratefully acknowledged. Temperature-dependent WAXS/SAXS measurements were performed at DESY Hamburg, HASYLAB, beamline A2 with financial support of DESY within a BMBF project. Many thanks to Dr. S. Funari and M. Döhrmann for support at the beamline A2. Furthermore, the authors thank M. Dittrich, K. Arnold, K. Pöschel, R. Keska and S. Kummer for valuable technical assistance.

[1] K. Song, J. R. Twieg, J. F. Rabolt, *Macromolecules,* **1990**, *23*, 3712.
[2] J. Höpken, M. Möller, *Macromolecules,* **1992**, *25*, 1461.
[3] E. Lindner, *Recent Dev. Biofouling Control*, **1994**, 305.
[4] C. P. Jariwala, L. J. Mathias, *Macromolecules,* **1993,** *26*, 5129.
[5] C. P. Jariwala, L. J. Mathias, *Macromolecules,* **1996,** *29*, 3186.
[6] Y. Katano, H. Tomono, N. Nakajima, *Macromolecules,* **1994**, *27*, 2342.
[7] R. Ramharack, *Polym. Prepr., Am. Chem. Soc.,* **1988**, *29*, 1, 146.
[8] J. Wang, G. Mao, C. K. Ober, E. J. Kramer, *Macromolecules*, **1997**, *30*, 1906.
[9] M. Xiang, X. Li, C. K. Ober, K. Char, J. Genzer, E. Sivaniah, E. J. Kramer, D. A. Fischer, *Macromolecules*, **2000**, *33*, 6106.
[10] E. Beyou, P. Babin, B. Bennetau, J. Dunogues, D. Theyssie, S. Boileau, *Polym. Int.,* **1994**, *32*, 1673.
[11] V. Percec, M. Lee, *J. Macromol. Scie., Pure Appl. Chem.*, **1992**, *A29(9)*, 723.
[12] D. Pospiech, D. Jehnichen, L. Häußler, D. Voigt, K. Grundke, C. K. Ober, H. Körner, J. Wang, *Polym. Prepr., Am. Chem. Soc.,* **1998**, *39*, 2, 882.
[13] D. Jehnichen, D. Pospiech, A. Janke, P. Friedel, L. Häußler, A. Gottwald, S. Kummer, W. Kollig, K. Grundke, *Materials Science Forum*, **2001**, 378-381, 378.
[14] D. U. Pospiech, D. E. Jehnichen, A. Gottwald, L. Häussler, U. Scheler, P. Friedel, W. Kollig, C. K. Ober, X. Li, A. Hexemer, E. J. Kramer, D. A. Fischer, *Polym. Mat.: Science & Engin.,* **2001**, *84*, 314.
[15] D. Y. Kwok, T. Gietzelt, K. Grundke, H. J. Jacobasch, A. W. Neumann, *Langmuir,* **1997**, *13*, 2880.
[16] D. Pospiech, L. Häußler, K. Eckstein, H. Komber, D. Voigt, D. Jehnichen, P. Friedel, A. Gottwald, W. Kollig, H. R. Kricheldorf, *High Performance Polymers,* **2001**, 13, 2, 275.

[17] D. Pospiech, L. Häußler, K. Eckstein, D. Voigt, D. Jehnichen, A. Gottwald, W. Kollig, A. Janke, K. Grundke, C. Werner, H.R. Kricheldorf, *Macromol. Symp.,* **2001**, 163, H. Cherdron, B. Sandner, H. U. Schenk, B. Voit (Eds.), "Tailormade materials", ISSN 1022-1360, pp. 113.
[18] T. Koch, H. Ritter, *Macromol. Chem. Phys.,* **1995**, *195,* 1709.
[19] O. Mitsunobu, J. Kimura, M. Kawashima, M. Sugizaki, N. Nemoto, *J. Chem. Soc., Chem. Commun.,* **1979**, 303.
[20] J. Moore, *Polycondensation 2002*, Lecture and Proc., Hamburg, **2002** (September 15-17, 2002).
[21] P. Friedel, D. Pospiech, D. Jehnichen, J. Bergmann, C. K. Ober, *J. Polym. Sci., Phys.,* **2000**, *38*, 1617.
[22] A. Gottwald, D. Pospiech, D. Jehnichen, L. Häußler, P. Friedel, J. Pionteck, M. Stamm, G. Floudas, *Macromol. Chem. Phys.*, **2002**, *203*, 854.
[23] D. Pospiech, L. Häußler, Dieter Jehnichen, Wolfram Kollig, Kathrin Eckstein, Karina Grundke, *Macromol. Symp.* (accepted 11/2002).

Synthesis of Well-Defined Poly(*p*-benzamide) from Chain-Growth Polycondensation and Its Application to Block Copolymers

*Tsutomu Yokozawa,**[1,2] *Miyuki Ogawa,*[1] *Atsushi Sekino,*[1] *Ryuji Sugi,*[1,2] *Akihiro Yokoyama*[1]

[1] Department of Applied Chemistry, Kanagawa University, Rokkakubashi, Kanagawa-ku, Yokohama 221-8686, Japan
[2] "Synthesis and Control", PRESTO, Japan Science and Technology Corporation (JST), Japan
Email: yokozt01@kanagawa-u.ac.jp

Summary: Poly(*p*-benzamide) with a defined molecular weight and a low polydispersity and block copolymers containing this well-defined aramide was synthesized. Phenyl 4-(4-octyloxybenzylamino)benzoate (**1b**) polymerized at room temperature in the presence of base and phenyl 4-nitrobenzoate (**2a**) as an initiator in a chain-growth polycondensation manner to give well-defined aromatic polyamides having the 4-octyloxybenzyl groups as a protecting group on nitrogen in an amide. It was confirmed by a model reaction that deprotection of this protecting group proceeded completely with trifluoroacetic acid (TFA) without breaking the amide linkage. The utility of this approach to poly(*p*-benzamide) with a low polydispersity was demonstrated by the synthesis of block copolymers of poly(*p*-benzamide) and poly(*N*-octyl-*p*-benzamide) or poly(ethylene glycol). The SEM images of the supramolecular assemblies of the former block copolymer showed μm-sized bundles and aggregates of flake structures.

Keywords: block copolymer, chain-growth polycondensation, living polymerization, polyamide, self-assembly

Introduction

The self-assembly of well-defined polymers containing condensation polymers or oligomers in part has recently received a great deal of attention due to their surprising properties or functionality.[1] The condensation polymer units in those polymers are prepared by conventional polycondensation and possess broad molecular weight distribution[2] except for several

 DOI: 10.1002/masy.200350916

examples.[3] General synthetic method for condensation polymers with low polydispersities would promote the development of new materials based on self-organization. We have recently succeeded in synthesizing *N*-alkyl aromatic polyamides having defined molecular weights and low polydispersities by chain-growth polycondensation of phenyl 4-(octylamino)benzoate (**1a**), where the monomer reacts with the polymer end group selectively, not with other monomers.[4] From the point of view of self-assemble materials and fabrication materials, well-defined *N*-unsubstituted aromatic polyamide, poly(*p*-benzamide), is very attractive. In this paper, we report the synthesis of poly(*p*-benzamide) having defined molecular weights and low polydispersities by chain-growth polycondensation of monomer **1b** bearing the *N*-protecting group, followed by deprotection, and also synthesis of block copolymers[5] of poly(*p*-benzamide) and *N*-alkyl aromatic polyamides or poly(ethylene glycol) (PEG).

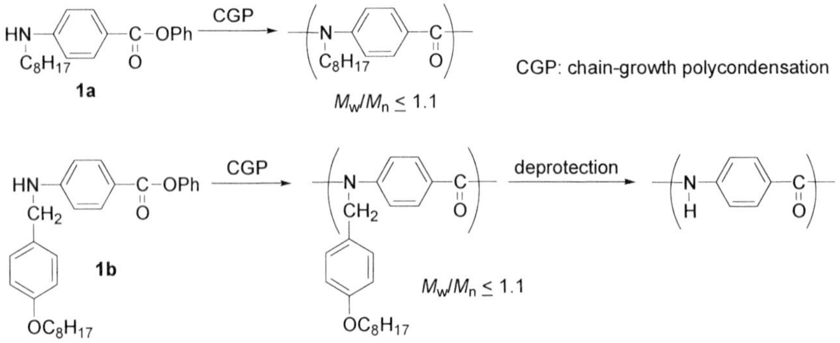

1a —CGP→ $M_w/M_n \leq 1.1$

CGP: chain-growth polycondensation

1b —CGP→ $M_w/M_n \leq 1.1$ —deprotection→

Polycondensation of 1c

We first attempted the polycondensation of phenyl 4-aminobenzoate (**1c**), which leads directly to poly(*p*-benzamide), in the presence of phenyl 4-nitorbenzoate (**2a**) as an initiator under the same conditions as the chain-growth polycondensation of **1a**.[4] However, not polyamide but the *N*-methylated 1:1 adduct **3** of **1c** and **2a** was obtained after quenching with iodomethane. The observed reaction is accounted for by involvement of the abstraction of the amide proton with the aminyl anion of **1c**. The anion generated on the amide bond between **2a** and **1c** is such a strong electron-donating group that the phenyl ester at the *para*-position of the anion is deactivated for polymerization, resulting in **3** after methylation of the amide anion. This result indicates that monomers having the primary amino group are not applicable to the chain-growth polycondensation for aromatic polyamides.

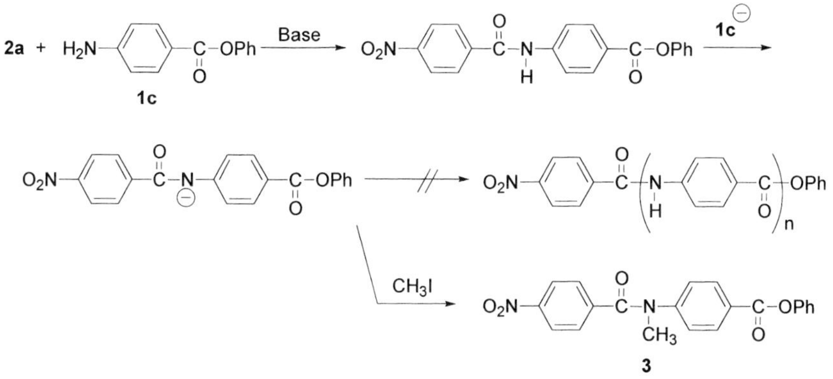

Polycondensation of 1b

The polymerization of monomers bearing the secondary amino group substituted by a protecting group was studied. On the basis of peptide chemistry, the amino group of monomer was protected as the *tert*-butyl carbamate (Boc), but this monomer did not polymerize. Monomer protected with the methoxyethoxymethyl (MEM) group was too unstable to be isolated. Monomer with the methoxybenzyl group was not freely soluble in the solvent for polymerization. Eventually, phenyl 4-(4-octyloxybenzylamino)benzoate (**1b**) was prepared and polymerized in the presence of 10 mol% of **2a** and base (*N*-triethylsilyl-*N*-octylaniline, CsF, and 18-crown-6)[4] in THF at room temperature. The polymerization proceeded homogeneously and was completed in 4 h to yield a polyamide with M_n of 3700[6] and M_w/M_n of 1.07. The calculated M_n value based on feed ratio of $[\mathbf{1b}]_0/[\mathbf{2a}]_0$ was 3600, and therefore the polyamide was synthesized in a controlled fashion. A living polymerization nature was also ascertained by a linear correlation between the M_n values and monomer conversion, retaining low polydispersities. Unfortunately, however, the polyamides were precipitated during polymerization when the feed ratios of $[\mathbf{1b}]_0/[\mathbf{2a}]_0$ were more than 10.

Deprotection

Deprotection of the 4-octyloxybenzyl groups on nitrogen in the polymer obtained was carried out in trifluoroacetic acid (TFA)[7] to precipitate polymer during reaction. This polymer was only soluble in H_2SO_4, and it was not confirmed whether deprotection took place quantitatively without scission of the amide linkages in polymer. For this reason, model compound **4** was treated with TFA at ambient temperature for 72 h, resulting in complete deprotection without breaking the amide and ester linkages.

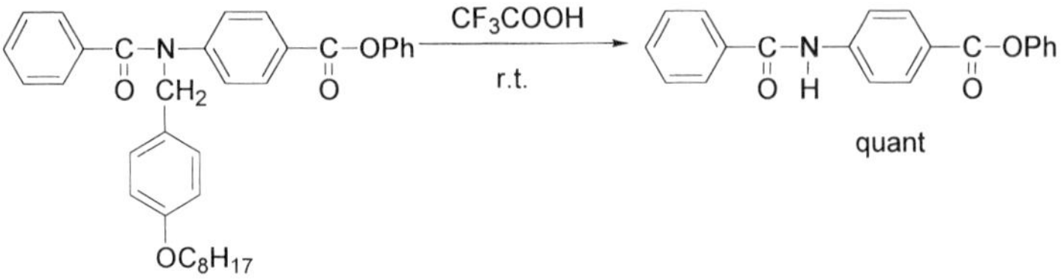

CF3COOH
r.t.
quant
4

Block Copolymers of *N*-H and *N*-Octyl Aromatic Polyamide

According to the above results, a soluble block copolymer of poly**1a** and poly(*p*-benzamide) was synthesized. Thus, **1a** was polymerized in the presence of **2a** and base in THF at room temperature to give a prepolymer.[4] A fresh feed of **1b** and *N*-triethylsilyl-*N*-octylaniline[4] was added to the prepolymer in the reaction mixture. The added **1b** feed was smoothly polymerized. The GPC chromatogram of the product (Figure 1A (b)) clearly shifted toward the higher molecular weight region, while retaining the narrow distribution, indicating a successful production of the block copolymer of **1a** and **1b** with a controlled molecular weight (Table 1).

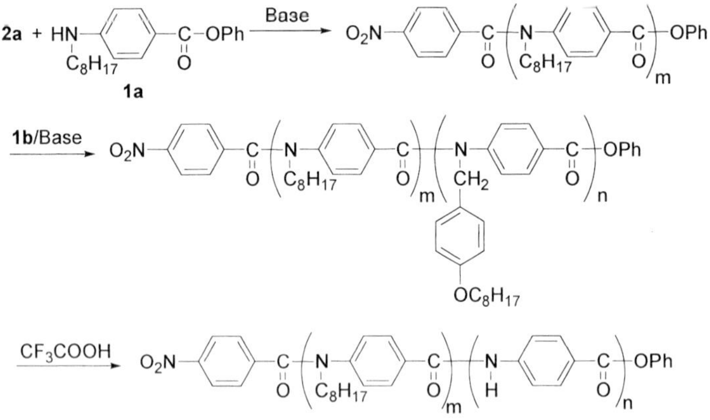

2a + 1a
Base
1b/Base
CF3COOH

Table 1. Synthesis of block coplymers of poly**1a** and poly**1b**.

1st segment [a)]			diblock copolymer		
M_n(calcd)	M_n [b)]	M_w/M_n [c)]	M_n(calcd)	M_n [b)]	M_w/M_n [c)]
1370	1170	1.13	3030	2000	1.11
1420	2410	1.13	4770	4630	1.10
2460	2900	1.10	7270	8130	1.08
2470	2490	1.10	5900	6420	1.10
2530	2410	1.14	4270	4150	1.12
4230	4730	1.08	7160	8890	1.06
4800	4600	1.09	6530	6070	1.08
4830	6100	1.08	11500	11900	1.08

Polymerization was initiated with **2a** in THF at 25 °C in all cases.
[a)] Poly**1a**.
[b)] Estimated by ^{1}H NMR.
[c)] Estimated by GPC based on polystyrene standards (eluent: THF).

The block copolymer obtained was then stirred in TFA at ambient temperature for 72 h, followed by purification with a preparative HPLC to yield a yellowish solid, which was soluble in THF, $CHCl_3$, CH_2Cl_2, DMF. In the ^{1}H NMR spectrum of the product, the signal of the benzylmethylene protons of poly**1b** units at 4.90 ppm completely disappeared. The GPC chromatogram of the product (Figure 1B) in the low molecular weight region slightly shifted toward the lower moleculer weight region, while keeping the low polydispersity. Consequently, deprotection of the 4-octyloxybenzyl groups proceeded quantitatively without scission of the amide linkage in polymer, and a soluble diblock copolymer of aromatic *N*-H polyamide and *N*-octyl polyamide was successfully synthesized.

It should be noted that the GPC chromatogram (eluent: THF) of the above block copolymer showed a large peak in the high molecular weight region as well as the peak corresponding to the block copolymer (Figure 1B). The observed high molecular weight region peak implies that this block copolymer was arranged in a supramolecular self-assembly in THF. Scanning electron microscopy (SEM) was used to visualize the supramolecular assemblies of the block copolymer after drying the THF solution on a silicon wafer. Surprisingly, the SEM images revealed that µm-sized bundles were formed as well as aggregates of flake structures;

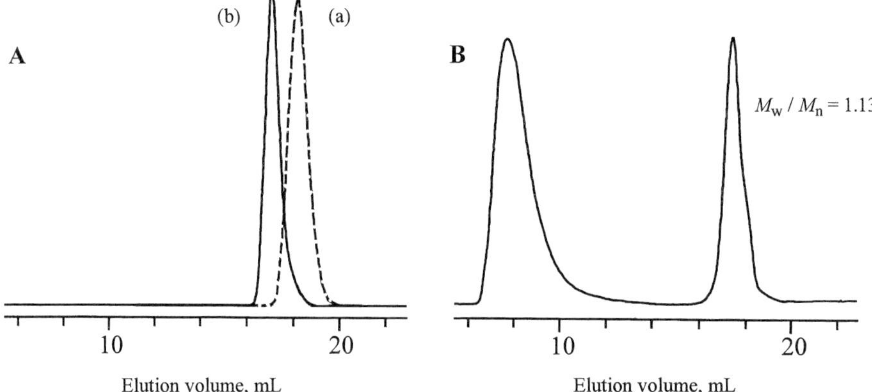

Fig. 1. GPC profile of polymer (eluent: THF). (A) Synthesis of the block copolymer of **1a** and **1b** by monomer addition method: (a) poly**1a** as a prepolymer ($[\mathbf{1a}]_0/[\mathbf{2a}]_0 = 9.6$), M_n = 2490, M_w/M_n = 1.10; (b) poly**1a**-*b*-poly**1b** as a postpolymer ([added $\mathbf{1b}]_0/[\mathbf{2a}]_0 = 10.2$), M_n = 6420, M_w/M_n = 1.10. (B) The block copolymer of poly(*N*-octyl-*p*-benzamide) and poly(*p*-benzamide) obtained by the deprotection of the 4-octyloxybenzyl groups on poly**1a**-*b*-poly**1b** with TFA.

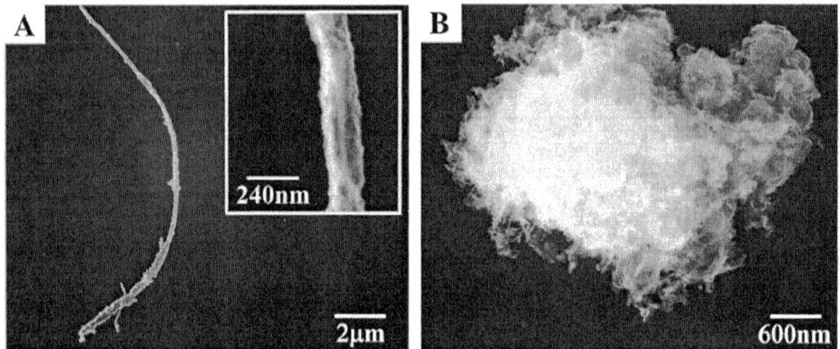

Fig. 2. SEM images (A, B) of the supramolecular assemblies of the block copolymer of poly(*N*-octyl-*p*-benzamide) and poly(*p*-benzamide) from the copolymer solution in THF dried at 25 °C on a silicon wafer and sputter coated with carbon. (A) is magnified in the inset in (A).

the reason for the formation of two kinds of structures is unclear at present time (Figure 2). The length of the bundles was in the range of 4-15 μm and their diameter was in the range of 150-250 nm. The DMF solution of the block copolymer, however, did not show the high molecular weight peak in the GPC chromatogram, and the bundle structures were not observed in the SEM, either. The SEM image of homopolymer of **1a** having the *N*-octyl groups also did not show the bundle structures even by use of the THF solution. Therefore, the hydrogen bondings of the poly(*p*-benzamide) unit of the block copolymer are probably responsible for supramolecular bundle structures. Further studies by X-ray diffraction methods will be essential to achieving a detailed understanding of the supramolecular structures of the block copolymer.

Block Copolymers of *N*-H Aromatic Polyamide and PEG

The polymer end group of poly**1b** is the phenyl ester, which would be applicable to the reaction with anionic living polymer end groups to yield other kind of block copolymers containing poly(*p*-benzamide) after deprotection. We tried the synthesis of block copolymers of poly(*p*-benzamide) and PEG. Phenyl benzoate (**2b**) was used instead of **2a** for the chain-growth polycondensation of **1b**, because the amide linkage of **2a** unit and **1b** was cleaved during the deprotection of 4-octyloxybenzyl group with TFA. The reaction of poly**1b** with monomethylated PEG in the presence of NaH yielded a block copolymer of poly**1b** and PEG, which was then treated with TFA to give a block copolymer of poly(*p*-benzamide) and PEG. Unfortunately, this block copolymer was poor solubility in common organic solvents, and it was difficult to characterize the polymer. For the production of a soluble block copolymer containing poly(*p*-benzamide) and PEG, the block copolymer of poly**1a** and poly**1b**, mentioned in the above section, was used. Thus, the block copolymer of poly**1a** and poly**1b** reacted with monomethylated PEG (M_n = 2000, M_w/M_n = 1.05) in the presence of NaH to give the triblock copolymer of poly**1a**, poly**1b**, and PEG (M_n = 3200, M_w/M_n = 1.15). Treatment of this block copolymer with TFA yielded the soluble triblock copolymer of *N*-octyl and *N*-H aromatic polyamide and PEG (M_n = 3380, M_w/M_n = 1.20). The GPC elution curves showed a peak in the higher molecular weight region both in the THF and DMF solution, implying that this block copolymer also self-assembles. Further studies about this supramolecular structure are currently under way.

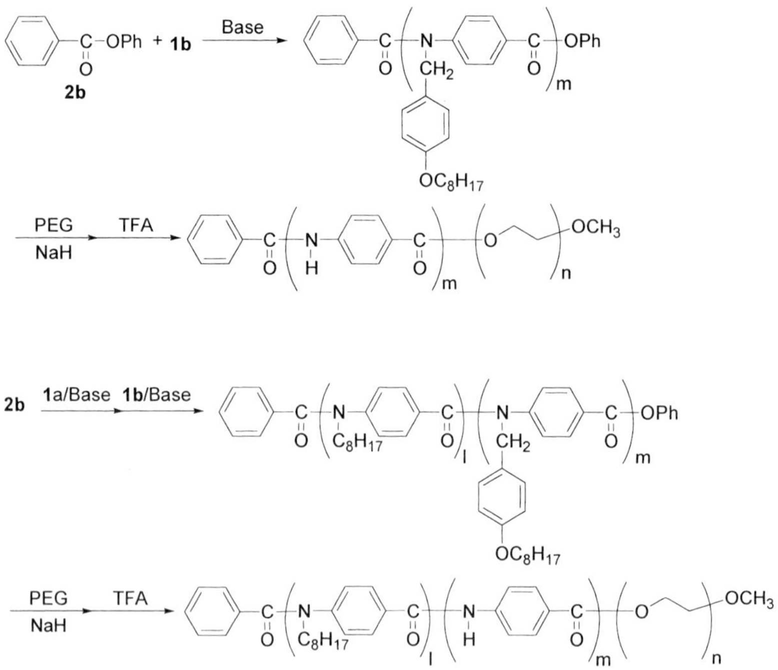

Conclusion

Our chain-growth polycondensation method for well-defined aromatic polyamides has been developed to the synthesis of poly(*p*-benzamide) with a low polydispersity by the polycondensation of **1b** followed by deprotection with TFA. By this method we synthesized soluble diblock copolymers of poly(*p*-benzamide) and poly(*N*-octyl-*p*-benzamide) and of poly(*p*-benzamide) and PEG, and a triblock copolymer of poly(*p*-benzamide), poly(*N*-octyl-*p*-benzamide) and PEG. The first block copolymer was arranged in a supramolecular self-assembly in THF to give intriguing μm-sized bundles probably by virtue of the multiple intermolecular hydrogen bondings of the poly(*p*-benzamide) unit. The self-assembly of block copolymers and star polymers containing the well-defined poly(*p*-benzamide) units will be a versatile protocol for nanoarchitectures of aromatic polyamides.

Acknowledgment

This work was supported in part by a Grant-in-Aid (12450377) for Scientific Research from the Ministry of Education, Science, and Culture, Japan.

[1] M. Lee, B.-K. Cho, W.-C. Zin, *Chem. Rev.* **2001**, *101*, 3869.
[2] For example, (a) S. G. Gaynor, K. Matyjaszewski, *Macromolecules* **1997**, *30*, 4241; (b) S. G. Gaynor, S. Z. Edelman, K. Matyjaszewski, *Polym. Prepr.* (Am. Chem. Soc., Polym. Div.) **1997**, *38* (1), 703; (c) S. A. Jenekhe, X. L. Chen, *Science* **1998**, *279*, 1903; (d) S. A. Jenekhe, X. L. Chen, *Science* **1999**, *283*, 372; (e) G. Klaeerner, M. Trollsås, A. Heise, M. Husemann, B. Atthoff, C. J. Hawker, J. L. Hedrick, R. D. Miller, *Macromolecules* **1999**, *32*, 8227; (f) D. Marsitzky, M. Klapper, K. Müllen, *Macromolecules* **1999**, *32*, 8685; (g) P. K. Tsolakis, E. G. Koulouri, J. K. Kallitsis, *Macromolecules* **1999**, *32*, 9054. (h) X. L. Chen, S. A. Jenekhe, *Macromolecules* **2000**, *33*, 4610; (i) U. Stalmach, B. Boer, A. D. Post, P. F. Hutten, G. Hadziioannou, *Angew. Chem. Int. Ed.* **2001**, *40*, 428.
[3] For block copolymers containing condensation polymer units with narrow molecular weight distributions, see: (a) T. J. Deming, *Nature* **1997**, *390*, 386; (b) J. N. Cha, G. D. Stucky, D. E. Morse, T. J. Deming, *Nature* **2000**, *403*, 289; (c) R. Prange, S. D. Reeves, H. R. Allcock, *Macromolecules* **2000**, *33*, 5763; (d) H. R. Allcock, R. Prange, *Macromolecules* **2001**, *34*, 6858; (e) X.-Z, Zhou, K. J. Shea, *Macromolecules* **2001**, *34*, 3111; (f) K. R. Brzezinska, T. J. Deming, *Macromolecules* **2001**, *34*, 4348; (g) J. Liu, E. Sheina, T. Kowalewski, R. D. McCullough, *Angew. Chem. Int. Ed.* **2002**, *41*, 329
[4] T. Yokozawa, T. Asai, R. Sugi, S. Ishigooka, S. Hiraoka, *J. Am. Chem. Soc.* **2000**, *122*, 8313. Recently, the same polyamides were synthesized by using a different leaving group of monomer and base: Y. Shibasaki, T. Araki, M. Okazaki, M. Ueda, *Polym. J.* **2002**, *34*, 261.
[5] For examples of diblock copolymers containing aromatic polyamides from conventional polycondensation, see: (a) E. Marsano, E. Bianchi, G. Conio, G. Mariani, S. Russo, *Polym. Commun.* **1991**, *32*, 45. (b) H.-H. Wang, M.-F. Lin, *J. Appl. Polym. Sci.* **1991**, *43*, 259. (c) G. Conio, E. Marsano, F. Bonfiglioli, A. Tealdi, S. Russo, E. Bianchi, *Macromolecules* **1991**, *24*, 6578. (d) T. Tagami, JP Pat. 05112651 A2, 1991; *Chem. Abstr.* **1993**, *119*, 226723. (e) T. Tagami, JP Pat. 05132563 A2, 1991; *Chem. Abstr.* **1993**, *119*, 227191. (f) G. Conio, A. Tealdi, E. Marsano, A. Mariani, I. Ponomarev, *Polymer* **1994**, *35*, 1115. (g) P. Cavalleri, N. N. Chavan, A. Ciferri, C. Dell'Erba, M. Novi, G. Marrucci, C. S. Renamayor, *Macromol. Chem. Phys.* **1997**, *198*, 797.
[6] The M_n value of polyamide was estimated by the ^{1}H NMR spectra based on the ratios of signal intensities of the repeating units to the initiator unit.
[7] For examples of the deprotection of the 4-methoxybenzyl group with TFA, see: (a) G. M. Brooke, S. Mohammed, M. C. Whiting, *Chem. Commun.* **1997**, 1511. (b) Y. Miki, H. Hachiken, Y. Kashima, W. Sugimura, N. Yanase, *Heterocycles* **1998**, *48*, 1. (c) A. Bouzide, G. Sauve, *Tetrahedron Lett.* **1999**, *40*, 2883.

Synthesis and Properties of Dendritic Polymers Based on Natural Amino Acids

Alexander Bilibin, Ivan Zorin, Sergey Saratovsky, Irina Moukhina, Galina Egorova, Nina Girbasova*

Saint-Petersburg State University, The Chemistry Institute, University ave., 26, Petrodvorez, 198504, Saint-Petersburg, Russia
E-mail: bilibin@w.spb.ru

Summary: Different approaches, including polycondensation, polymerization, polymer analog condensation, and ionic binding have been investigated for synthesis of amino acid-based dendritic polymers. It was shown that a growth of dendrons generation prevents obtaining of products with high polymerization degree in polycondensation and polymerization procedures. In polymer analog condensation a growth of dendrons generation leads to considerable decrease of polymer analog reaction rate as well as substitution degree. Degree of ionic binding depends on a strength of ionogenic groups and dendrns generation.

Keywords: α–amino acids, dendritic polymers, solubility, synthesis, viscosity

Introduction

Dendritic polymers are composed of linear macromolecular core to which dendrons have been attached in any way. They display a complex of unusual properties due to which they attract attention to their synthesis and study. First they were reported in 1987.[1] Their physical properties as well as conformational properties of their macromolecules are under thorough investigation.[2-4] The polymers consisting of dendritic macromolecules are considered as very promising objects, especially for biomedical purposes,[5] but complexity of their structure and difficulties and ambiguity of their synthesis as well as "inadequate analytical methodology precluded unequivocal characterization until now".[6] First dendritic lysine-based polypeptides were reported by Denkewalter in 1981.[7]

At present the most studied group of dendrimers are those, which have small molecule with a few functional groups as a core. Usually molecules of these dendrimers have a shape close to spherical. Replacement of the small molecule core by macromolecule leads to formation of new type of polymer molecules – dendritic macromolecules, whose shape is assumed to be

 DOI: 10.1002/masy.200350917

close to cylindrical. The bigger and the more complex the core molecule, the more complex the final dendritic macromolecule is arranged.

Several parts of such molecule determine final properties of dendritic polymer. Firstly, it is the core chain, its rigidity, its skeletal chemical bonds and other structural features. Secondly, they are "anchor" groups, which join pendent dendrons to the core. Thirdly, they are the pendent dendrons proper, which constitute the main part of inner volume of the macromolecule. And finally, they are terminal groups of the dendrons located at periphery of the macromolecule. Each of these parts contributes to properties and behavior of the resulting dendritic polymer depending on their chemical nature and mass portion. Several teams successfully work in this area since early 90-th.[1-6]

Most of described denderitic polymers were prepared by chain polymerization, whereas information dealing with other routes is insufficient. Many approaches and their combination may be used for design of dendritic molecules.

In the present work we have investigated synthesis of amino acids based dendritic polymers with use of various approaches. The choice of trifunctional amino acids as monomers for design of pendent dendrons and sometimes core macromolecule was determined by their very attractive properties. These compounds are biocompatible, which makes them very convenient for synthesis of polymers for biomedical purposes. A presence of chiral carbon atom enables to use them as chirality carriers. Due to the progress in biosynthesis of α-amino acids they became accessible monomers at the moderate price. Besides, they may be easily transformed into polymers via polymerization as well as polycondensation procedures. Endo gave the detailed consideration of use of amino acids as monomers for polymer synthesis.[8]

The present paper gives a brief overview of our recent works in dendritic polymers.

Objectives of the Work

i. Development of synthetic procedures to preparing dendritic polymers based on amino acids

ii. Preparation of dendritic polymers samples for their detailed study

iii. Investigation of synthesis and properties of the polymers.

Approach: Gradual transition in the study of synthesis and properties of the polymers from relatively simple structures of their macromolecules (zero-generations of pendent dendrons) to more and more complicated ones.

Methods: In the present work we used all available methods of polymer and peptide synthesis. The main method for characterization and investigation of the products was study of behavior of macromolecules in (dilute) solutions. Special attention in the present work was paid to solubility and viscosity behavior of the products.

Different approaches were used to prepare the polymers:

1. Polycondensation
2. Polymerization
3. Polymer analog condensation
4. Ionic binding of polyelectrolyte macromolecule with ionogenic focal groups of dendrons

L-aspartic acid and L-glutamic acids were used for design of dendrons similar to those described in ref.[9]

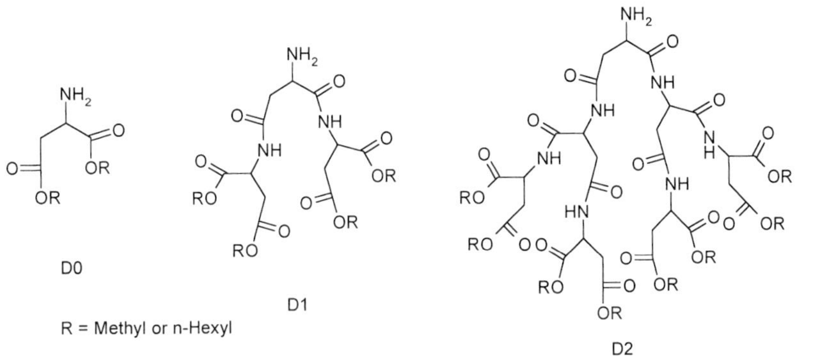

D0 D1 D2

R = Methyl or n-Hexyl

Scheme 1

Acrylic chains were used as core macromolecules for polymerization and polymer analog condensation. Lysine was used as monomer for formation of the core macromolecule in polycondensation procedure. Usually, polycondensation is not applied for synthesis of poly-α-amino acids because of competitive cyclization reactions. Amino acids may be inserted in polymer chains by polycondensation in some specific routes.[8]

We performed polycondensation with use of lysine as a monomer for formation of a core macromolecule (see the scheme below). Aspartic dendrons bound to carboxylic group were used in one route, whereas zero-generation "Frechet-type" dendron was used in another route.

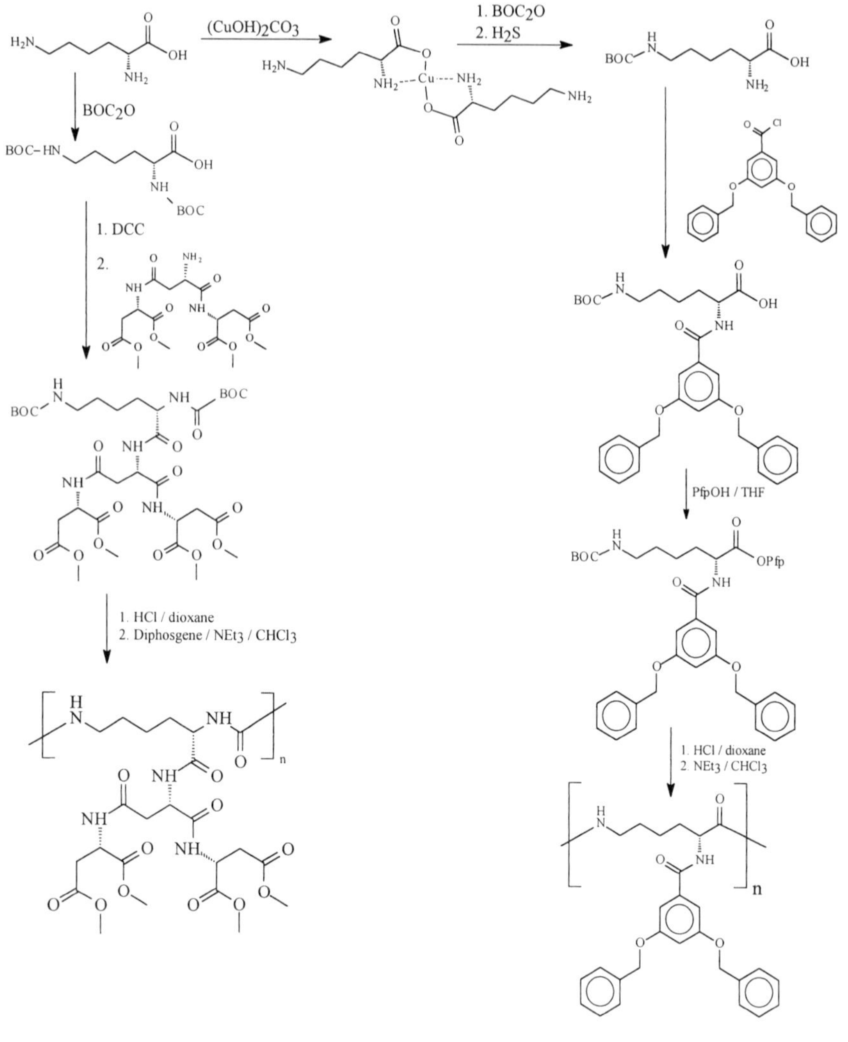

Scheme 2.

The resulting polymers were isolated as white powders, their structure was confirmed by ^1H NMR. Relatively low intrinsic viscosity (0.08 dl/g) may indicate rather low polymerization degree. At the same we showed that rather high molecular weight dendritic polymers might have low [η] values (see below).

Free-radical polymerization of acrylic monomers bearing aspartic and glutamic dendrons of different generations as well as conformational properties of the macromolecules are described in ref.[10,11] in detail. Structures of monomers used for the polymerization are given below:

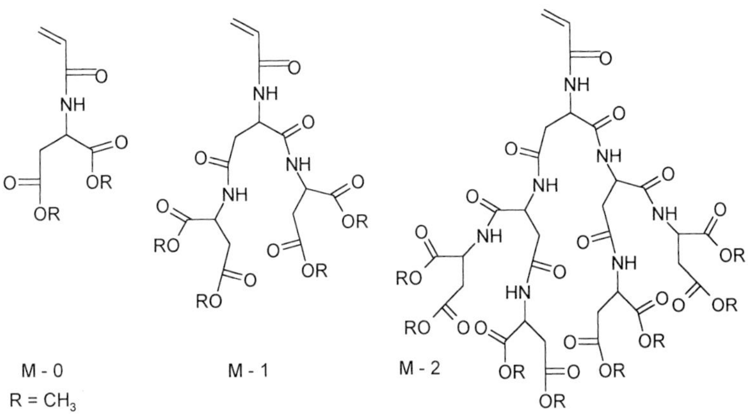

M - 0
$R = CH_3$

M - 1

M - 2

Scheme 3.

The mostly interesting feature of the polymerization is high M_w of the polymers derived from M_0 and M_1 (about 10^6). Similar facts were mentioned for polymerization of N-acryloyl amino acids by Endo.[8] Polymerization of M_2 gave oligomers only. Some of the prepared polymers were selectively hydrolyzed at methoxy group carboxy-terminated dendritic polymers.

In order to perform polymer analog condensation of core macromolecule with dendrons we have synthesized activated polyacrylates p-nitrophenyl and N-hydroxysuccinimide esters of polyacrylic acid, which were introduced in a reaction with aspartic dendrons with focal amino group. The main series of the syntheses were carried out with use of poly-p-nitrophenyl acrylate (PNPA) with intrinsic viscosity 0.14 dl/g. In order to determine its polymerization degree we have transformed it to polyacrylamide, whose molecular weight was calculated by

Mark-Kuhn equation $[\eta] = 0.631*10^{-4}M_\eta^{0,8}$.[12] The polyacryamide had intrinsic viscosity 0.17 dl/g, which corresponds to molecular weight 19000 and polymerization degree about 270. Figure 1 shows that both reaction rate as well as substitution degree decrease considerably for higher generation dendrons. It should be noted that the content of p-nitrophenyl units in polyacrylic chains decreased to higher extent as compared with increase of the bound dendrons. We explain this fact by formation of succinimide and glutarimide cycles in which two carboxlic groups are bound with one focal group of dendrons. NMR spectra confirm a presence of such structures.

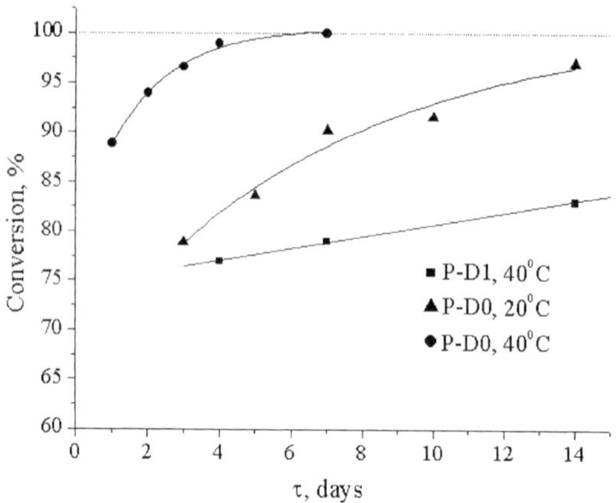

Fig. 1. Conversion degree of p-nitrophenyl groups determined by UV-spectrometry during reaction of PNPA with aspartic dendrons.

Assuming that polymerization degree doesn't change after polymer analog condensation molecular weight (M) of polymers with higher generations dendrons should increase. But we didn't observe increase in intrinsic viscosity of the products in chloroform at 30° (Table 1). Moreover, it decreased from 0.16 dl/g for P-D0 (acrylic polymer with attached D0; M = 53000) to 0.12 dl/g for P-D1 (acrylic polymer with attached D1; M = 115000) and to 0.11 dl/g for P-D2 (polymer with attached D2; M = 230000). The same tendency was observed for polymers prepared from N-oxysuccinimide ester of polyacrylic acid. More detailed characteristics of the condensation process and the products are given in Table 1.

Table 1. Characteristics of the process of polymer analog condensation of PNPA with aspartic dendrons of different generations.

Polymer	[η], dl/g	M calculated	Conversion of NP groups, %	Substitution degree, % (NMR)	Dendrons attached via imides, %
PNPA	0.14 (Ph-NO_2)	52000	-	-	-
P-D0	0.16 ($CHCl_3$)	53000	100	87	15
P-D1	0.12 ($CHCl_3$)	115000	83	~83	~2
P-D2	0.11 ($CHCl_3$)	230000	86	~86	0

Table 1 shows that complete conversions of p-nitrophenyl groups takes place in a reaction with dimethyl ester of aspartic acid only. Some of dendrons were bound by formation of not only amide bond, but by imide bond too. Growth of the generation number of dendron doesn‘t lead to decrease of substitution degree with respect to attached dendrons, but decreases a number of imide cycles formed, which may be explained by steric factors.

One of the most important directions of the area is self-assembling of complex macromolecular structures based on non-covalent interactions.[5] Hydrogen bonding of dendrons bearing terminal amino groups with poly(vinylpyridine) is described in.[13] We have studied an opportunity to form proper dendritic macromolecules due to self-assembly based on ionic binding of polyelectrolyte macromolecules with dendrons bearing oppositely charged groups at their focal point.

Aspartic dendrons have basic amino group at their focal point thus they are able to form ionic bond with polyanionites. We have studied ionic interaction of these dendrons with macromolecules of weak and strong polyacids. Polyacrylic and polystyrene sulfonic (PSS) acids were used as those. Ionic binding of weakly basic focal amino group of dendrons with weakly acidic carboxylic group of polyacrylic acid proceeds with binding degree not above 60% not depending on dendron generation. As to strongly acidic polystyrene sulfonic acid the binding of dendrons proceeds with binding degree close to 100%, which was confirmed by IR, NMR and elemental analysis data (N/S ratio).

Solubility and Viscosity Properties

We have studied properties of the synthesized dendritic polymers with use of different methods. Conformational properties of macromolecules of the polymers were investigated by flow and electrical birefringence as well as by diffusion and sedimentation.[10] In the present publication we will briefly discuss unusual solubility and viscosity behavior of some polymers.

Solubility of the polymers with acrylic chain and methoxycarbonyl terminal groups decreases in common organic solvents with growth of generation number due to increased contribution of intramolecular hydrogen bonding. These polymers after selective hydrolysis of methyl ester group display pronounced polyelectrolyte effect.[11] The most interesting solubility and viscometry data were obtained for ionic complexes of polystyrene sulfonic acid and aspartic denrons bearing methoxycarbonyl and n-hexyloxycarbonyl terminal groups. Solubility of these complexes is influenced by a number of factors. Hydrophobicity and high flexibility of a core polystyrene macromolecule should promote solubility in organic solvents with moderate polarity. Anchor sulfoammonium salt groups should promote solubility in highly polar water and methanol. Dendritic part of a macromolecule should form a network of predominantly intramolecular hydrogen bonds, which should prevent solubility in aprotic organic solvents. And, finally terminal alkoxycarbonyl groups should promote solubility of the products in organic solvents with low polarity. Table 2 displays solubility of ionic complexes of PSS and aspartic dendrons bearing two (C^M-1 and C^H-1); four (C^M-2 and C^H-2) and eight (C^M-3 and C^H-3) methoxy and hexyloxy terminal groups respectively (dendrons D0; D1 and D2).

Table 2. Solubility of ionic complexes of PSS and aspartic dendrons.

	Hexane	Toluene	$CHCl_3$	MeOH	Dioxane	DMSO	Water
C^M-1	-	-	+	+	-	+	-
C^M-2	-	-	-	+ *	-	+	-
C^M-3	-	-	-	+ *	-	+	-
C^H-1	-	-	+	+	-	+	-
C^H-2	+ *	+	+	+	+	+	-
C^H-3	+ *	+*	+	+	+	+	-

* - At heating above 40°C

These data allows evaluating relative contributions of each component of ionically-assembled dendritic macromolecule to its solubility. Remarkable fact is the solubility of PSS polysalts in solvents with very low polarity.

Solubility of the complexes in a wide range of solvents made it possible to determine intrinsic viscosity values in different solvents. Viscometry data were used by some researchers for evaluating conformational properties of dendritic macromolecules.[14]

We have measured viscosity of ionic complexes of PSS and aspartic dendrons in different solvents. Some of the results are given at Figure 2 for hexyloxy-terminated dendritic polymers. The broad range of solvents allows drawing some preliminary conclusion about behavior of the macromolecules in these solvents. C^{H}-1 and C^{H}-2 display pronounced polyelectrolyte effect in polar methanol. At that reduced viscosity values for C^{H}-2 are lower than those for C^{H}-1, although molecular weight of the latter is much lower (165000 for C^{H}-1 and 300000 for C^{H}-2). In non-polar solvent such as toluene and chloroform the complexes do not display polyelectrolyte effect, but their intrinsic viscosity values look to be very small for the complexes of rather high molecular weights ($[\eta] = 0.11$ dl/g in chloroform for C^{H}-3 with $M = 570000$).

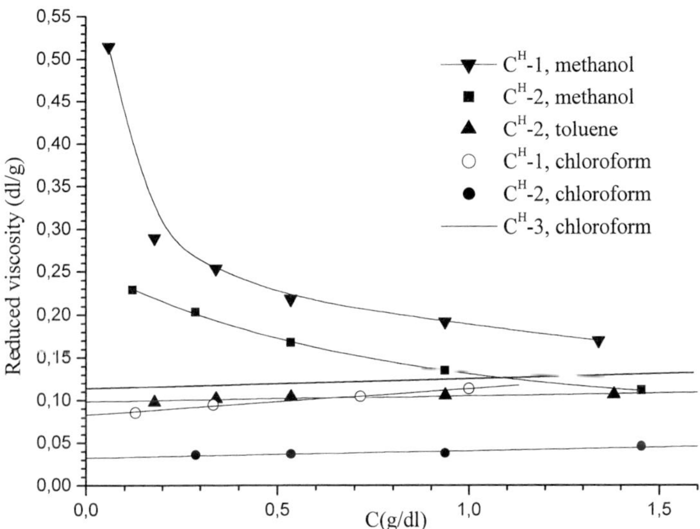

Fig. 2. Dependencies of reduced viscosity for the hexyloxy-terminated complexes on concentration in different solvents at 30° C.

The tendency of lowering intrinsic viscosity with growth of molecular weight (dendrons generation numbers at the same polymerization degree of a core macromolecule) seems to be common for the polymers under investigation (see above). It may indicate significant compactization of macromolecules in some solvent. At present we study conformational properties of dendritic molecules in different solvents by methods hydrodynamics, electrooptics, diffusion and sedimentation.

Experimental Part

All solvents were purified by distillation. BOC-amino acids and hydrochlorides of diesters of aspartic and glutamic acids were prepared according to standard methods. NMR spectra were recorded by "Bruker DPX-300" instrument in $CDCl_3$ or DMSO-D_6. IR spectra were recorded by "Specord IR-75" spectrometer.

Experimental details of polymerization and polymer analog condensation procedures are given in [11] and [15] respectively. Dendritic derivative of di-BOC-lysine was synthesized by reaction of di-BOC-lysine with aspartic dendron D-1 with use of dicyclohexyl carbodiimide. NMR, elemental analysis and thin-layer chromatography confirmed structure, composition, and purity of all derivatives.

Typical procedure for synthesis of condensation polymer with pendent aspartic dendronG-1 is given below:

2.5 ml of 3 M HCl solution in dry dioxane was added to 400 mg of di-BOC-lysine dendronized derivative. After a few minutes colorless oil of lysine dendronized derivative dihydrochloride was precipitated. Dioxane and excess of HCl were removed in vacuum and resulting oil was dissolved in 2 ml of chloroform containing 400 mg of triethylamine. 0.1 ml of diphosgene in 3 ml of chloroform was added dropwise to this mixture for 15 min. Reaction mixture was poured in 50 ml of hexane, the precipitate was filtered off, washed with water, dissolved in chloroform and poured into hexane to give 120 mg of white solid. Intrinsic viscosity in chloroform at 30°C was 0.08 dl/g. NMR spectra and elemental analysis confirm its composition and structure.

Polystyrene sulfonic acid was obtained from its sodium salt (Aldrich, M_W = 70000) using ion exchange chromatography column filled with Dowex 50x8.

Dendrons were synthesized by method similar to those described in,[11] but hexan-1-ol was used instead of methanol.

Ionic complexes were obtained via mixing equimolar amounts of dendrons and polyacids in methanol solution followed by precipitation in ethylacetate and drying in a vacuum. The composition of complexes was confirmed by elemental analysis (see Table 3), NMR and IR spectroscopy. The presence of absorption bands located near 1600 cm^{-1} and 1500-1520 cm^{-1} are assigned to H_3N^+ and shows the formation of ionic bonds. It should be noted that the first band could be seen for C^M-1 and C^H-1 only. In all other complexes it is overlapped by broad and intensive band of amide groups.

Table 3. Elemental analysis data for ionic complexes of polystyrene sulfonic acid and aspartic dendrons esters.

Ionic Complex	Elemental analysis (%)								N/S Ratio	
	Found				Calculated					
	C	H	N	S	C	H	N	S	Found	Calculated
C^M-1	47.9	5.5	3.6	8.4	48.7	5.5	4.1	9.3	0.4	0.4
C^M-2	46.9	5.4	6.5	5.1	47.8	5.5	7.0	5.3	1.3	1.3
C^M-3	46.0	5.7	8.3	2.9	47.2	5.5	8.8	2.9	2.9	3.0
C^H-1	58.5	7.5	2.5	6.3	59.4	8.0	2.9	6.6	0.4	0.4
C^H-2	59.8	8.3	4.8	3.6	59.8	8.3	4.8	3.6	1.3	1.3
C^H-3	58.9	8.1	5.7	2.0	60.0	8.4	5.8	1.9	2.9	3.0

Conclusion

A set of dendritic polymers with aspartic dendrons and various core macromolecules was synthesized by polycondensation, polymerization, polymer analog condensation and ionic binding methods. The polymers were investigated in dilute solutions and some unusual features of their solubility and viscosity behavior were found.

Acknowledgement

The authors thank Russian Foundation for Basic Researches for financial support of the work (Grant 00-03-33199).

[1]. D. A. Tomalia, P. M. Kirchoff, **U.S. Pat.** 4,694,064, **1987.**

[2]. D. A. Tomalia, A. M. Naylor, W. A.Goddard, III *Angew. Chem.***1990**, *102* (2), 119-157; *Angew. Chem., Int. Ed. Engl.* **1990**, *29* (2), 138-175.

[3]. L. Shu, A.D. Schluter, Ch. Ecker, N. Severin, J.P. Rabe *Angew. Chem. Int. Ed.* **2001.** Vol.40. № 24. P. 4666.

[4]. V.Percec, C.-H. Ahn, V.D.Cho, A.M. Jamieson, J.Kim, T.Leman, *J.Am.Chem.Soc.* **1998.** V.120. P. 8619.

[5]. I.Goessl, L. Shu, A.D. Schlueter, J.P. Rabe, *J. Am. Chem. Soc.* **2002**, V. 124, p. 6860

[6]. R.Yin, Y. Zhu, D. A. Tomalia *J. Am. Chem. Soc.* **1998,** V.120, p. 2678

[7]. R.G. Denkewalter, J. Kolc, W.J. Lukasavage, **U. S. Pat.** 4,289,872, **1981**

[8]. F. Sanda, T. Endo, *Macromol. Chem. Phys.*, **1999,** V. 200, p. 2651

[9]. M. Niggeman, H. Ritter *Acta Polymer.* **1996**. B. 47. p. 351.

[10]. S.V. Bushin, N.V. Girbasova, E.V. Belyaeva, M.A. Bezrukova, L.N. Andreeva, A.Yu. Bilibin, *Polymer Science,* **2002**, V. 44, No 6, Series A, p. 632

[11]. N.V. Girbasova, I.I. Migunova, G.V., Raspopova, E.I., A.Yu. Bilibin, *Polymer Science*, in press.

[12]. Encyclopedia of Chemical Technology, **1963,** N.Y. 2-nd Ed., V. 1, p. 274

[13]. M. Jikei, T. Kouketsu, M. Kakimoto, in "Polycondensation 2002", Book of Abstracts, Hamburg, **2002,** p. 152

[14]. S. Jahromi, B. Coussens, N. Meijerink, A.W.M. Braam, *J. Am. Chem. Soc.*, **1998,** V. 120, p. 9753

[15]. A.Yu. Bilibin, G.G. Egorova, N.V. Girbasova, S.V. Saratovsky, I.V. Moukhina, *Polymer Science*, in press.

Cyclizations in Hyperbranched Aliphatic Polyesters and Polyamides

*Linda Chikh, Xavier Arnaud, Céline Guillermain, Martine Tessier, Alain Fradet**

UMR 7610 - Chimie des Polymères, Université Pierre-et-Marie Curie
Courrier 184, 4 Place Jussieu, 75252 Paris Cedex 05, France

Summary: Hyperbranched aliphatic polyesters of 2,2'-bis-(hydroxymethyl) propanoic acid and hyperbranched aliphatic polyamides obtained from new carboxy- and amino-functionalized caprolactams were studied by NMR spectroscopy and MALDI-TOF mass spectrometry. Ring-chain equilibria taking place through intramolecular hydroxy-ester, carboxy-amide or amine-amide interchanges and leading to the formation of cyclic branches or end-groups were found to exert a predominant influence on the molar mass of these hyperbranched polymers. A number of intra- or intermolecular side reactions, such as the formation of ethers in polyesters and the formation of anhydrides, imides, amidines and secondary amines in polyamides were also detected and resulted in polymer crosslinking on prolonged heating. The existence of such ring-chain equilibria and side-reactions make the control of hyperbranched polymer structure much more difficult than generally accepted.

Keywords: aliphatic polyamides, aliphatic polyesters, hyperbranched polymers, ring-chain equilibrium, side reactions

Introduction

The existence of intramolecular side reactions is a common feature of virtually all polycondensation or polyaddition reactions and results in the competitive formation of cyclic molecules along with linear polymer chains.[1] For most polycondensation polymers, the content in cyclics is generally low (a few %) and their presence plays only a minor role in polymer properties. The situation is somewhat different in AB_f-type hyperbranched condensation polymers, since intramolecular reactions result in the formation of ring-terminated hyperbranched macromolecules and limit the maximum attainable molar mass to values well below those predicted by Flory's theory, as shown by Monte-Carlo simulations.[2] The existence of intramolecular reactions has been detected in many hyperbranched polymers, such as polyesters[3,4] and polysiloxanes,[5] and are likely to exert an important influence on

 DOI: 10.1002/masy.200350918

the properties of nearly all hyperbranched condensation polymers. The presence of cyclic branches or rings in AB_f-type hyperbranched polymers is generally assigned to the existence of A+B intramolecular reactions during the synthesis.[3] Since end-group concentration is very high in such polymers - ca. one B end-group per monomer unit for AB_2-type polycondensations - "back-biting" reactions, however, are likely to play an important role in the ring formation taking place in these polymers.

This work was devoted to the study of cyclization reactions taking place in hydroxy-terminated aliphatic hyperbranched polyesters and in amino- and carboxy-terminated aliphatic hyperbranched polyamides under ring-chain equilibrium conditions, i.e. thermodynamically controlled cyclizations rather than kinetically controlled ones.

Experimental

Hyperbranched Aliphatic Polyesters

Trimethylolpropane (TMP) or pentaerythritol (PE) core molecules were reacted in the bulk with 2,2-bis(hydroxymethyl)propanoic acid (BMPA) at 140°C under nitrogen in the presence of 0.5 mass % p-toluenesulfonic acid (PTSA) catalyst[6]. Polyesters of pseudo-generation 2 or 3 (G2 or G3 respectively) were obtained using 1:9 or 1:21 TMP:BMPA mol ratio and 1:12 or 1:28 PE:BMPA mol ratio (Scheme 1).

*γ-carboxyethyl-ε-caprolactam (**1**), γ-aminoethyl-ε-caprolactam (**2**) and 4-aminoethyl-1,7-heptanedioic acid (**3**)*

Carboxy- and amino-functionalized ε-caprolactams **1** and **2** were prepared by a multi-step synthesis starting from methyl 3-(4-hydroxyphenyl)propanoate and consisting of (i) hydrogenation of the phenyl group by H_2/Raney-Ni[7], (ii) CrO_3 oxidation of the resulting cyclohexanol to the corresponding cyclohexanone[7] and (iii) conversion of this compound to γ-carboxyethyl-ε-caprolactam **1** by reaction with O-sulfonic hydroxylamine in 95% formic acid (Schmidt-Beckmann reaction).[8] γ-Aminoethyl-ε-caprolactam **2** was synthesized by reacting **1** with sodium azide in a 2:1 vol $CHCl_3/H_2SO_4$ mixture.[8] 4-Aminoethyl-1,7-heptanedioic acid (**3**) was prepared by hydrolysis of **1** in refluxing 35% aqueous HCl.[8]

Hyperbranched Aliphatic Polyamides

Samples of **1** or **2** (100 mg) were placed in a 5mL tube and heated to 250°C under nitrogen for a predetermined reaction time (Scheme 2).

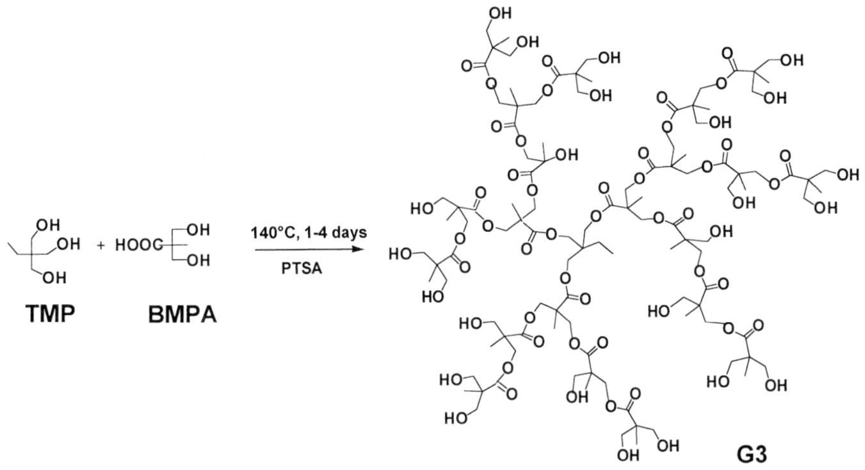

Scheme 1. Synthesis of TMP-BMPA hyperbranched polyester of generation 3.

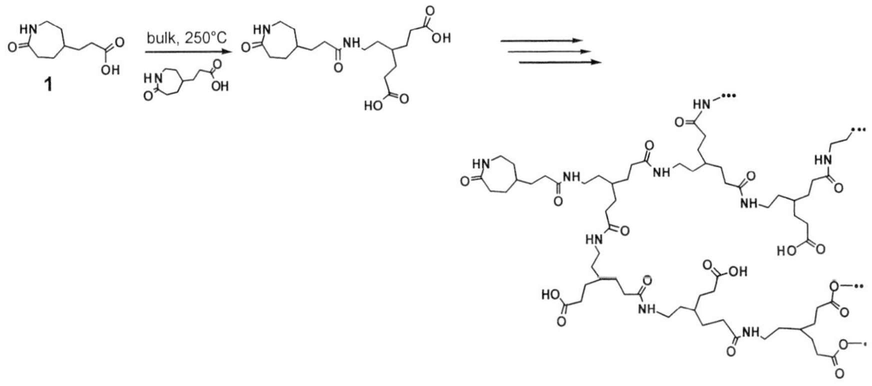

Scheme 2. Synthesis of carboxy-terminated hyperbranched polyamides.

Results and Discussion

Hyperbranched Aliphatic Polyesters

Hyperbranched polyesters were synthesized according to the procedure described by Malmström et al.[6] by the reaction of 2,2-bis(hydroxymethyl)propanoic acid (BMPA) with pentaerythritol (PE) or trimethylolpropane (TMP) core molecules (Scheme 1). No carboxy end-groups were detected in the ^{13}C NMR spectra of the hyperbranched polyesters after 24 h reaction (Fig. 1) and the chemical titration of carboxy end-groups indicated that the conversion of COOH groups was above 98.7 %. The degree of branching (DB) was in the range 0.46-0.48 for all synthesized polyesters, close to the theoretical DB value (0.5) expected for hyperbranched AB_2 condensation polymers.

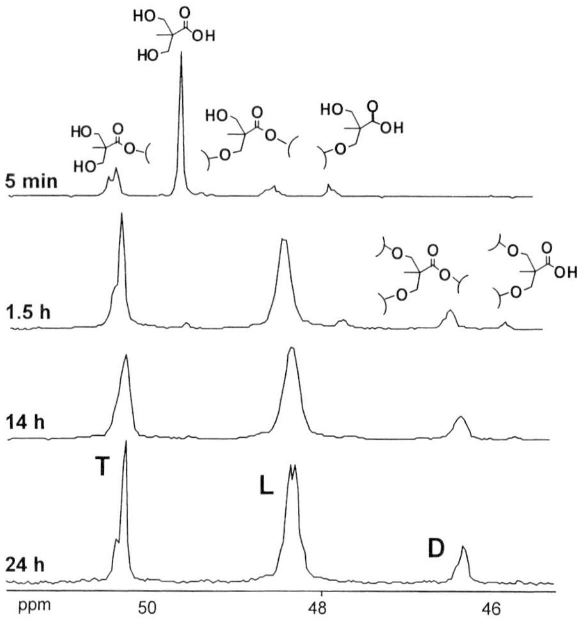

Fig. 1. ^{13}C NMR spectra (62.9 MHz, DMSO d-6) of G2-TMP:BMPA (1:9 mol ratio) polyesters after various reaction times (bulk reaction, 140°C, 0.5 mass % PTSA). T, D, L: terminal, dendritic and linear units, respectively.

The MALDI-TOF mass spectra of samples taken after 24 h reaction show, however, the presence of a series of small peaks at -18 Th from the main ones for either TMP- or PE-based hyperbranched polyesters, reflecting the existence of water-forming intramolecular side reactions. The 1250-1350 Th m/z region (DP = 11 oligomer molecules) of the mass spectrum of a G2 PE-based hyperbranched polyester obtained after 24 h reaction is given in Fig. 2.

The main peaks correspond to the expected core-unit-containing molecules (sodium cationized PE-M_x series where PE is the core unit and M_x monomer units) while the series of small M_x peaks corresponds to COOH-terminated molecules i.e. hyperbranched molecules that do not contain a core-unit. The presence of such molecules is not unexpected since conversion is not complete at this reaction time.

The (PE-M_x - 18) series is assigned to the formation of cyclic branches by intramolecular etherification - one cyclic branch per molecule. Etherification is a well-known side reaction of the synthesis of aliphatic and aliphatic-aromatic linear polyesters and results in the formation of a small proportion of oxyalkylene units. Due to the high OH functionality of PE-BMPA and TMP-BMPA hyperbranched polyesters, intramolecular etherification takes place to a much higher extent than in linear polyesters. On the other hand, intermolecular etherification cannot be discriminated from intermolecular esterification (the normal reaction) by mass spectrometry. The presence of an increasing amount of etherified units was observed at t > 24h in the ^{13}C NMR spectra of PE-BMPA hyperbranched polymers (Fig. 3). Since crosslinking was observed at t > 96 h, intermolecular etherifications should obviously be involved. It is worth mentioning that the presence of structures resulting from etherifications has been recently reported in similar polyester systems.[9]

The (M_x - 18) series is assigned to cyclic-terminated hyperbranched molecules resulting from intramolecular esterifications. Such reactions produce ring-containing molecules without core unit. Model cyclization studies were carried out on ester-, hydroxy- or carboxy-terminated linear aliphatic polyesters. The results obtained showed that the formation of cyclics is favored in hydroxy-terminated polyesters, and that hydroxy-ester interchange is the predominant reaction. The (M_x-18) series may be assigned, therefore, to compounds resulting from hydroxy-ester interchange or carboxy-hydroxy reaction. The hydroxy-ester equilibrium, well known in high temperature bulk polyesterifications, results in chain scission and in the formation of hyperbranched molecules containing one cyclic ester branch. The maximum attainable molar mass, therefore, depends on the cyclization equilibrium constant K (Scheme 3).

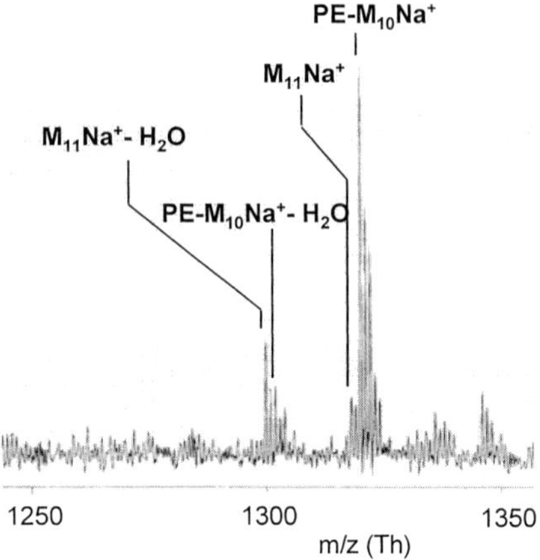

Fig. 2. MALDI-TOF MS spectrum (undecamer region) of G2-PE:BMPA (1:12 mol) polyester. Bulk reaction, 24 h, 140°C, 0.5 mass % PTSA.

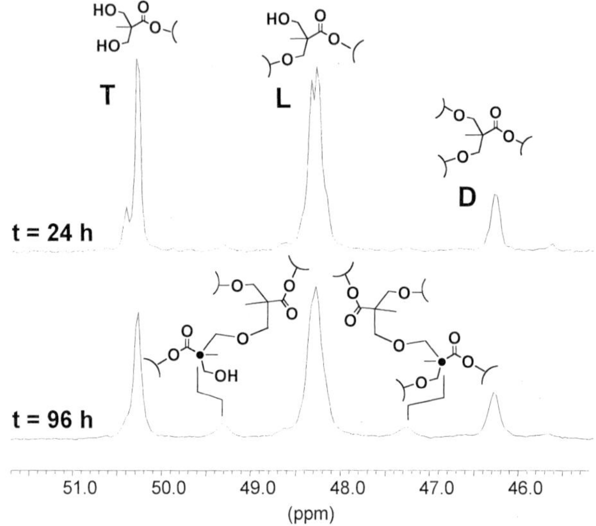

Fig. 3. ^{13}C NMR spectra (62.9 MHz, DMSO d-6) of G2 PE:BMPA (1:12 mol) polyester at 24h and 96h reaction. Bulk reaction, 140°C, 0.5 mass % PTSA.

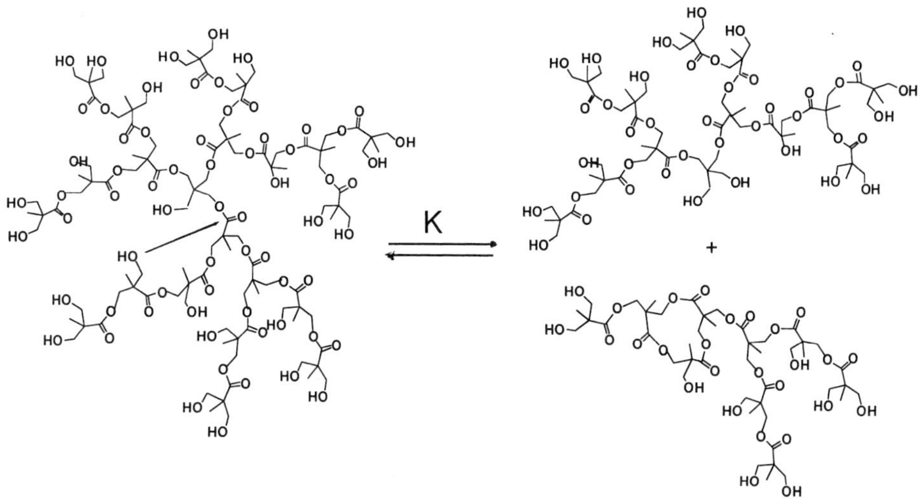

Scheme 3

Hyperbranched Aliphatic Polyamides

Polymerization

Carboxy-terminated hyperbranched polyamides were obtained by polymerizing γ-carboxyethyl-ε-caprolactam **1** in the conditions of the acidolytic polyamide-6 synthesis (250°C, bulk reaction) (Scheme 2). The course of polymerization was followed by integrating the ^{1}H NMR resonances of caprolactam and polyamide CON$\underline{H}$ at 7.41 and 7.76 ppm respectively.

A rapid polymerization took place, but conversion reached a plateau value close to 0.53 after 3h reaction. The addition of polyamidation catalysts such as phosphorous and hypophosphorous acids (1 mass %) accelerated the first steps of reaction ($t < 1h$), but did not change the conversion plateau value (Fig. 4).

The NMR spectra of all samples fitted with the expected structure and exhibited resonances corresponding to both residual lactam and carboxylic acid end-groups. Similarly, the MALDI-TOF MS spectra were in agreement with the expected structure. Each hyperbranched molecule gives multiple peaks, assigned to H^+-Na^+ or H^+-K^+ exchanges taking place on COOH end-groups (Fig. 5). This confirms the high COOH functionality of these polyamides. According to these structural studies, it is obvious that the reaction did not stop because of reactive group consumption by possible side reactions. In the same way, kinetic factors cannot be involved since polyamidation catalysts did not change the conversion plateau value. The existence of the conversion limit can, therefore, be assigned to a ring-chain equilibrium, similar to the caprolactam-polyamide equilibrium taking place during the synthesis of linear polyamide-6. This equilibrium involves carboxy-amide interchange reactions (Scheme 4).

Similar observations were made during the polymerization of γ-aminoethyl-ε-caprolactam (**2**). Lactam CONH conversion rapidly reached a plateau value close to 0.57 after only 30 min reaction. This was assigned to a ring-chain equilibrium, analogous to that depicted in Scheme 4, in which amine end-groups are involved instead of carboxy end-groups.

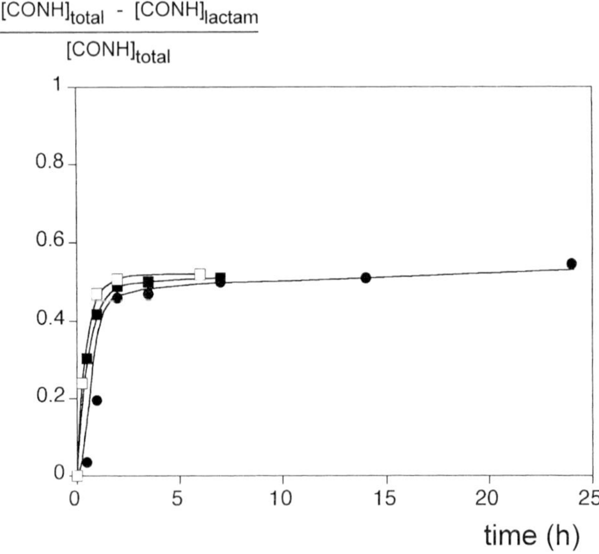

Fig. 4. Polymerization of caprolactam **1** (250°C): Variation of caprolactam CONH conversion versus reaction time for uncatalyzed (●), H_3PO_3- (■) and H_3PO_2-catalyzed(❑) reactions.

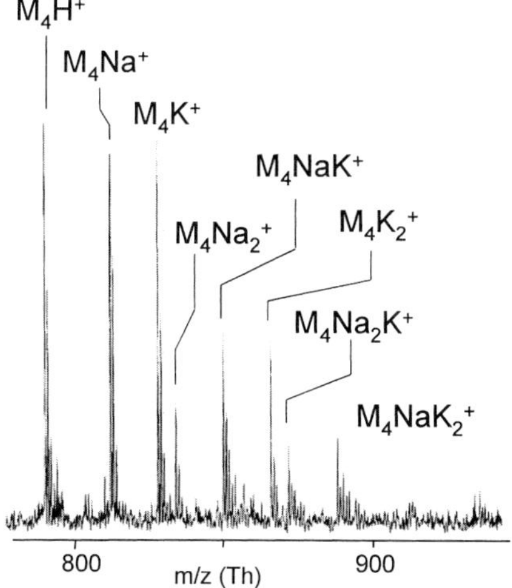

Fig. 5. Polymerization of capro-lactam **1** (250°C, 3.5h): MALDI-TOF MS spectrum of the 800-900 Th region.

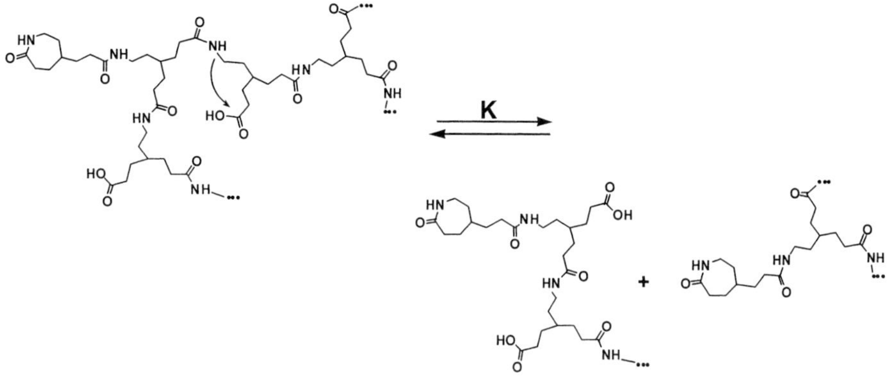

Scheme 4

The polycondensation of 4-aminoethyl-1,7-heptanedioic acid, obtained from caprolactam **1** hydrolysis, was carried out at 170°C for 3h. It resulted in the formation of a lactam-terminated low-molar mass compound. Carboxy-amide interchange reaction is very slow at 170°C and cannot be involved in the formation of lactam end-groups during the polycondensation of this aminodiacid. In this case ring formation is clearly under kinetic control, which means that the intramolecular $COOH+NH_2$ reaction is faster than the intermolecular reaction.

Side reactions on prolonged heating

In both carboxy- and amino-terminated hyperbranched polyamides, the formation of an insoluble, crosslinked fraction was observed on prolonged heating, 48h for the carboxy system and 4 h for the amine system. Crosslinking clearly involves intermolecular side reactions of end-groups, amine or carboxy, which are present in large quantity in the hyperbranched polymers. In the case of carboxy-terminated hyperbranched polyamides, IR and NMR spectra indicated the probable formation of imide and anhydride linkages. MALDI-TOF MS study showed that multiple dehydration reactions take place on the same molecule, since peaks corresponding to the loss of one and two water molecules, respectively M_x - 18 and M_x - 36 series, are present in the spectra (Fig. 6). In the same way, multiple dehydrations were observed in the MS spectra of amino-terminated polyamides after prolonged heating (M_x - 18, M_x - 36 and M_x - 54 series). These dehydrations are attributed to amidine formation by reaction of -NH_2 end-groups on amide -CO-. Moreover, NH_3 loss reflecting the formation of secondary amines by deamination reaction of two amine end-groups was also observed in the spectra. The extent of these various side reactions is quite high since no peak corresponding to the expected series (M_x) can be seen in the spectra after 2h reaction (Fig. 7). In this case, mass spectrometry cannot discriminate intermolecular- from intramolecular dehydration or deamination side reactions. Since crosslinking is observed, intermolecular reactions must obviously take place, but the existence of intramolecular side reactions leading to the formation of cyclic branches cannot be excluded.

Study of ring-chain equilibrium

In order to examine the influence of a high reactive group concentration on the lactam/polymer equilibrium, model reactions were carried out on linear polyamide-6, namely (i) the acidolysis of high molar mass polyamide-6 by dodecanoic acid and (ii) the acidolytic polymerization of ε- caprolactam in the presence of dodecanoic acid. Both reactions were carried out in equimolar COOH:CONH ratio at 250°C, i.e. in reaction conditions similar to those used for the polymerization of γ-carboxyethyl-ε-caprolactam **1**. ε-Caprolactam concentration was monitored by ^1H NMR spectroscopy (Fig. 8). Caprolactam was consumed during the acidolytic polymerization, and formed during the acidolysis of polyamide-6. In both reactions, however, lactam concentration reached the same plateau value after a few hours of reaction, corresponding to lactam amide conversion = 0.75 or [lactam amide]/[total amide] = 0.25. This indicate that carboxyl groups, caprolactam amides and linear amides take part in a ring-chain equilibrium through amide-carboxylic acid interchange reactions (Scheme 5).

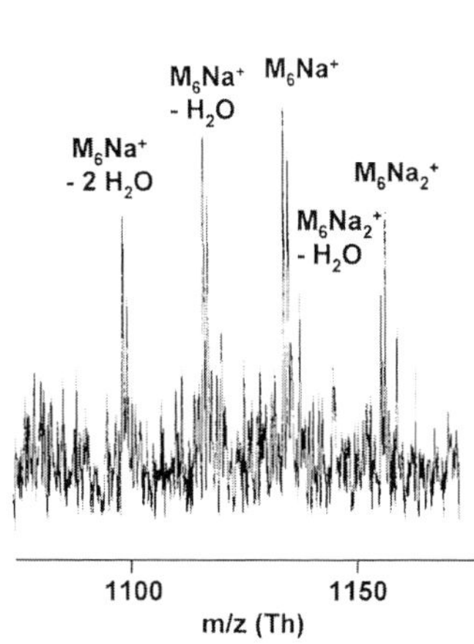

Fig. 6. MALDI-TOF MS spectrum (hexamer region, Na^+ cationization) showing the existence of dehydration reactions during carboxy-functionalized caprolactam **1** polymerization (250°C, 14 h).

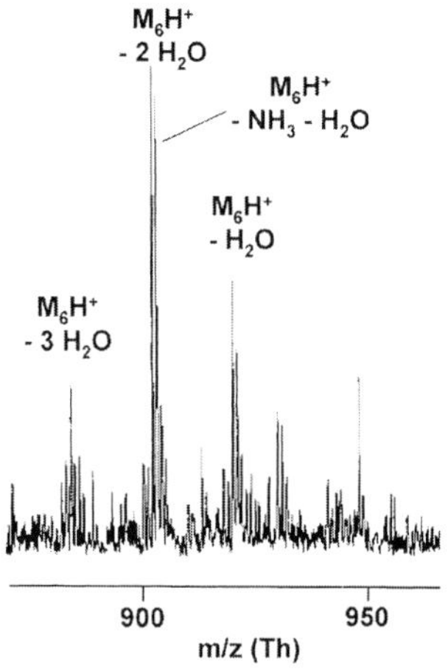

Fig. 7. MALDI-TOF MS spectrum (hexa-mer region, H^+ cationization) showing the existence of dehydration and deamination reactions during amino-functionalized capro-lactam **2** polymerization (250°C, 2 h).

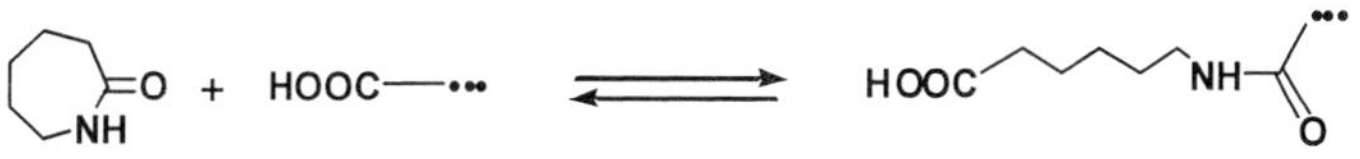

Scheme 5

The acidolytic polymerization of γ-carboxyethyl-ε-caprolactam **1** presents features quite similar to those of the linear polymerization (Fig. 8 and Scheme 4), the only noticeable difference being the lower conversion plateau value, 0.53 instead of 0.75 in the case of the linear polymerization. Such a large equilibrium lactam concentration, however, is not in agreement with Jacobson-Stockmayer's theory of ring-chain equilibrium for which the equilibrium concentration of a given cyclic species is a constant and does not depend on polymer or end-group concentrations. Since the equilibrium lactam conversion usually observed in linear polyamides is ca. 0.92, values of[lactam amide]/[total amide] close to 0.08 were expected instead of the value (0.25) found for the model linear polymerizations. The model reactions show that the formation of rings is favored in the presence of high end-group concentration. This may explain to a certain extent the high equilibrium [lactam amide]/[total amide] value (0.47) found in the acidolytic polymerization of carboxyethylcaprolactam **1**. Some other parameters should be considered to explain this value. The first one is the high "internal" concentration of pairs of reactive groups available for intramolecular reactions in hyperbranched polymers with respect to linear ones[10] and the second one is the lower polymerizability generally observed for substituted lactams with respect to non-substituted ones.[11]

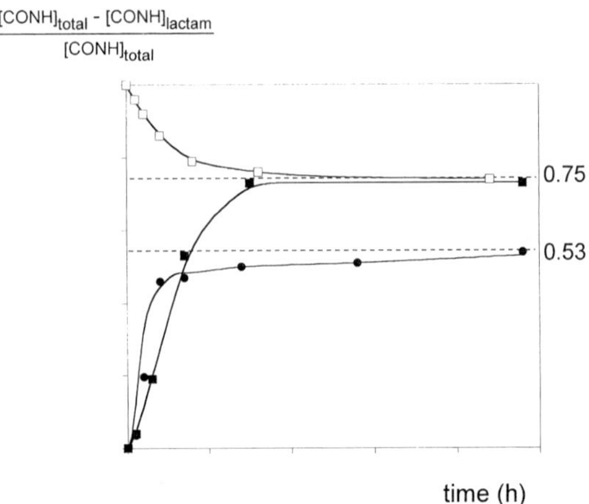

Fig. 8. Variation of lactam conversion versus time during polyamide-6 dode-canoic acid reaction (❑), ε-caprolactam dodecanoic acid reaction (■) and γ-carboxyethyl-ε-caprolac-tam polymerization (●). (1:1 COOH: CONH mol ratio, bulk reaction, 250°C).

Conclusions

The synthesis of high molar mass aliphatic AB_f-type hyperbranched condensation polymers is hindered by the existence of a competition between cyclization and polymerization, which exert a predominant influence on polymer structure. Hydroxy-ester, carboxy-amide and amine-amide back-biting ring-forming reactions taking place during the synthesis result in the formation of cyclic branches or cyclic end-groups such as lactams, and lead to a noticeable lowering of the molar mass of final polymers. The preparation of high molar mass aliphatic hyperbranched polymers should, therefore, involve either a fractionation step to eliminate low molar mass oligomers or reaction conditions in which back-biting reactions are very slow, i.e. polymerization under kinetic control rather than thermodynamic control - although cyclizations have also been reported in hyperbranched polymer syntheses in these conditions.[12]

The presence of a large number of end-groups exert also a deleterious effect on polymer structure and properties. On prolonged heating, a number of side reactions, negligible in linear polymers due to very low end-group concentration at high conversion, become quite important and may result in the formation of crosslinked material: The formation of ethers was detected in hyperbranched polyesters, and the formation of anhydrides, imides, amidines and secondary amines in hyperbranched polyamides.

Ring-chain equilibrium and side-reactions make the control of hyperbranched polymer structure obviously much more difficult than generally accepted.

[1] H. R. Kricheldorf, S. Böhme, G. Schwarz, *Macromolecules* **2001**, *34*, 8879.
[2] C. Cameron, A. H. Fawcett, C. R. Hetherington, R. A. W. Mee, F. V. McBride, *Macromolecules* **2000**, *33*, 6551
[3] A. Burgath, H. Sunder, H. Frey, *Macromol. Chem. Phys.* **2000**, *201*, 782.
[4] D. Parker, W. J. Feast, *Macromolecules* **2001**, *34*, 2048.
[5] C. Gong, J. Miravet, J. M. J. Fréchet, *J. Polym. Sci. Polym. Chem.* **1999**, *37*, 3193.
[6] E. Malmström, A. Hult *Macromolecules* **1995**, *28*, 1698.
[7] F. Leyendecker, G. Mandville, J. M. Conia *Bull. Soc. Chim.* **1970**, 556.
[8] X. Arnaud, C. Guillermain, M. Tessier, A. Fradet, *to be published.*
[9] H. Komber, A. Ziemer, B. Voit, *Macromolecules* **2002**, *35*, 3514.
[10] S. B. Roos-Murphy, R. F. T. Stepto, "*Macromolecular Cyclizations*" in "*Large Ring Macromolecules*", Vol. 1., *Macrocyclic Coupounds*, Ed. J. A. Semlyen, Wiley, New-York (1996).
[11] R. C. P. Cubbon *Makromol. Chem.* **1964**, *80*, 44.
[12] L. Vakhtangishvili, G. Schwarz, H. R. Kricheldorf, Poster P56, "Polycondensation 2002", Hamburg, Sept. 2002.

Synthesis and Characterization of Hyperbranched Aromatic Polyamide Copolymers Prepared from AB_2 and AB Monomers

Mitsutoshi Jikei, Ken Fujii, Masa-aki Kakimoto*

Department of Organic and Polymeric Materials, Tokyo Institute of Technology
O-okayama, Meguro-ku, Tokyo 152-8550, Japan
Email: mjikei@o.cc.titech.ac.jp

Summary: A series of hyperbranched aromatic polyamide copolymers has been prepared and characterized from direct polycondensation of AB_2 and AB monomers. Structure of the monomers and the molar ratio of AB_2/AB showed strong influence on the properties of resulting copolymers. A small amount of AB_2 branching unit improved markedly the solubility of the resulting copolymer. Mark-Houwink parameters of the copolymers were essentially independent of the mole ratio of the monomers. The physical and mechanical properties of resulting copolymers were influenced not only by the mole ratio of monomers, but also by the structure of the monomers employed.

Keywords: copolymerization, hyperbranched polyamide, mechanical properties, solution properties, thermal properties

Introduction

It is well-known that architecture of polymers affects strongly polymer properties even if the polymers are composed of the same repeating units. Figure 1 shows that the architecture of branched macromolecules depends on the branching density. Random and irregular branching leads to a crosslinked material (**a** in Figure 1). On the other hand, regular and consecutive branching results in the formation of a highly branched material (**b** and **c** in Figure1), called dendritic macromolecules. It is known that properties of dendritic macromolecules are very different from those of crosslinked materials and linear polymers. Unique properties of dendritic macromolecules, which include good solubility, low viscosity, encapsulation effect, and multi-functionality, are attributed to globular shape of the molecules and the presence of many functional groups on the periphery. Dendritic macromolecules are usually classified into dendrimers and hyperbranched polymers. One of the advantages of hyperbranched polymers is that they can be prepared from one-step polymerization of AB_x type monomers. This is in sharp contrast to dendrimers which must be prepared by multi-step reactions. Although hyperbranched polymers usually contain insufficient branching units, which are called linear

 DOI: 10.1002/masy.200350919

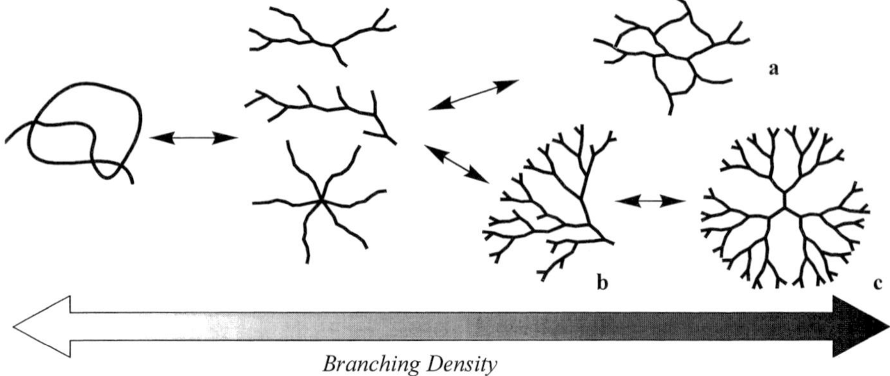

Fig. 1. Architecture of branched macromolecules.

units, the properties of hyperbranched polymers are often very similar to those of the corresponding dendrimers.[1-4] Therefore, hyperbranched polymers are attractive new materials for industrial applications because of their simple production process. Branching density of a hyperbranched polymer can be controlled by the mole ratio of AB_2 and AB monomers. The effect of branching point on polymer properties of resulting polymers can be quantitatively evaluated by the mole ratio of AB_2 and AB monomers. The concept of the copolymerization was first proposed by Flory about 50 years ago,[5] and then, Kricheldorf reported the first attempt to introduce AB_2 type branching unit for synthetic polymers in 1982.[6] We have recently reported the copolymerization of 3,5-bis(4-aminophenoxy)benzoic acid (AB_2) and 4-aminophenoxybenzoic acid (AB).[7,8] In this paper, we focus on the effects of the number of branching unit and the structure of branching component on the physical and mechanical properties of resulting copolymers prepared from direct copolycondensation of AB_2 and AB monomers.

Experimental Section

Materials. 3,5-Bis(4-aminophenoxy)benzoic acid (AB_2-1) was prepared from 4-fluoronitrobenzene and 3,5-dihydroxybenzoic acid, as described in the literature.[7] 3,5-Diaminobenzoic acid (AB_2-2) was purified by recrystallization in ethanol. 3-(4-Amino phenoxy)benzoic acid (AB-1) was prepared from 4-fluoronitrobenzene and 3-hydroxybenzoic acid, as described in the literature.[7] 4-(4-Aminophenoxy)benzoic acid (AB-2) was preparedfrom condensation reaction of 4-hydroxybenzoic acid and 4-fluoronitrobenzene and subsequent hydrogenation in the presence of palladium on acitivated carbon as a catalyst. 4-Aminobenzoic acid (AB-3) was purified by recrystallization in ethanol. 3-Aminobenzoic acid

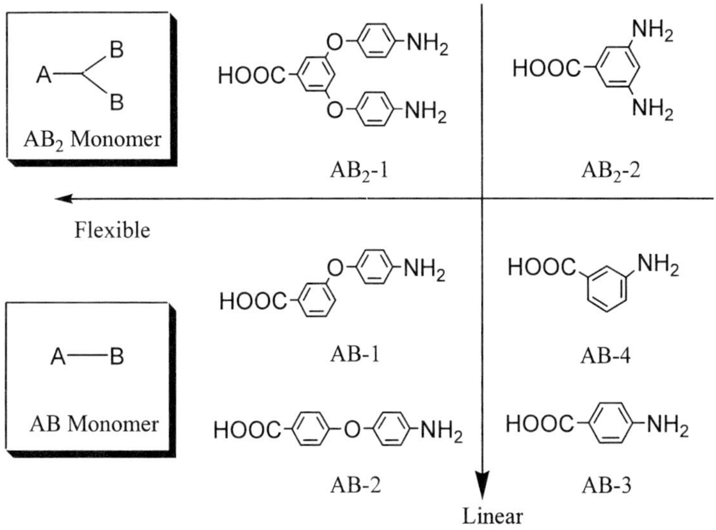

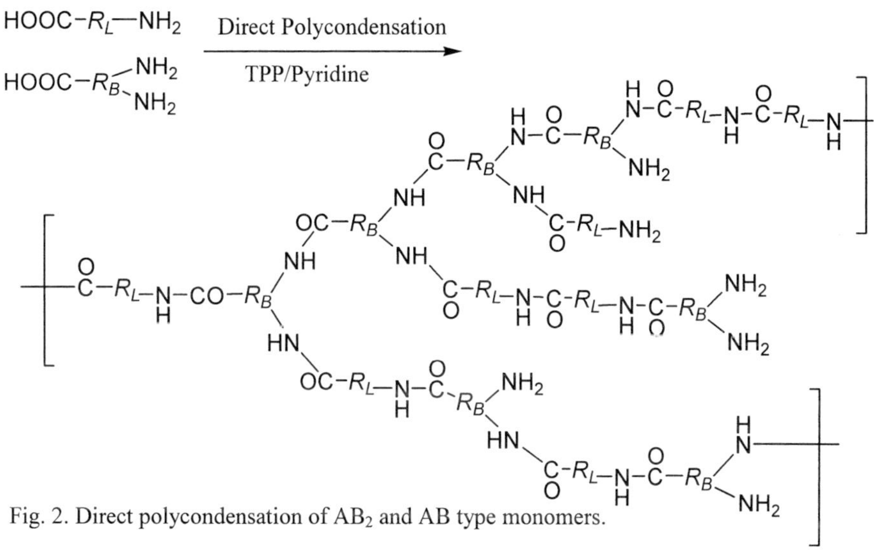

Fig. 2. Direct polycondensation of AB_2 and AB type monomers.

(AB-4) was purified by recrystallization in water. N-Methyl pyrrolidone (NMP) and pyridine were used after distillation under reduce pressure from calcium hydride. Lithium chloride was dried at 230°C overnight before use. Other solvents and reagents were used without further purification.

Measurements. Infrared (IR) spectra were recorded using a Shimadzu FTIR-8100 fourier transform infrared spectrophotometer. ^{1}H and ^{13}C-NMR spectra were recorded using a JEOL JNM-AL 300 spectrometer. Thermo gravimetric analysis (TGA) was carried out with a Seiko TGA 6200 at a heating rate of 10 °C/min under nitrogen. Differential scanning calorimetry (DSC) was carried out with a Seiko DSC 6200 at a heating rate of 10 °C/min in nitrogen. Gel permeation chromatography (GPC) measurements were carried out using a Viscotek TDA302 Triple Detector Array with polystyrene-divinylbenzene columns (two Shodex KD-806M and KD-802). DMF containing 0.01 mol/L of lithium bromide was used as an eluent. Tensile tests were carried out at room temperature with a Toyo Baldwin Tensilon UTM-II-20 using a crosshead speed of 4.0 mm/min. The width and thickness of film strips were 5.0 mm and about 40 μm, respectively, and the gauge length was 10 mm.

Copolymerization of AB_2-1 and AB-1 (50/50). In a three-necked flask, 0.84 g (2.5 mmol) of AB_2-1 and 0.57 g (2.5 mmol) of AB-1 were dissolved in 10 mL of NMP, and then, 0.5g of lithium chloride, 2.5 mL of pyridine and 1.43 mL (5.5 mmol) of triphenylphosphite (TPP) were charged into the flask. The solution was heated to 100 °C and stirred under nitrogen. After 3h, the solution was poured into 1000 mL of methanol to precipitate the polymer. The precipitate was collected by filtration and dried in vacuum at room temperature. The crude product was purified by reprecipitation from DMF solution into methanol containing 0.1 wt% of lithium chloride. The product was finally filtered and washed in boiling mixture of methanol/ water (1: 1), and dried in vacuum at 100 °C. Yield 93 %. IR(KBr): 1659, 1586, 1505, 1482, 1435, 1406, 1318, 1267, 1211, 1167, 1127, 1003, 835 cm^{-1}. ^{1}H-NMR (DMSO-d_6, ppm): 10.03 (amide), 7.76, 7.61, 7.41, 7.36, 7.25, 7.15, 7.07, 6.79, 6.63, 6.55, 4.71 (amine). Anal. Found: C, 71.95; H, 4.64; N, 7.70.

Copolymerization of AB_2-1 and AB-3 (50/50). AB_2-1 (1.261 g: 3.75 mmol) and AB-3 (0.514 g:3.75 mmol) were copolymerized in the same manner as that of AB_2-1 and AB-1. Benzoyl chloride (4 mL) was added to the reaction mixture before isolation of crude product in order to modify end amino groups. Yield 94 %. IR(KBr): 1657, 1601, 1505, 1435, 1406, 1318, 1267, 1208, 1167, 1127, 1003, 841 cm^{-1}. ^{1}H-NMR (DMSO-d_6, ppm): 10.07, 7.97, 7.80, 7.51, 7.39, 7.10, 6.79.

Spectorscopic data for other copolymers. AB_2-2-AB-1 (50/50): IR(KBr): 1657, 1601, 1539, 1447, 1273, 1225, 855 cm^{-1}. ^{1}H-NMR (DMSO-d_6, ppm): 10.22, 8.47, 8.10, 8.02, 7.80, 7.52, 7.19, 7.07. AB_2-2-AB-3 (50/50): IR(KBr): 1659, 1599, 1525, 1514, 1449, 1323, 1282, 1244, 1186, 850 cm^{-1}. ^{1}H-NMR (DMSO-d_6, ppm): 10.23, 7.97, 7.80, 7.51, 7.39, 7.10, 6.79. AB_2-2-AB-4 (50/50): IR(KBr): 1659, 1599, 1545, 1485, 1447, 1327, 1304, 880 cm^{-1}. ^{1}H-NMR (DMSO-d_6, ppm): 10.24, 8.49, 8.38, 8.11, 8.01, 7.75, 7.52.

Results and Discussion

Preparation of Copolymers. Hyperbranched aromatic polyamide copolymers were prepared by direct polycondensation of AB_2 and AB monomers in the presence of triphenylphosphite (TPP) and pyridine acting as condensation agents, as shown in Figure 2. Two AB_2 type monomers, 3,5-bis(4- aminophenoxy)benzoic acid (AB_2-1) and 3,5-diaminobenzoic acid (AB2-2), and four AB type monomers were examined for the copolymerization. Table 1 shows the results for copolymerization of AB2-1 and AB-1. The functional groups in AB-1 were designed to give equal reactivity with AB2-1. The direct polycondensation proceeded successfully and the resulting copolymers were isolated in good yield. The amount (mol%) of incorporated AB_2-1 in the copolymers, determined by ^{1}H NMR measurements, was almost consistent with the formulated mole ratio of monomers. Absolute molecular weight determined by a triple-online detector was in the range 10^4-10^5 with relatively large polydispersity.

Table 1. Preparation of hyperbranched polyamide copolymers from AB_2-1 and AB-1.

Initial mole ratio AB_2:AB	Yield (%)	r_{AB2} [a)]	η_{inh}(dL/g) [b)]	M_w [c)]	M_w/M_n [c)]
100 : 0	95	100	0.22	3.60 x 10^4	2.08
87.5 : 12.5	82	87	0.42	7.57 x 10^4	3.68
75 : 25	93	76	0.27	5.91 x 10^4	3.10
50 : 50	93	50	0.40	6.67 x 10^4	3.19
12.5 : 87.5	85	13	0.68	11.7 x 10^4	3.68
0 : 100	88	-	0.77	-	-

[a)] Incorporated mol% of AB_2-1 in the copolymer determined by integration ratio of ^{1}H NMR measurements.
[b)] Measured in NMP at a concentration of 0.5 g/dL at 30□C.
[c)] Absolute molecular weight determined by GPC equipped with a triple-online detector (refractive index, low-angle laser light scattering, and viscosity) in DMF containing lithium bromide of 0.01 mol/L.

Table 2. Preparation of hyperbranched polyamide copolymer from AB_2-1 and AB-3.

Initial mole ratio AB_2 : AB	r_{AB2} [a)]	Yield (%)	η_{inh} [b)] (dL/g)	M_w [c)]	M_w/M_n [c)]
80:20	80	96	0.21	3.56×10^4	1.86
50:50	50	94	0.23	3.66×10^4	1.67
20:80	25	79	0.24	0.822×10^3	1.18
0 :100	-	96	1.05[d)]	-	-

a) Incorporated mol% of AB_2-1 in the copolymer determined by ^{1}H-NMR measurement.
b) Inherent viscosity measured in NMP at a concentration of 0.5 g/dL at 30 °C.
c) Absolute molecular weight determined by GPC equipped with a triple-online detector (refractive index, low-angle laser light scattering, and viscosity) in DMF containing lithium bromide of 0.01 mol/L.
d) Inherent viscosity measured in concentrated sulfuric acid at a concentration of 0.5 g/dL at 30 °C.

A rigid AB type monomer, 4-aminobenzoic acid (AB-3), was also copolymerized with AB_2-1. The reactivity of functional groups may be different from that of AB_2-1. The copolymerization was carried out in the presence of TPP and pyridine in NMP. In order to improve the solubility of the resulting copolymers, amino end groups were reacted with benzoyl chloride before isolation. As shown in Table 2, the copolymers were isolated in good yield. When the mole ratio of AB_2-1 was 20 %, the resulting copolymer contained 25% AB_2-1 component, which is greater than the initial mole ratio (20%). The data suggested that AB-3 has lower reactivity than AB_2-1.

Table 3. Copolymerization of AB_2-2 and AB monomers to form hyperbranched polyamide copolymers.[a)]

Monomer [b)]		Yield (%)	η_{inh} [c)] (dL/g)	M_w [d)]	M_w/M_n [d)]
AB_2-2	-	93	0.26	15.4×10^4	1.42
AB_2-2	AB-1	91	0.32	7.09×10^4	1.72
AB_2-2	AB-3	94	0.20	9.08×10^4	1.35
AB_2-2	AB-4	85	0.17	7.71×10^4	1.31

a) The direct polycondensation was carried out in the presence of triphenylphosphite and pyridine at 100□C for 3h. End amino groups were reacted with benzoyl chloride before isolation of the product.
b) Initial molar ratio of AB_2-2 and an AB monomer was 1:1.
c) Inherent viscosity measured in NMP at a concentration of 0.5 g/dL at 30 °C.
d) Absolute molecular weight determined by GPC equipped with a triple-online detector (refractive index, low-angle laser light scattering, and viscosity) in DMF containing lithium bromide of 0.01 mol/L.

Copolymerization of AB_2-2 and various AB monomers was carried out in the same manner. Resulting copolymers having amino end groups were insoluble or partially soluble in aprotic polar solvents after isolation. The lower solubility compared with the copolymers prepared from AB_2-1 most likely is caused by the more rigid structure of AB_2-2. Therefore, end amino groups were reacted with benzoyl chloride in order to improve the solubility. The results for the copolymerizations are summarized in Table 3. All copolymerizations proceeded successfully to form corresponding copolymers in high yield. Their molecular weights determined by GPC with the triple-online detector were reasonablly high.

Properties of Copolymers. The solubility of copolymers was markedly improved by incorporation of AB_2 type branching unit. In the case of AB_2-1 and AB-1, the addition of AB_2-1 in 5mol% affected the solubility of the resulting copolymer.[7] The solubility in m-cresol and methoxyethanol depended on the mole ratio of the monomers. As shown in Table 4, the copolymers from AB_2-1 and AB-3 having amino end groups were insoluble in most of organic solvents after isolation. However, the solubility was improved after the amino end groups were reacted with benzoyl chloride. It is clear that the end-capping reaction reduces solute-solute interaction.

Table 4. Solubility of hyperbranched polyamide copolymers from AB_2-1 and AB-3.

AB_2: AB	End group	Solubility					
		NMP	DMAc	DMSO	DMF	*m*-Cresol	Methoxy ethanol
20:80	$-NH_2$	−	−	−	−	−	−
50:50	$-NH_2$	−	−	−	−	−	−
100:0	-NHCOPh	+	+	+	+	+	+
80:20	-NHCOPh	+	+	+	+	+	−
50:50	-NHCOPh	+	+	+	+	+	−
20:80	-NHCOPh	−	+	+	+	−	−
0:100	-NHCOPh	−	−	−	−	−	−

+: Soluble
-: Insoluble

As shown in Table 1, inherent viscosity of the resulting copolymers increased with increasing the feed amount of AB-1. The relationship between weight average molecular weight and intrinsic viscosity of the copolymers from AB_2-1 and AB-1 was investigated by GPC with a triple-online detector: the triple-online detector is capable of monitoring low angle laser light scattering, viscosity, and reflactive index all at the same time. Figure 3 shows Mark-Houwink plots of the copolymers. A linear relationship was observed for each copolymer and the lines closely overlapped with each other. As shown in the inserted figure, all of the slopes observed were around 0.45 even if the mol% of AB_2-1 was 12.5%. It is well-known that most of linear polymers have a value of the slope between 0.5 to 0.8 and dendritic macromolecules often

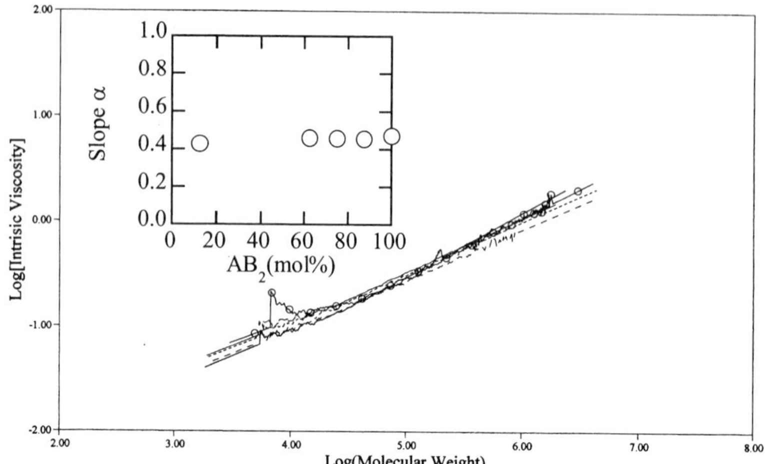

Fig. 3. Mark-Houwink plot of the hyperbranched aromatic polyamide copolymers.

show the slope smaller than 0.5. In this work, the slope was slightly lower than 0.5 and seemed to be independent on the initial molar ratio of the monomers in the range studied. The slope for the AB-1 homopolymer could not be determined because of their poor solubilitiy in the GPC solvent used.

Glass transition temperature (T_g) of the copolymer also varied showing a dependence on the structure and mole ratio of the monomers used, as shown in Figure 4. T_g of the copolymer decreased with increasing the feed ratio of AB_2-1, whereas the incorporation of AB_2-2 unit increased T_g of the copolymer. It is interesting to note that all of the relationship showed unique curvature shape. The T_gs of the copolymers were lower than the value expected by Fox's equation,[9] which is commonly applied for linear copolymers. In literatures, Kricheldorf observed unexpected minimum peaks of T_g for hyperbranched polyester copolymers.[10,11] There are two possible reasons. One is that the number of end functional groups is directly ralated to the mole ratio of monomers. The other is that the degree of intermolecular

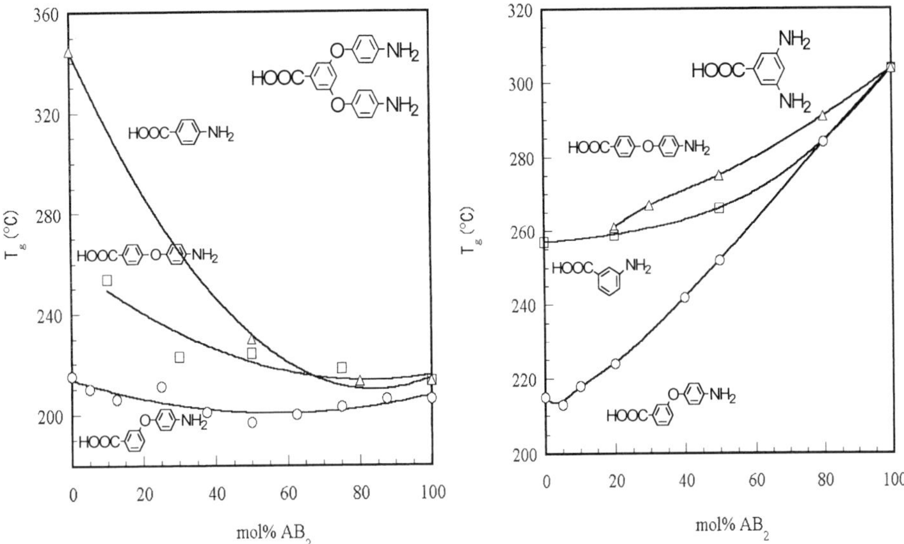

Fig. 4. Glass transition temperature (T_g) of hyprabranched polyamide copolymers prepared from AB_2-1 (a) and AB_2-2 (b).

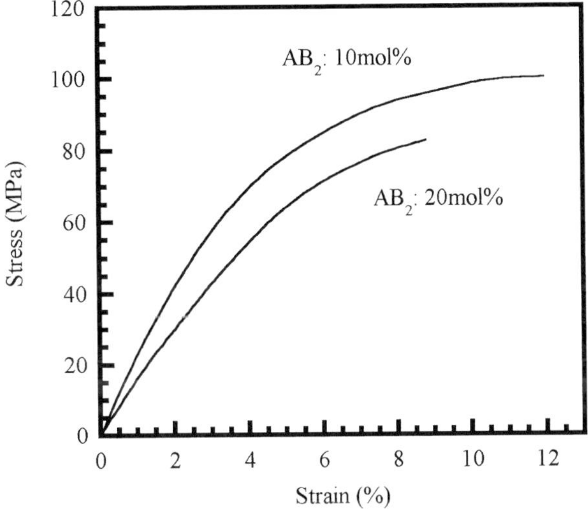

Fig. 5. Stress-strain curve of the hyperbranched polyamide copolymers prepared from AB_2-2 and AB-1.

entanglement is highly influenced by the ratio of monomers. We are unable to discern which of the two factors governs the Tg behavior, based on the results obtained. Further study is needed.

It was extremely difficult or impossible to cast films from the hyperbranched polyamide homopolymers due to low solution viscosity. The linear aromatic polyamides were practically insoluble in organic solvents after isolation. Only some of the copolymers gave flexible films by casting copolymer solution onto glass plates. Tensile tests were carried out to evaluate the effect of the branching units on the mechanical properties. The stress-strain (S-S) curve of the copolymer prepared from AB_2-2 and AB-1 is shown in Figure 5. As expected, Young's molulus determined by the initial slope of S-S curves decreased when the feed ratio of AB_2-2 was increased. The same trend was also observed for the copolymers from AB_2-1/AB-1[7] and AB_2-2/AB-2. This suggests that intermolecular entanglement decreases by the incorporation of the AB_2 branching units. The decrease in maximum strain or elongation may be caused by the incorporation of the rigid structure of AB_2-2 in the polymer backbone.

Conclusion

Hyprabranched aromatic polyamide copolymers were successfully prepared by direct polycondensation of AB_2 and AB monomers in the presence of TPP and pyridine as condensation agents. Properties of resulting copolymers, such as solubility, viscosity, glass transition temperature, and mechanical properties, were strongly dependent on the structure and the mole ratio of monomers. A small amount of AB_2 branching unit markedly affected solubility and viscosity of the copolymers. Flexible cast films could be prepared from the copolymers due to inreased intermolecular entanglement by the incorporation of AB unit.

[1] E. Malmström, A. Hult, *J. Macromol. Sci.-Rev. Macromol. Chem. Phys.*, **1997**, *C37*, 555.
[2] Y. H. Kim, *J. Polym. Sci.: Part A: Polym. Chem.*, **1998**, *36*, 1685.
[3] B. Voit, *J. Polym. Sci.: Part A: Polym. Chem.*, **2000**, *38*, 2505.
[4] M. Jikei, M. Kakimoto, *Prog. Polym. Sci.*, **2001**, *26*, 1233.
[5] P. J. Flory, *J. Am. Chem. Soc.*, **1952**, *74*, 2718.
[6] H. R. Kricheldorf, Q.-Z. Zang, G. Schwarz, *Polymer*, **1982**, *23*, 1821.
[7] M. Jikei, K. Fujii, G. Yang, M. Kakimoto, *Macromolecules*, **2000**, *33*, 6228.
[8] M. Jikei, K. Fujii, M. Kakimoto, *J. Polym. Sci.: Part A: Polym. Chem.*, **2001**, *39*, 3304.
[9] $1/T_g = W_a/T_{ga} + W_b/T_{gb}$.
[10] H. R. Kricheldorf, G. Löhden, *J. Macromol. Sci., Pure Appl. Chem.*, **1995**, *A32*, 1915.
[11] H. R. Kricheldorf, O. S. Stöber, M. Lübbers, *Macromol. Chem. Phys.*, **1995**, *196*, 3549.

Macromol. Symp. **2003**, *199*, 233-241

Hyperbranched Polyimides Prepared by Ideal A_2+B_3 Polymerization, Non-Ideal A_2+B_3 Polymerization and AB_2 Self-Polymerization

*Jianjun Hao, Mitsutoshi Jikei, Masa-aki Kakimoto**

Department of Organic & Polymeric Materials, Tokyo Institute of Technology, 2-12-1, O-okayama, Meguro-ku, Tokyo 152-8550, Japan

Summary: In this paper, hyperbranched polyimides having the same repeating unit were synthesized by employing ideal A_2+B_3 polymerization, non-ideal A_2+B_3 polymerization and AB_2 self-polymerization methods. The polymerization behavior, polymer properties were compared for three methods. Hyperbranched polyimides by ideal A_2+B_3 polymerization, non-ideal A_2+B_3 polymerization and AB_2 self-polymerization methods show apparent difference in many physical properties, such as inherent viscosity, glass transition temperature, and film formation behavior etc. The hyperbranched polymers by the non-ideal A_2+B_3 polymerization are suitable for smooth, flexible and self-standing film preparation, which provides useful information for hyperbranched polymers toward self-standing materials.

Keywords: A_2+B_3 polymerization, film, hyperbranched, polyimides, preparation

Introduction

Hyperbranched polymers have received considerable attention for ten more years.[1-7] Hyperbranched polymers are generally prepared by one-pot self-polymerization of the AB_x monomers.[1-7] Since the AB_x monomers are not always commercially available and their preparation sometimes involves in synthetic effort, a facile A_2+B_3 approach was put forward recently.[8] Hyperbranched polymers, such as polyamide,[8] polyether,[9] polyimide[10] and poly (sulfone-amine)[11] etc., were successfully prepared by this approach.

An A_2+B_3 polymerization has many advantages over AB_2 self-polymerization, such as: (1) monomers are commercially available and easy to obtain; (2) polymerization is convenient to scale up; and (3) the polymer structure is easy to tailor due to various choice of monomers for a certain polymerization. However, it must be noted that A_2+B_3 polymerization has an intrinsic problem, i.e. the gelation is unavoidable over a certain conversation in 1:1 mol monomer feed ratio, as pointed out by Flory over 50 years ago.[12] Thus the major concern of the A_2+B_3 polymerization focuses on how to avoid the gelation. An ideal A_2+B_3 polymerization system toward gelation, as described by Flory,[12] is based on three

 DOI: 10.1002/masy.200350920

assumptions: (1) equal reactivity of all A or B groups at any given stage of the reaction, (2) the neglect of intramolecular cyclization, and (3) the condensation being restricted to the reaction between an A and a B group. However, if an A_2+B_3 polymerization did not obey these assumptions, gelation would be probably avoided.

In our previous paper,[13] a new strategy for preparing hyperbranched polyimides by employing a non-ideal A_2+B_3 polymerization has been reported. The unique A_2+B_3 direct polycondensation by using the DBOP as a condensation agent was found to deviate from the ideal A_2+B_3 polymerization. Therefore gelation was effectively avoided and soluble hyperbranched polyimides with high molecular weight were successfully prepared. In this paper, hyperbranched polyimides having the same repeating unit were synthesized by employing ideal A_2+B_3 polymerization, non-ideal A_2+B_3 polymerization and AB_2 self-polymerization methods. The polymerization behavior, polymer properties were compared for three methods.

Results and Discussions

In this paper, two novel B_3 monomers, i.e. tri(phthalic anhydride) and tri(phthalic acid methyl ester), were employed to carry out A_2+B_3 polymerization. The molecular design of two B_3 monomers was aimed at high thermal stability, good solubility and equal reactivity of each functinal group. As shown in Fig.1, each B_3 monomer is designed to have three independant functional groups with the same reactivity. To carry out AB_2 self-polymerization, a new AB_2 monomer was designed from a template of a repeating unit of the hyperbranched polyimides by A_2+B_3 polymerization.

Hyperbranched polyimides were synthesized by A_2+B_3 polymerization from both B_3 monomers (tri(phthalic anhydride) and tri(phthalic acid methyl ester)), and 1,4-phenylene diamine (A_2) in molecular ratio of 1:1, respectively. The ideal A_2+B_3 polymerization of the tri(phthalic anhydride) and 1,4-phenylene diamine afforded poly(amic acid) (**PAA**) precursor, and the non-ideal A_2+B_3 polymerization of the tri(phthalic acid methyl ester) and 1,4-phenylene diamine gave poly(amic acid methyl ester) (**PAAME**) precursor. The **PAA** and **PAAME** precursors were end-capped with 4-toluidine by adopting the same reaction conditions as the precursor synthesis. The 4-toluidine end-capped poly (amic acid) (**TE-PAA**) and 4-toluidine end-capped poly (amic acid methyl ester) (**TE-PAAME**) were then converted into 4-toluidine end-capped polyimides (**TEPI**) by cyclodehydration in the presence of acetic anhydride and pyridine. The anhydride-terminated polyimides (**ATPI**) were also prepared

ideal A_2+B_3 polymerization
non-ideal A_2+B_3 polymerization
AB_2 self-polymerization
DMAc
DBOP NMP
DBOP NMP
hyperbranched poly(amic acid) (PAA)
hyperbranched poly(amic acid methyl ester) (PAAME)
Ac_2O/pyridine 115°C/NMP
Ac_2O/pyridine 115°C/NMP
terminal unit
linear unit
dendritic unit
anhydride-terminated hyperbranched polyimide (ATPI)
DBOP NMP
toluidine end-capped hyperbranched poly(amic acid) (TE-PAA)
toluidine end-capped hyperbranched poly(amic acid methyl ester) (TE-PAAME)
Ac_2O/pyridine 115°C/NMP
Ac_2O/pyridine 115°C/NMP
toluidine end-capped hyperbranched polyimide (TEPI)

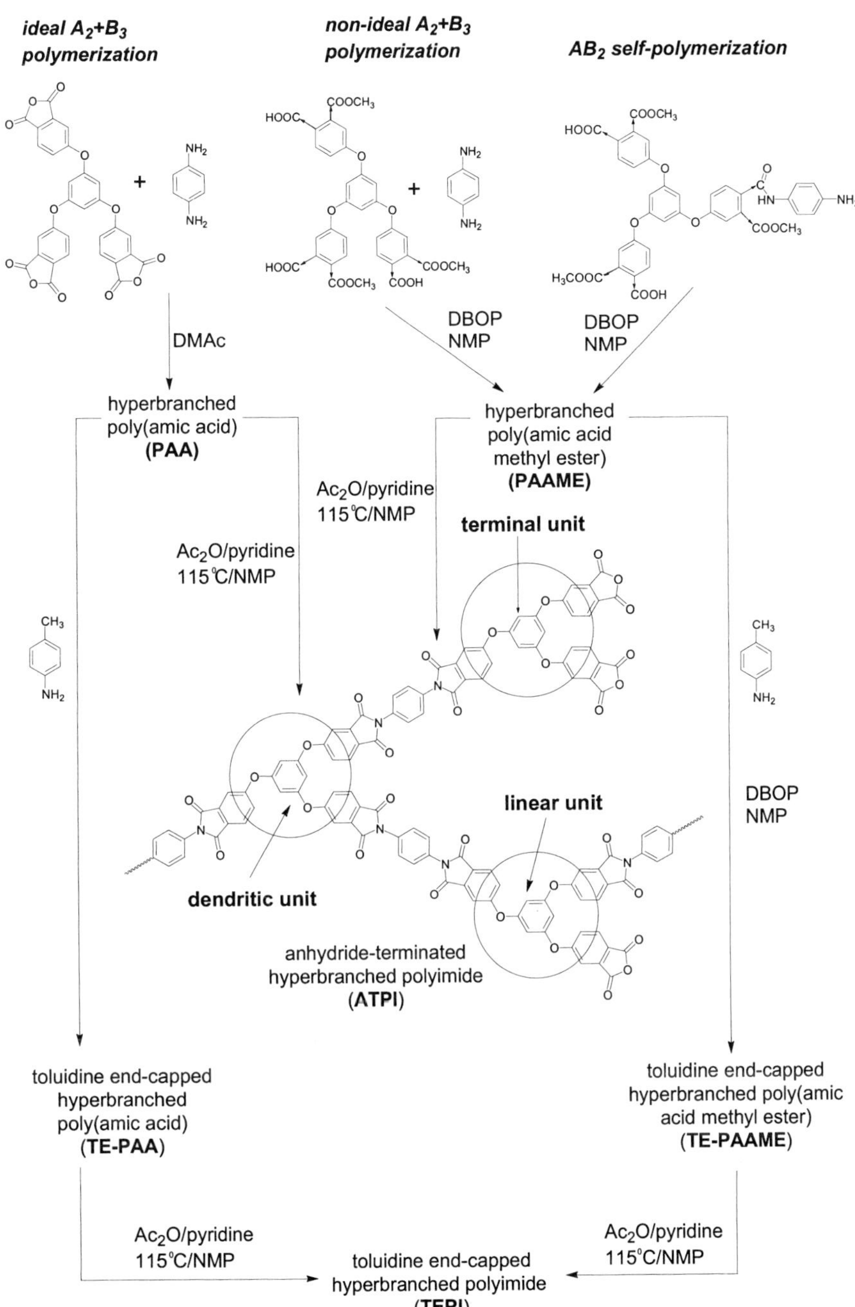

Figure 1

from **PAA** and **PAAME** under the same reaction conditions as **TEPI** synthesis. It can be seen that **TEPIs** and **ATPIs** by both A_2+B_3 polymerizations show the same chemical structure in the repeating unit. On the other hand, hyperbranched polyimides were prepared by self-polymerization of the new AB_2 monomer. As-prepared poly(amic acid methyl ester) (**PAAME**) precursors were end-capped with 4-toluidine by adopting the same reaction conditions as the precursor synthesis. The 4-toluidine end-capped poly(amic acid methyl ester)s (**TE-PAAMEs**) were then converted into 4-toluidine end-capped polyimides (**TEPI**) by cyclodehydration in the presence of acetic anhydride and pyridine. The anhydride-terminated polyimides (**ATPI**) were also prepared from **PAAME** under the same reaction conditions as **TEPI** synthesis. It is apparent that hyperbranched polymers by both AB_2 self-polymerization and non-ideal A_2+B_3 polymerization have the same chemical structure for the repeating unit at every synthetic stage of **PAAMEs**, **TE-PAAMEs**, **ATPIs** and **TEPIs**.

The polymerization conditions and results are summarized in Table 1. It is clear that A_2+B_3 polymerization of the tri(phthalic anhydride) and 1,4-phenylene diamine is a typical ideal A_2+B_3 polymerization, which always causes gelation. Thus, a dropwise addition way by dripping A_2 monomer into B_3 monomer was employed (see table 1, ideal A_2+B_3 polymerization, entry 3). It is noted that the weight-average molecular weight of the **TEPI** is unexpectedly high, but its number-average molecular weight is much low (1.31×10^4 g/mol). The molecular weight distribution attains 23, indicating that the **TEPI** is a mixture of oligomers and high molecular weight polymers. Since the **PAA** and **TEPI** are only soluble in organic solvents upon heating, their poor solubility demonstrate that the high molecular part of the **PAA** and **TEPI** is indeed a slightly crosslinking microgel.

The A_2+B_3 polymerization of tri(phthalic acid methyl ester) (B_3) and 1,4-phenylene diamine (A_2) was carried out through a 'one-step' procedure as literature mentioned, [14] which can successfully achieve the high molecular weight and soluble polymers. A weight-average molecular weight ranging from 33,600 to 125,000 is achieved for this polymerization approach. It was pointed by Flory that, the A_2+B_3 polymerization in 1:1 mol generally results in gelation over a certain conversation of functional groups.[12] This prediction is based on the exclusive reactivity between functional group A and B. However in our case, the functional group B is not restricted to react only with functional group A. The *in-situ* activation of carboxyl groups in a B_3 monomer by DBOP affords an intermediate functional group B′ at first. Moreover a by-product from DBOP is also formed. The B′ can either react with a functional group A to carry out a step of polymerization, or react with a by-product (as

Table 1. Polymerization conditions and results for the synthesis of hyperbranched polyimide precusors via ideal A_2+B_3 polymerization, non-ideal A_2+B_3 polymerization and AB_2 self-polymerization.

method	entry	concent. [a] (g/ml)	temp.[b] (^{0}C)	yield %	η_{inh} [c] (dL/g)	M_w [d]	M_w/M_n [d]
ideal A_2+B_3 polymerization	1	0.025	r.t.		gel		
	2	0.017	0		gel		
	3	0.017	0/r.t.[f]	95	0.28	3.02×10^5	23.0
non-ideal A_2+B_3 polymerization	1	0.19	r.t.		gel		
	2	0.11	r.t.		gel		
	3	0.097	r.t.	97	0.97	1.25×10^5	2.63
	4	0.073	r.t.	90	0.25	6.74×10^4	2.08
	5	0.058	r.t.	86	0.23	3.76×10^4	1.84
	6	0.032	r.t.	78	0.17	3.36×10^4	2.17
AB_2 self-polymerization	1	0.06	r.t.	70	0.12	1.14×10^4	1.2
	2	0.16	r.t	71	0.12	1.11×10^4	1.3
	3	0.32	r.t	75	0.13	2.55×10^4	1.5
	4	0.06	50	82	0.15	5.13×10^4	1.9
	5	0.08	50	94	0.17	1.73×10^5	2.3

a) Concentration, calculated by (the total mass monomers)/ (the volume of the solvent).
b) Temperature.
c) Inherent viscosity measured at a concentration of 0.5g/dL at 30^{0}C.
d) Determined by GPC measurement with a laser light scattering detector in DMF containing lithium bromide (0.01 mol/L) as an eluent. The samples for GPC dertermination were 4-toluidine end-capped polyimide (**TEPI**) for ideal A_2+B_3 polymerization and 4-toluidine end-capped poly (amic acid methyl ester)s (**TE-PAAME**s) for non-ideal A_2+B_3 polymerization as well as AB_2 self-polymerization. The specific refractive increments (d*n*/d*c*) were 0.1278mL/g for entry 3 of ideal A_2+B_3 polymerization; 0.150, 0.172, 0.188 and 0.196 for entry 3, 4, 5 and 6 of non-ideal A_2+B_3 polymerization; 0.139, 0.143, 0.136, 0.140 and 0.143 mL/g for entry 1, 2, 3, 4 and 5 of AB_2 self-polymerization.

mentioned above) to result in an intermediate functional group B″. Although B″ can also react with an A to carry out a step of polymerization, the competition between B′/A reaction and B′/by-product reaction prevents B′ from reacting with A entirely. Such an A_2+B_3 polymerization in 1:1 mol deviates from Flory's ideal A_2+B_3 polymerization toward gelation. Thus the A_2+B_3 polymerization of tri(phthalic acid methyl ester) (B_3) and 1,4-phenylene diamine (A_2) is a non-ideal A_2+B_3 polymerization. This may be the main cause for avoiding the gelation. Due to such a unique polymerization characteristic, it is assumed that a macromolecular structure with a low branching density would be formed at the early polymerization stage, whose topology was retained at the later polymerization stage.

The self-polymerization of the new AB_2 monomer was also carried out through a 'one-step' procedure,[14] which is the same as that in the non-ideal A_2+B_3 polymerization. However low molecular oligomers with a molecular weight around 1.1×10^4 g/mol were obtained at room temperature. It is observed that the polymerization at 50^0C could afford a polymer with a molecular weight high up to 1.7×10^5 g/ml. Gelation was not observed under the given conditions listed in Table 1.

As seen in Table 1, a comparision of the inherent viscosities for as-prepared hyperbranched polyimide precursors reveals that polymers by non-ideal A_2+B_3 polymerization show the highest inherent viscosity, indicating the existence of stronger chain-entanglement. The formation of chain-entanglement is due to the intermolecular penetration, which is resulted from the low branching density topology of polymers by non-ideal A_2+B_3 polymerization. Apparently, hyperbranched polymers by AB_2 self-polymerization is lack of chain-entanglement, resulting in a low inherent viscosity despite of high molecular weight (table 1, AB_2 self-polymerization, entry 5). Just like the general hyperbranched polymers by AB_2 self-polymerization, as-prepared hyperbranched polyimide precursors have a compact highly branching structure, leading to a impenetrable molecule. In addition, hyperbranched polyimide precursors by ideal A_2+B_3 polymerization also exhibit a low inherent viscosity. This is perhaps due to its slightly crosslinking structure, preventing the molecules from penetration one another.

The degree of branching (DB) of hyperbranched polymer was defined as the ratio of the sum of dendritic and terminal units vs total units (dendritic, linear and terminal units). It was found that **ATPI**s gave a clear difference in chemical environment among dendritic, linear and terminal units, as shown in Fig. 1. The aromatic protons of the central aromatic ring in a given unit can be clearly distinguished with the aid of ^{1}H NMR measurement. Thus the DBs for hyperbranched polyimides by ideal A_2+B_3 polymerization, non-ideal A_2+B_3 polymerization and AB_2 self-polymerization were estimated to be 0.54, 0.52-0.55 and 0.50, respectively. The DB for hyperbranched polyimides by ideal A_2+B_3 polymerization and non-ideal A_2+B_3 polymerization is seriously deviated from the statistically predication value (0.50), which is different from that by AB_2 self-polymerization.

The thermal properties of hyperbranched polyimides by chemical imidization are summarized in Table 2. **TEPI** by ideal A_2+B_3 polymerization shows slightly higher glass transition temperature than that by non-ideal A_2+B_3 polymerization. The glass transition temperatures of **TEPI**s by non-ideal A_2+B_3 polymerization are in the range of 212-235^{0}C, which are far higher than that (160^{0}C) of hyperbranched polyimide by AB_2 self-polymerization. The big difference

Table 2. Thermal properties of hyperbranched polyimide via ideal A_2+B_3 polymerization, non-ideal A_2+B_3 polymerization and AB_2 self-polymerization.

method	entry [a]	T_g (^{0}C) [b]	T_5 (^{0}C) [c]	T_{10} (^{0}C) [c]
ideal A_2+B_3 polymerization	3	235	500	535
non-ideal A_2+B_3 polymerization	3	230	505	545
	4	223	480	520
	5	219	480	525
	6	212	485	535
AB_2 self- polymerization	4	160	450	510
	5	161	460	520

[a] The entry code in Table 1.
[b] Glass transition temperature (T_g) measured by DSC under nitrogen, heating rate 10^{0}C/min
[c] 5% and 10% weight loss temperature measured by TGA under nitrogen, heating rate 10^{0}C/min

in glass transition temperatures was caused by the different molecular structure, i.e. the slightly crosslinking structure for ideal A_2+B_3 polymerization, the low branching density topology for non-ideal A_2+B_3 polymerization and the compact highly branching structure for AB_2 self-polymerization. As given in Table 2, the thermal stability for hyperbranched polyimides are also different for three polymerization methods. The 5% weight loss temperatures of **TEPI**s by non-ideal A_2+B_3 polymerization were in the range of 480~505^{0}C. **TEPI** by ideal A_2+B_3 polymerization showed a 5% weight loss at 500^{0}C, close to one of **TEPI**s (entry 3) by non-ideal A_2+B_3 polymerization. The 5% weight loss temperatures of **TEPI**s by A_2+B_3 polymerization methods surpasses that by AB_2 self-polymerization method (450-460^{0}C), although both have the same chemical structure for the repeating unit. This may be caused by the different structure topology. By the way, the residue trace unreacted amine group in hyperbranched polyimides by AB_2 self-polymerization, of which at least one unreacted amine group was theoretically existed in every molecules, should be another reason. Hyperbranched polymers are generally unsuitable for the preparation of self-standing films due to lack of chain entanglements. The film preparation was attempted for hyperbranched polyimide precursors by three polymerization methods. Hyperbranched polyimide films from three kinds of precursors were successfully prepared by casting the DMAc solutions onto glass plates upon heating. Film from **TE-PAA** by ideal A_2+B_3 polymerization was prepared directly from a condensed original reaction solution due to poor solubility of **TE-PAA** precursor after precipitation. The films from **TE-PAAME**s by non-ideal A_2+B_3 polymerization and **TE-PAA** by ideal A_2+B_3 polymerization were self-standing, while that from **TE-PAAME**s by AB_2 self-polymerization were brittle, fragile and not self-standing. The different film formation behavior is related to their molecular structure. The formation of a self-standing film by ideal A_2+B_3 polymerization may be due to the chain extension reaction among the oligomers, and that by non-ideal A_2+B_3 polymerization should be caused by chain-entanglement due to their low branching density structure. The failure in self-standing film formation for **TE-PAAMEs** by AB_2 self-polymerization should be resulted from chain-entanglement lack due to the compact impenetrable moleculr structure. The film from **TE-PAA** by ideal A_2+B_3 polymerization is heterogeneous and rough despite of its flexibility, suggesting existence of microgels. However films from **TE-PAAMEs** by non-ideal A_2+B_3 polymerization are flexible and smooth in transparent yellow appearance. The mechanical properties of as-prepared polyimide films are listed in Table 3. The hyperbranched polyimide film by non-ideal A_2+B_3 polymerization (N-3), whose precursor has the highest inherent viscosity, showed the highest tensile strength. This suggests the existence of the strongest chain-entanglement in this film. Although the chain extension reaction among the oligomers affoded a self-standing film, the tensile strength of the film by ideal A_2+B_3 polymerization was the lowest one. This indicates that the chain extension reaction is limited in enhancing the mechanical properties. As-prepared films showed a elongation at break around 1.0%. Their tensile modulus are above 2.3 GPa, which is close to their linear analogues.

Table 3. Mechanical properties of hyperbranched polyimide films by A_2+B_3 polymerization.

properties	AEPI				
	I-3 [a]	N-3 [b]	N-4	N-5	N-6
tensile strength (MPa) [c]	18 ± 2	29 ± 1	27 ± 2	21 ± 4	18 ± 2
elongation at break (%)	0.8 ± 0.1	0.9 ± 0.2	1.1 ± 0.1	0.9 ± 0.1	0.8 ± 0.1
tensile modulus (GPa) [d]	2.3	3.2	2.6	2.3	2.3

a) I, ideal A_2+B_3 polymerization; 3, entry code in table 1.

b) N, non-ideal A_2+B_3 polymerization; 3, entry code in table 1.

c) Tensile test was carried out at room temperature with a film specimen at a dimension of 40 × 5 × 0.02 mm. The tensile rate is 4 mm/min.

d) Calculated from dividing the average tensile strength by the average elongation at break

Conclusions

The hyperbranched polyimides by ideal A_2+B_3 polymerization, non-ideal A_2+B_3 polymerization, and AB_2 self-polymerization have different topological structure one another, although they have the same repeating unit. The polyimides by ideal A_2+B_3 polymerization have a three-dimensional network (crosslinking) structure, and that by AB_2 self-polymerization have a compact highly branching architecture without chain-entanglement. However the polyimides by non-ideal A_2+B_3 polymerization have a low branching density topology with apparent chain-entanglement, although they are hyperbranched polymers with a *DB* above 0.5. Due to their difference in topological structure, hyperbranched polyimides by ideal A_2+B_3 polymerization, non-ideal A_2+B_3 polymerization and AB_2 self-polymerization methods show apparent difference in many physical properties, such as inherent viscosity, intrinsic viscosity, glass transition temperature, and film formation behavior etc. The hyperbranched polymers by the non-ideal A_2+B_3 polymerization are suitable for smooth, flexible and self-standing film preparation, which provides useful information for hyperbranched polymers toward self-standing materials.

[1] (a) Y. H. Kim, O. W. Webster, *Am. Chem. Soc., Div. Polym. Chem. Polym. Prepr.* **1988**, *29(2)*, 310. (b) Y. H. Kim, O. W. Webster, *J. Am. Chem. Soc.* **1990**, *112*, 4592.
[2] D. A. Tomalia, H. Durst, *Top. Curr. Chem.* **1993**, *165*, 193.
[3] J. S. Moore, *Curr. Opin. Solid State Mater. Sci.* **1996**, *1*, 777.
[4] B. Voit, *J. Polym. Sci.: Part A: Polym. Chem.* **2000**, *38*, 2505.
[5] A.Sunder, J. Heinemann, R. Haag, H. Fery, *Chem. Eur. J.* **2000**, *6*, 2499.
[6] S. S. Sheiko, M. Möller, *Topics in Current Chemistry* **2001**, *212*, 137.
[7] M. Jikei, M. Kakimoto, *Progress in Polymer Science* **2001**, *26*, 1233.
[8] M. Jikei, S. Chon, M. Kakimoto, S. Kawauchi, T. Imase, J. Watanebe, *Macromolecules* **1999**, *32*, 2061
[9] T. Emrick, H. Chang, J. M. J. Fréchet, *Macromolecules* **1999**, *32*, 6380.
[10] J. Fang, H. Kita, K. Okamoto, *Macromolecules* **2000**, *33*, 4639.
[11] D. Yan, C. Gao, *Macromolecules* **2000**, *33*, 7693.
[12] P. J. Flory, *Principles of Polymer Chemistry*; Cornell University Press, Ithaca, NY, 1953, Chapter 9.
[13] J. Hao, M. Jikei, M. Kakimoto, *Macromolecules* **2002**, *35*, 5372.
[14] M. Ueda, A. Kameyama, K. Hashimoto, *Macromolecules* **1988**, *21*, 19.

Synthesis and Characterisation of Branched Polyarylethers

Martin Weber, Piyada Charoensirisomboon*

BASF Aktiengesellschaft, Polymer Research, Engineering Plastics, D-67056 Ludwigshafen, Germany

Summary: Polyarylethers (Polysulfone: PSU, Polyethersulfone: PES) belong to the group of high performance polymers, having high glass transition temperature as well as high continuous use temperature. As a consequence of their high glass transition temperature, these polymers display high melt viscosity, which limits the number of accessible applications. Since the conventional methods to improve the flow characteristics are limited, the influence of branching by incorporation of the tri-functional monomer 1,1,1-Tris-(4-hydroxyphenyl)ethane (THPE) on the flow and the mechanical performance of PES was studied.
Branching enhances the flow of Polyethersulfone significantly, but has a deleterious effect on the toughness, especially Charpy impact and tensile elongation.

Keywords: branching, mechanical properties, melt flow, polyarylethers

Introduction

Polyarylethers (Polysulfone: PSU, Polyethersulfone: PES) belong to the group of high performance polymers, having high glass transition temperature as well as high continuous use temperature.[1-4] These materials are widely used in high end applications, like dialysis membranes, head lamp reflectors, and microwave cookware.[5-7]

As a consequence of their high glass transition temperature (PSU: 186°C, PES: 225°C), these polymers have comparatively high melt viscosity. Subsequently, processing of large parts or of parts with low wall thickness very often causes severe problems. This issue limits the number of accessible applications significantly.

As known from the literature, branched polymers usually display better flow performance than linear products with comparable molecular weight.[8] Especially in the area of Polyolefines, branching is a powerful tool to tailor flow and processing behaviour.[9] Also in the case of Polyamides, branching nowadays is used to improve the processing behaviour, especially the cycle time in injection moulding processes.[10] The improvement of the melt

 DOI: 10.1002/masy.200350921

flow without changing other positive properties of Polyarylethers would enhance the market potential of these materials significantly.

The common ways to improve the melt flow of thermoplastic materials, like the addition of flow promoters e.g. waxes, mineral oils or stearates, have limited impact in the case of polyarylethers, due to problems like dispersion of low viscosity additives in the highly viscous melt, or reduction of the transparency, or thermal degradation of the additive and subsequent discoloration of the product. blending of polyarylethers with liquid crystalline polymers (LCP) gives nice improvements of the processing behaviour, but the obtained mixtures suffer from bad toughness, even at very small loading of LCP.[11,12]

The melt flow of polyarylethers can be significantly improved by blending with engineering plastics like Polyamides or polycarbonates.[13,14] But in all cases, the obtained blends are opaque as a consequence of their phase separated morphology.

As already discussed, branching should improve the melt flow of polyarylethers. By incorporation of a small amount of a heat resistant molecule as branching unit, no deleterious influence on the thermal stability and transparency should be observed. Since the molecular weight for entanglements (M_e) is quite low in the case of polyarylethers (appr. 2000 g/mol),[15] the mechanical properties of branched products are expected to remain within the limits known for linear products. Thus, branched polyarylethers were synthesised and characterised.

The synthesis of such products is already described in the patent literature, but limited information about the influence of branching on the mechanical performance of polyarylethers is available up to now.[16]

Generally, polyarylethers are prepared via nucleophilic polycondensation starting from di-chloro and di-hydroxy-components.[4,17] For the synthesis of polyethersulfone, dichloro-diphenylsulfone (DCDPS) and dihydroxydiphenylsulfone (DHDPS) are used as monomers. Using a tri-functional unit, like 1,1,1-Tris-(4-hydroxyphenyl)ethane (THPE, Figure 1) as co-monomer, branching is possible. This product (THPE) is commercially available and soluble in the reaction mixture. The reaction scheme for the synthesis of the branched polyethersulfone is depicted in Figure 1. This procedure can also be used to synthesise unique branched polysulfones, having anhydride end-groups as compatibilizer for PSU/PA-alloys.

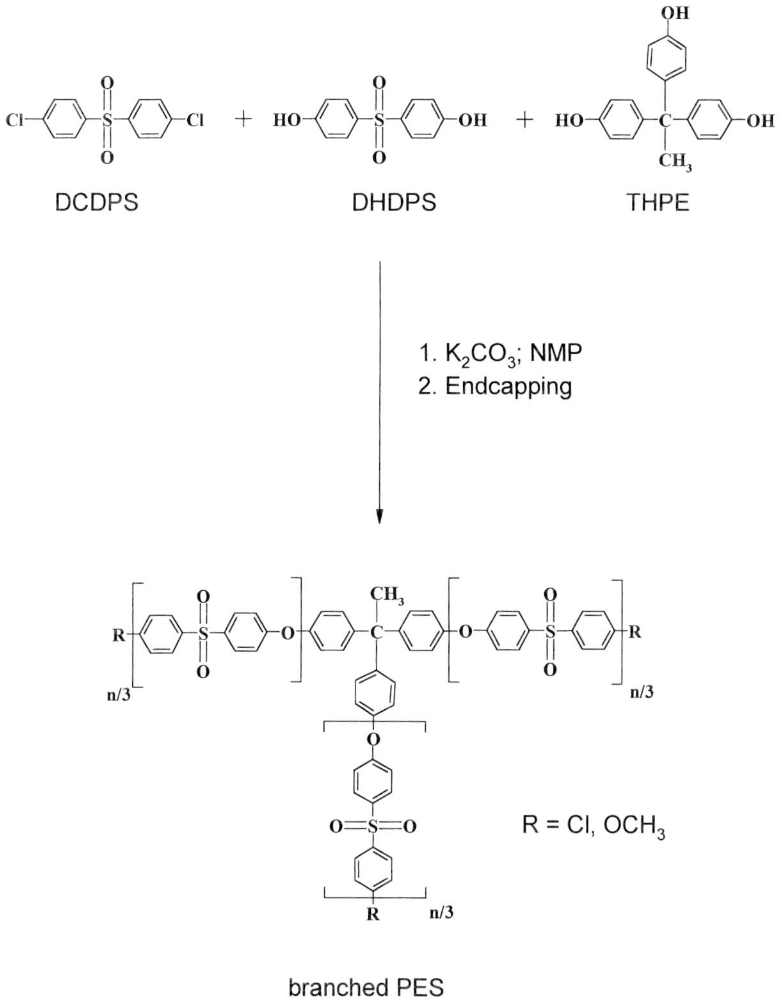

DCDPS DHDPS THPE

1. K_2CO_3; NMP
2. Endcapping

$R = Cl, OCH_3$

branched PES

Fig. 1. Reaction scheme for the synthesis of branched polyethersulfone.

Results and Discussion

Branched Polyethersulfones

Using a standard process for the preparation of Polyethersulfone (1 mol monomers in 1000 ml N-methylpyrrolidone (NMP), 1 mol K_2CO_3, 6h 195°C), products with THPE amounts from 0,5 to 3 mol-% were prepared. In the first experiments, the ratio between OH- and Cl-end-groups was not adjusted. After a condensation time of 6 h the products were isolated by precipitation in water and characterised. As can be seen from Figure 2, the viscosity number (V.N., determined at 25°C, 0,5 g product in 100 ml NMP), as a measure for the molecular weight of the obtained products, decrease with increasing amount of charged THPE as a consequence of the stochiometric ratio. The scattering of the data might be due to differences in the reaction conditions (monomer purity, heating rate, etc.).

The amount of incorporated THPE in the obtained branched polyethersulfones (PES-b) can be determined by ^{1}H-NMR-spectroscopy (Figure 3). The protons of the methyl group of THPE can be used to calculate the amount of incorporated THPE. The obtained data (Table 1) indicate, that only 70 % of the charged THPE is incorporated into the branched polyethersulfons.

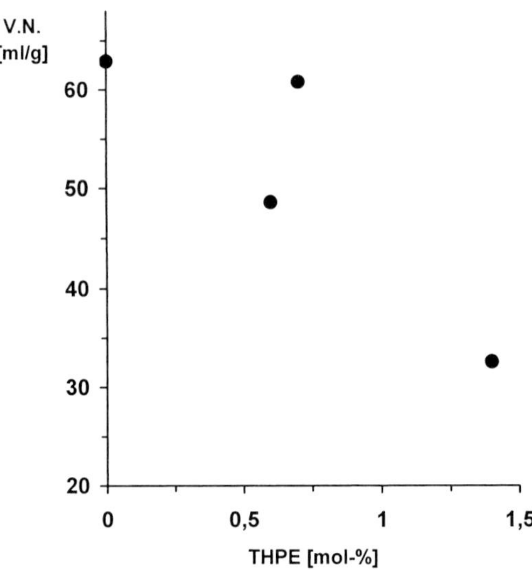

Fig. 2. Viscosity number as a function of incorporated THPE in branched PES.

The product with the THPE-content of 0,7 mol-% and the linear PES have almost the same viscosity number (Figure 2), subsequently these two samples were used for rheological measurements. Using a capillary rheometer, the apparent melt viscosity as a function of shear rate was determined for these two products at 350°C (Figure 4).

Table 1. Results of the characterisation of branched PES.

Product*	THPE [mol-%]		V.N.	Tg
	charged	found	[ml/g]	[°C]
PES	---	---	62,9	229
PES-b1	0,5	0,6	48,6	223
PES-b2	1	0,7	60,8	226
PES-b3	2	1,4	32,6	215
PES-b4	3	n.b.	non soluble	n.b.

* 6h condensation time

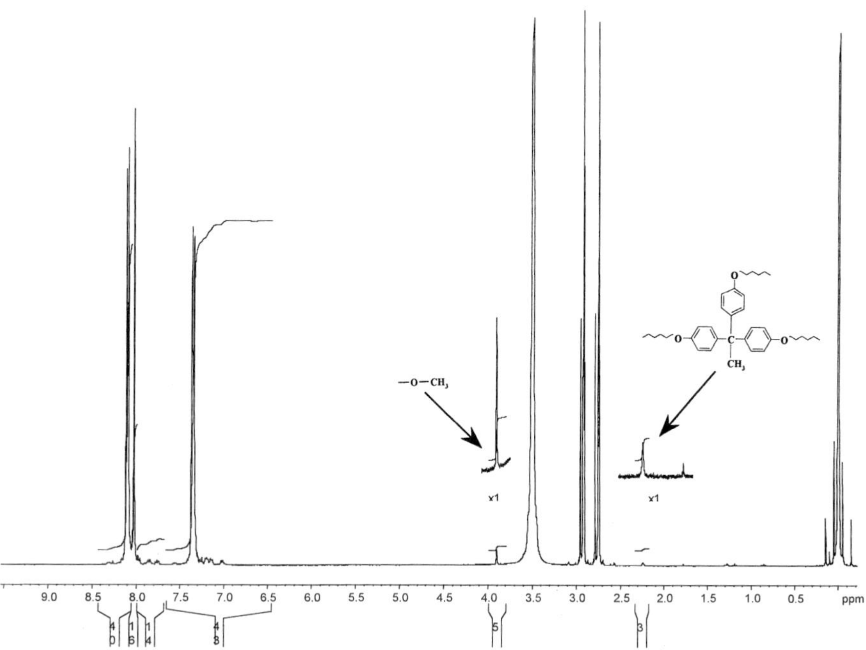

Fig. 3. ¹H-NMR-spectrum of PES-b2 (d_6-DMF as solvent), δ= 2,3 ppm: CH_3 of THPE unit, δ=3,9 ppm: CH_3 of Methoxy-endgroups .

As can be clearly seen from Figure 4, the branched product has a lower melt viscosity as the linear product with the same viscosity number. From these small-scale trials it was concluded, that the desired reduction of the melt viscosity could be achieved by incorporation of THPE.

In order to have enough material for mechanical testing, branched PES-samples were prepared on a 1 kg-scale. To have a better control of the reaction, the stochiometric ratio between the di-chloro and the hydroxy-monomers was adjusted to be 1/1 in these experiments. The amount of charged THPE was again varied from 0 to 3 mol-%. The reaction time was adjusted according to the THPE-content. The obtained polymers were isolated by precipitation in water. After drying, the powders were compounded on a Prizm PTW 16 extruder to yield granulates. The barrel temperature was set at 350 °C; the throughput was 1 kg/h at a screw speed of 200 rpm.

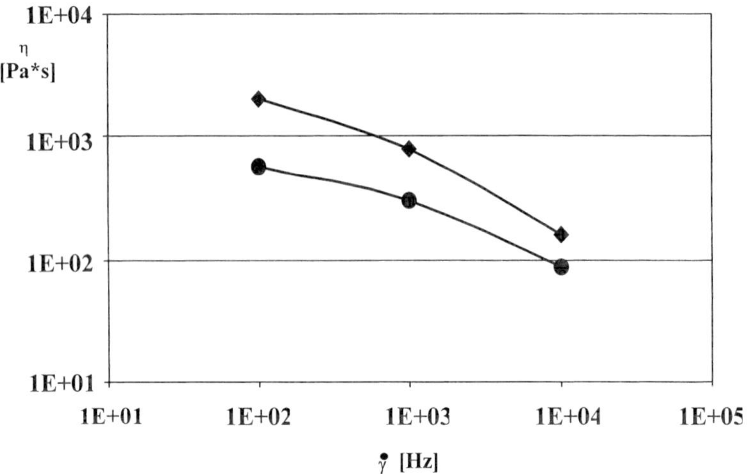

Fig. 4. Flow curves of a linear ▲ (PES) and a branched • (0,7 mol-% THPE, PES-b2) Polyether-sulfone at 350°C.

From the granulates ISO-samples for impact (Charpy impact: ISO 179 1eU, Charpy notched impact: ISO 179 1eA), tensile testing (ISO 527) and Vicat B (ISO 306) measurements were injection moulded at a melt temperature of 350°C and a mould temperature of 120°C.

The viscosity number and the THPE-content of the obtained products are summarised in Table 2.

Table 2. Results of the characterisation of branched PES.

	Condensation time [h]	THPE [mol-%] charged	THPE [mol-%] found	V.N. [ml/g]
PES	3,5	0	0	44,8
PES-b5	3,75	1	0,7	44,7
PES-b6	3	2	1,5	44,6
PES-b7	3,5	2	1,5	64,8
PES-b8	4	2	1,6	88,9

Three products obtained after different reaction times have almost the same viscosity number (45 ml/g). The properties of these three samples can be compared to study the influence of the amount of incorporated THPE on the mechanical performance. It can be clearly seen, that the incorporation of THPE has almost no influence on the Vicat B softening temperature and the E-modulus (Figure 5a and b), while the toughness is dramatically reduced by the incorporation of the branching units (Figure 6a and b). Especially the Charpy impact shows an immediate drop to a level below the 50% value of the linear product with the same viscosity number. In the case of the notched impact, a decrease of the toughness with increasing amount of THPE was also observed.

From comparative trials with linear PES the influence of the viscosity number on the toughness of PES was known. Below a viscosity number of 40 ml/g a tremendous reduction of the toughness occurs (Figure 7a). Since the samples discussed have only a slightly higher viscosity number than 40 ml/g, the influence of the viscosity number on the mechanical performance at a fixed THPE-amount was studied as well.

For the products with about 1,5 wt.% THPE (Table 2), only a slight dependence of the Charpy impact with increasing viscosity number is visible (Figure 7a). The level of the linear products is not achieved, even at very high viscosity number.

For the Charpy notched impact a significant increase of the toughness with increasing viscosity number was observed (Figure 7b). For a high viscosity number (> 65 ml/g), the toughness level of linear products was achieved.

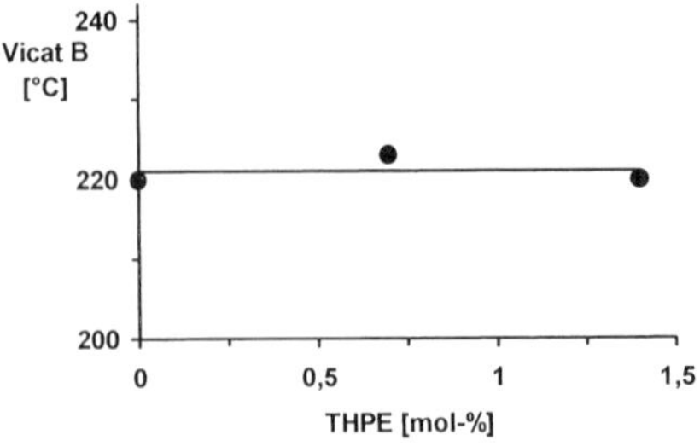

Fig. 5a. Vicat B softening temperature as a function of the THPE-content for PES.

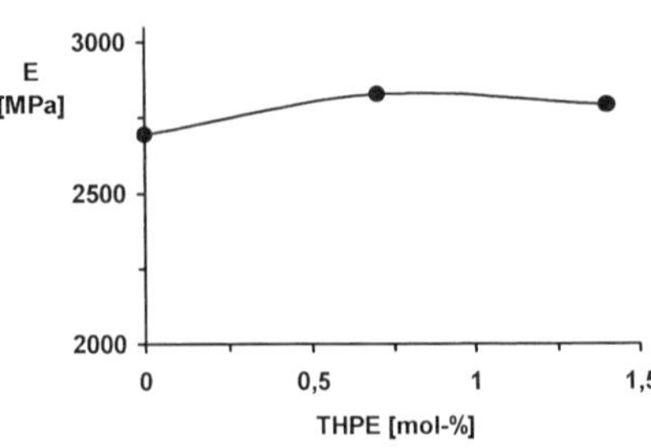

Fig. 5b. E-Modulus as a function of the THPE-content for PES.

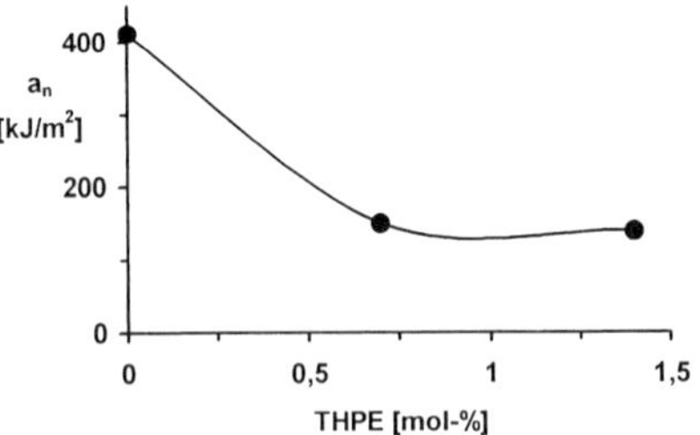

Fig. 6a. Correlation between the Charpy impact toughness and the THPE-content for PES.

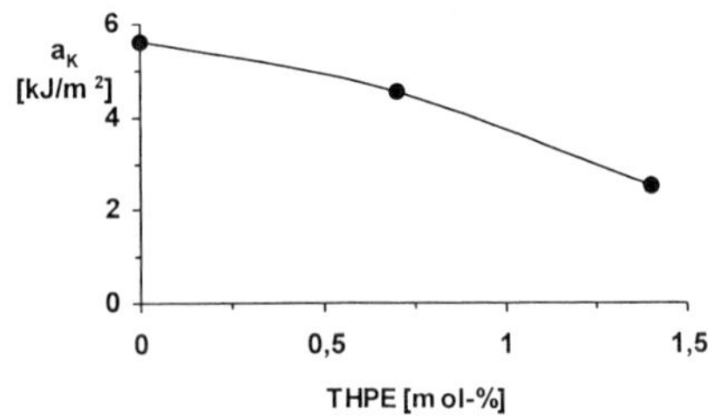

Fig. 6b. Correlation between the Charpy notched impact toughness and the THPE-content for PES.

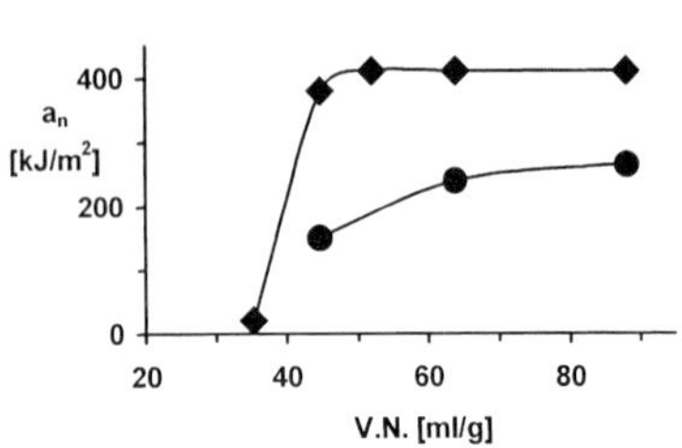

Fig. 7a. Relationship between Charpy impact toughness and viscosity number for linear and branched PES (1,5 mol-% THPE).

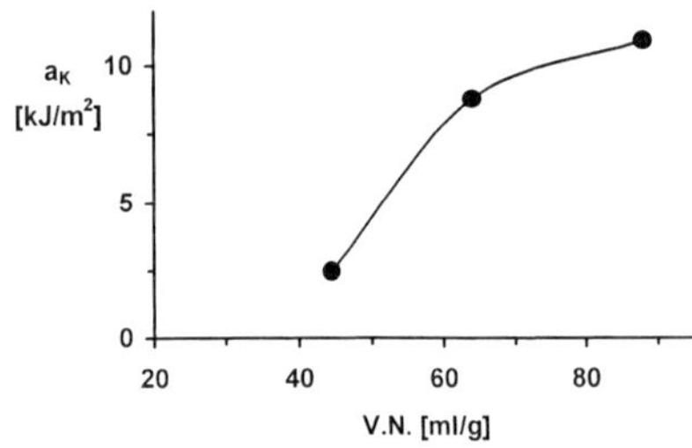

Fig. 7b. Relationship between the Charpy notched impact and viscosity number for branched PES (1,5 mol-% THPE).

In the Charpy impact test without notch, the energy to initiate a crack as well as the energy to drive the crack through the material is characterised. Since the difference between linear and branched PES is less pronounced in the Charpy notched impact, the observed behaviour in the Charpy impact is mainly related to the crack initiation. Polyarylethers usually belong to the group of pseudo-ductile polymers having a high entanglement density[15]. The observed reduction in toughness can thus be explained in terms of a reduced entanglement density as a consequence of the branching and a higher M_e-value. On the other hand it has to be mentioned, that due to the low molecular weights of these products (M_n appr. 6000 - 12000 g/mol) and the low amounts of incorporated THPE, not every chain has one branching unit.

Conclusions

The incorporation of 1,1,1-Tris-(4-hydroxyphenyl)ethan (THPE) in the Polyethersulfone chain leads to branched materials with reduced melt viscosity compared to linear material with a comparable viscosity number. ^{1}H-NMR reveals that only about 70 % of the THPE is incorporated in the obtained product. As a consequence of the THPE-incorporation, the toughness of the branched PES is significantly lower than for linear material. Especially the Charpy impact and the tensile elongation of branched PES are pretty low. Branched PES with high viscosity number on the other hand has the same notched impact strength as linear products. Branching obviously disturbs the entanglement of the PES-chains, enhancing the sensitivity for crack formation.

Acknowledgements

The contributions of P.Liebig (mechanical testing) and D. Garau (synthesis) are acknowledged. The authors also wish to thank B. Cunningham for her technical assistance.

[1] R.N. Johnson, A.G. Farnham, R.A. Clendinning, W.F. Hale, and C.N. Merriam, *J. Polym. Sci. A 1* 1967, *5*, 2375.
[2] B.E. Jennings, M.E.B. Jones, and J.B. Rose, *J. Polym. Sci. C* 1968, *16*, 715.
[3] M.E.A. Cudby, R.G. Feasy, B.E. Jennings, M.E.B. Jones, and J.B. Rose, *Polymer* 1965, *6*, 598.
[4] J.P. Critchley, G.J. Knight, W.W. Wright, *"Heat-Resistant Polymers"*, Plenum Press, New York and London 1983.
[5] E.M. Koch, H.-M. Walter, *Kunststoffe* 1990, *80*, 1146.
[6] E. Doering, *Kunststoffe* 1990, *80*, 1149.
[7] J. Queisser, M. Geprägs, R. Bluhm, G. Ickes, *Kunststoffe* 2002, *92*, 90.
[8] M Pahl, W. Gleißle, H.-M. Laun, *"Praktiche Rheologie der Kunststoffe und Elastomere"*, VDI-Verlag 1991, p. 347.
[9] H.M. Laun, *Progr. Coll. Polym. Sci.* 1987, *75*, 111.
[10] P. Guinot, *Proc. New Plastics 2001 Conference* 2001.

[11] G. Kiss, *Polym. Eng. Sci.* 1987, *26*, 410.
[12] M.H.B. Skovby, J. Kops, and R.A. Weiss, *Polym. Eng. Sci.* 1991, *31*, 954.
[13] M. Weber in: *"POLYMER BLENDS AND ALLOYS"*, G.O. Shonaike, G.P. Simon, Eds., Marcel Dekker, New York, Basel 1999, p. 270.
[14] M. Weber, U.Eichenauer, J. Queisser, *Kunststoffe*, 1998, *88*, 1472.
[15] S.Wu, *Polymer Int.*, 1992, *29*, 229.
[16] DE 27 35 092 (1977), Bayer AG, R. Binsack, E. Reese, J. Wank.
[17] R.N. Johnson, A.G. Farnham, R.A. Callendinning, W.F. Hale, C.N.J. Merian, *J.Polym. Sci., Polym. Chem. Ed.*, 1967, *5*, 2375.

Lyotropic Para-Linked Aromatic Poly(amic ethyl ester)s

*Christian Neuber, Reiner Giesa, Hans-Werner Schmidt**
Makromolekulare Chemie I und Bayreuth Center for Colloids and Interfaces (BZKG), Universität Bayreuth, D-95440 Bayreuth, Germany
E-mail: hans-werner.schmidt@uni-bayreuth.de

Summary: The synthesis, characterization and application of lyotropic precursor polymers for polyimides, poly(amic ethyl ester)s (PAE) are presented. By the use of non-coplanar and para-linked monomers, rigid-rod PAEs are synthesized in a typical polycondensation reaction yielding liquid crystalline solutions at concentrations of 30-40 wt% in NMP at 100 °C. These lyotropic solutions permit for the first time the fabrication of orientation PAE layers by shear which are thermally converted to the corresponding polyimide. The bulk orientation in the thin films is characterized by spectroscopic methods. In addition the lyotropic solution is spun into fibers in a dry-jet wet spinning process. The obtained fiber orientation is characterized by tensile tests and wide angle X-ray scattering.

Keywords: nanolayers, orientation, polyimides, solution properties, structure-property relations

Introduction

The macroscopic orientation of polyimides and their precursor polymers in bulk and at the surface is important for a technical as well as an academic interest. In the course of our research we are investigating the development of intrinsically anisotropic precursor systems, their orientation and application possibilities.

Since several years poly(amic ethyl ester)s (PAE) are used increasingly instead of poly(amic acid)s (PAA) as precursors for polyimides.[1-4] PAEs have improved solubility, better resistance to hydrolytic degradation, yield high molecular weight products, but require higher and broader imidization temperatures. For technical applications based on PMDA, such as insulating layers in computer chips, only the meta isomer is employed since the good solubility of the resulting precursor polymer permits the preparation of concentrated solutions. PAEs based on the para isomer only should exhibit similar behavior as it is well established for lyotropic para-linked polyaramides.[5-7] As shown in Scheme 1 the incorporation of flexible elements such as meta or oxygen links results in flexible precursor systems. In contrast the usage of para links in

 DOI: 10.1002/masy.200350922

combination with linear non-coplanar diamines results in an extended rod-like precursor polymer. For PAAs no lyotropic phases have been found for PMDA/4,4'-ODA up to 30wt% solvent content which was attributed to the existence of the para and meta isomer within the PAA backbone. Indications of lyotropic behavior was described for the first time in the PAA of PMDA/4,4'-MDA. At room temperature this system turns opaque over 15wt% solid content in NMP and shows at 20wt% spherulitic crystals.[8] A few years ago we reported[9] the first para-linked aromatic PAE that exhibits a lyotropic liquid crystalline phase. By use of THF as cosolvent in combination with NMP, at 35-55wt% in NMP lyotropic solutions were observed. after evaporation of the volatile cosolvent.[10]

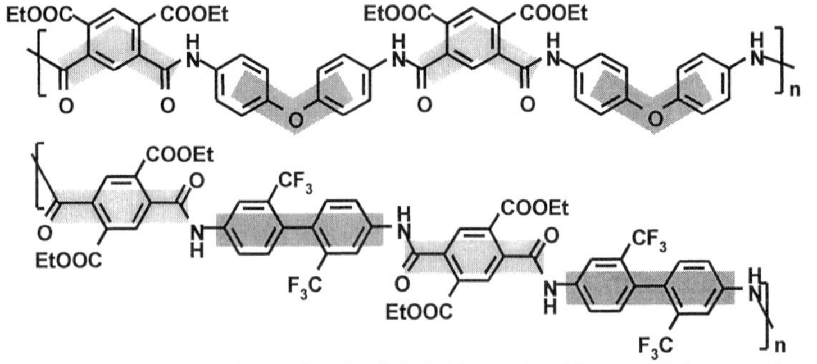

Scheme 1. Chain conformations of a flexible PAE (top) with meta-linked amid units and kinked, flexible diamines compared to an extended PAE (bottom) with para-linked amid units and rod-like non-coplanar diamines.

Polyimides are commonly used as orientation layers in nematic liquid crystal displays (LCD). The planar alignment of the LC director is achieved by rubbing (buffing) the polyimide layer.[11] The mechanism of the orientating effect is still subject of many investigations. Most polyimides used for orientation layers possess a more or less flexible structure. The development of lyotropic rigid-rod PAEs permits for the first time the investigation of PAE and polyimide layers obtained additionally by orientation methods such as shear. This opens up also a method of preparing PI orientation layers without a separate buffing processing step. Furthermore structure-property relationships comparing rod-like (para) to flexible (meta) PAEs can be developed.

Another area of utilizing the orientation development in PAE and/or PI is the processing into fibers and subsequent drawing. The spinning of lyotropic solutions of polyaramids into fibers and their orientation was thoroughly investigated for the commercialization of Kevlar® fibers.[5] In the case of para-linked PAEs the high aspect ratios of fibers can be advantageously used to induce orientation by i) starting with nematic lyotropic solutions and ii) drawing the obtained fibers before and during imidization.

Synthesis of Para-linked Poly(amic ethyl ester)s

All PAEs were synthesized according to literature procedures[1-4, 9] as shown in Scheme 2. PMDA was refluxed in excess ethanol to form an isomeric mixture of para (**1a**) and meta (**1b**) isomers. The key step of separating the isomers[4] is based on the lower solubility of **1a** which can be isolated and purified by a series of extraction and recrystallization to 96% (^{1}H-NMR). **1a** can be easily converted to the acid dichloride **2a** with excess oxalyl chloride in chloroform. The polycondensation to the PAEs was performed in anhydrous NMP at 5wt% polymer concentration as described in ref. [9] An in-situ activation of the diamine **3** was achieved by adding molar amounts of chlorotrimethylsilane (TMSCl). The acid dichloride **2a** was added to this mixture and after 4 h the PAEs **4a-e** were precipitated in water, extracted with acetone and dried, yielding almost quantitative conversions.

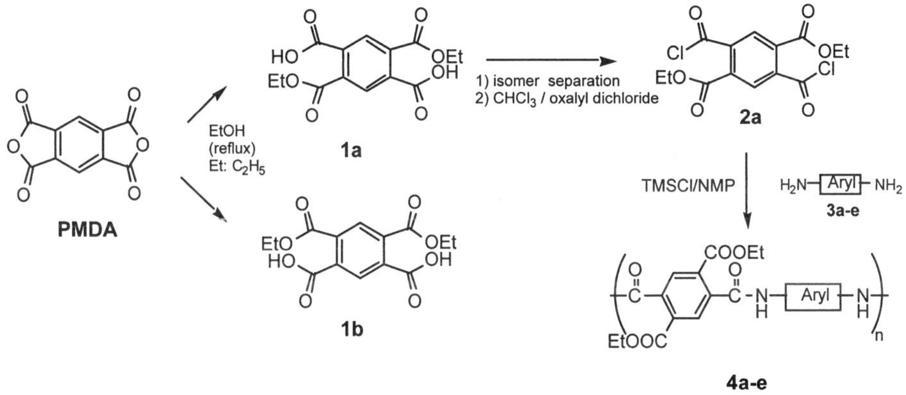

Scheme 2. Synthesis of para-linked rigid-rod PAEs **4a-e**.

Solution Properties and Lyotropic Behavior

All synthesized PAEs form isotropic solutions up to 20wt% and are completely soluble in NMP without the addition of inorganic salts such as LiCl. The inherent viscosities in NMP at 30°C are listed in Table 1 and range between 0.30-1.88 dl/g. The GPC (gel permeation chromatography) in DMAc/LiCl at 70°C revealed number average molecular weights in the range of 30000-50000 g/mol except **4e**. However, the calibration with polystyrene standards results in too high values since polyaramids are considerably more rigid. The low inherent viscosity/molecular weight of the polymer **4e** seems to be the consequence of the high degree of fluorination and the reduced reactivity of the diamine. For the formation and examination of lyotropic systems the polymers **4a-e** were dissolved in NMP at 10wt% without a co-solvent and then concentrated by evaporation at 100°C. The final concentration was determined gravimetrically. All PAEs turned opaque above 30wt% and in some cases liquid crystalline behavior at elevated temperatures or under strong shear was found.

Table 1. Molecular weight characterization of PAEs **4a-e** by viscosimetry and GPC

PAE 4	a	b	c	d	e
–[Ar]–:	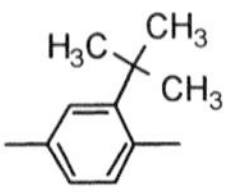 H_3C, CH_3, CH_3	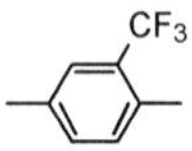 CF_3	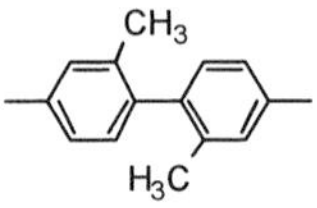 CH_3, H_3C	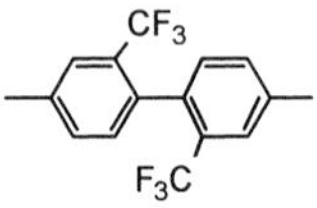 CF_3, F_3C	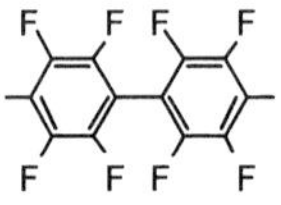 F F F F, F F F F
η_{inh}[a)] [dl/g]	1.00	1.02	1.88	1.86 (para) 0.66 (meta)	0.30
M_n (GPC)[b)]	29000	34000	39000	47000 (para) 49000 (meta)	18000
PD[c)]	2.2	2.1	2.5	2.7 (para) 2.1 (meta)	1.9

a) In NMP at 30°C and 0.5 g/dl.
b) In DMAc/0.05M LiCl at 70°C, polystyrene calibration
c) Polydispersity (M_w/M_n)

Rod-like aromatic polyamides and consequently the ones investigated in this study have the strong tendency to form gels. Table 2 summarizes the temperature ranges for the melting of the PAE-gels either to an isotropic (iso) or a lyotropic (lyo) solution. The observation under a polarized light microscopy is possible up to 140°C before major solvent evaporation occurs and imidization progresses. In most cases for concentrations below 30wt% the birefringent gels melt to an isotropic non-birefringent solution. However at higher concentrations around 40wt% the gels melt into a liquid crystalline solution. After annealing a typical nematic Schlieren texture is observed. The transition of the gel state to the isotropic or lyotropic solution is not sharp but extended over a wide temperature range of 20-30°C and coexistence of a gel and isotropic, or gel and lyotropic biphase is observed (Fig. 1).

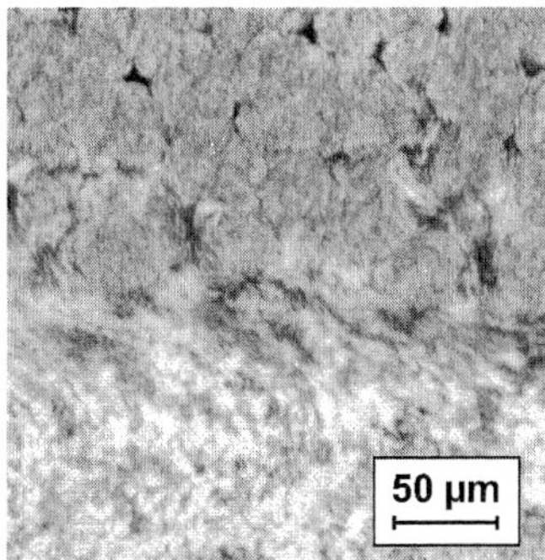

Fig. 1. Spherulitic gel (top) and lyotropic (bottom) domains coexisting in PAE **4d** at 40wt% in NMP at the transition of gel to lyotropic solution (70°C).

Table 2. Temperature ranges in °C of the melting of **4a-e** either to an isotropic (iso) or a lyotropic (lyo) solution determined at a polarization light microscope up to 140°C.

	Concentration/wt%				
4	10	20	30	40	50
a	iso	80-95⇨iso	90-110⇨iso	**105-125⇨lyo**	gel[a)]
b	iso	iso	30-50⇨iso	gel[a, b)]	gel[a, b)]
c	50-80⇨iso	80-110⇨iso	**100-120⇨lyo**	**110-130⇨lyo**	**120-140⇨lyo**
d	iso	iso	60-80⇨iso	**50-80⇨lyo**	**60-100⇨lyo**
e	iso	iso	gel[a, c)]	gel[a, c)]	gel[a, c)]

a) No melting of the gel up to 140°C.
b) After strong shearing a lyotropic phase was observed which gels again within minutes.
c) After strong shearing a lyotropic phase was observed which is stable for months.

For **4d** which contains a non-coplanar 2,2'-di(trifluoromethyl)biphenylene diamine unit the gel formation was first observed at 30wt%. The gel melts between 60 and 80°C into an isotropic solution. At 40 and 50wt% above the gel melting temperature in the range of 50-80°C and 60-100°C, respectively, a lyotropic solution with a typical nematic texture was found. A hypothesis to explain the unusual gelation behavior of the solutions of **4b** and **4e** could be the substitution position of the fluorine atoms at the diamine unit. In both cases it seems to be possible that under shear the fluorine atoms form predominately *intra*molecular hydrogen bonds with the N-H-groups of the adjacent amid linkage, thus supporting the formation of a lyotropic solution. Whereas from isotropic solutions and without shearing a substantial fraction of *inter*molecular hydrogen bridges is formed between the polymer chains resulting in macroscopic gelation. In conclusion, it is evident that the capability of para-linked aromatic PAEs to form stable lyotropic solutions is depending on the particular diamine used. The critical concentration for lyotropism for such polymers is in the range of 30 to 40wt%. The lyotropic phase of **4d** was characterized in great detail by polarization microscopy, DSC, temperature dependent X-ray diffraction, and FT-IR spectropscopy, the results are summarized in ref.[15] Here only a comparison of the surface morphology by SEM is depicted in Figure 2. The meta-linked PAE **4d** forms an isotropic solution and the dried gel shows no distinct surface morphology (Fig. 2, a). However the para-linked **4d** forms a lyotropic solution and here the dried gel forms a distinct 'ground beef' morphology and worm-like structural features are visible (Fig. 2, b).

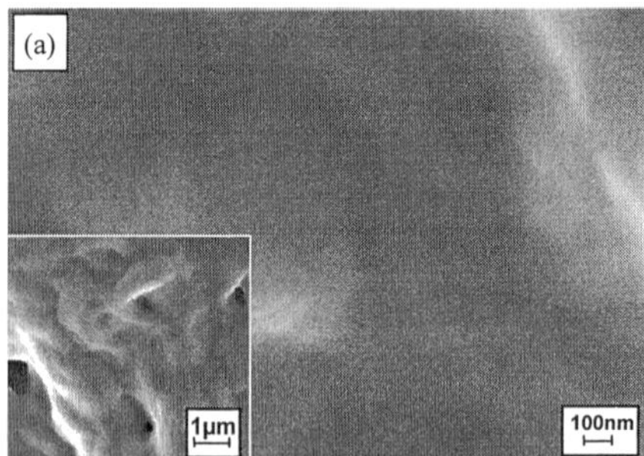

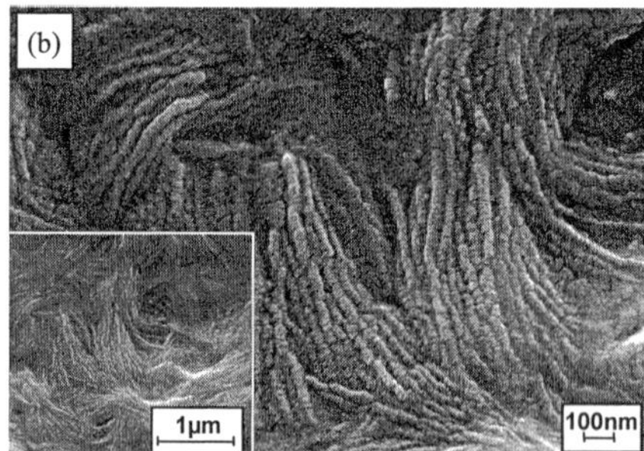

Fig. 2. Scanning electron microscopy images obtained of dried gels of PAE **4d** at a concentration of 50wt%. (a): meta-**4d**,(b): para-**4d.**

Orientation of PAEs in Thin Films by Shear

One objective of the conducted research is to compare the orientation behavior of PAE layers originated from lyotropic para-**4d** solutions with layers obtained from isotropic solutions, e. g. the meta isomer of **4d** or commercial isotropic PI kits[12] for the preparation of orientation layers. For layer preparation the PAE are concentrated up to 50wt% in NMP and heated into the lyotropic phase around 100°C. Then the lyotropic solution was sheared onto a cleaned glass substrate with a doctor blade constructed of a bundle of parallel glass rods which were pressed on the glass substrate while moving with 25 mm/s. This technique permits the fabrication of highly oriented PAE thin films in a thickness range of 20 to 60 nm with high reproducibility. The thus obtained PAE layers were dried at 100°C, characterized, and thermally converted to the PI.

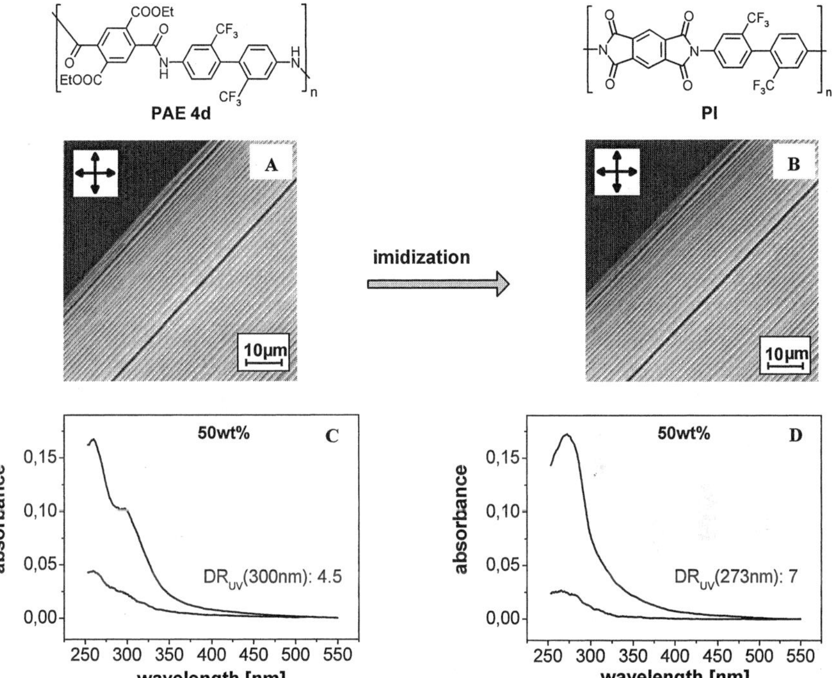

Fig. 3. Polarized light microscopy of para-**4d** (A) and the corresponding imidized structure (B). The layers were characterized by polarized UV/Vis spectroscopy (C, D). The resulting dichroic ratios are plotted in the graphs.

To characterize the obtained orientation the layers were investigated with polarizing techniques, such as light microscopy, FT-IR and UV/Vis spectroscopy.[13] Figure 3 shows the polarized light micrograph of the layer prepared with this shearing technique using the lyotropic solution of para-**4d**. The shearing (and orientation) axis of the layer is rotated at 45° since complete extinction is observed at 0 and 90°. No differences are visible after imidization. Polarized UV/Vis spectroscopic measurements (Figure 4) show that the orientation is fully preserved and the dichroic ratio, as indicator for orientation, is increased after imidization. In the polarization micrographs shown in Figure 3 the layers show striations in shearing direction. These striations where further characterized by a mechanical scanning profiler and SEM (Figure 4 and 5).

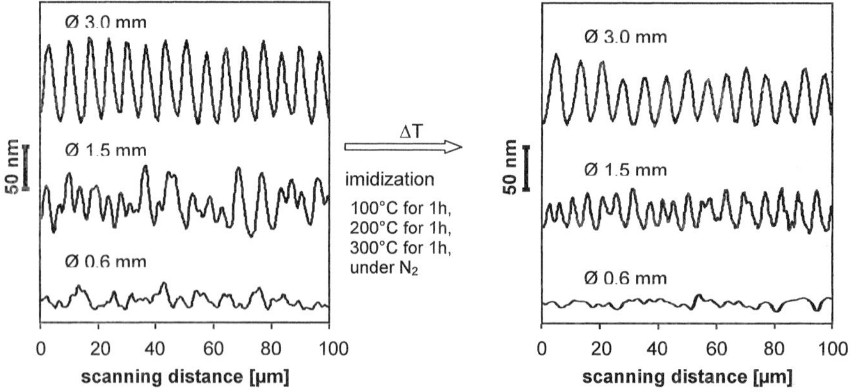

Fig. 4. Mechanical scanned surface profiles of para-**4d** orientation layers prepared by the shearing technique using glass rods with different diameters before (left) and after imidization (right).

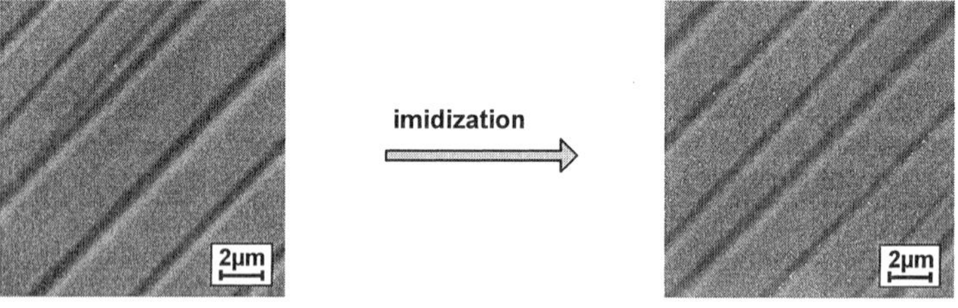

Fig. 5. Scanning electron microscopy (SEM) of oriented layers prepared by shearing with 3 mm glass rods. Left: precursor polymer para-**4d**; right: after imidization (1h at 100°C, 1h at 200°C, 1h at 300°C, 10 K/min; under nitrogen). The darker diagonal lines represent the raised profile.

The distance between the maxima of the striations is around 2 μm. The film shows no dewetting and completely covers the glass slide. The average film thickness below the profile is around 25 nm and the profile height is in the range of 40 nm.

Fiber Spinning from Lyotropic Solutions

PAE monofilaments of para-**4d** were spun from a 40 wt% lyotropic solution in NMP in a dry jet-wet spinning process.[14] At this time no air gap was used and spinning rates were low i. e. 3m/min. In preliminary spinning experiments several coagulation solvents and spinning conditions were tested (Fig. 6) and mechanical fiber properties were evaluated as function thereof.

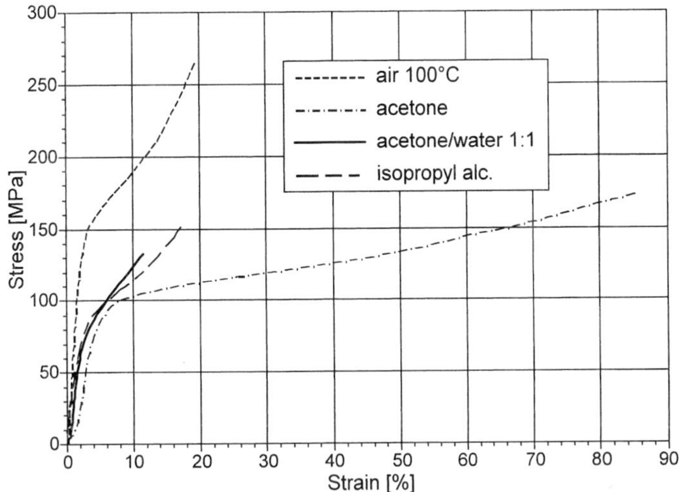

Fig. 6. Stress-strain curves for para-**4d** fibers coagulated in different solvents.

Water is not suitable and fibers with extremely poor mechanical properties are obtained. Acetone yields fibers with lower modulus but very high elongation probably due to a plasticizing effect. An acetone/water 1:1 mixture reduces the elongation dramatically, even at lower water contents. Dry spinning results in fibers with high tenacities and a twofold increase in modulus. Figure 7 shows SEM micrographs of an acetone coagulated PAE fiber. The fiber surface is smooth and unstructured, the fracture surface shows no fibrillation and a very porous morphology is revealed. Based on these results spinning was first attempted as dry spinning. However the spinning set-up available which is optimized for dry-jet wet-spinning did not yield sufficient amounts of continuous fibers. Therefore acetone was used as coagulation medium and in a continuous spinning homogenous fibers with sufficient wet-strength were obtained. Also the fibers could be drawn during spinning in the acetone bath with draw ratios of 2-3. Table 3 lists the obtained mechanical properties of fibers spun from para-**4d** at 100°C and 40wt% with acetone as coagulation bath. All samples have been dried at 50°C in vacuum for 24h before testing. The fibers just coagulated in acetone show high elasticity and thus the highest elongation. Drawing at 100°C in a continuous process with 2.5 cm/sec in air does not change the fiber properties dramatically thus all solvent was removed during drying. Successive heat treatment at 200°C toughens the fiber shown by the increased modulus, tenacity and decreased elongation. This indicates a progressing imidization. After the last treatment process at 300°C the imidization is more complete, modulus and tenacity increased again, the elongation is dramatically decreased indicating an embrittlement of the material. The fiber which was directly drawn in acetone shows prior to any heat treatment high mechanical performance, however accompanied with a low elongation which impede further drawing steps.

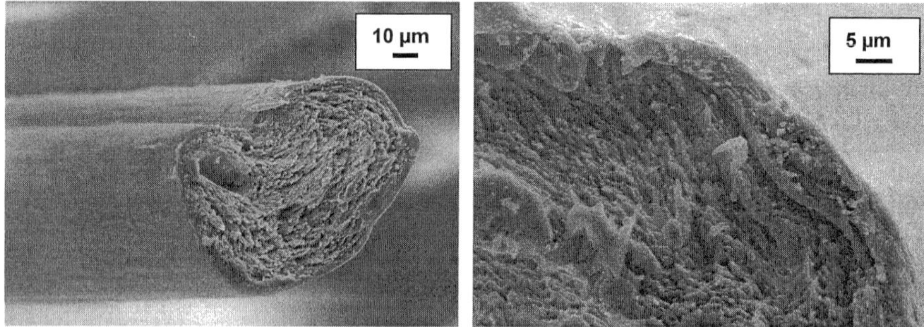

Fig. 7. Fracture surface of a para-**4d** fiber coagulated as-spun in acetone. The left micrograph shows a detailed view of the layered skin-core morphology of the fiber.

Table 3. Mechanical properties of para-**4d** fibers spun in a dry-jet wet-spinning process at 100°C. Data are the average of eight single measurements.

Fiber sample	**(1)**	**(2)**	**(3)**	**(4)**	**(5)**	**(6)**
Process	as spun and coagulated in acetone	sample (1) drawn at 100°C	sample (2) drawn at 200°C	sample (3) drawn at 300°C	drawn in acetone during spinning	sample (5) imidized
cross-head speed [mm/min]	4	4	4	2	2	1
Fiber diameter [µm]	96±8	96±7	84±9	81±9	59±3	54±2
Modulus E [GPa]	6.4±1	6.21±0.7	10.62±2.1	15.8±2.9	23.3±1.8	37.5±5
Tenacity [MPa]	212±37	203±30	286±67	307±51	539±62	547±136
Elongation at break [%]	23.8±2.6	26.0±4.1	13.1±2.9	3.4±0.6	3.4±0.2	1.7±0.4

In Figure 8 the WAXS fiber patterns of sample (1), left, and (5), center, and (6), right, are depicted. The intrinsically much higher orientation of sample (5) is evident by sharper equatorial reflexes whereas the increased meridional reflexes indicate also a better alignment of the polymer chains in fiber axis. This orientation is further improved during imidization in fiber (6). To study the improvement of orientation during drawing more experiments with the imidized fiber are in progress.

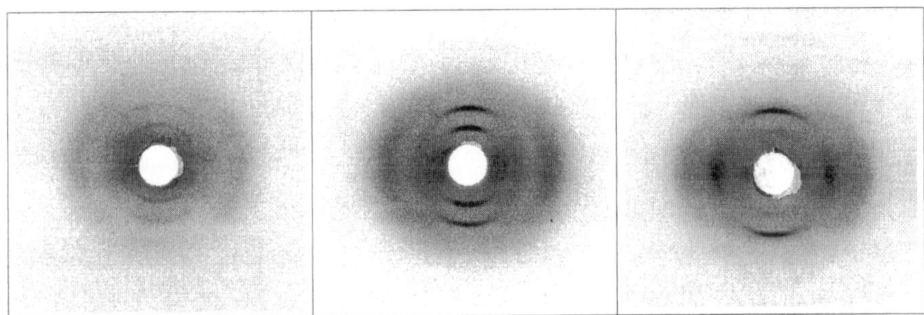

Fig. 8. Wide-angle X-ray diffraction pattern of para-**4d** fibers coagulated in acetone (1) (left) and the same polymer coagulated and drawn in acetone (5) (center). Fiber (5) was finally imidized, forming fiber (6) (right).

Experimental

The synthesis, detailed analytical and mesomorphic characterization of PAEs **4a-e** is published in ref.[15] The orientation of the lyotropic solution of **4d** was performed on oxygen plasma etched glass slides with a DACA Instr. Tribotrak machine. A pressure of 1 MPa at a moving speed of 25 mm/sec at 100°C was used. Polarized optical microscopy was performed on a Leitz Laborlux 12-Pol equipped with a Mettler FP 82 hot stage. UV/Vis spectra were recorded on quartz glass slides on a Hitachi U-3000 spectrophotometer with Glan-Thompson polarizers. Layer thickness and profile were scanned with a Dektak 3030 Step Profiler. Fiber spinning experiments were conducted with a solution fiber spinning line similar to the equipment described in ref.[14] A 100 μm spinnerette was used. For heat setting experiments a custom built drawing rig was employed. With this frequency controlled equipment the fiber can be drawn at high reproducibility. Imidization was performed on a metal spool at 100°C for 1h, 200°C for 1h and 300°C for 1h with a ramping time of 10K/min. Fibers were tested according to ASTM D3379 in an Instron 5565 universal tester equipped with a 10N load cell. The complete fiber spinning process and fiber characterization will be published elsewhere.[16] SEM micrographs were recorded on a LEON 1530 scanning electron microscope after sputtering with Pd/Pt. WAXS diffraction pattern were obtained on a Seifert Isoflex 3000 (CuKα at 1.5kW) equipped with a fiber set-up in a Laue camera.

Conclusions

Lyotropic precursor solutions for polyimides based on para-linked aromatic poly(amic ethyl ester)s were prepared and characterized. Lyotropic nematic phases in NMP were observed at concentrations above 40 to 50wt%. Then the lyotropic solutions were sheared on a substrate and oriented layers in the thickness range of 20-40 nm were obtained. This technique permits the fabrication of bulk orientation layers without a separate buffing/rubbing step. The UV/Vis measurements resulted in dichroic ratios of 4.5 for the PAE orientation layer and an increase to 7 for the imidized layer. Furthermore the lyotropic solutions were spun into fibers in a dry-jet wet-spinning process. The mechanical properties of the fibers were investigated as function of processing conditions and thus the fiber orientation. By drawing in acetone a fiber with three times higher modulus and doubled tenacity was obtained.

Acknowledgement

This work was supported by the German Science Foundation (SFB 481, Projekt A6). We are indebted to C. Abetz (BIMF) for assistance with the SEM micrographs.

[1] M. Konieczny, H. Xu, R. Battaglia, S. L. Wunder, W. Volksen, *Polymer* ***38***, 2969 (1997).
[2] W. Volksen, D. Y. Yoon, J.L. Hedrick, D. Hofer, *Mat. Res. Soc. Symp. Proc.* ***227***, 23 (1991).
[3] W. Volksen, D.Y. Yoon, J.L. Hedrick, *IEEE Trans. of Comp., Hybrids, and Manuf. Techn.* ***15***, 107 (1992).
[4] W. Volksen, *Adv. Polym. Sci.* ***117***, 111 (1994).
[5] H. H. Yang, *Kevlar aramid fiber*, J. Wiley & Sons, Chichester, 1993.
[6] W. Hatke, H.-W. Schmidt, *Makromol. Chem. Phys.* ***195***, 3579 (1994).
[7] W. Hatke, H.-W. Schmidt , W. Heitz, *J. Polym. Sci. Polym. Chem.* ***29***, 1387 (1991).
[8] W. T. Whang, S. C. Wu, *J. Polym. Sci. Part A*, ***26***, 2749 (1988).
[9] K. H. Becker, H.-W. Schmidt, *Macromolecules* ***25***, 6784 (1992).
[10] K. H. Becker, H.-W. Schmidt, *ACS Polymer Preprints* ***35***(1), 349 (1994).
[11] K. Sakamoto, R. Arafune, N. Ito, S. Ushioda, Y. Suszuki, S. Morokowa, *Jpn. J. Appl. Phys.* ***33***, L1323 (1994); C.L.H. Delvin, S.D. Glab, S. Chiang, T.P. Russell, *J. Appl. Polym. Sci.* ***80***, 1470 (2001).
[12] Merck Liquicoat® PI-Kit ZLI-2650
[13] C. Neuber, R. Giesa, H.-W. Schmidt, *Adv. Materials*, submitted.
[14] G. C. Rutledge, U. W. Suter, C. D. Papaspyrides, *Macromolecules* ***24***, 1934 (1991).
[15] C. Neuber, R. Giesa, H.-W. Schmidt, *Macromol. Chem. Phys.* ***203***, 598 (2002).
[16] C. Neuber, R. Giesa, H.-W. Schmidt, *J. Appl. Polym. Sci.* (2003), to be published.

Liquid Crystalline Thermosets from Dimeric LC Epoxy Resins Cured with Primary and Tertiary Amines

*David Ribera, Ana Mantecón, Angels Serra**

Dept. Q. Analítica i Q. Orgànica. Universitat Rovira i Virgili.
Plaça Imperial Tàrraco 1 43005 Tarragona, Spain
E-mail: serra@quimica.urv.es

Summary: Liquid crystalline thermosets (LCTs) were prepared by curing difunctional LC dimeric epoxy monomers with imine moieties in the mesogenic core and central spacers of different length. Primary diamines or tertiary amines were used as curing agents obtaining materials with different characteristics. The results obtained were related to the mesogen structure, since dipolar moments in the mesogenic cores affect the ability to form ordered networks.

Keywords: crosslinking, epoxy resins, high performance polymers, liquid-crystalline polymers (LCP), networks

Introduction

There are two different ways to obtain liquid crystalline thermosets (LCTs). One of them is the thermal, chemical or photochemical crosslinking of high molecular LC polymers having reactive groups. This route is versatile, because many reactive groups can be crosslinked with several curing agents.[1] The second procedure, proposed by De Gennes,[2] consists of polymerizing a difunctional monomer, which reacts with a curing agent to form a tridimensional network. The advantage of this method is that monomers have a low viscosity, which facilitates processing and makes it possible to work at lower temperatures. Epoxy networks are widely applied as engineering thermosets because they have good mechanical and thermal properties. They have been extensively used to prepare LCTs.[3] The first references to epoxy LCTs appeared in the 1986, when large industrial groups patented their results.[4] It has been described[5,6] that the best characteristics of epoxy LCTs are their high order parameter and their high birefringence. Moreover, the ordering in the liquid crystal state facilitates a rapid transition to the solid phase with little molecular movement. This means that there is a reduction in the thermal expansion coefficient and the

 DOI: 10.1002/masy.200350923

shrinkage upon curing, which are problems inherent to the technological applications of epoxy resins.

In the nematic mesophase, monomers are easier to orient because of their low viscosity. The anisotropic networks obtained by fixing nematic mesophases are interesting as structural materials or as insulation for microelectronics. They are also suitable for electro-optical applications.[7] Although smectic networks have not been studied so much, they are described to be exceptionally tough.

The monomers studied in the present work are epoxy resins with dimeric or twin architectures consisting of two calamitic mesogens and central aliphatic spacers of different lengths. This architecture is chosen to lower the melting and isotropization temperatures and to increase the mesophase stability. Thus, the crosslinking can be carried out at relatively low temperatures. The mesogenic moiety contains aromatic imine groups, with a high mesogenic character and a high thermal stability.[8] This unit is easy to synthesize from commercially available products with high yields. At the end of the mesogenic unit, ether or ester groups link the mesogen to the aliphatic spacer and the glycidylic reactive group as is shown in the following scheme.

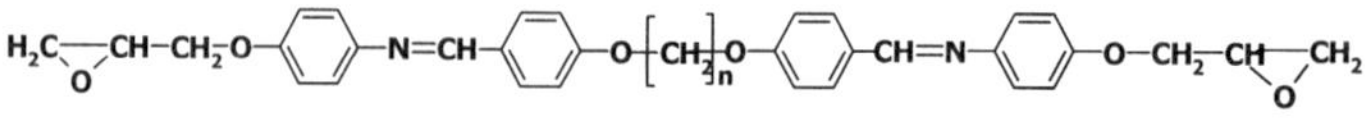

IAn

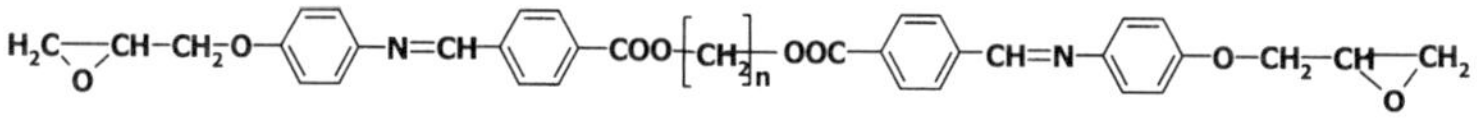

IBn

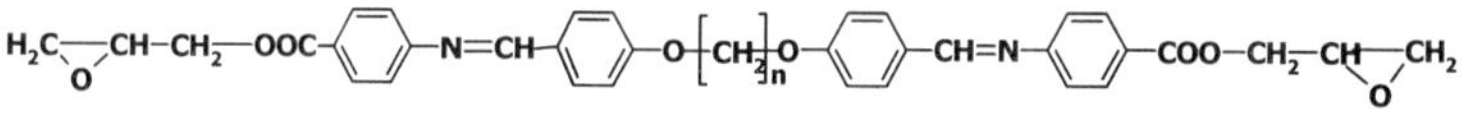

IIAn

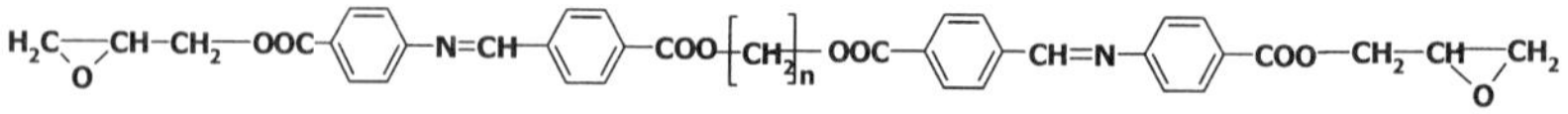

IIBn

Scheme 1

The aim of the present paper is to show how the ether or ester groups directly attached to the mesogen and the length and parity of the central spacer affect the ability to form ordered phases and the nature of these mesophases. We have also studied the influence of the structure of these monomers in the formation of LCTs by crosslinking.

Results and Discussion

The monomers were synthesized as shown in Scheme 2. The first step is the substitution of bromine atoms of a dibromoalkane by a p-formyl phenolate or benzoate in the presence of potassium carbonate and 18-crown-6. These compounds were condensed with p-aminophenol and p-aminobenzoic acid[9,10] to produce the imino compounds with high yields and purities. The reactive glycidylic groups are introduced by reacting these compounds with an excess of epiclorohydrin and a quaternary ammonium salt. The products obtained were characterized by standard methods.

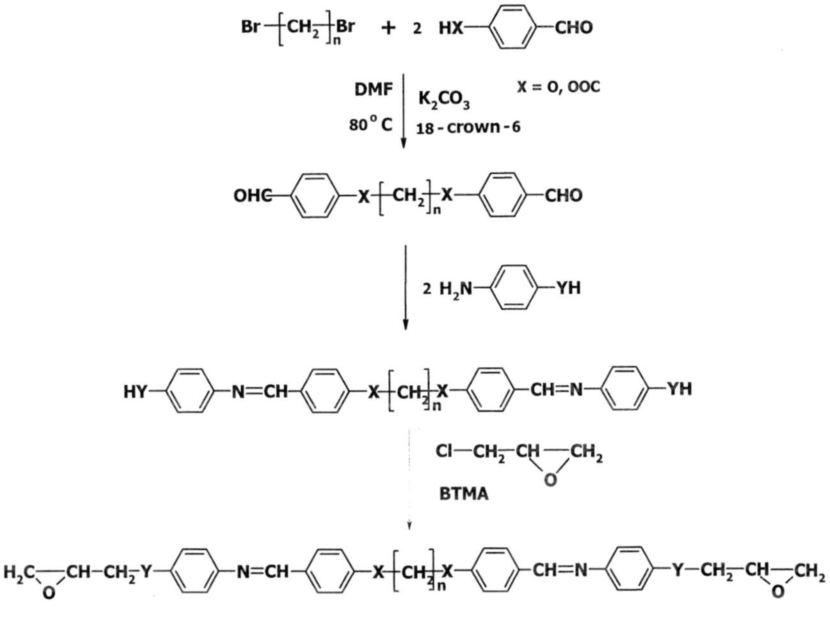

Scheme 2

Table 1 shows the mesogenic behaviour of the IB series and some structural characteristics. As we can see, monomers IB3, IB5 and IB7, which have spacers with an odd number of methylenes form smectic C mesophases while those that have an even number of methylenes form smectic A mesophases. Some of these monomers also show nematic mesophases. Generally, the smectic A range decreases and nematic mesophase stabilizes slightly as the spacer length increases. These results confirm that the spacer of the monomers has a strong odd-even effect on both the melting and isotropization temperatures and on the type of smectic mesophase formed. Both the melting and isotropization temperatures in each series of odd- and even-membered spacers decrease as the spacer length increases.

Table 1. Characteristics of dimeric epoxy compounds of the IB series.

Comp	Transition Temperatures (°C)	Mesophase ranges S/N (°C)	Half spacer length (Å) [b]	Monomer length (Å) [b] l	Layer spacing (Å) [c] d	d/l
IB3[a]	K 149 S_C 159 I	10	3.66	34.46	33.04	0.96
IB4	K 176 S_A 210 N 212 I	34/2	4.30	39.16	20.27	0.52
IB5	K 107 S_C 149 I	42	4.92	37.21	18.96	0.51
IB6	K 151 S_A 181 I	30	5.56	41.23	21.72	0.53
IB7	K 114 S_C 127 I	13	6.19	39.85	20.41	0.51
IB8	K_1 133 K_2 139 S_A 157 N 162 I	18/5	6.82	43.27	22.89	0.53
IB10[a]	K 131 S_A 138 N 145 I	7/7	8.09	44.88	21.81	0.49

a. Monotropic mesophase;
b. calculated by Cerius2;
c. determined by WAXS

The spacings of the smectic layers were evaluated by X-Ray diffraction. If we compare these results with the length of the monomer calculated by Cerius 2, the d/l ratio is around 0.5, which indicates that the even monomers are intercalated. Imrie[11] described that symmetric calamitic dimers have a strong tendency to form monolayer smectic phases. Our results can be explained if we consider the structural characteristics of the mesogens. So, we can formulate a resonance structure, in which the ether assumes a positive charge and the ester a negative one. Because of the polarization of the mesogenic core favourable electrostatic interactions appear, which favours an intercalated structure, as can be seen in the Scheme 3. These interactions disfavour monolayer arrangements.

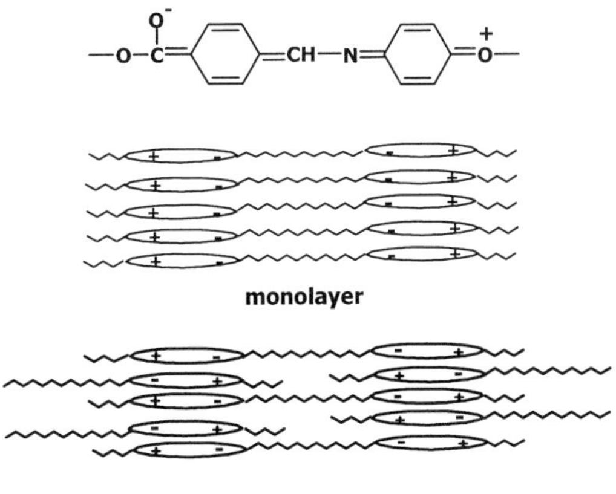

Scheme 3

Odd-membered spacer monomers show smectic C phases. In this case intercalated organizations were also confirmed by the d/l ratio for monomers with spacers of 5 and 7 methylene carbons. However, a monolayer arrangement was formed in the case of monomer IB3, because the terminal chain (4.60 Å) was longer than half of the spacer. Thus, in this case, steric hindrance of the aliphatic chains is predominant over electronic interactions between mesogens.

The liquid crystalline behaviour of the other series of diglycidylic compounds is collected in Table 2. In all monomers only nematic mesophases were detected. Of the compounds with spacers of 10 or 12 methylene carbons only IIA compounds show LC behaviour. This observation, in addition to the broader mesophase range, leads to the conclusion that IIA series have a higher mesomorphic character.

If we compare the mesomorphic characteristics of the compounds synthesized in this study, only compounds of the IB series form smectic mesophases, while the others present only nematic ones. Imrie reported that an ester linkage connecting the central spacer to the mesogen promotes smectic behaviour. Compounds in IIB series also have ester linkages in the same position, but only nematic mesophases are formed. In addition to ester groups linking the central spacer to the mesogens, it seems that the nature of both groups plays a significant role in the LC behaviour. Strong lateral interactions between mesogens facilitate smectic arrangements, since the smectic mesophases are produced by a microseparation of the phases, in which the mesogens are close in a definite region, while the alkyl chains constitute another well differentiated. The interactions between mesogens may be of electrostatic nature, because electron-donating and electron-withdrawing groups at their ends, cause high dipolar moments.

Table 2. Mesomorphic characteristics of compounds of IA, IIA and IIB series.

Comp.	Transition temp. (°C)	Mesophase range (°C)	ΔH (kJ/mol)
IA6	K 208 N 224 I	16	70 / 6
IA8	K 201 N 209 I	8	90 / 7
IA12	K 188 I	-	67
IIA3	K 142 I	-	38
IIA4	K 162 N 207 I	45	35 / 4
IIA5	K 106 N 130 I	24	21 / 1
IIA6	K 178 N 189 I	11	69 / 4
IIA8	K 114 N 164 I	50	43 / 5
IIA10	K 119 N 158 I	39	55 / 5
IIA12	K 122 N 149 I	27	55 / 5
IIB4	K 164 N 167 I	3	45 / 4
IIB6[a]	K 99 N 138 I	39	37 / 4
IIB8[a]	K 93 N 119 I	26	46 / 5

[a] Monotropic mesophase

Monomers of the IA and IIB series have functional groups of the same character at the end of the mesogens and, therefore, electrostatic interactions are low. However, IB and IIA monomers have groups of different characters at the end of the mesogens, although IB monomers have a maximum dipolar moment, such is represented in Scheme 4. The relative orientation of the imine group to the final groups in the IIA series allows an additional resonant structure to be formulated. This reduces the positive charge in the ethereal oxygen. Therefore, these results confirm that dipolar moments have considerable influence on the smectogenic character of this type of dimeric compounds.

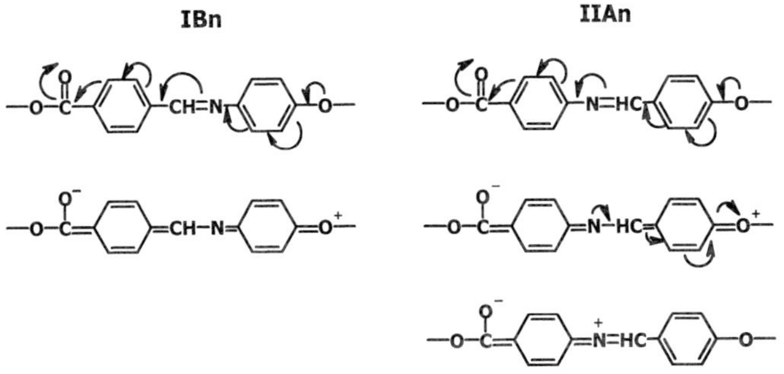

IBn **IIAn**

Scheme 4

To obtain LCTs, we studied the reaction of these epoxy compounds with primary aromatic diamines. In a first step, the primary amine reacts with epoxide to form linear oligomers with an extended chain that enhances the molecular anisotropy. In a second step, the secondary amine groups react to fix the anisotropic network. So, from an isotropic initial melt, ordered networks can be formed.

Shiota and Ober[12] used primary aromatic diamines to study the crosslinking of epoxy resins with dimeric architectures. They obtained smectic networks in spite of the initial mesophase. They suggested that the crosslinking sites at the end of the mesogens help to fix the spacing distances between crosslinks and favour smectic layer organizations.

Of the various primary aromatic diamines, the most suitable for getting from our epoxy monomers ordered networks were 2,4-diaminotoluene (DAT) and p-amino-acetophenone azine (NA2). It should be pointed out that NA2 was more effective than DAT at fixing ordered networks, although thermogravimetry always detected a weight loss, which can be due to a release of nitrogen.

We managed to fix ordered networks only when mesogens have both ether and ester linkages at the end. This confirmed the significance of the polarization of the mesogenic cores. As we can see in Table 3, in the most cases the ordering was nematic even from the isotropic melt. Smectic C organizations were locked only from monomers with long spacers. The monomers not included in the table led to isotropic networks. The mesophases fixed in the networks were identified by polarized optical microscopy and XR-diffraction.

Table 3. Isothermal curing conditions and characteristics of the LCTs obtained from mixtures of monomers IB and IIA with primary aromatic diamines.

Compound	Amine	Curing temperature (°C)	Curing time (min)	Initial mesophase	Network	Tg (°C)
IB4	DAT	160	30	I	N	-
IB5	DAT	110	57	I	N	120
IB6	DAT	150	36	I	N	-
IB8	DAT	150	47	I	Sc	-
IB10	DAT	140	70	I	N	108
IB4	NA2	170	40	N	N	-
IB5	NA2	130	120	N	N	102
IB6	NA2	180	40	N	N	-
IB8	NA2	180	70	N	S_C	-
IB10	NA2	180	120	N	S_C	-
IIA4	NA2	190	35	N	N	131
IIA5	NA2	190	40	N	N	118
IIA6	NA2	190	30	N	N	122
IIA8	NA2	190	40	N	N	110
IIA10	NA2	190	40	N	N	89

If mesogenic epoxy resins are cured with primary aromatic diamines, a relative large proportion of amine is required, which can disturb the organization of the monomer in the mesophase. In order to achieve highly ordered structures, we have studied the catalytic crosslinking of liquid crystalline compounds with tertiary amines. In this way, curing agent remains at the beginning of the crosslinked chain and does not disturb the inner structure. Another advantage of catalytic systems is the possibility to select the appropriate proportion of catalyst, so the crosslinking process can be carried out at different temperatures and reaction rates.

The used reaction conditions and the obtained results when the curing was catalyzed by a 3%.-w/w. of 4-(N,N-dimethylamino)pyridine (DMAP) are collected in Table 4.

Table 4. Isothermal curing conditions and characteristics of the LCTs obtained from mixtures of monomers IB and IIA with 3 phr. of DMAP as curing agent.

Compound	Curing temperature (°C)	Curing time (min)	Initial state	Network	ΔH (KJ/mol)	Tg (°C)
IB4	180	26	S_A	S_C	112	-
IB5	120	240	S_C	S_C	103	-
IB6	160	32	S_A	S_C	114	-
IB7	120	300	S_C	S_C	114	-
IB8	150	140	S_A	S_C	97	-
IB10	148	50	S_A	S_C	72	-
IIA4	180	100	N	N	150	110
IIA5	120	160	N	N	100	73
IIA6	180	130	N	N	143	98
IIA8	150	180	N	N	141	86
IIA10	150	160	N	N	148	79
IIA12	145	180	N	N	163	72

As we can see, monomers of the series IB led to smectic C-like networks. On the other hand monomers IIA led to nematic organizations in all cases. Also in this case, the behaviour of both series is different, although LCTs are obtained from all these monomers. However, when the groups linked at the end of the mesogens have identical nature only isotropic materials were always obtained. These results confirm the influence of the dipolar moments on the mesogens in the obtention of materials with different degree of order and the suitability of tertiary amine as the curing system to obtain liquid crystalline thermosets.

Acknowledgments

The authors would like to thank the CICYT (Comisión Interministerial de Ciencia y Tecnología) (MAT1999-1113) and CIRIT (Comissió Interdepartamental de Recerca i Innovació Tecnològica) SGR 000318 for their financial support.

[1] L.Strezelecki, L. Liebert, *Bull. Soc. Chim. France* **1973**, 597
[2] P. De Gennes, , *Phys. Letters* **1969**, *A8*, 725
[3] C. Carfagna, E. Amendola, M. Giamberini, *Prog. Polym. Sci.* **1997**, *22*, 1607
[4] Jpn. 58.206.579 Jpn. Kokai Tokkyo Koho C.,invs.:Agency of Industrial Sciences and Technology; *Chem. Abstr.* **1984**, 100, 138934x. Ger. Patent 36.22.610 **1986**, invs. : H.P. Mueller, R. Gipp, H. Heine, *Chem. Abstr.* **1988**, 109, 38835h. *Ger. Patent* 36.22.613, **1986**, invs. : R. Dhein, H.P. Mueller, M. Meier, R. Gipp, *Chem. Abstr.* **1988**, 109, 7136b
[5] G.G. Barclay, C.K. Ober, *Prog. Polym. Sci.,* **1993**, *18,* 899
[6] A. Shiota, C.K. Ober, *Prog. Polym. Sci.,* **1997**, *22,* 975
[7] S. Jahromi, J. Lub, G.N. Mol, *Polymer* **1994**, *35,* 622
[8] A.C.N. Mija, Gh. Cascaval, D. Shioca, B.C. Simionescu, *Eur. Polym. J.* **1996**, *32*, 779
[9] D. Ribera, A. Mantecón, A. Serra, *Macromol. Chem. Phys.* **2001**, *202*, 1658
[10] D. Ribera, A. Mantecón, A. Serra, *J. Polym. Sci., Part A: Polym. Chem.*, in press
[11]R.W. Date, C.T. Imrie, G.R. Luckhurst, J.M. Seddon, *Liq. Cryst.* **1992**, *12*, 203
[12] A. Shiota, C.K. Ober, *Polymer* **1997**, *38,* 5857

Liquid Crystalline Properties of Unsegmented and Segmented Polyurethanes Synthesised from High Aspect Ratio Mesogenic Diols

Kalathur Sabdham Vangepuram Srinivasan, Tallury Padmavathy*

Polymer Division, Central Leather Research Institute, Adyar, Chennai 600 020, India
E-mail: srinivasan1@hotmail.com

Summary: A series of novel tetrad high aspect ratio mesogenic diol monomers 4-{[4-(n-hydroxyalkoxy)-phenylimino]-methyl}-benzoic acid 4-{[4-(n-hydroxyalkoxy)-phenylimino]-methyl}-phenyl ester were prepared with varying alkoxy spacer length (n=2,4,6,8,10) by reacting 4-formylbenzoic acid 4-formylphenyl ester and 4-(n-hydroxyalkoxy) anilines. Two series of thermotropic main chain liquid crystalline unsegmented polyurethanes (PUs) were obtained by the polyaddition of the mesogenic diols with hexamethylene diisocyanate (HMDI) and methylene bis(cyclohexylisocyanate) (H_{12}MDI) in dimethylformamide respectively. The effect of the incorporation of a third component namely polyol on the liquid crystalline properties of the polyurethanes was also studied. Linear segmented PUs were synthesised by a two-step block copolymerisation method. The PUs synthesised were based on six spacer mesogenic diol chain extender, soft segments poly(tetramethylene oxide)glycol (PTMG) (M_n= 650,1000,2000) and polycaprolactone diol (PCL) (M_n=530,1250,2000) of varying molecular weights and different diisocyanates including HMDI, H_{12}MDI and methylene bis(phenylene isocyanate) (MDI). Structural elucidation was carried out by elemental analysis, fourier transform infra red (FT-IR), nuclear magnetic resonance (^{1}H NMR and ^{13}C NMR) spectroscopy. Inherent viscosity of the unsegmented polymers measured in methanesulphonic acid at 26°C was in the range of 0.13 - 0.65 dL/g while the molecular weights and molecular weight distribution of the segmented polyruethanes was determined using gel permeation chromatography (GPC). Mesomorphic properties were studied by differential scanning calorimetry (DSC) and hot stage polarising optical microscopy and the thermal stability was determined by thermogravimetric(TG)analysis. The monomeric diols and the polyurethanes exhibited nematic texture and good mesophase stability. It was observed that the partial replacement of the mesogenic diol by the polyol of varying molecular weights influenced the phase transitions and the occurrence of mesophase textures. The phase transition temperatures of the investigated polyurethanes showed dependence on the chain length of the soft segment and on the content of the mesogen moiety. A higher content of mesogenic moiety was needed to obtain

 DOI: 10.1002/masy.200350924

liquid crystalline property when the soft segment length was increased as observed in the case of PTMG. Grained and threaded textures were observed depending on the molecular weight of the soft segment, the mesogen content and the diisocyanate. The stress-strain analyses showed that the polymers based on high molecular weight PTMG soft segment have elastomeric property while the PCL based PUs displayed no elastomeric property.

Keywords: chain extender, high aspect ratio mesogenic unit, liquid crystalline polymers, polyurethane

Introduction

Polyurethanes constitute one of the most versatile classes of polymeric materials known today. Thermoplastic liquid crystalline polyurethanes (TLCPUs) are being investigated with much interest owing to their application potential.[1-3] Due to the strong intermolecular interaction arising from hydrogen bonding of the urethane linkages, the synthesis of the TLCPUs seems more difficult than the synthesis of the other LC polymers. To prepare a LCPU, many methods of minimizing the hydrogen bonding effect have been developed, such as (i) inserting flexible spacers into the mesogenic units,[4] (ii) introducing substituted mesogenic segments[5] or (iii) using secondary amines reacted with dichloroformates.[6]Among these methods, the introduction of long flexible spacers such as PTMG-1000 and 2000 into the polymer main chains was effective in decreasing the phase transition temperature, increasing solubility, reducing hydrogen bonding[7] and showing elastic properties. However, the use of long flexible spacers as modifying agents could result in partial or complete loss of liquid crystalline properties. Hence, we envisage that the incorporation of a high axial ratio mesogenic unit into the polymer backbone could help to balance the mesogenic property and the elastic nature of the polyurethane. Also it is important to investigate as to what extent a modifying agent would decrease the phase transition temperatures without destroying the liquid crystalline properties.

Experimental Part

Materials

4-Hydrxoybenzaldehyde, 4-formylbenzoic acid (Spectrochem, India), 4-hydroxyacetanilide, 6-chloro-1-hexanol (Fluka, Switzerland), PTMG (M_n=650, 1000 and 2000) and PCL (M_n= 530, 1250, 2000) was purchased from Aldrich, USA. The polyol was dried and degassed at

90-100°C under vacuum for 6-7 hours before use. 4-formyl benzoic acid, 4-hydroxy benzaldehyde, HMDI, $H_{12}MDI$ and MDI (Aldrich), dibutyl tin dilaurate (DBTDL) (Merck) were used as received. The solvent N, N'-dimethylformamide (DMF) was purified by distillation under reduced pressure over calcium hydride before use.

Characterization Techniques

Elemental analysis was done in a Heraues CHN RAPID, CHN analyzer. The FTIR spectra of the polymers were recorded using NICOLET Impact 400 by casting film in KBr disks. ^{1}H and ^{13}C NMR spectra were recorded on a Bruker MSL 300P, 300 MHz NMR spectrometer. Visual observation of liquid crystal transitions were studied by using OLYMPUS BX 50 polarizing microscope equipped with a LINKAM THMS 600 heating stage. Thermogravimetry was performed on a Seiko model SSC 5200H system attached to a TG/DTA 220 module at a heating rate of 10°C/min in nitrogen atmosphere. A Seiko model SSC 5200H attached to a DSC module was used to determine the second heating thermograms. The equipment was calibrated using indium and tin as standards. The measurements were carried out at a heating rate of 10°C/min in nitrogen atmosphere. The molecular weight determination was carried out using GPC with polystyrene standard in tetrahydrofuran solvent using an ultrastyragel column (Waters). Stress–strain measurements of polyurethane films were measured by INSTRON Universal Testing Machine at a constant speed of 50 mm/min. The specimens were conditioned before testing and the measurements were performed at room temperature with dumb bell shaped specimens with dimensions of 3.5x2.5x0.16mm.

Synthesis of Mesogenic Diols 4-{[4-(n-hydroxy alkoxy)-phenylimino]-methyl}-benzoic acid 4-{[4-(n-hydroxy alkoxy)-phenylimino]-methyl}-phenyl ester (3a-e)

0.01 mol (**1**) and 0.02 mol of the corresponding 4-(n-hydroxyalkoxy) aniline (n=2,4,6,8,10) (**2**) was refluxed in 20 ml ethanol for 1 hour with a catalytic amount of acetic acid. The precipitated product (**3**) was filtered and dried. All the compounds were recrystallised from DMF/methanol solvent mixture. Thus, 2.54g (0.01 mol) of 4-formyl benzoic acid-4-formyl phenyl ester and 4.18g (0.02 mol) of 4-(6-hydroxyhexyloxy) aniline were refluxed in 20 ml of ethanol for 1hour. The precipitated product **3c** was filtered, dried and recrystallised from DMF/methanol. Yield = 80%.

^{1}H NMR($CDCl_3$+Trifluoroacetic acid): δ 8.98 (s, 1H, C$\underline{H}$=N), 8.84 (s, 1H, C$\underline{H}$=N), 8.47 (d, 2H, Ar-$\underline{H}$), 8.28 (d, 2H, Ar-$\underline{H}$), 8.23 (d, 2H, Ar-$\underline{H}$), 7.61 (d, 2H, Ar-$\underline{H}$), 7.64-7.69 (m, 4H, Ar-$\underline{H}$), 7.1-7.27 (m, 4H, Ar-$\underline{H}$), 4.43 (t, 2H,-C$\underline{H}_2$), 4.08 (m, 4H, -C$\underline{H}_2$OH), 3.85 (t, 2H, -OC$\underline{H}_2$), 1.40-1.60 (m, 8H, -C$\underline{H}_2$), 1.52 (p, 4H, -C$\underline{H}_2$$CH_2$OH), 1.71 (p, 4H, -ArO$CH_2C\underline{H}_2$); FT-IR (KBr): 3306 (OH), 1736 (C=O of ester), 1623 (C=N), 1248 cm^{-1}(C-O).

Synthesis of Unsegmented Polyurethanes from HMDI (4a-e) and H_{12}MDI (5a-e)

Unsegmented polyurethanes (**4a-e**) and (**5a-e**) were synthesized by adding 0.0052 mol of the diisocyanate to a stirred solution of 0.005 mol of **3a-e** in 50 ml DMF at 60°C in nitrogen atmosphere. The reaction mixture was then stirred at 90°C. The progress of the reaction was followed by the disappearance of the NCO peak at 2260 cm^{-1} by FT-IR. Pouring the reaction mixture into methanol, the polyurethane was precipitated, filtered, and dried in vacuum. Phase separation was observed in the reactions with HMDI. The polyurethanes obtained were pale green solids.

Synthesis of Segmented Liquid Crystalline Polyurethanes

The polyurethanes were synthesised by the polyaddition reaction of HMDI with the dihydroxy compound at a molar ratio of 1:1. The mesogenic diol / PTMG ratio was changed according to the compositions as described in Table 1. A typical procedure for the synthesis of PU is as

Table 1. Composition of segmented polyurethanes.

Mesogenic Diol	Polyol	Diisocyanate
0.5	0.5	1
0.4	0.6	1
0.3	0.7	1
02	0.8	1

follows: PTMG 1000 (0.0005 M) was reacted with HMDI (0.001 M) at 80°C in inert atmosphere to prepare the prepolymer with end capped NCO group. When the theoretical NCO content was reached (confirmed by dibutylamine titration), the prepolymer was chain extended with the mesogenic diol (0.0005 M) in DMF with a catalytic amount of DBTDL at 70 °C for 12 h. The polymer solution was then poured into a mold and films were cast.

Results and Discussion

Mesogenic Diols

The synthesis of the mesogenic diols is outlined in Scheme 1. The high aspect ratio tetrad mesogenic diols (**3a-e**) containing four phenyl rings para linked by ester and azomethine linkages and terminal spacers were synthesized by condensing 4-formylbenzoic acid-4-formyl phenyl ester with 4-(n-hydroxyalkoxy) anilines of varying spacer length (n=2,4,6,8,10).

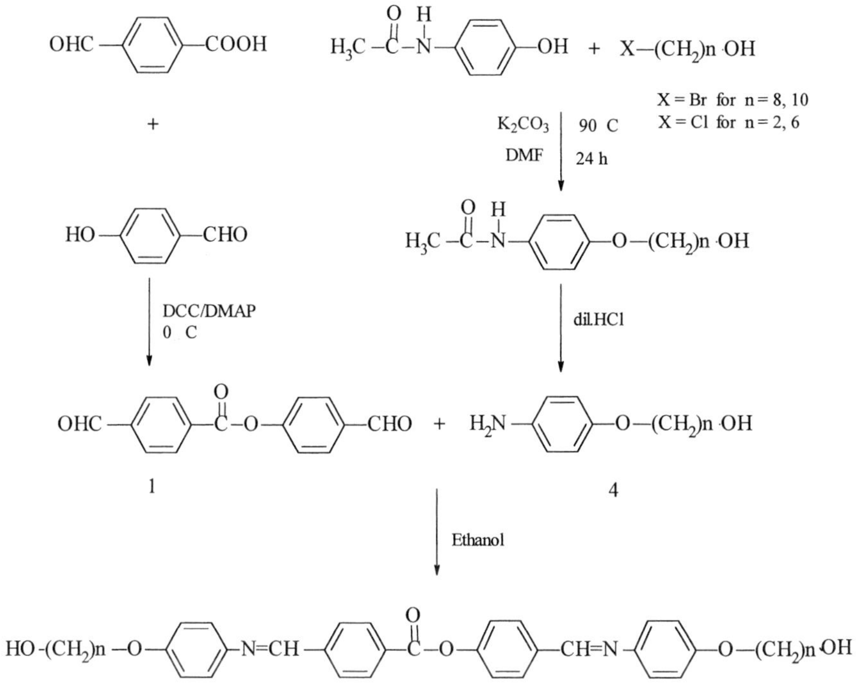

	3a	3b	3c	3d	3e
n	2	4	6	8	10

Scheme 1. Synthesis of mesogenic diols.

The yields and elemental analysis results are summarized in Table 2. The ester and azomethine linkages contribute to the linearity and the overall polarisability while the terminal spacers induce flexibility to the molecule. The diols were insoluble in all common solvents at room temperature but were soluble in aprotic solvents such as dimethylsulphoxide and DMF at elevated temperatures. FT-IR and ^{1}H NMR spectroscopy confirmed the chemical structure of the mesogenic diols.

DSC of the diols shows two endotherms corresponding to melting and isotropization transition, respectively. The transition temperatures of the mesogenic diols are summarized in Table 2. Though all the diols exhibited liquid crystalline phases in the first heating, no transition could be detected in the DSC cooling cycle, which could be due to the partial decomposition of the diols near their clearing temperatures. The transition temperatures decrease as the number of methylene spacers increase. All the diols exhibit nematic mesophase.

Table 2. Properties of mesogenic diols.

Diol	Yield (%)	Elemental Analysis (%)						LC phase[#] °C	Transition Temperature[§]		
		Calculated			Found						
		C	H	N	C	H	N		T_m °C	T_i °C	ΔT °C
3a	77	70.97	5.38	5.34	70.75	5.31	5.22	228 - 310	228.4	312.9	84.5
3b	82	72.39	6.25	4.86	72.18	6.17	4.91	222 - 300	225.0	303.0	78.0
3c	80	73.56	6.97	4.40	73.52	6.84	4.33	202 - 269	196.7	270.3	73.6
3d	74	74.54	7.56	4.04	74.31	7.38	4.12	190 - 235	186.1	243.7	57.6
3e	76	80.12	2.29	3.97	79.86	2.23	3.94	180 - 230	180.2	226.9	46.7

$\Delta T = T_i - T_m$
[#] As observed under the polarising microscope
[§] From DSC, first heating, 10 °C /min heating rate

Figure 1a shows the first heating DSC thermogram of the diol **3c**. A sharp endotherm at 196.7°C and a weak endotherm at 270.3°C was observed corresponding to the melting and the isotropization transitions respectively. At these temperatures, the mesophase was observed to be nematic schlieren texture. Figure 2 represents a typical polarized microphotograph of the diol **3c** on heating at 212 °C.

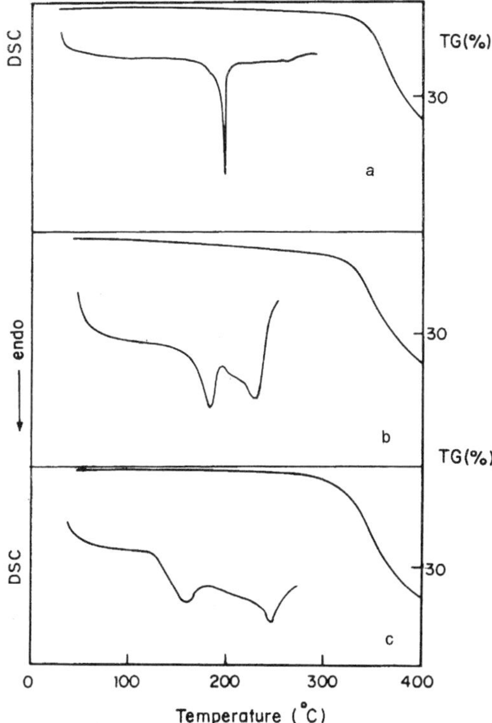

Fig. 1. Overlapped TG/DSC thermograms of mesogenic diol a) **3c**, polyurethanes b) **4c** and c) **5c.**

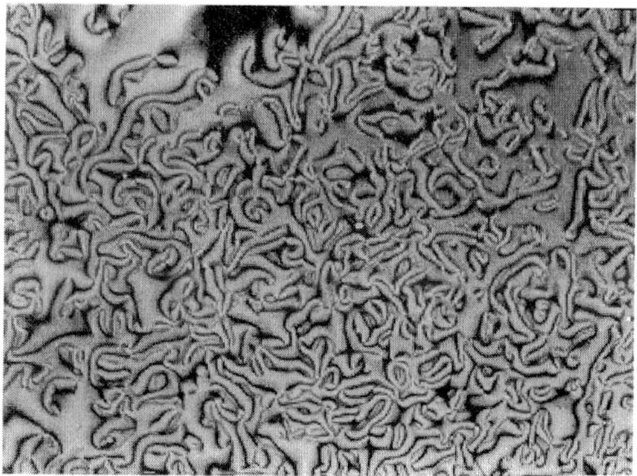

Fig. 2. Optical photomicrograph of mesogenic diol **3c** at 212° C.

Unsegmented Liquid Crystalline Polyurethanes

Polyurethanes (**4a-e, 5a-e**) were synthesized by the polyaddition of the tetrad mesogenic diol (**3a-e**) with HMDI and H_{12}MDI respectively. Progress of the reaction was monitored by FT-IR. Phase separation was noticed in the polyurethanes prepared from HMDI. The polyurethanes were insoluble in common solvents at room temperature. Due to their insoluble nature, structural characterization of the polyurethane was restricted to FT-IR spectroscopy alone. The inherent viscosity of the polymers was measured in methanesulphonic acid at 26°C as a function of time and is shown in Table 2. Decomposition of the polyurethanes occurred when they were dissolved in methanesulphonic acid. It is documented in the literature that schiff-bases undergo extensive degradation in the presence of protonic acids due to the hydrolysis of -CH=N- linkages.[8,9] The relationship between inherent viscosity (η_{inh}) and time is shown in Figure 3. As the time is increased, η_{inh} of **5c** decreased from 0.652 dL/g to 0.014 dL/g in 60h. As shown in Table 2, **4a-e** polyurethanes have low inherent viscosity in the range 0.18 - 0.2 dL/g while **5a-e** polyurethanes have higher viscosity in the range 0.42 – 0.65 dL/g. than the HMDI based polyurethanes, which precipitated during the course of the reaction.

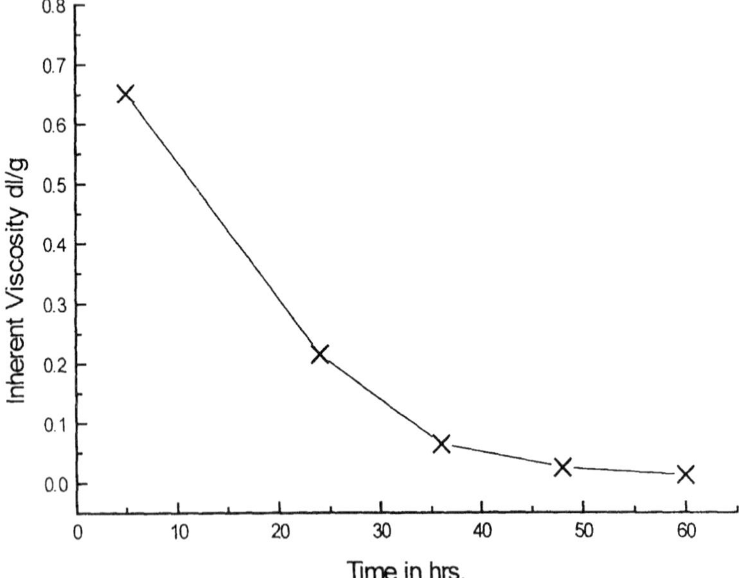

Fig. 3. Relationship between inherent viscosity and time of polyurethane **5c.**

The difference in the viscosity values could be attributed to the structure of the polyurethanes, which probably influences their solubility characteristics. The $H_{12}MDI$ used in this present study is a mixture of isomers (trans–trans, cis-cis, cis-trans). As a result of these isomers, the solubility of $H_{12}MDI$ based polyurethanes could have been enhanced and therefore these polymers have higher values of inherent viscosity.

Table 3. Properties of unsegmented Polyurethanes.

Polymer	Yield (%)	η_{inh}*(dL/g)	LC phase[#] °C	Transition Temperature[§]			5% Decomp Temp.
				T_m °C	T_i °C	ΔT °C	°C
4a (n=2)	78	0.184	232 - 270	231.2	274.2	43.0	310
4b(n=4)	79	0.163	230 - 251	224.0	245.5	21.5	308
4c(n=6)	75	0.139	175 - 258	185.4	231.6	46.2	312
4d(n=8)	72	0.168	185 - 231	186.2	223.5	37.3	323
4e(n=10)	70	0.202	175 - 228	180.2	226.9	46.7	320
5a(n=2)	77	0.426	226 - 263	220.1	252.7	32.6	316
5b(n=4)	76	0.473	214 - 255	212.3	249.0	36.7	314
5c(n=6)	80	0.652	191-269	162.2	246.2	84.0	319
5d(n=8)	73	0.546	175-250	155.3	245.5	90.2	322
5e(n=10)	71	0.482	145- 228	137	220.3	83.3	312

$\Delta T = T_i - T_m$
*Measured at a concentration of 0.5 dL/g in methanesulphonic acid at 26 °C.
[#] As observed under the polarising microscope
[§] From DSC, first heating, 10 °C /min heating rate

The thermal stability was determined by thermogravimetry and the thermotropic liquid crystalline property was evaluated by DSC and polarizing microscopy. The results are summarized in Table 3. DSC traces show two endothermic transitions corresponding to melting and isotropisation. In both the homologous series of the polyurethanes, the transition temperatures decrease as the length of the flexible spacers is increased. These polymers have good mesophase stability, which is attributed to the high length to breadth ratio of the mesogen.[10] On cooling, no transition was detected in both the polyurethane series, which could be due to the partial decomposition of the polymers. Figures 1b and 1c show the DSC heating thermograms of the polyurethanes **4c** and **5c** respectively.

As shown in Table 3, the phase transition temperatures of the polyurethanes measured by DSC were confirmed by observation with a polarised microscope. Both the series of polyurethanes based on HMDI and $H_{12}MDI$ showed a nematic texture

Segmented Liquid Crystalline Polyurethanes

The thermotropic LCPUs were synthesised by the polyaddition of the diisocyanate with polyol of various molecular weights and chain extended with the mesogenic diol in the presence DBTDL catalyst by a two-step method. In order to study the effect of the soft segment on the properties of the LCPUs, three different molecular weights of the soft segment namely PTMG (M_n 650, 1000, 2000) and PCL (M_n 530, 1200, 2000) were used. The mesogen content was also varied by 20-50 mol % in order to study the influence of the hard segment on the thermal and physical properties. The structure of the polyurethanes synthesised was characterized by FT-IR spectroscopy.

Table 4. Molecular weights and molecular weight distribution of polyurethanes.

Polymer code	M_n	M_w	M_w/M_n
PG-6-H-50	15000	25800	1.72
PG-1-H-50	20400	37800	1.61
PG-2-H-20	81000	14000	1.87
PG-1- H_{12}-40	40300	67200	1.66
PC-5-H-50	-	-	-
PC-1-H-50	-	-	-
PC-2-H-50	-	-	-
PC-5-H_{12}-50	10600	17700	1.67
PC-1-M-30	29000	54000	1.85

[PG=PTMG, PC=PCL,1=M_n 1000, 5 = M_n 530, 6 = M_n 650, 2 = M_n 2000, H=HMDI, H_{12}=H_{12}MDI, M = MDI, the numbers 20, 30, 40 and 50 stand for mesogen content in mol%].

The molecular weight of the polymers was determined by gel permeation chromatography. The number average molecular weight of the segmented polyurethanes is in the region of 10,000 – 80,000. Table 4 shows the molecular weight and molecular distribution of segmented polyurethanes. This is in contrast to unsegmented polyurethanes wherein the molecular weight data could not be obtained due to the insoluble nature of the polymers. Thus it is evident that the incorporation of soft segment improves the solubility characteristics of the polymers. Polyurethanes prepared from PCL M_n530 with HMDI and MDI were insoluble in tetrahydrofuran and the 50 mol % composition as in the case of PCL 1250 and PCL 2000 with HMDI were also insoluble.

The DSC heating cycle showed phase transition temperatures for the polyurethanes, but in the cooling cycle no transitions were observed which could be due to the partial decomposition of the polymers but the polarizing optical microscope showed the formation of a mesophase on cooling. Representative DSC thermograms of PG-1-H-50 and PG-6-H-40 are shown in Figures 4a and 4b.

PG-1-H-50 exhibited a transition at 14.7°C corresponding to the glass transition temperature of the PTMG soft segment and another transition at 104.5°C corresponding to the glass transition of the hard segments which is derived from isocyanate and mesogenic units. The thermogram shows four endotherms positioned at 133.6, 151.7, 169.3 and 190°C .The endotherm appearing at the lower temperature might be associated with a crystal-crystal transition since the polarizing microscope did not reveal any crystal-mesophase transition in this region. A similar observation was made by Lee *et al.*[11] The endotherms at 151.7 and 169.3°C were the result of crystal- mesophase transition temperature and isotropisation temperature respectively as confirmed by polarizing optical microscope.

The higher temperature melting endotherm at 190°C is a result of melting of microcrystalline hard segments, which is commonly observed in materials having longer aromatic urethane segments. The polymers prepared from HMDI containing 30, 40 and 50 mol% mesogen exhibited the crystal–crystal transition in the range 133-139 °C. In the MDI series, some of the polymers displayed very weak T_m and T_i transitions and in most of the polymers, these transitions could not be detected in DSC and were determined by optical polarizing microscope. In all the thermograms, the most prominent transition is that of the hard segment melting.

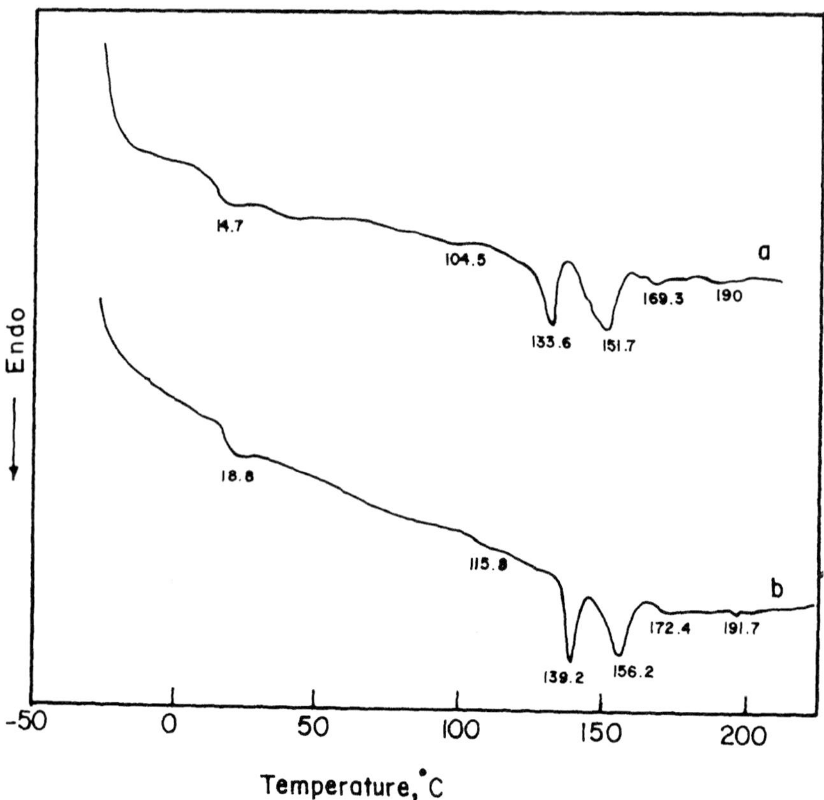

Fig. 4. DSC thermograms of a) PG-1-H-50 and b) PG-6-H-40.

Effect of Soft Segment

The phase transition temperatures of the segmented polyurethanes shift to lower values as a result of the variation in the flexible soft segment. Figure 5 shows the representative DSC thermograms of PCL 530, 1250 and 2000 with HMDI containing 50 mol % mesogen content. An increase in the molecular weight of the PCL soft segment from 530 to 2000 results in the lowering of glass transition, melting and isotropisation temperatures.

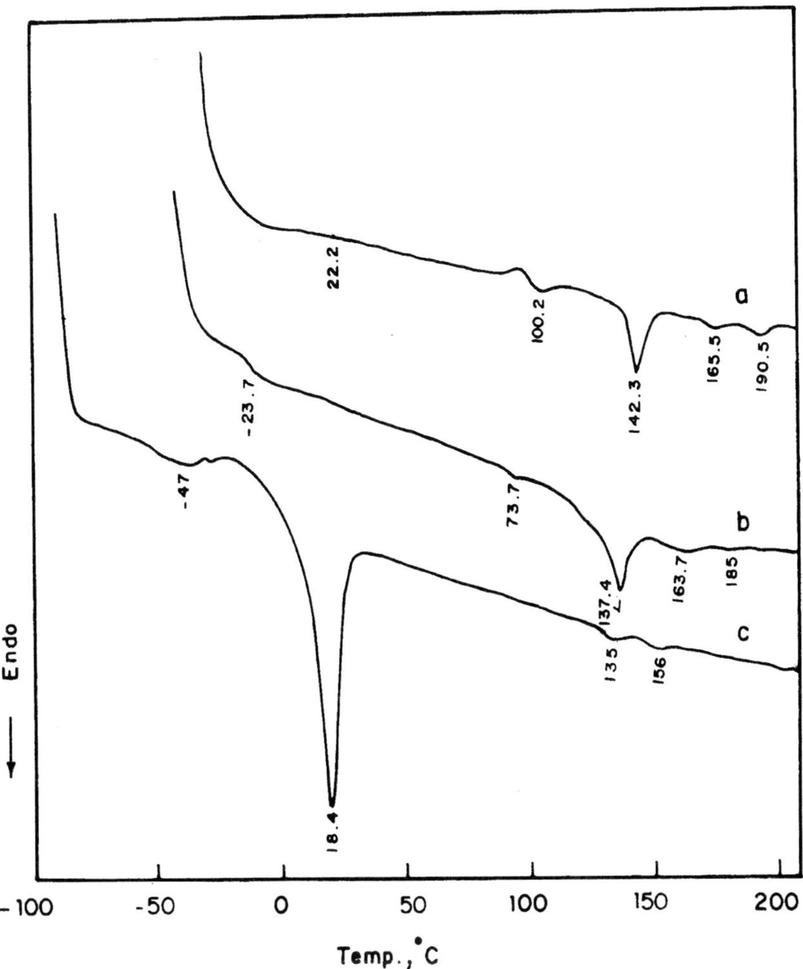

Fig. 5. DSC thermograms of a) PC-5-H-50 b) PC-1-H-50 and c) PC-2-H-50.

Effect of Diisocyanate

The LC phase transition is in the order of $PU_{MDI} > PU_{HMDI} > PU_{H12MDI}$. This could be due to the MDI and HMDI hard segments exhibiting a more symmetric and rigid structure than $H_{12}MDI$. Similar observations were made by Jia *et al.*[12] The PUs based on MDI have wider LC transition region than the polymer based on HMDI and H_{12}MDI. Mesophase texture variation is observed when the diisocyanate is changed from aliphatic to aromatic. In the PUs containing the aromatic diisocyanate MDI, grained texture is observed in all the compositions and in the PUs containing the aliphatic diisocyanates HMDI and H_{12}MDI, the lower mesogen contents exhibit grained texture while higher mesogen contents exhibit threaded texture.

Effect of Mesogen

The effect of partial replacement of the mesogenic diol by the polyols resulted in the increase of transition temperatures increase with increasing mesogen content. Visual observation of mesophase transitions in optical microscope revealed that the minimum amount of mesogen content required to impart LC varies with PTMG soft segment molecular weight while a minimum amount of 20 mol % mesogen to impart LC property for the PCL containing polyurethanes.

Table 5. Appearance of LC Phase in polyetherurethanes.

Diisocyanate	Polyol M.W	Minimum of Mesogen Content to impart LC (mol%)
HMDI	650	20%
	1000	30%
	2000	40%
H_{12}MDI &	650	20%
MDI	1000	20%
	2000	40%

Apart from the transition temperatures, the mesophase texture of the polymers also varied with mesogen content. Lower mesogen contents displayed grained texture and thread like texture was observed as the mesogen content was raised in the case of HMDI and H_{12} MDI while only grained texture for MDI based Polyurethanes.

Table 6. Variation in Mesophase Texture of polyetherurethanes.

Diisocyanate	Mesogen Content	Mesophase Texture
HMDI	Low	Grain
	High	Thread
H_{12} MDI	Low	Grain
	High	Thread
MDI	All Compositions	Grain

Table 7. Variation in Mesophase Texture of polyesterurethanes.

Diisocyanate	Mesogen Content	Mesophase Texture
HMDI	Low	Grain
(M.W 530)	High	Thread
(M.W.1250)	All Compositions	Grain
(M.W. 2000)	All Compositions	Grain
H_{12} MDI	Low	Grain
	High	Thread
MDI	All Compositions	Grain

Stress-strain Properties of the Polyurethanes

Polymers prepared from PTMG 650 and PTMG 1000 formed very brittle films and could not be subjected to stress-strain analysis while those prepared from PTMG 2000 exhibited elastomeric property. In the polyetherurethanes, as the mesogen content increased from 20 – 40 mol %, the tensile strength increased and the elongation at break decreased. In contrast to the polyetherurethanes, the polyesterurethanes did not exhibit elastomeric property. This could be due to the presence of a number of aromatic rings (in the hard segment domains) that stiffens the polymer chains coupled with the crystalline nature of the polyol leading to enhanced rigidity and brittleness. On the other hand, the presence of flexible bonds in the ether polyol imparts elasticity to PU chains, which is lacking in the case of polyesterurethanes.

Conclusion

High aspect ratio tetrad mesogenic diols containing four phenyl rings para linked by ester and azomethine linkages and terminal spacers varying from 2-10 were synthesized. These mesogenic diols were reacted with HMDI and $H_{12}MDI$ resulting in thermotropic liquid crystalline main chain unsegmented polyurethanes. All monomeric diols and polyurethanes showed liquid crystalline behaviour. The diols and the unsegmented polyurethanes exhibited nematic mesophase. The mesogenic diol was partially replaced by polyol of varying molecular weights, which influenced the phase transitions and the occurrence of mesophase textures. The phase transition temperatures of the polyurethanes showed dependence on the chain length of the soft segment and on the content of the mesogen moiety. A higher content of mesogenic moiety was needed to obtain liquid crystalline property when the soft segment length was increased as observed in the case of PTMG. Grained and threaded textures were observed depending on the molecular weight of the soft segment, the mesogen content and the diisocyanate. The stress-strain analyses showed that the polymers based on high molecular weight PTMG soft segment have elastomeric property.

Acknowledgement

One of the authors, T.P., thanks the Council for Scientific and Industrial Research (CSIR), New Delhi, India for the award of Senior Research Fellowship.

[1]G. Smyth, E.M. Valles, S.K. Pollack, J. Grembowicz, P.J. Stenhouse, S.L. Hsu, W.J. Macknight, *Macromolecules* **1989**, *22*,1467.
[2] T.Padmavathy, K.S.V.Srinivasan, *J. Macromol. Sci. Polym. Rev.*, **2003**, 43 (1) (to appear)
[3] T.Padmavathy, S.Sudhakar, K.S.V.Srinivasan, *Liquid Crystals*, **2001,** *28*, 1475.
[4] P.J.Stenhouse, E.M. Valles, S.W. Kantor ,W.J. MacKnight, *Macromolecules*, **1989,** 22, 1467
[5] K. Imura, N. Koida, H. Tanabe, M. Takeda, *Makromol. Chem.* **1981**, *182*, 2569.
[6] H.R. Kricheldorf, J. Awe, *Makromol Chem.* **1989**, *190*, 2579.
[7] F. Higeshi, F, K. Nakajima, M. Watabiki, W.X. Zang, *J. Polym. Sci., Polym. Chem.* **1993**, *31*, 2929.
[8] C.J. Yang and S.A. Jenekhe, *Chem. Mater.*, **1991,** *3*, 878.
[9] S. Banerjee, P.K. Gutch ,C. Saxena, *J. Polym. Sci, Polym. Chem.*, **1995**, *33*, 1719.
[10]G.W.Gray, "*Molecular structure and the properties of liquid crystals,*" Academic Press, New York 1962, p. 158.
[11] T.J. Lee, D.J. Lee, J., H.D.Kim, *J. Appl. Polym. Sci.*, **2000**, *77*, 577.
[12] X. Jia, X. He, X. Yu, *J. Appl. Polym. Sci.* **1996**, *62*, 465.

Designing Aromatic Polyamides and Polyimides for Gas Separation Membranes

Javier de Abajo, José G. de la Campa, Angel E. Lozano, Jorge Espeso, Carolina García*

Instituto de Ciencia y Tecnología de Polímeros (Consejo Superior de Investigaciones Científicas). Juan de la Cierva, 3. 28006 Madrid, Spain
Fax:+34 91 5644853. E-mail: deabajo@ictp.csic.es

Summary: Aromatic polyamides and polyimides with improved gas permselectivity, can be designed and prepared by systematically changing structural elements that affect these properties. Indeed, a conscientious choosing of the chemical changes may still provide a promising approach to get better and better polymers for selective filtration of gases.
The results of this work, in which novel monomers have been used, have confirmed that gas permeability through aromatic polyamides and polyimides much higher than that of conventional polyamides and polyimides can be achieved. It has been done by introducing bulky side groups, using non-planar monomers, and combining these elements on both monomers: diamines and dianhydrides or diamines and diacids. A theoretical study has also been made to explain the behaviour of some individual polymers, comparing experimental and calculated values of density and free volume.

Keywords: aromatic polyimides, gas separation, theoretical calculations

Introduction

Aromatic condensation polymers have got special importance in advanced technologies, mainly because of their excellent balance of mechanical, electrical and thermal properties. Thus, aromatic polyamides and polyimides have achieved particular importance for applications in a number of industrial fields such as electrical and electronic industry, aerospace industry and automotive industry.[1-3] For these applications, thermal resistance has been a crucial factor, but new applications, such as selective filtration through semipermeable membranes, are imposing other requirements, that are dependent to a great extent on chemical composition. For these

 DOI: 10.1002/masy.200350925

applications, polymers have to be designed with a favourable balance of properties such as solubility, crystallinity, glass transition temperature, density or free volume.

Many efforts have been done at this respect on engineering thermoplastics so far, for instance on polysulfones, polycarbonates, or cellulose derivatives, and more recently on thermoplastic polyimides specially designed for these applications.[4-6] A characteristic of all these polymer materials is that they are amorphous and show relatively high glass transition temperatures (Tg), so that they behave as actual glassy materials at room temperature. Although significant advances have helped for the application of these materials in separation technologies, involving operation such as dialysis, in particular hemodialysis, ultrafiltration, or reverse osmosis, gas separation and purification are operations where polymers have not yet achieved the degree of performance and disposal enough to be competitive against conventional technologies, like chemical absorption or cryogenic distillation.[7-9]

At this respect, it is worthy remarking that there is a growing demand for novel polymers to be used as permselective membranes in the separation of gas and vapour mixtures in refineries and in the petrochemical industry in general, and also for the separation of carbon dioxide and water from natural gas, the separation of nitrogen oxides and sulphur oxides from industrial gas streams and polluted atmospheres, the enrichment of synthesis gas, the recuperation of hydrogen from ammonia synthesis, the separation of nitrogen and oxygen from air, etc. Pairs of gases that are intensively studied at present are: O_2/N_2, H_2/CH_4, CO_2/CH_4, CO_2/N_2, He_2/N_2, H_2O/CH_4 and C_2H_2/ C_2H_4 and in general the mixtures olefin/paraffin.

This paper deals with the preparation and evaluation of novel aromatic polyamides and polyimides, especially designed and developed to be tested as barrier materials for the separation of gases. A study has been done of their general properties, particularly thermal transitions and solubility (processability), and a comparative evaluation of their properties as membranes has been carried out based on permeation measurements and theoretical calculations of macroscopic properties, such as density and fractional free volume.

Experimental

The synthesis of monomers and the preparation of polymers by conventional polycondensation methods have been previously reported.[10-12]

Glass transition temperatures (Tg) were determined by differential scanning calorimetry (DSC), using a Perkin-Elmer DSC-7 device, in nitrogen atmosphere, on approx. 10mg samples, at a heating rate of 20°/min from 50 to 400°C.

Solubilities were investigated by mixing 20 mg polymer with 1 mL solvent at room temperature for 24 h, shaking the mixture from time to time. Samples that did not dissolve in 24 h were heated up to the boiling temperature of the solvent if needed to get a clear solution.

The permeability to pure gases was studied on polymer films, made by casting of polymer solutions from appropriate solvents. A barometric method was used, applying a constant feed pressure of 3 bars, and an initial pressure in the expansion chamber less than 0.01 mbar.

Simulations of the amorphous structures of the polymers were performed using the computer program Cerius2, version 4.2[13] on a Silicon Graphics Octane workstation. A combination of molecular mechanics and molecular dynamics was used to obtain low energy structures. For these computations, 10 bulk amorphous chains of each polymer were built using the amorphous builder incorporated in the Cerius2 package. The bulk amorphous state was simulated using periodic boundary conditions. The amorphous polymer chains were generated using a random method without any rotational restriction.

The potential energy of each bulk polymer chain was minimized using the Dreiding [14] force field, version 2.21. Partial atomic charges, computed by the Rappé-Goddard method [15], were included in the non-bonded energy calculations. An external stress of 0.0001 GPa (1 atm) was applied to the cell. Ewald summation[16] was used to determine the Van der Waals and Coulomb energy terms, using a non-bond list with a 2.000 Å buffer in both cases.

Molecular dynamics simulations were performed to relax the minimized structures (annealing cycles at constant pressure P (1 atm) and moles N, starting and ending at 300 K, mid-cycle temperature of 1000 K). The structures obtained after each cycle were minimized again as described previously. Successive cycles of molecular mechanics and molecular dynamics were performed until energy and density convergence was achieved.

Results and Discussion

A major goal of this investigation was to achieve aromatic polyamides and polyimides with a number of specific properties, more precisely: good solubility in organic solvents, absence of crystallinity, high glass transition temperatures that assure glassy state at temperatures below 100°C, good mechanical properties and high fractional free volume, and film-forming properties.

Considering the poor permeability of conventional wholly aromatic polyimides and polyamides, monomers were designed and synthesized following empirical rules concerning chemical modification oriented to the development of processable aromatic polymers. Thus, dianhydrides, diamines and diacids were prepared that incorporate some of the following elements:

Bulky side substituents (pendent groups)

Combination of *meta/para* substitution

Enlarged monomers

Non planar chemical structures

Polyimides were prepared by conventional condensation methods, particularly by low temperature solution polycondensation and chemical imidation. Polyamides were prepared by direct polyamidation from diamines and diacids using the method of Yamazaki-Higashi (phosphorylation) at about 100°C. All polymers were soluble in organic media, showed high molecular weight, and exhibited good film-forming properties.

Polyimides

The polyimides of this work were prepared from dianhydrides containing three benzene rings. The general formula for them has been depicted in Fig 1. The substituent R may be H, phenyl, nitrophenyl or *tert*-butyl, and the benzene rings are directly joined in a *m*-terphenyl fashion or they are joined through carbonyl bonds, corresponding to an isophthaloyldiphthalic moiety. Most of them are original polymers, prepared from the novel dianhydrides and selected aromatic diamines.

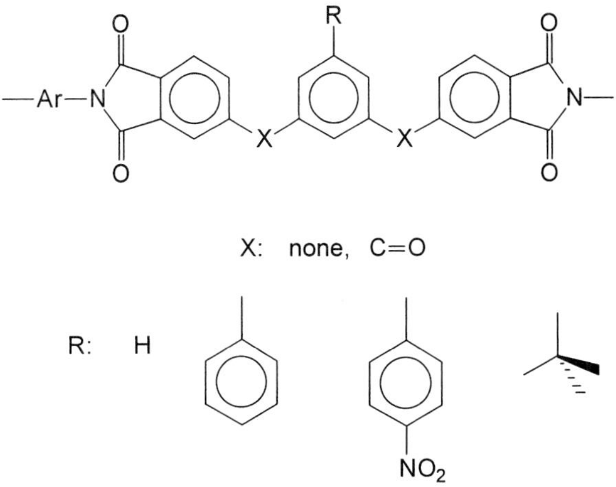

Fig. 1. Chemical structure of novel polyimides.

Polyimides of this family showed glass transition temperatures around 300° C, and fairly good solubility in organic solvents. The good solubility of polyimides from isophthaloyldiphthalic anhydrides is to be mainly attributed to the torsional mobility provided by the carbonyl groups. For the polymers containing *m*-terphenyl moieties, the good solubility should be attributed to the non-planar structure of the anhydride. The presence of pendent groups in both series greatly helps for an improvement of the solubility and free volume.

In dense membranes, the transport of gases is conveniently described by the solution-diffusion mechanism, which involves three steps: 1) solution of the permeant in the polymer, 2) diffusion through the polymer matrix and 3) desorption. Thus, the permeability P of a solid polymer is defined by the product of a diffusion coefficient D and a solubility coefficient S. Hence, the permeate flux J can be correlated with the permeability by

$$J = P\frac{\Delta p}{l}$$

where Δp is the pressure difference and l is the membrane thickness.

The ideal separation factor α, which is an index of the overall selectivity of the polymer, can be expressed in the following terms:

$$\alpha_{A/B} = \frac{P_A}{P_B}$$

Where P_A and P_B are the permeabilities for gases A and B. In Table 1 are listed the gas permeation properties for these polyimides

Furthermore, the diffusion of gases through dense polymer membranes directly depends on the free volume (FV), as it has been established by Fujita[17] and Cohen and Turnbull.[18] The FV can be calculated by using additive contributions combined with experimental measurements of the polymer specific weight, and more recently by methods of computational chemistry. A comparison of the results obtained by both methods for a set of polyimides is presented in Fig. 2.

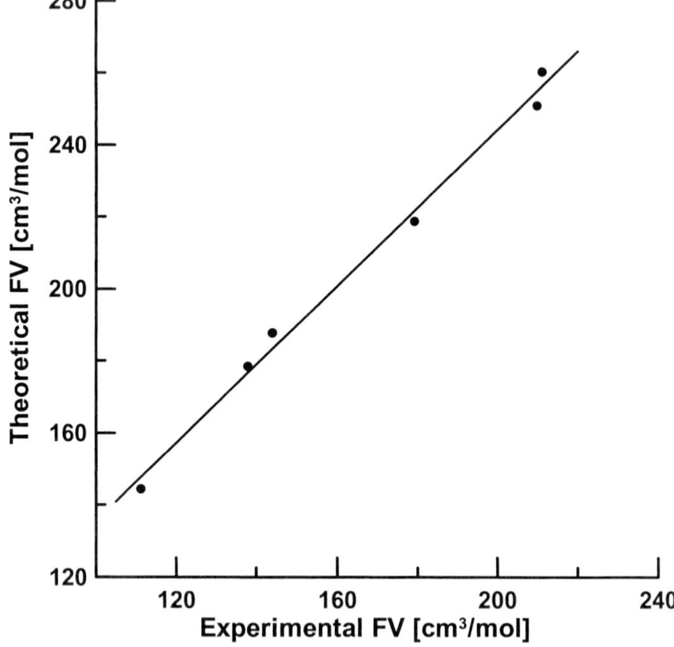

Fig. 2. Plot of experimental FV vs. calculated FV for novel polyimides.

The good correlation observed permits a theoretical treatment of molecular structures, and an *a priori* estimation of some macroscopic properties of condensation polymers specially designed for specific applications. In our case, they were foreseen to be useful barrier materials for gas separation, so that calculations were made of the density of the foreseen polyimides structures by a combination of molecular mechanics and molecular dynamics. The procedure implies the generation of chain conformations by a Monte Carlo method, and the minimization of energy states by iterative calculations that lead to average minimum energy conformations. Polymer chains containing not more than eleven repeating units are visualized as confined in a box of defined size, as it is shown in Fig. 3. for one polyimide containing *tert*-butyl pendent groups.

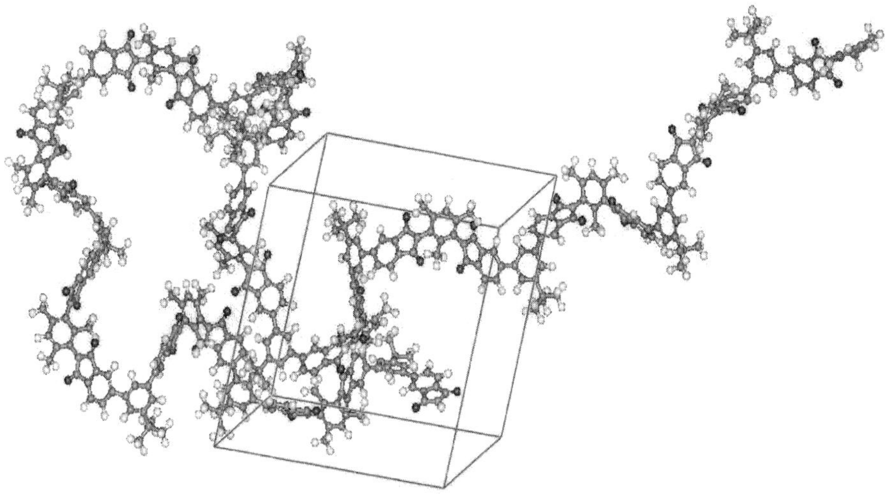

Fig. 3. Periodic cell of a chain of polyimide from TMMPD (trimethyl-*m*-phenylenediamine).

Hence, the theoretical density can be calculated with reasonably acceptable accuracy (s. Fig. 4), even for polymers with a complex structure and strong intermolecular attractive forces as polyimides. At this respect, it is worthy mentioning that, apart from density, extreme-extreme distance or mean square radius can also be calculated, what gives additional information on polymer flexibility.

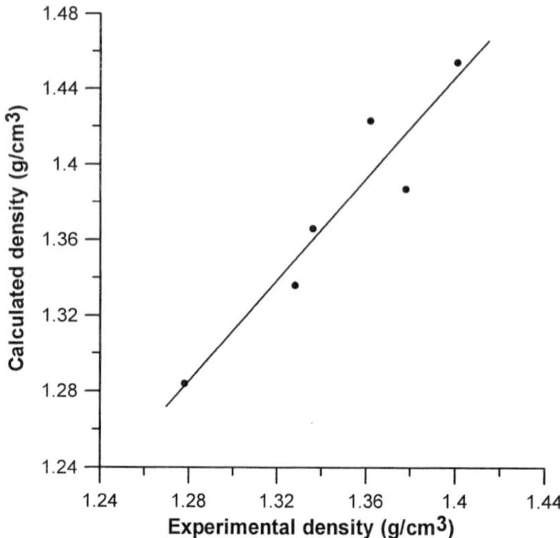

Fig. 4. Plot of measured density vs. calculated density of polyimides.

The theoretical permeabilities agreed fairly well with the experimental permeation measurements, and both coincide on the good permeation properties that could be achieved by using monomers that hinder molecular packing and three dimensional order, and improve solubility and free volume (s Table 1). As a matter of fact, upon comparing the permeability to pure oxygen and nitrogen of some of these novel polyimides with conventional aromatic

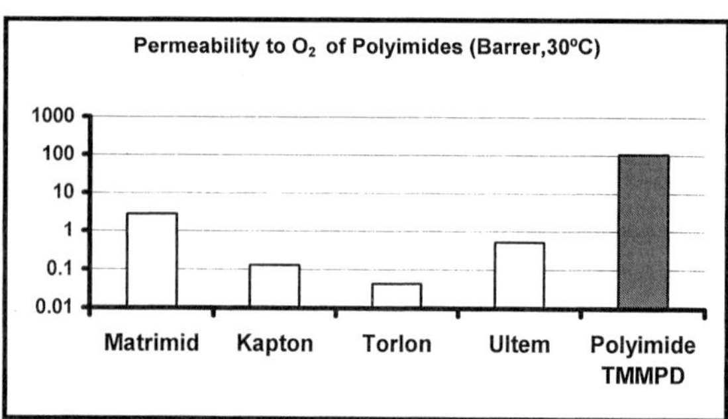

Fig. 5. Comparison of permeability to O_2 for technical polyimides and the experimental polyimide containing TMMPD.

polyimides like Kapton®, or even with thermoplastic polyimides like Ultem® or Matrimid®, a great gain of permeability can be observed, and that without a significant loss of selectivity (s. Fig. 5).

Table 1. Permeation properties of novel polyimides.

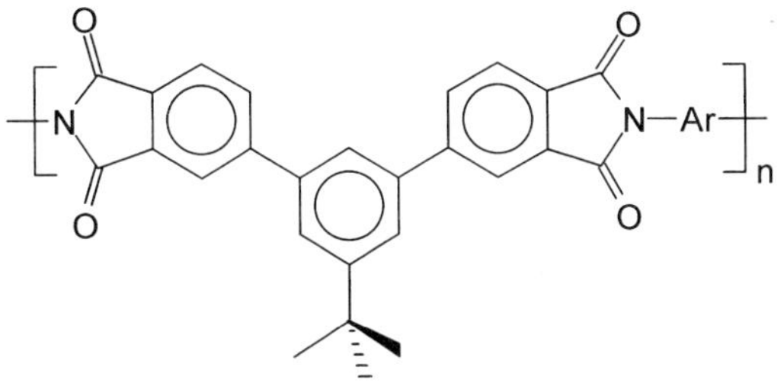

	Permeability (Barrer)					Ideal separation factor		
Ar	**He**	**CO_2**	**O_2**	**N_2**	**CH_4**	**He/CH_4**	**CO_2/CH_4**	**O_2/N_2**
6F	115	114	24.2	5.8	5.0	23.0	22.9	4.2
TMMPD	270	600	130	35.1	47.6	5.7	12.6	3.7
IMDDM	130	196	43.1	10.8	14.7	8.8	13.4	4.0
ODA	33	25	5.1	0.97	0.96	34.0	25.6	5.3

6F: Hexafluoroisopropylidene dianiline.
TMMP: Trimethyl *m*-phenylene diamine.
IMDDM : Methylene bis(3-isopropyl-5-methyl-4-aminobenzene).
ODA: Oxydianiline.

Polyamides

The same strategy was applied to aromatic polyamides. Traditionally, aromatic polyamides have been considered as very efficient barrier materials, with a high selectivity for gas separation, however, the high packing density of conventional wholly aromatic polyamides, such as poly-p-phenylene terephthalamide (PPTA) or poly-m-phenylene isophthalamide (PMIA), provides very small values of gas permeability. Rigid, rod-like polyamides are among the less permeable polymers, showing permeability to oxygen some orders of magnitude lower than that of common glassy polymers. The high density of cohesive energy of polyamides is caused by the strong associative forces through hydrogen bonds, so that most of the chemical modifications outlined to improve the solubility (processability) of aromatic polyamides, are directed to effectively prevent interchain hydrogen bonds, for instance by introducing bulky pendent groups.[19-22]

Chemical modification of the primary structure can provide, also in this case, a considerable improvement of free volume and permeability. For instance, the family of polyamides derived from a monomer specially designed for this application, that is shown in Table 2, exhibited an outstanding permeability to gases, comparable or even higher than that of glassy, engineering thermoplastics, and with an acceptable selectivity to such pairs of technical gases as O_2/N_2, H_2/CH_4 or CO_2/CH_4.

Table 2. Permeation properties of polyamides.

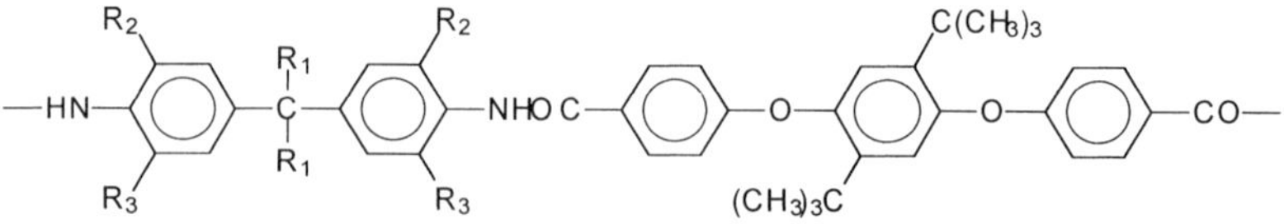

Key	**R_1**	**R_2**	**R_3**	**$P(O_2)$**	**$P(N_2)$**	**$P(CO_2)$**	**$P(CH_4)$**	**$\alpha(O_2/N_2)$**	**$\alpha(CO_2/CH_4)$**
PA1	CH_3	H	H	4.59	0.83	21.52	0.91	5.53	23.64
PA2	CH_3	CH_3	H	2.85	0.48	11.78	0.48	5.93	24.54
PA3	CH_3	CH_3	CH_3	9.32	1.83	43.00	2.04	5.09	21.07
PA4	H	CH_3	$CH(CH_3)_2$	6.25	1.10	25.21	1.20	5.68	21.00
PA5	CF_3	H	H	7.52	1.46	33.58	1.33	5.15	25.24

Theoretical calculations corroborated that the chemical modification can actually help for an efficient improvement of the conformational freedom and free volume. The results are consistent with a gain of permeability that agrees fairly well with experimental measurements. In Fig. 6, the permeability/selectivity relationships of one of them are compared with that of a technical, amorphous polyamide. As it can be observed, gases permeate much faster through polyamides of the family shown in Table 2 than through Trogamid®, which is probably the commercial polyamide that offers higher permeability to gases.

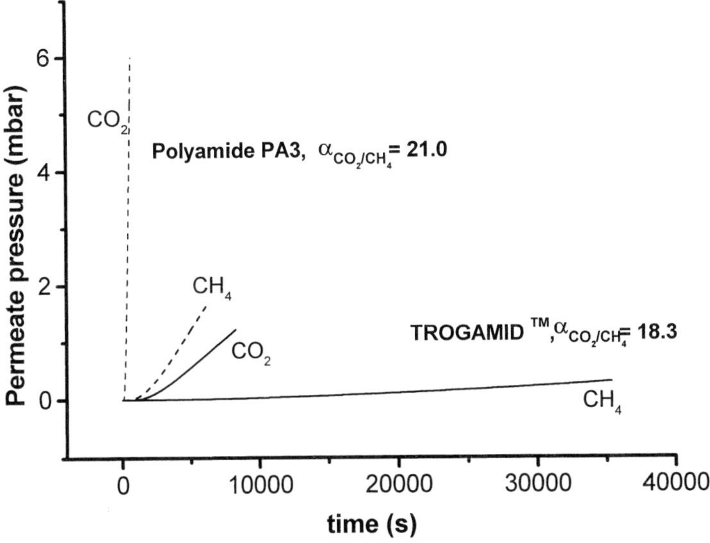

Fig. 6. Permeation curves of Trogamid® and polyamide PA4 for O_2, N_2, CO_2 and CH_4.

Here again, a good agreement between theoretical calculations and experimental measurements of permeability could be observed. (Fig. 7), is quite illustrative at this respect. It was confirmed that there is a direct relationship between calculated FV and experimental values of permeability through aromatic polyamide dense membranes, and that the presence of a growing number and growing size of substituents may substantially improve the FV and the permeability. An additional advantage of these polyamides is their unusually good solubility in organic solvents, so that they could be fabricated into defect-free films, of excellent mechanical properties, by casting from solutions of common solvents, such as chloroform or tetrahydrofuran.

As a conclusion, it can be stated that methods of computational chemistry and molecular modelling, have revealed exceptional potential as a tool to investigate novel chemical compositions for aromatic polymers, more precisely aromatic polyamides and polyimides, that may offer better and better performances as barrier materials for gas separation and purification.

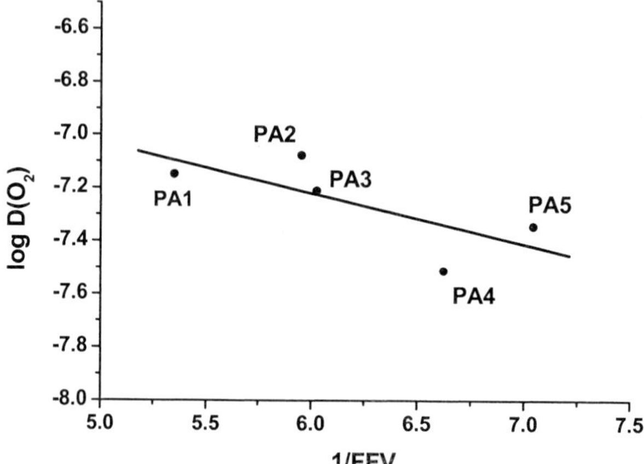

Fig. 7. Plot of diffusion coefficient vs. calculated FFV for polyamides.

[1] D. Wilson, H. D. Stenzenberger, P. M: Hergenrother, eds. "Polyimides", Chapman & Hall, New York **1990**
[2] M. K. Ghosh, K. L. Mittal, eds., "Polyimides. Fundamentals and Applications", Dekker, New York, **1996**
[3] G. Rabilloud, "High-Performance Polymers", Editions Technip, Paris **1999**
[4] J. de Abajo, J. G. de la Campa, *Adv. Polym. Sci.* **1999**, *140*, 23
[5] K. Tanaka, H. Kita, M. Okano, K._I. Okamoto, *Polymer*, **1992**, *23*, 585
[6] M. Langsam, W. F. Burgoyne, *J. Polym. Sci. Polym. Chem.*, **1993**,*31*, 909
[7] W. J. Koros, G. K. Fleming, *J. Membrane Sci.*, **1993**, *83*, 1
[8] S. A. Stern, *J. Membrane Sci.*, **1994**, *94*, 1
[9] R. Prasad, R. L. Shaner, K. J. Doshi, eds. "Polymeric Gas Separation Membranes", CRC Press, Boca Raton **1994**, pp. 513-614
[10] D. Ayala, A. E. Lozano, J. de Abajo, J. G. de la Campa, *J. Polym. Sci. Polym. Chem.*, **1999**, *37*, 805
[11] J. E. Espeso, J. G. de la Campa, A. E. Lozano, J. de Abajo, *J. Polym. Sci. Polym. Chem.*, **2000**, *38*, 1014
[12] J. E. Espeso, E. Ferrero, J. G. de la Campa, A. E. Lozano, J. de Abajo, *J. Polym. Sci. Polym. Chem.*, **2001**, *39*, 475
[13] Cerius2, version 4.2, Molecular Simulations, Inc., San Diego, CA, **2000**
[14] S. L. Mayo, B. D. Olafson, W. A. Goddard III, *J. Phys. Chem.*, **1990**, *94*, 8897
[15] A. K. Rappe, W. A. Goddard III, *J. Phys. Chem.*, **1991**, *95*, 3358
[16] N. Karasawa, W. A. Goddard III, *J. Phys. Chem.*, **1989**, *93*, 7320
[17] H. Fujita, *Forschr.Hochpolym. Forsch.*, **1961**, *3*, 1

[18] M. H. Cohen, D. Turnbull, *J. Chem. Phys.,* **1959**, *31*, 1164
[19] M. A. Kakimoto, M. Yoneyama, Y. Imai, *J. Polym Sci. Polym. Chem.,* **1988**, *26,* 149
[20] K. Ghosal, B. D. Freeman, R. T. Chern, J. C. Alvarez, J. G. de la Campa, A. E. Lozano, J. de Abajo, *Polymer*, **1995**, *36*, 793
[21] Y. T. Chern, H. C. Shiue, S. C. Kao, , *J. Polym. Sci. Polym. Chem.,* **1998**, *36*, 785
[22] J. M. García, F. García, R. Sanz, J. G. de la Campa, A. E. Lozano, J. de Abajo, *J. Polym. Sci. Polym. Chem.,* **2001**, *39*, 1825

New Reactive Polymeric Systems for Use as Waveguide Materials in Integrated Optics

Christian Dreyer,[*1] Jürgen Schneider,[1] Karin Göcks,[1] Brigitte Beuster,[1] Monika Bauer,[1] Norbert Keil,[2] Huihai Yao,[2] Crispin Zawadzki[2]*

[1] Fraunhofer Institute for Reliability and Microintegration, Branch Lab Polymeric Materials and Composites, Kantstr. 55, D-14513 Teltow, Germany
E-mail:dreyer@epc.izm.fraunhofer.de
[2] Heinrich-Hertz Institut für Nachrichtentechnik, Berlin GmbH Einsteinufer 37, D-10587 Berlin, Germany

Summary: Cross-linked polymeric materials are used in a wide range of applications. Special compositions are used in micro-system technologies and its utilization is extending to integrated optics. We have been working for several years in the field of polymers for optical applications.
Highly fluorinated polycyanurate systems have been proven as promising waveguide materials in integrated optics. The refractive index can be adjusted reproducibly in a wide range almost continuously. The layer quality was optimized. Low optical losses of less than 0.3 dB/cm@1550nm were obtained and working optical prototypes were developed. The birefringence had been a major problem, but this was solved by adjusting the coefficient of thermal expansion of substrate and film.
We will report in this paper on the polycyanurate ester resins and the new triazine containing polymeric systems.
Keywords: integrated optic, optical loss, polycyanurate ester resins; triazine, waveguide

Introduction

Today, in the age of telecommunications and the internet, the amount of data that is transferred is growing continuously. So called killer-applications like video-on-demand, virtual reality, huge electronic databases, file-sharing and online-games require greater and greater bandwidth. However, it is not only these "fun" applications that require fast data-transfer. The modern information-society has the need for worldwide fast access to e-mail, important files, bank accounts, scientific literature, stock exchanges, product ranges of companies, etc.

To cope with this huge amount of data, optical telecommunications technology is a sufficient tool. Due to its huge bandwidth, with more than 40 Tbps, the glass fiber has almost completely replaced the copper wire for long-range data transmission. The potential of the glass fiber is

 DOI: 10.1002/masy.200350926

such that all telephone-calls worldwide could conceivably be transferred at once via one single fiber. And the use of the glass fiber now extends more and more to short-distance applications, for example in the automotive or aeronautic fields. Especially for these last two applications the insensitivity of glass fiber to electromagnetic fields, as produced by, for example, cellular phones; and its lower weight in comparison to the copper wire are among the advantages of glass fiber. In about ten years we will likely see an all-optical network with glass fiber connections to every single house or even to every single computer.

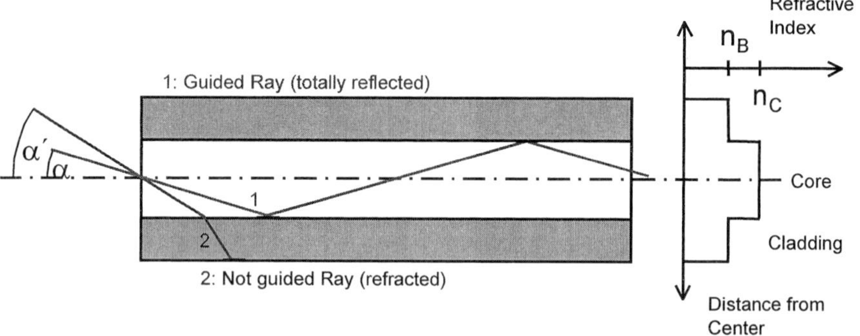

Fig. 1. Waveguiding in the glass fiber or in the waveguide of an optical device. To be guided, the rays must be launched inside of an acceptance angle (1), otherwise the ray is refracted into the cladding.

However, an all-optical network not only consists of glass fibers, but also of a number of different integrated optical devices, such as multiplexers, demultiplexers, thermooptic switches, attenuators, etc. that are needed to run such a high-speed network.

Currently, most optical components are realized by using silica-based optical waveguides, but polymers have many advantages compared with these established silica-based devices. Polymeric materials can provide a significantly higher refractive index-contrast between waveguide core and cladding, which results in smaller devices. Furthermore polymers posses a thermooptic coefficient that is ten times higher than that of silica. Additionally the heat capacity of many polymers is one order of magnitude higher than silica. Hence, a thermooptic switch made from polymer requires a switching power that is only about one percent of that of inorganic switches. Particularly big switching matrices with thousands or more switches can be realized without the well known problem (in microprocessor-technology) of waste heat. Last but not least, polymeric devices can be produced at lower costs, through techniques established in semiconductor production.

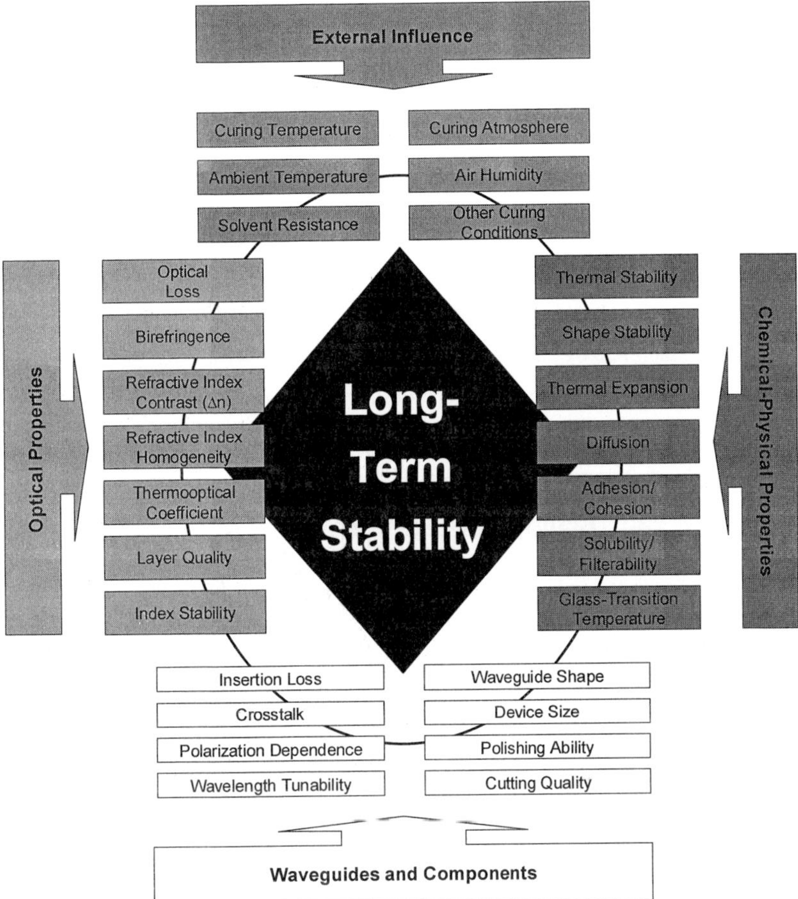

Fig. 2. Requirements on polymers for use as waveguide materials, external influence and device properties.

The dimensions of these waveguides are in the range of the light guiding core of the glass-fiber, about 5-7 μm. The surrounding cladding material must show a slightly lower (about 0.01) index of refraction, to allow waveguiding by total internal reflection at the interface between both media, as shown in Figure 1.

Demands on Polymers for Use in Integrated Optics

In order to be used as a waveguiding material, the polymers used in integrated optics must meet a number of different demands, as shown in Figure 2. The optical loss must be very low, which is hindered by the fact that -NH, -OH, and weaker C-H groups absorb in the region around 1.55 μm, the wavelength of choice for optical data transfer. Thus, the content of these groups in the polymer needs to be kept very low. The refractive index and the index contrast must be adjustable in a reproducible way, as the design of the devices is determined delicately by these parameters. The birefringence of the waveguiding materials must also be very low, to ensure an error free data-transfer. Good adhesion of the different polymer layers to each other and to the substrate is necessary. Finally, the thermal, mechanical and optical stability must be guaranteed for several years, along with the other properties mentioned in Figure 2, as they all influence device performance. Also, external parameters must be taken into account.

Polycyanurate Ester Resins

Polycyanurate ester resins are a lesser known class of high-performance-polymers, developed in the late 1960s as base materials for printed circuit boards. They show high thermal and mechanical stability and good adhesion properties on various surfaces. They are used as encapsulants and adhesives, in flame stable composites, in laminates, as friction materials, in lightweight construction, in automotive engineering, and in the aircraft industry.[1]

Scheme 1 shows some basic cyanate monomers, from which we synthesized several polycyanurate copolymers used as waveguide materials.

The two dicyanates DCBA and DCHFBA were purchased from Lonza AG, Switzerland, and used without further purification. The other mono- and difunctional cyanates were synthesized using the cyanogen-bromide method as reported by Martin and Bauer.[2]

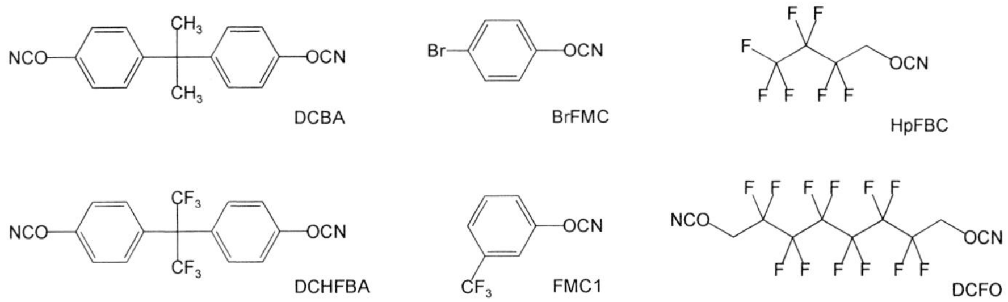
NCO
CH3
CH3
OCN
DCBA
Br
OCN
BrFMC
F
F F
F F F F
OCN
HpFBC
NCO
CF3
CF3
OCN
DCHFBA
OCN
CF3
FMC1
NCO
F F F F F F
F F F F F F
OCN
DCFO

Scheme 1. Some basic cyanate monomers.

The basic reaction used to obtain polycyanurates from the corresponding monomers is the cyclotrimerisation of aromatic or fluoroaliphatic cyanate esters, as shown in the reaction scheme in Figure 3.

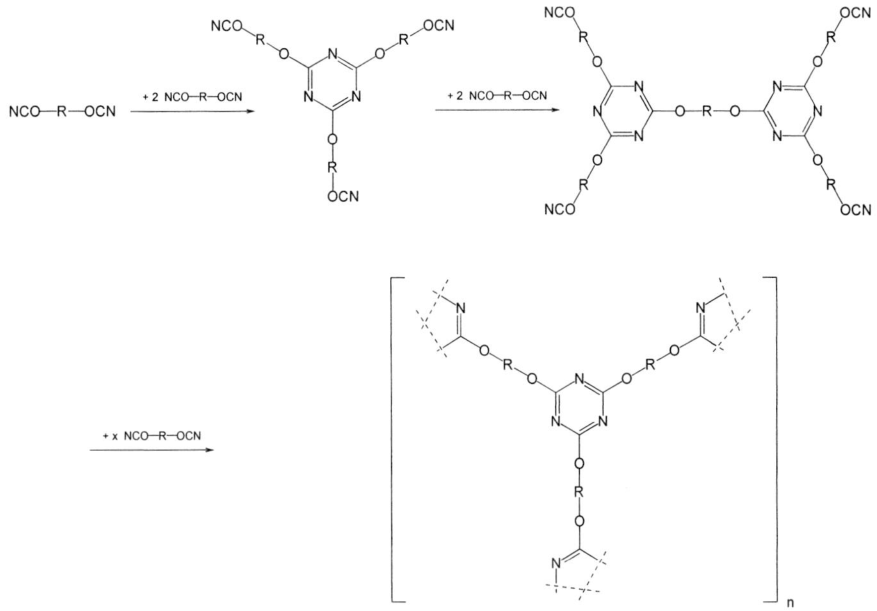
NCO—R—OCN
+ 2 NCO—R—OCN
+ 2 NCO—R—OCN
+ x NCO—R—OCN
n

Fig. 3. Polycyclotrimerisation and network formation.

A difunctional cyanate molecule reacts with two other cyanate molecules and forms a triazine. This triazine has three free reactive groups left; therefore it can react with further monomers and forms the pentamer, the heptamer and higher oligomers, until finally a highly branched three-dimensional network is built. At sufficiently high temperatures and appropriate reaction times an almost complete transformation of the cyanate-groups can be obtained.

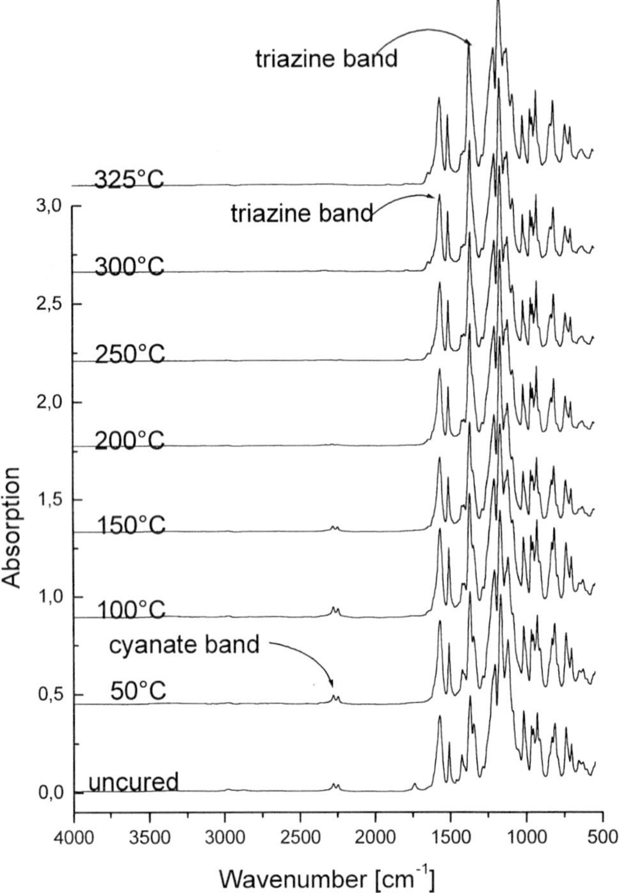

Fig. 4. IR-spectra of the ternary copolymer DCHFBA:DCFO:HpFBC (61,99:6,28:31,74 mole-%). Offline-measurement of thin films on silicone, cured at different temperatures.

This reaction has the great advantage that no by-products, neither gaseous nor liquid, are formed during the reaction. The syntheses of the (pre-)polymers were performed by polymerization of difunctional cyanates in bulk in sealed vials at temperatures between 140 and

190°C. The reaction was stopped rapidly at conversion rates between 40 and 48-% using liquid nitrogen. The prepolymers were dissolved in 2-ethylethoxyacetate and the solutions (45-65 wt.%) were filtered under cleanroon conditions through 0.2 μm PTFE-membrane-filters. Thin films were obtained by spin coating and thermal curing.

The progress of the curing can be examined by the decrease of the cyanate band in the wavenumber region around 2250 cm^{-1} and by the increase of the triazine bands at 1370 and 1558 cm^{-1} (Figure 4).

Optical Properties

The first step to obtaining a polymeric waveguide material was to adjust the refractive index and find a material pair with a sufficient index-contrast (about 0.010, depending on the design of the integrated optical device and the waveguide dimensions). This can easily achieved with polycyanurates as shown in Figure 5. Almost all monomers can be combined with one another, so the refractive index can be adjusted over a broad range. It can be seen that almost every refractive index between 1.48 and 1.59 can be realized using these materials.

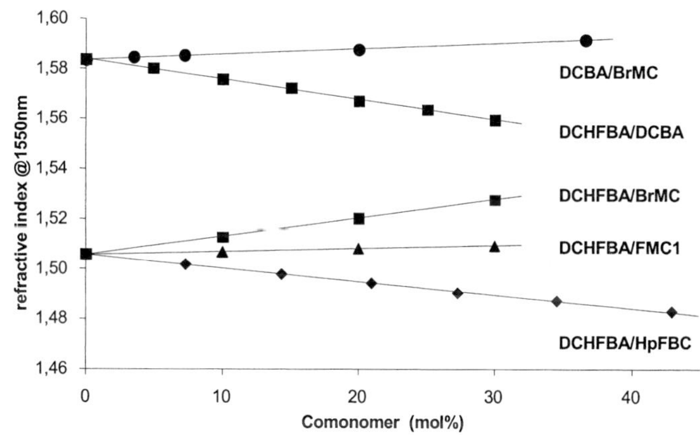

Fig. 5. Refractive index of different copolymers versus comonomer-content.

The optical loss of polycyanurate ester resins should be quite low, as an intrinsic property of the polymer is the absence of OH- and NH- groups. Even after the cyclotrimerisation, to form the network, none of these undesireable groups should be present.

The optical loss of different polycyanurate copolymers is shown in Figure 6. The neat DCBA and DCHFBA show a high increase in the loss during crosslinking: The DCBA monomer shows 0.58 dB/cm and DCHFBA 0.48 dB/cm, whereas the fully cured layers show losses around 1.0 dB/cm. The copolymerization with monofunctional cyanates (BrMC, FMC1, HpFBC) results in an amazing reduction of the optical loss with increasing comonomer content down to 0.4 dB/cm. In part this is due to the increasing halogen content of the resulting copolymer, but this effect is only minor, as the binary copolymers consisting of DCHFBA and FMC1 have a constant content of fluorine, and because both DCHFBA and FMC1 have almost identical fluorine contents. The higher contents of monofunctional cyanates result in a decrease of the network density and in a decrease of the glass transition temperature (T_g). First results indicate as previously reported,[3] that side reactions (especially the reverse reaction with the formation of the phenol from the cyanates) which cause high optical losses are now inhibited, because the chains are more mobile and the cyclotrimerisation is not hindered by rigid chains.

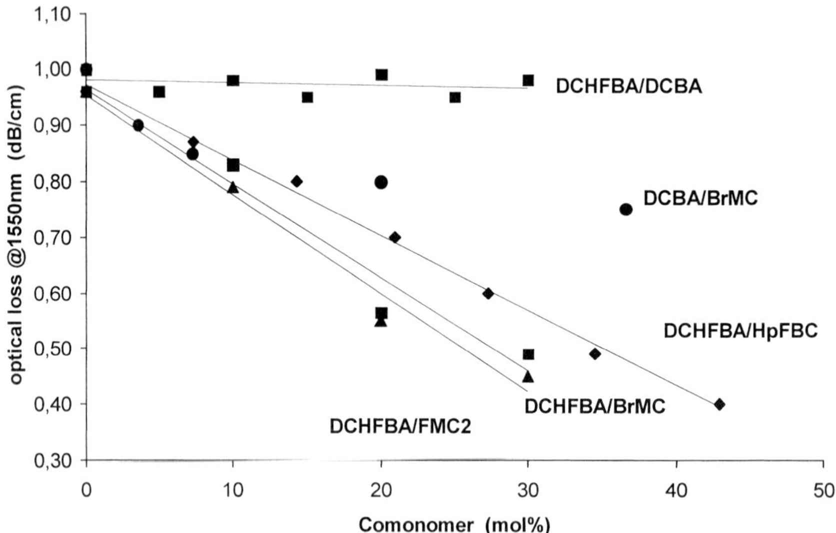

Fig. 6. Optical loss of different copolymers versus comonomer-content.

Further optimization of the copolymers and incorporation of higher fluorine content using the bisfunctional DCFO (Fig. 2) resulted in ternary copolymers with optical losses down to 0.28 dB/cm in the cured layer. Figure 7 shows the optical loss of different copolymers vs. their fluorine content. The increase in the fluorine content also results in a significant decrease of the optical loss. However, it must be considered that the introduction of a higher amount of fluorine into these copolymers is always coupled with a reduction in the network density and T_g, as the carrier of high amounts of fluorine are aliphatic monofunctional cyanates (HpFBC) or difunctional aliphatic cyanates (DCFO) with low intrinsic T_g.[4]

Optical multiplexers and demultiplexers are especially sensitive to birefringence. Therefore extensive investigations on the reduction of the birefringence of the polycyanurates were done. The birefringence of the polycyanurate layers is a result of the different coefficients of thermal expansion (CTE) between the polymeric layer and the substrate. As previously reported[3, 5, 6] we were able to reduce the birefringence by invention of polymeric substrates and proper adjustment of the substrate-CTE; even negative values were obtained.

As previously reported embedded optical waveguides in good quality could be obtained from the synthesized polycyanurates.[3]

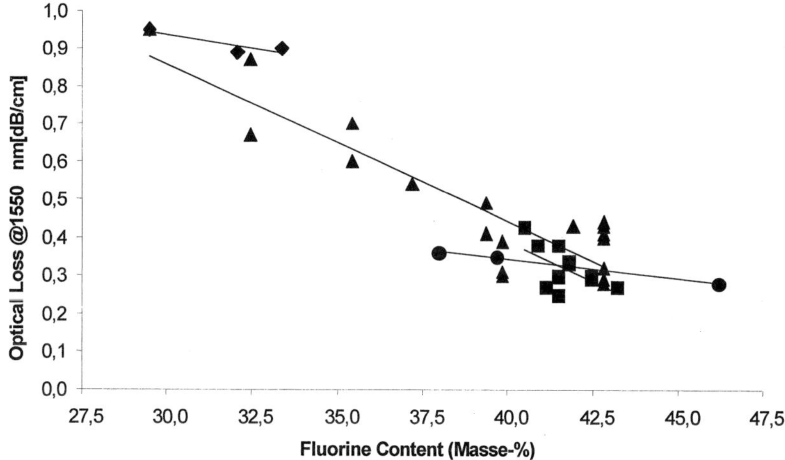

Fig. 7. Optical loss of different polycyanurate copolymers vs. fluorine content (mass-%). ■ -HpFBC = constant (≈ 31.7 mol-%), DCFO-content variable. ◆-DCFO-content variable, 0 % HpFBC. ▲ - HpFBC-content variable, 0 % DCFO. ●-other mixtures.

New Triazine Containing Polymers

Our present work is also focused on triazine-polymers with alternate structures with the goal of solving some of the disadvantages of the polycyanurate systems described above, like high curing temperatures and high birefringence on inorganic substrates. Figure 8 exhibits our strategy for obtaining these new materials. In the first step we incorporate fluorinated or partially fluorinated chains (R^1, R^2, R^3) into the triazine-branchend network, which is still soluble and therefore not fully crosslinked. In a second step we are linking additional reactive groups (X) with the network. These give us the possibility to crosslink the network even further, which is then insoluble and stable. The final crosslinking can be done thermally or by UV-radiation, depending on the particular reactive group.

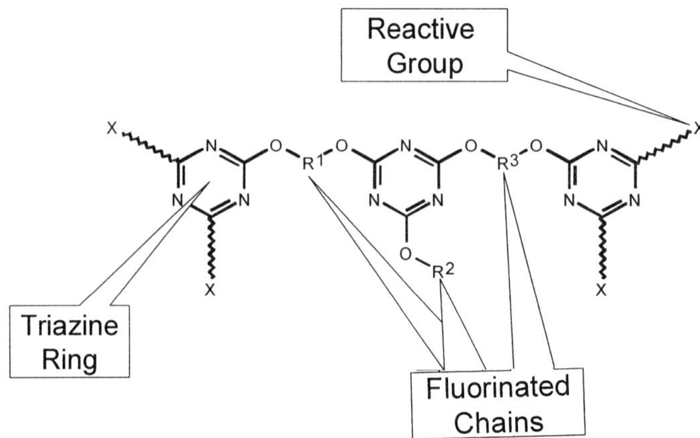

Fig. 8. Alternate triazine containing structures.

We synthesized several polymers using the route described above. The first results appear to be very promising. The films obtained from these new polymers show a good quality and low optical loss. The refractive index is reproducible (with regard to different batches and different curing runs). By varying the chains (R^1, R^2, R^3) and their compositions we can change the fluorine content and thus adjust the refractive index accurately. The chemical structure is only marginally changed.

Using these materials a core/cladding system with a refractive index-contrast of about 0.011 is available. The optical loss of the cladding material is about 0.41 dB/cm; the core material shows a slightly higher absorption of 0.44dB/cm, due to its lower fluorine content. The loss measurements were carried out by using the sliding prism method.

The new system is cured 60 minutes at 180°C under nitrogen atmosphere. The obtained layers are highly crosslinked and unsoluble in common organic solvents.

The best loss-values of our established polycyanurate systems were 0.28 dB/cm - after several years of optimization - thus we are optimistic about obtaining loss values better than 0.3 dB/cm with our new triazine-containing polymers.

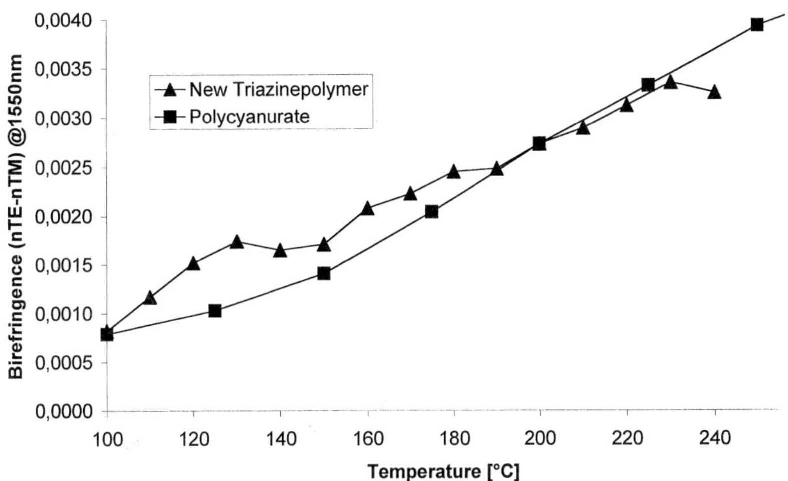

Fig. 9. Dependence of the birefringence of the core material and of a fluorinated polycyanurate on silicone.

Also, an important part of our research is the investigation of birefringence. Figure 9 shows the dependence of the birefringence of the new core material and the ternary polycyanurate copolymer (DCHFBA:DCFO:HpFBC, 61,99:6,28:31,74 mole-%). The birefringence is increasing almost linearly with temperature, as an effect of the different CTE-values of silicone substrate and polymeric layer. It is remarkable that despite the different structures of the polymers the birefringence is almost identical for both systems. Thus, birefringence could not be reduced by this change of the chemical structure. However, as the curing temperature to obtain stable films of the new triazine containing polymers (180°C) could be lowered in comparison to the polycyanurates (250°C), the birefringence of the fully cured layers has been reduced from 0.0039 to 0.0025. This is still too high for polarization sensitive devices, like multiplexers. Therefore further improvement is required.

Comparision of Polycyanurates and the New Triazine Polymers

Table 1 shows some selected properties of the ternary polycyanurate copolymer and the new triazine polymers. These preliminary results are very promising. The lower curing temperatures allow combinations with temperature sensitive materials, which could not be used together with the polycyanurates.

Table 1. Selected properties of the ternary polycyanurate copolymer (DCHFBA:DCFO:HpFBC, 61,99:6,28:31,74 mole-%) and the new triazine polymers (core and cladding material).

Property	Polycyanurate (DCHFBA:DCFO:HpFBC)	Triazine Polymer
Optical Loss @1550nm	0.28 dB/cm	0.44 dB/cm (cladding)
		0.41 dB/cm (core)
Birefringence on Silicone	0.0039	0.0025
Birefringence on (adjusted) Polymer	< 0.0001	work in progress
Thermal Stability	> 250°C	> 180°C
Curing Temperature	250°C	180°C
Structurizability	good	work in progress
Layer Quality	good	good

Our work is now focused on the minimizing of the birefringence and the reduction of the optical loss.

Acknowledgements

We thank the German government for funding our work on the new triazine-based polymers. BMBF FKZ: 01BP151 "Innovative Kommunikationsnetze (KomNet) - Teilvorhaben: Polymer-system für einen integriert-optischen Add/Drop-Multiplexer (OADM) in Polymertechnologie".

[1] I. Hamerton, *"Chemistry and Technology of Cyanate Ester Resins"*, Blackie Academic & Professional, London 1994
[2] D. Martin, M. Bauer, *Org. Synth.* **1983**, *61*, 35
[3] C. Dreyer, M. Bauer, J. Bauer, N. Keil, H. Yao, C. Zawadzki, *Microsystem Technologies* **2002**, *7*, 229
[4] A. W. Snow, J. P. Armistead: *"NRL Mem. Rep. 6848"* Washington, DC: Naval Research Laboratory, 1991
[5] C. Dreyer, M. Bauer, J. Bauer, N.Keil; H.H. Yao, C. Zawadzki, *SPIE Conf. on Linear and Non-Linear Optics of Organic Materials", San Diego, CA (USA), Conf. Dig.* **2001**, *4461*, 188
[6] C. Dreyer, M. Bauer, J. Bauer, N.Keil; H.H. Yao, C. Zawadzki, *Polytronic 2001, Conf. Dig.* **2001**, 188

Polymer Design for Thermally Stable Polyimides with Low Dielectric Constant

Kohei Goto, Toshiyuki Akiike, Yasutake Inoue, Minoru Matsubara*

Tsukuba Research Laboratories, JSR Corporation, 25 Miyukigaoka, Tsukuba, Ibaraki 305-0841, Japan
E-mail: kouhei_gotou@jsr.co.jp

Summary: We successfully prepared a series of thermally stable polyimides (PIs) with low dielectric constant (*k*) by introducing bulky diphenyl fluorenylidene moieties in backbone. The lowest *k* was found to be 2.77 among non-fluorinated PIs and 2.35 among fluorinated ones. In order to prove the lowest limit of *k* in PIs, we prepared soluble and thermally stable polyarylenes (PArs) without polar imide linkage with the same aromatic moieties by coupling polymerization. The lowest *k* was 2.7 without fluorine (F) and 2.2 with F atom, which showed also promising for low *k* materials. From these results, PIs we prepared were estimated to the lowest *k* values among PIs. On the basis of statistics on these results, we could express contour lines of *k* as a function of imide concentration and F content with high correlation factor (r= 0.96) in PIs and PArs.

Keywords: dielectric constant, 9,9-diphenyl fluorine, polyarylenes, polyimides, structure-property relations,

Introduction

A challenge for thermally stable polymers in the fields of microelectronics is lowering their dielectric constant (*k*), since at higher wiring density the signal propagation speed depends on the *k* values of the substrate used as interlayer dielectrics.

Aromatic polyimides (PIs) are among the candidate materials with low *k* and thermal stability. Although the most of the thermally stable PIs used in the electronics industry show *k* over 3.0, they have some advantages to apply for this purpose. That is a diversity of monomer synthesis including aromatic diamines and tetracarboxylic dianhydrides on the standpoint of polymer design.

 DOI: 10.1002/masy.200350927

Our polymer design for PIs with low *k* is based on reducing polar imide concentration (weight fraction of imide linkage by molecular weight (Mw) in repeating unit) which contribute to make *k* higher, as shown examples of the difference of *k* values among Kapton® of Du Pont (k = 3.5; imide concentration= 36.6 wt%), Upilex R® of Ube (k = 3.2; imide concentration= 27.7 wt%), and Ultem® of GE (k= 3.15; imide concentration= 23.6 wt%).

We choose a bulky non-polar aromatic substituent, diphenyl fluorenylidene moieties, which has a possibility to lower *k* with keeping thermal stability, because it combines short aromatic π-conjugation length with a cardo aromatic structure.

In this article we report, 1) synthesis, *k* values and thermal properties of a series of PIs based on substituted diphenyl fluorenylidene moieties in diamines and/or tetracarboxylic dianhydrides, 2) a comparative study of polyarylenes (PArs) without polar imide linkage with the same aromatic moieties in order to prove the lowest limit *k* of PIs, and 3) the relationship between *k* and chemical structure of PIs.

Experimental

Monomer

Among diamines with diphenyl fluorenylidene moieties, a series of diamines, 9,9-bis(4-aminophenyl)fluorene (**1a**), its phenyl substituted derivative (**1b**), 9,9-bis[4-(amino-phenoxy)phenyl]fluorene (**2a**) and its methyl, phenyl and/or trifluoromethyl substituted derivatives (**2b-2g**) except for commercially available **1a**, were prepared.

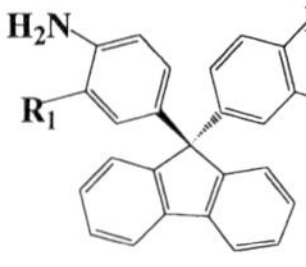

1a: R_1=H
1b: R_1=C_6H_5

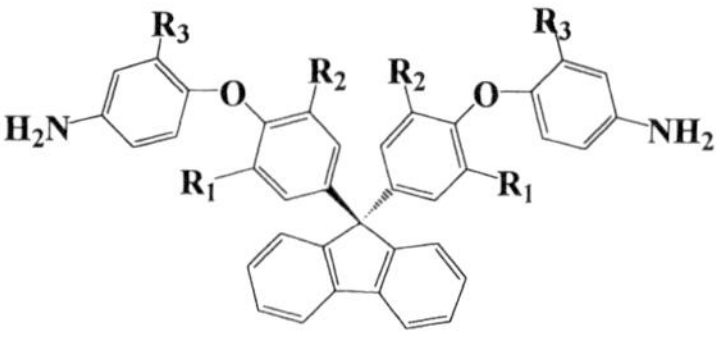

2a: R_1=R_2=R_3=H
2b: R_1=C_6H_5, R_2=R_3=H
2c: R_1=H, R_2=R_3=CH_3,
2d: R_1=R_2=R_3=CH_3
2e: R_1=H, R_2=CH_3, R_3=CF_3
2f: R_1= R_2=CH_3, R_3=CF_3
2g: R_1=R_2=C_6H_5, R_3=CF_3

Among tetracarboxylic dianhydrides, four commercially available monomers, pyromellitic dianhydride (PMDA), 3,3’,4,4’-biphenyltetracarboxylic dianhydride (s-BPDA), 4,4’-oxydiphthalic anhydride (ODPA), and 4,4’-(hexafluoroisopropylidene)diphthalic anhydride (6FDA), were used. Two dianhydrides, 9,9-bis[(3,4-dicarboxyphenoxy)phenyl] dianhydride (**3a**) and its 3-phenyl substituted derivative (**3b**) were synthesized.

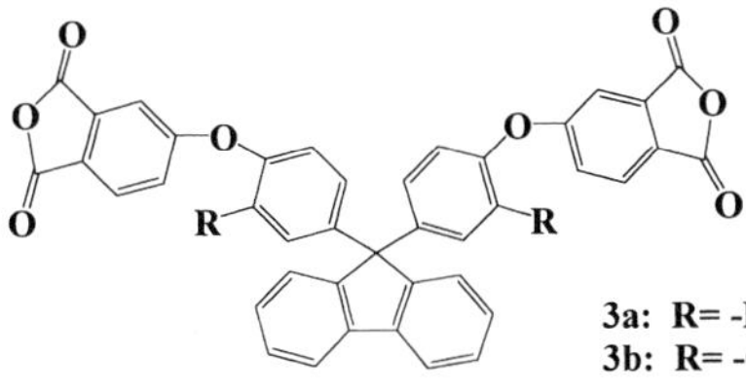

3a: R= -H
3b: R= -C_6H_5

Physical Properties

A conventional polymerization method was used to prepare poly(amic acid)s (PAAs) by polyaddition of corresponding diamine and tetracarboxylic dianhydride in aprotic solvent, such as N,N-dimethylformamide (DMF). The PI film was prepared from the PAA solution by thermal imidization with stepwise heating at 80 °C for 20 min., 140 °C for 20 min., 200 °C for 20 min., 250 °C for 20 min., and 300 °C for 20 min.

Relative Mws of PAAs were estimated from inherent viscosities in DMF at a concentration of 0.5 dL/g at 30 °C.

$Td_{5\%}$ (thermal decomposition temperature at 5 % weight loss) was measured by thermogravimetric analysis (TGA) under nitrogen atmosphere with TGA 320 (Seiko Denshi Kogyo Co.) at heating rate of 10 °C/ min. Glass transition temperature (Tg) was measured by DSC 910 (Du Pont Co.) at heating rate 20 °C/ min. under nitrogen flow.

For the measurement of *k* the polymers were spin-coated onto flat conductive substrates (As-doped Si wafer). Aluminum electrode was deposited *in vacuo* through a mask on top of the polymer films. All measurements of *k* were performed at 1 MHz and room temperature in air, using a LCR meter (HP 4284, Hewlett Packard). The film thickness was measured by elipsometry (DVA-36LH, Mizojiri).

Results and Discussion

PIs with Low k[1] [2]

Table 1 shows the characteristics of the PIs derived from **1a**, **1b** and commercially available four tetracarboxylic dianhydrides, PMDA, s-BPDA, ODPA and 6FDA. A series of these PIs showed the lower k with the lower imide concentration as expected. The PIs without F atom showed relatively low k= 3.0-3.3. PIs with F atom showed k= 2.6–2.8, with the lowest value, k= 2.63, being obtained from **1b** and 6FDA.

Table 1. Characteristics of PIs with 9,9-bis(4-aminophenyl)fluorine (1).

Diamine	Dianhydride	Imide (wt%)	F (calcd) (wt%)	PAA	PI		
				η_{inh} (dL/g)	$Td_{5\%}$ (°C)	Tg (°C)	k
1a	PMDA	26.4	0	0.95	548	>450	-
	s-BPDA	23.1	0	0.58	575	420	3.27
	ODPA	22.5	0	0.57	553	345	3.03
	6FDA	18.5	15.1	0.72	533	366	2.81
1b	PMDA	20.5	0	0.60	529	366	-
	ODPA	18.1	0	0.66	543	315	3.00
	6FDA	15.4	12.5	0.41	536	297	2.63

In order to lower k by reducing imide concentration, **2a** or **2b** were used instead of **1a** or **1b**. A series of these PIs also showed lower k with lower imide concentration, as shown in Table 2. The lowest k of the PI without F atom was showed by **2b**/ODPA, to be 2.80. The lowest k of the PI with F atom was 2.65, from **2a**, or **2b** and 6FDA. It seems possible to approach the lower limit of k, around 2.65, by using the fluorenylidene diamine, **1a**, **1b**, **2a**, **2b**, in combination with commercial aromatic dianhydrides.

PIs with much lower imide concentration were prepared by using some methyl- and/or trifluoromethyl-substituted fluorenylidene diamines, **2c**, **2d**, **2e**, **2f**, and **2g**. In these PIs syntheses PMDA and 6FDA were used as a typical teteracarboxylic dianhydride. Table 3 shows the characteristics of PIs with a series of substituted fluorenylidene diamines.

Table 2. Characteristics of PIs with 9,9-bis(4-aminophenoxyphenyl)fluorene (2).

Diamine	Dianhydride	Imide (wt%)	F (cald) (wt%)	PAA η_{inh} (dL/g)	PI $Td_{5\%}$ (°C)	Tg (°C)	k
2a	PMDA	19.6	0	0.92	546	358	2.91
	s-BPDA	17.7	0	0.61	567	302	2.95
	ODPA	17.4	0	0.78	540	293	2.93
	6FDA	14.9	12.1	1.31	534	293	2.65
2b	PMDA	16.2	0	1.15	544	293	2.82
	s-BPDA	14.9	0	0.54	543	272	2.86
	ODPA	14.6	0	0.61	539	253	2.80
	6FDA	12.8	10.4	0.89	539	268	2.65

PIs with methyl-substituted fluorenylidene diamines showed inferior thermal stability, exhibiting $T_{d5\%}$ of *ca.* 100 °C lower than that of unsubstituted fluorenylidene diamine, owing to aliphatic substituent. On the other hand, PIs with trifluoromethyl-substituted fluorenylidene diamines kept superior thermal stability, which reflects the difference in bond energy between C-F and C-H bond. While the lower imide concentration led to the lower k, it was also observed that F content from CF_3 groups in both diamines and tetracarboxylic dianhydrides contributed to reduce k.

Table 3. Characteristics of PIs with substituted 9,9-bis(4-aminophenoxyphenyl)fluorene (2).

Diamine	Dianhydride	Imide (wt%)	F (calcd) (wt%)	PAA η_{inh} (dL/g)	PI $Td_{5\%}$ (°C)	Tg (°C)	k
2c	PMDA	18.2	0	0.99	454	312	2.77
	6FDA	14.0	11.4	0.97	446	278	2.62
2d	PMDA	17.5	0	0.62	434	347	2.81
	6FDA	13.7	11.1	0.54	436	307	2.70
2e	PMDA	15.9	13.0	0.77	460	307	2.62
	6FDA	12.7	20.6	0.73	454	281	2.51
2f	PMDA	15.4	12.6	0.69	447	362	2.70
	6FDA	12.4	20.1	0.58	439	309	2.53
2g	PMDA	14.0	11.4	0.60	550	290	2.69
	6FDA	11.4	18.5	0.67	538	269	2.46

In the F-free PIs, methyl-substituted PIs, **2c**/PMDA and **2d**/PMDA, which reached k= 2.77-2.81, had higher imide concentration than that of phenyl-substituted **2b**/PMDA. The reason of reversed relation is thought to originate from larger bulkiness of phenyl substituent compared with methyl ones. Introducing plural substituents into **2c** and **2d** resulted in increasing free volume in the PI. In a series of these PIs with F atom, the lowest *k* of the PI, **2g/6FDA**, with thermal stability was 2.46 ($Td_{5\%}$= 538 °C, Tg= 269 °C). *K* values approached *k*= 2.5 more and more closely, which is the lowest *k* value among PIs reported before.[3] [4]

In order to achieve lower *k* by using a bulky fluorenylidene skeleton effectively as diluent to polar imide bond, we tried to introduce bulky substituents not only into diamine but also into tetracarbxylic dianhydride moieties. PI with fluorenylidene moieties both in diamine, **2a** (without F), **2g** (with F), and tetracarbxylic dianhydride, **3a**, and its 3-phenyl substituted, **3b**, were investigated. The results are listed in Table 4.

Table 4. Characteristics of PIs with 9,9-bis[(3,4-carboxyphenoxy)phenyl]fluorene dianhydride (3).

Dianhydride	Diamine	Imide (wt%)	F(calcd) (wt%)	PAA	PI		
				η_{inh} (dL/g)	$Td_{5\%}$ (°C)	Tg (°C)	*k*
3a	**2a**	12.6	0	0.47	542	284	2.80
	2g	10.0	4.1	0.34	558	263	2.61
3b	**2a**	11.1	0	0.51	525	266	2.77
	2g	9.0	3.7	0.35	554	253	2.60

2a/3b with fluorenylidene units in both monomers shows the lowest imide fraction, 11.1 wt%, resulting in the lowest *k* among the F-free PIs which we studied, *k*= 2.77, with keeping thermal stability ($Td_{5\%}$= 554 °C, Tg= 266 °C). Since **3a** and **3b** contain no F atoms, it does not contribute to higher F content in PI by using these tetracarboxylic dianhydrides. It shows the limitation of reducing imide concentration to get lower *k*. These results also suggested that PI with *k*= 2.5 required both reducing imide concentration and the increasing F content.

By introducing diphenyl fluorenylidene moieties as bulky cardo structure to reduce polar imide concentration into PI backbone, we successfully prepared non-fluorinated thermal stable **PI (A)**, **2a/3b,** with the lowest *k*, 2.77, and fluorinated PI, **2g/6FDA,** with the lower *k*, 2.46.

To obtain PI with the lowest *k* in fluorinated PI, we prepared **PI (B)** with the lowest *k*, 2.35, by lowering imide concentration from 11.4 to 9.9 wt% by using diamine **2g** and 2,2-bis[(3,4-

dicarboxyphenoxy)]hexafluoropropane dianhydride instead of 6FDA, with keeping 16.1 wt% of F content. By considering imide concentration, in other words, Mw of repeating unit, the lowest imide concentration is limited to around 10 wt%. This PI with the lowest k showed high thermal stability $Td_{5\%}$= 549 °C, Tg = 238 °C.

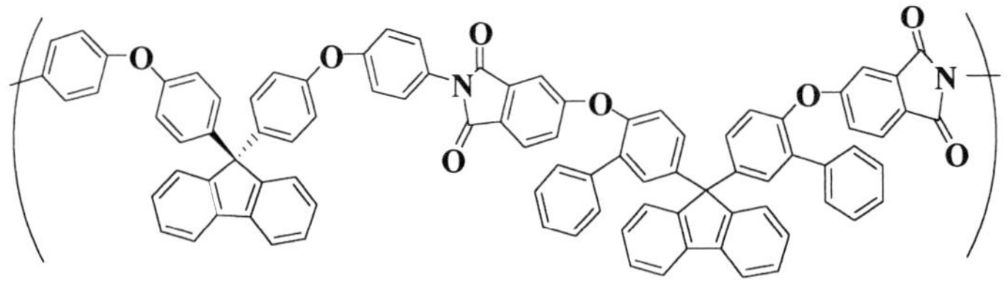

PI (A)

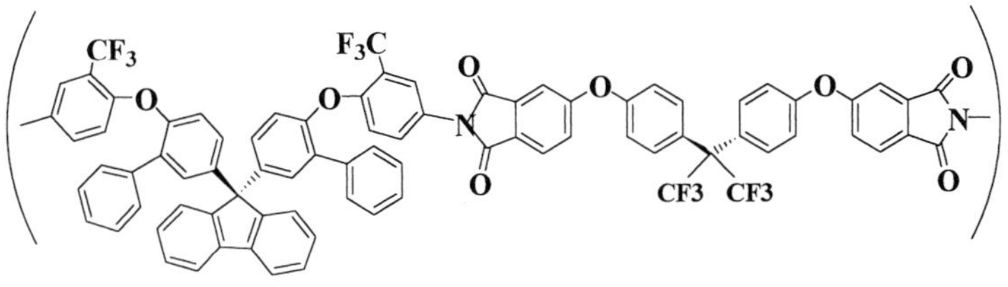

PI (B)

K of PArs as the Lowest Limit of PIs[5]

We obtained a series of PIs with lower k as mentioned above. However, we have a question whether k value of obtained PI is the lowest limit among known PI. In order to solve such a question, we compared k values between PI and PAr without polar imide linkage with the same aromatic moieties.

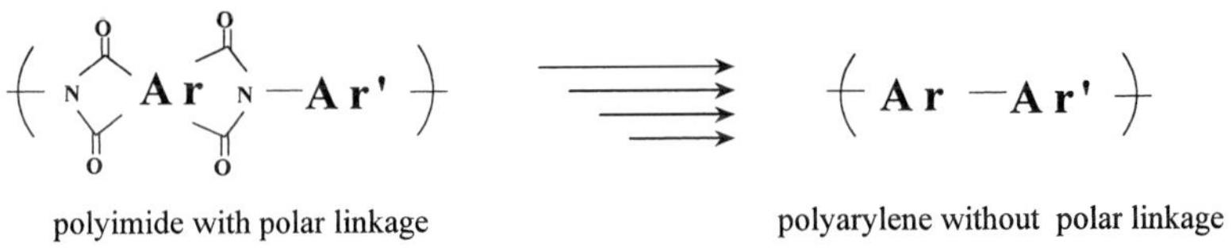

polyimide with polar linkage

polyarylene without polar linkage

In case of F-free PI with the lowest *k*, 2.77, a major chemical structure which contributes to lower *k* in PI is fluorenylidene moiety. Poly(diphenylfluorenylidene-diyl) is corresponding **PAr (C)** to this PI in more simple structure. It may show *k* lower than 2.77.

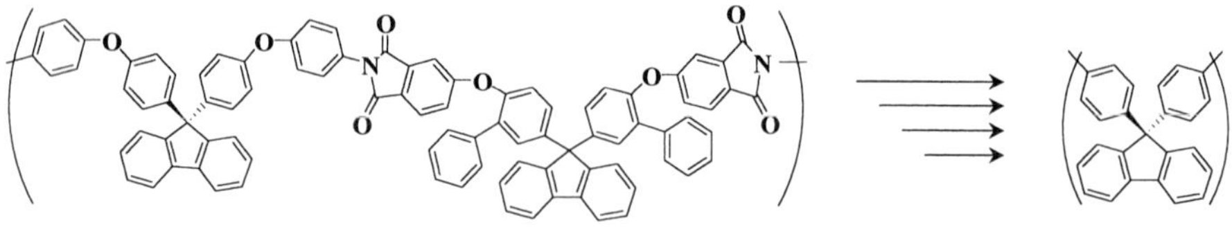

PI(A) **PAr (C)**

Moreover, in case of fluorinated PI with the lowest *k*, 2.35, a major chemical structure which contributes to lower *k* in PI, a main structure of 6FDA unit. Poly(2,2-diphenyl-hexafluoro-propane-diyl), **PAr (D)**, is corresponding PAr. It may show k lower than 2.35.

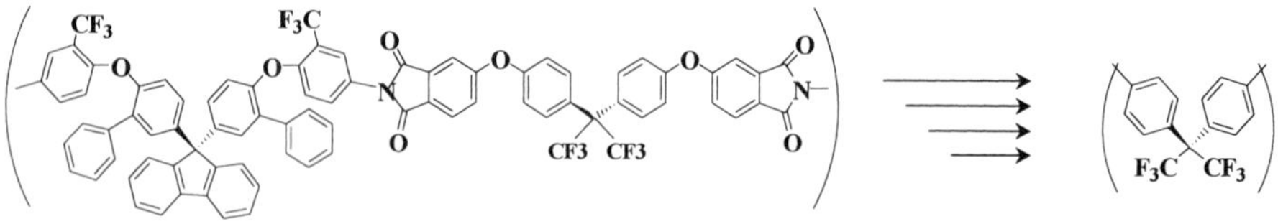

CF_3 F_3C O O O N O O N O CF3 CF3 O F_3C CF_3

PI(B) **PAr (D)**

Anyway the comparison of *k* value between PIs and PArs is significant to prove the lowest limit of *k* values among PIs.

There are some methods to prepare PArs by coupling polymerization. A typical synthesis of PArs is using Zn and nickel-phosphine catalyst. Aryl dichlorides and/or bis-methansulfonates (so-called bis-mesylates) of bisphenols were used as functional groups for coupling reaction according to the method by Percec *et al.*[6] [7] We used bisphenols as a starting monomer unit, because of availability of a variety of commercialized bisphenols.

Poly(diphenylfluorenilydene-diyl), **PAr (C)**, was successfully prepared with superior thermal properties, Tg higher than 300 °C and $Td_{5\%}$= 562 °C with high Mw (>13000 by GPC). However, *k* of this PAr was impossible to measure in accuracy owing to insufficient solubility for spin-coating processing.

To improve solubility of this PAr, we tried copolymerization with bis-mesylates of 1,1-bis(4-hydroxyphenyl)diphenylmethan, a similar structure to bis(4-hydroxyphenyl)fluorene. Copolymers (**PAr (E)**) showed improved solubility, to be soluble in NMP, cyclohexanone and even THF with keeping high thermal stability.

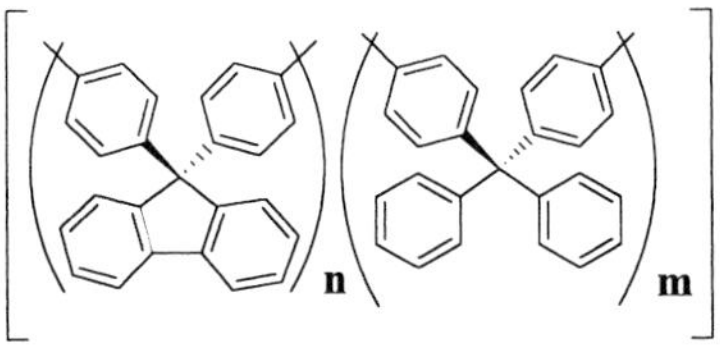

PAr (E)

K of copolymer with composition of molar ratio 60 to 40 mol% could be measured as 2.7. This copolymer can be also considered as a candidate of low *k* polymers with thermal stability ($Td_{5\%}$=549 °C, Tg>300 °C). Since the difference in *k* values between the PI and the PAr is less than 0.1, *k* of our non-fluorinated PI showed the lowest limit of *k* among PIs.

In case of fluorinated PAr, PAr from bismesylate of Bisphenol AF could be prepared with high Mw (27000 by GPC). It showed a good solubility, to be soluble in NMP, cyclohexanone, THF, chloroform and even in ethyl lactate recognized as a safe solvent in IC industry. However, thermal properties ($Td_{5\%}$= 508 °C, Tg= 249 °C) were not better than those of PI. *K* showed 2.2, the lowest value among our PIs and PArs. It has a little difference between that of the PI (*k*= 2.35). It is owing to the difference of F content between the PAr and the PI. The PAr has higher F content with 37.7 wt%. On the other hand, PI with *k*= 2.35 has 16.1wt% of F. Considering the different F content in polymers, *k* of our fluorinated PI estimated to be the lowest limit of *k* among PIs.

Relationship between *k vs.* Chemical Structure in PIs and PArs[5]

Next issue is to prove the relationship *k vs.* chemical structure of PIs and PArs. Contour lines of *k* could be obtained as a function of imide concentration (wt%) and F content (wt%) by our

31 experimental points of PIs and PArs on the basis of statistics with high correlation factor (r= 0.96) (Figure 1).

This figure tells that; 1) *K* is estimated by chemical structure of repeating unit. 2) At higher imide concentration, *k* is mainly affected by imide concentration. 3) The lowest *k* of PI with F-free PI is estimated to be *ca.* 2.7 judging from imide concentration, in other words Mw of repeating unit.

To achieve PI with *k* lower than 2.5, imide concentration has to be lower than 12 wt% (*ca.* 1170 or higher Mw of repeating unit) and F content higher than 12 wt%.

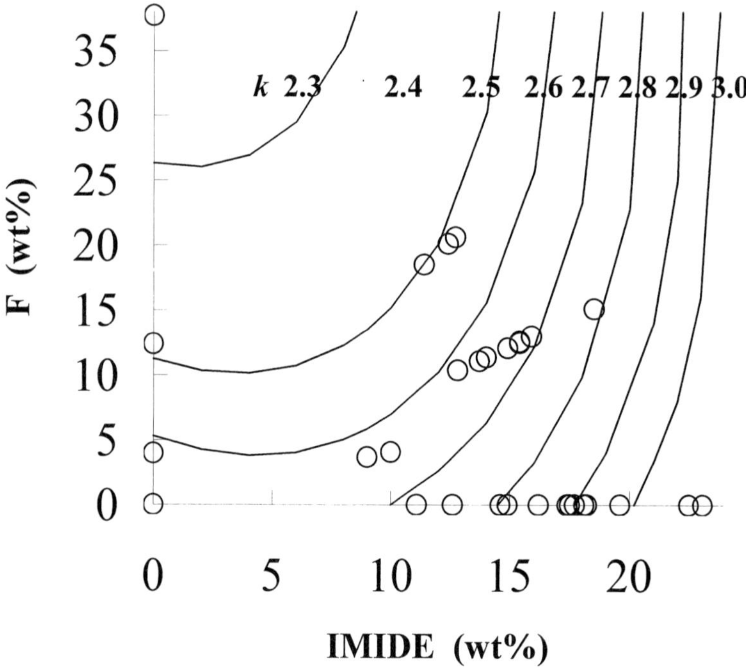

Fig. 1. Plots of *k* *vs*. imide (wt%) and F (wt%); (o; experimental data).

Conclusions

1) By introducing a bulky cardo structure, fluorenylidene moieties, into PI backbone, we successfully prepared F-free PI with the lowest k= 2.77, ($Td_{5\%}$= 525 ^oC, Tg= 266 ^oC), and fluorinated PI with the lowest k= 2.35 ($Td_{5\%}$= 549 ^oC, Tg= 238 ^oC).
2) A series of the thermally stable and low k PArs were obtained, which contained the similar main moieties of above PIs. F-free with fluorenylidene moiety showed k=2.7, ($Td_{5\%}$= 549 ^oC, Tg>300 ^oC) fluorinated showed k= 2.2 ($Td_{5\%}$= 508 ^oC, Tg= 249 ^oC). Such soluble PArs also proved to be promising as low k materials with thermal stability.
3) Based on the results of the low k PAr with corresponding structure of the PIs, we could estimate that PI with fluorenylidene moieties showed the lowest limit of k among PIs.
4) Using 31experimental data set of k values of our PIs and PArs, contour lines of k as a function of imide (wt%) and F (wt%) could be obtained on the basis of statistics with high correlation factor. We can estimate k values from the chemical structure of repeating unit of PI and PArs.

Acknowledgment

The authors would like to thank to Dr. M. Takahashi, Dr. M. Patz, Dr. I. Rozhanskii, Dr. K. Konno and Ms. M. Kakuta-Shinoda for their valuable comments through discussions.

[1] K. Goto, M. Kakuta, Y. Inoue, M. Matsubara, *J. Photopolym. Sci. Tech.,* **2000**, *13*, 313.
[2] K. Goto, Y. Inoue, M. Matsubara, *J. Photopolym. Sci. Tech.,* **2001**, *14*, 33.
[3] S. Hayashi, T. Nishikimi, M. Yamamoto, K. Yamamoto, Nitto Giho, **1990**, *28* (No.2), 49.
[4] B. C. Auman, and S. Trofinenko, *Polym. Mater. Sci. Eng.*, **1992**, 60, 253.
[5] K. Goto, T. Akiike, K. Konno, T. Shiba, M. Patz, M. Takahashi, Y. Inoue, and M. Matsubara, *J. Photopolym. Sci. Tech.*, **2002**, *15*, 223.
[6] V. Percec, S. Okita, and R. Weiss, *Macromolecules*, **1992**, *25*, 1816.
[7] V. Percec, J-Y. Bae, M. Zhao, and D. H. Hill, *Macromolecules*, **1995**, *28*, 6726.

Low Viscous Unsaturated Polyester Resin for Monomer Free UP-Resins

Günter Hegemann

Schenectady-Beck Electrical Insulating Systems Hamburg, Germany
Email: guenter.hegemann@schenectady.de

Summary: It was possible to develop comonomer free unsaturated polyester resins for the application in electrical industry. By testing model compounds and mixtures of model compounds it was possible to find the optimised structure of the resin molecule with respect to the position of the copolymerizable double bonds. By choosing the proper composition of the other components of the resin, e.g. the glycols and modifying dicarboxylic acids, it was also possible to manufacture low viscous resins.

Keywords: crosslinking, curing of polymers, esterification, polycondensation, polyesters, reactivity, resins, unsaturated polyester resin, viscosity

Introduction

The insulating system of an electrical machine consists of the primary insulation and the secondary insulation. The primary insulation is the wire enamel fulfilling the main insulation purpose of the whole system. The secondary insulation has only a small contribution to the electrical insulation itself. The main purposes are

- Mechanical reinforcing the magnet winding, especially in rotating parts of the machine
- Improving heat transfer resulting from the ohmic resistance of the windings
- Protection against mechanical and chemical attack on the primary insulation
- Last but not least supporting the electrical insulation power of the primary insulation.

As a standard today electrical machines normally run at temperatures of 155°C, some special machines at even higher temperatures. Therefore the secondary insulation has to be able to withstand such temperatures without losing the ability to fulfil the above mentioned requirements during the lifetime of the equipment.

 DOI: 10.1002/masy.200350928

For a long time, and in many parts of the world even today, the secondary insulation was built by varnishes curing at relatively high temperatures. Since the 1960ies in Europe and especially in Germany these varnishes were gradually replaced by unsaturated polyester resins. The advantages of such systems over varnishes are:

- No need to use solvents which have to be removed by time and energy consuming evaporation processes
- The filling of the gaps in the windings is much better as there is no material loss during the curing as with a varnish in which about 50% of the material in the winding has to be removed
- As a radically initiated curing process the reactivity is much higher as in varnishes resulting in much shorter curing cycles at lower temperatures

As there is no need to use a solvent to have the insulating resin in an applicable form UP resins theoretically are 100% systems. Therefore the impact on environment should be neglectable.

In reality the possibility of adjusting the viscosity of the system by addition of an excess of comonomer, mainly styrene, and the necessity to cure at elevated temperatures to have stable resins at ambient temperature leads to losses of styrene during curing. In order to reduce the VOC in technical processes it is necessary to clean the exhaust air from the evaporated styrene. To avoid the costly installation of such cleaning systems many efforts have been made to reduce the evaporation, mainly by using other comonomers as styrene with lower vapour pressures, for instance by using diallylphthalate or highly boiling acrylates. For various reasons such products did not gain a considerable market share.

Another strategy to reduce the VOC content to nearly zero would be the use of comonomer free unsaturated polyesters. The total amount of unsaturated polyester resins used in electrical insulation is very small compared to the huge amounts going into other business, but nevertheless it is useful to reduce the impact on environment even in such comparatively small areas of use.

Requirements for a Monomer Free Unsaturated Polyester Resin

An unsaturated polyester resin consists of a polyester containing unsaturated double bonds usually introduced by the unsaturated dicarboxylic maleic or fumaric acid. This unsaturated base resin normally is highly viscous or even solid at ambient temperature. Therefore it cannot be used for impregnating any object. On the other hand this base resin is hardly curable at normal conditions because the tendency of the ester double bonds to homopolymerize is quite low. Therefore a copolymerizable monomer is added, in which the base resin is soluble. By adding various amounts of this comonomer, mostly styrene, the viscosity can be adjusted as it is needed.

An unsaturated polyester without a copolymerizable monomer has to fulfil the following requirements:

- It has to be curable under economic conditions
- It has to be low viscous to allow the application
- The resin should be a one pot system with long potlife at ambient and slightly increased temperatures
- The potlife of the resin activated by the hardener has to be in a range allowing a production without problems

From the beginning of the work there was no doubt that it would not be possible to develop a polyester fulfilling the legal requirements of a polymer with viscosities in the range of the normal monomer containing UP resins. Otherwise it would have been necessary to pass the whole time consuming and expensive procedure for the testing of new chemicals. It was believed from the beginning that the resin had to be heated to reduce the viscosity for application. Therefore the question of potlife at increased temperature up to 40°C to 60°C became more important.

Solution of the Problem of Copolymerisation; Reactivity of Model Compounds

It had been possible to find two types of double bonds which are able to copolymerize. One of them is the standard unsaturation of UP resins, the maleic/fumaric unsaturation. The other one results from an adduct of cyclopentadiene with a functional group allowing a condensation reaction. It is possible to incorporate these two types of double bonds into one resin without a copolymerisation during the polycondensation of the polyester. The copolymerisation of the double bonds can be started by radicals, for instance from peroxides or UV radiation.

There are two strategies possible: It is possible to incorporate both copolymerizable double bonds in one resin. The other possibility is to make two resins, one containing the double bond from the cyclopentadiene adduct and the other resin containing the double bond from the maleic/fumaric acid alone. This strategy resembles the well known mixing of resin and hardener for instance in polyurethane or epoxy resins with the difference that in the case of monomer free UP-resins an additional initiator has to be added.

It became very quickly obvious that the reactivity strongly depends on the number of double bonds, the position of them in the molecule and on the ratio of them. Therefore it was necessary to find out these parameters in order to get molecules with the right reactivity on one hand and the good stability needed for the pot life on the other hand. So we decided to synthesise model compounds and mixing them to find the optimum in reactivity by measuring their reactivity by determining the time of gelation (as the aim is the production of a duroplastic system) and the reaction enthalpy of curing by thermal analysis.[1,2]

The model compounds contain

- the double bond of the adduct of the cyclopentadiene symbolised by 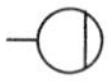
- the double bond of maleic/fumaric acid symbolised by ═══
    - in direct neighbourhood to the adduct 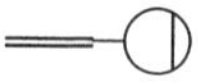
    - in a certain distance to the adduct 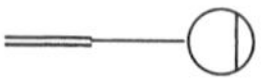
    - in different number in the molecule

Table 1 shows the first group of model compounds and their behaviour under curing conditions after activation with tert.-butyl perbenzoate.

Table 1. Model compounds with no or low reactivity determined by gel time measurement and DSC measurement.

Number	Model compound	Reactivity by Gel Time	DSC Measurement [J/Mol]
1	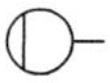	**No crosslinking**	**8600**
2		**No crosslinking**	**8700**
3	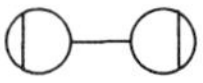	**No crosslinking**	**11500**
4		**Very low reactivity**	**52500**
5		**Good reactivity**	**84300**
6		**No crosslinking**	**28000**
7		**Very low reactivity**	**23100**

The double bonds of the cyclopentadiene adduct show no reactivity by gel time measurements, even when two groups are in one molecule. The addition of one maleic/fumaric double bond makes the molecule slowly curable, while the addition of two such double bonds increases the reactivity considerably. Two maleic/fumaric double bonds without the addition product show a very low reactivity. The measurement of the reaction enthalpy of all model compounds shows that they are reactive, but the reaction does not lead to the desired crosslinking in every case.

Table 2 shows the results of the reactivity measurements of the second group of model compounds. In these compounds the maleic/fumaric double bonds are in the direct neighbourhood of the cyclopentadiene adduct.

Table 2. Reactivity of model compounds with maleic/fumaric double bonds directly linked to the cyclopentadiene adduct.

Number	Model compound	Reactivity by Gel Time	DSC Measurement [J/Mol]
1	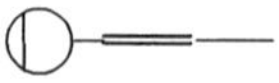	No crosslinking	31500
2		Low reactivity	52300
3	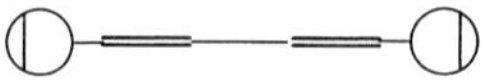	Increasing reactivity	90200
4			112000
5			120500

It is necessary to fix the maleic/fumaric double bond directly to the cyclopentadiene adduct to get a good reactivity in the sense that the material is crosslinked by the curing process. In this sense it is necessary to have at least one adduct with linked maleic/fumaric double bond plus one more maleic/fumaric double bond to have crosslinking. Good reactivity in the sense of crosslinking is obtained with two adduct groups with linked maleic/fumaric double bond as in 3. Further increase of the number of maleic/fumaric double bonds increases the probability of crosslinking and therefore the reactivity characterised by reduced gel time.

Reactivity of Mixtures of Model Compounds

The second strategy to mix model compounds which are not readily homopolymerizable is interesting in the respect of resin manufacture. If both copolymerizable groups are present in one resin there is the danger of uncontrolled reaction during the manufacturing of the resin. Table 3 shows such mixtures. In this case there was no measurement of reaction enthalpy made. Only the possibility of crosslinking was of interest and therefore determined by gel time measurement.

Table 3. Reactivity of mixtures of model compounds which do not crosslink alone.

Number of Mixture	Ratio molar	Model compounds	Reactivity by Gel Time
1	1 1	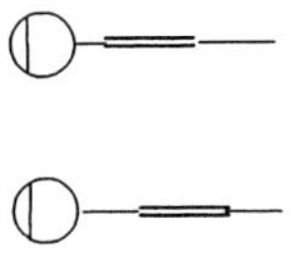	No crosslinking
2	1 1	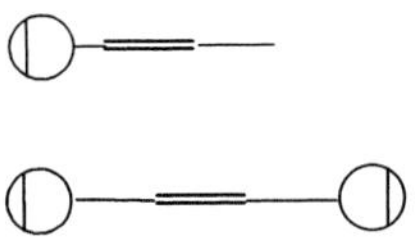	No crosslinking
3	1 2	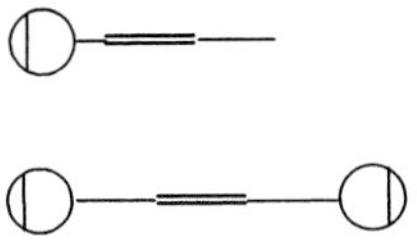	No crosslinking
4	1 1	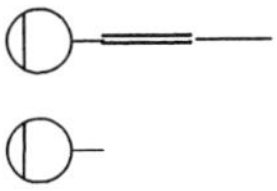	No crosslinking
5	1 1		No crosslinking
6	1 1	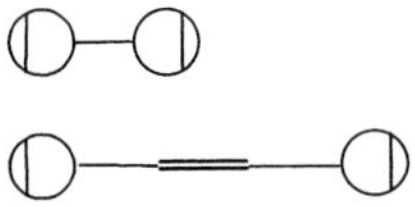	No crosslinking
7	1 1	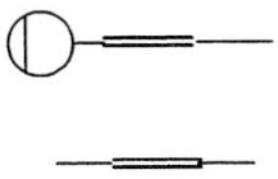	No crosslinking

As the table shows this strategy is not working. Therefore it was necessary to incorporate both reactive types of double bonds in one resin.

Optimisation of the Polyester

As the previous investigations had shown it is necessary to have the double bond of the cyclopentadiene adduct directly linked to a maleic/fumaric double bond and additional maleic/fumaric double bonds in the resin to get a good crosslinking characteristics. Fortunately the system showed such a good stability that under certain precautions it is possible to make polyesters which fulfil the legal requirements for a polymer (number average molecular weight >1000 and a molecular weight distribution) and which can be crosslinked by the addition of radical producing compounds like peroxides. The cured resin also fulfilled the requirements for the application in the manufacturing of electrical machines and the resin has an interesting curing characteristics. The Arrhenius plot shows a good linearity in the temperature range above about 100°C and also below about 80°C. In the temperature range between 80°C and 100°C there is a change in the curing mechanism so that the energy of activation below 80°C is much higher than above 100°C. Therefore it is possible to apply the resin at elevated temperatures up to 80°C to reduce the viscosity without getting problems with the pot life.

Nevertheless it was necessary to try to reduce the viscosity of the resin at ambient temperature by choosing the right composition of the polyester. This can be done by changing the glycols and by introducing modifying dicarboxylic acids in place of parts of the maleic/fumaric acids. Of course this will lower the reactivity of the resin as the probability for crosslinking is reduced. Nevertheless it could be shown that partial exchange of the maleic/fumaric acid by adipic acid greatly reduces the viscosity of the resin.

The second way of influencing viscosity is to vary the glycol used in the polycondensation. Unfortunately it was not possible to use a statistic test plan to reduce the number of tests. There are too many possibilities. In the many test batches it was possible to find an optimised composition of different glycols in one resin which led to a resin that was applicable at even slowly raised temperature. Fig. 1 shows the viscosity range vs. temperature which is usable.

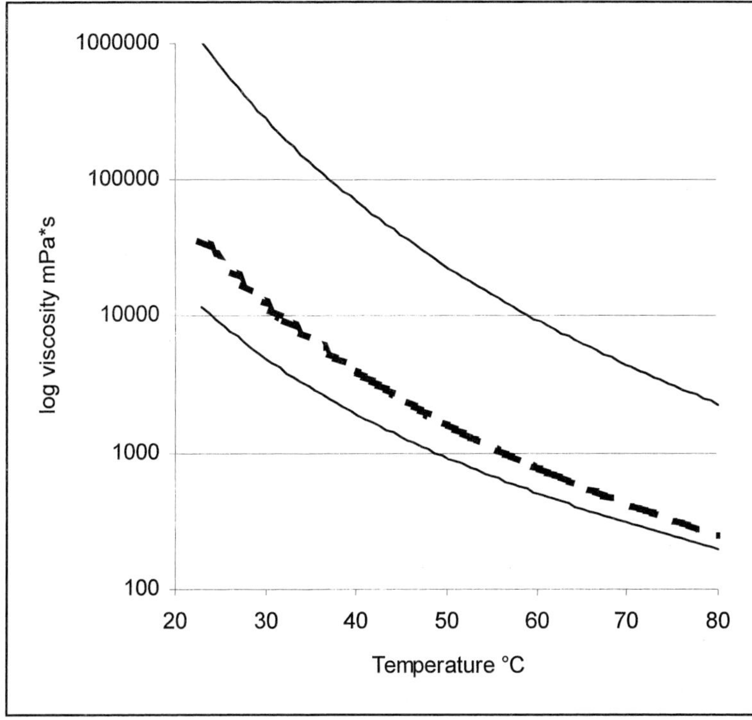

Fig. 1. Usable viscosity range of monomer free unsaturated polyester resins. Dotted line: Upper viscosity limit for standard applications.

A further reduction of viscosity was possible by reducing the acid number. The standard acid number of the polyesters is in the range of about 20 to 25 mg KOH/g. A further reduction by improving the degree of condensation leads to a rapid increase in viscosity as the molecular weight is increased. Therefore the carboxylic end group is esterified with methanol or ethanol until the acid number becomes lower as 5 mg KOH/g. This leads to a further reduction in viscosity of about 20%.

Conclusion

It was possible to develop a monomer free unsaturated polyester which

- can be manufactured by standard polycondensation procedures without reacting although the resin contains two types of copolymerizable double bonds
- can be processed in activated state at elevated temperatures without premature gelling
- can be cured by standard conditions to produce a material fulfilling the requirements for the construction of electrical machines
- has very low emissions during the curing process compared to standard UP resins.

To do this it was necessary to synthesise model compounds to find out which configuration of the two different double bonds gives the best results in reactivity. In this special case reactivity means the possibility to form a duroplastic resin. The other task was to optimise the composition of the polyester with regard to viscosity.

Due to the much higher viscosity compared with the standard UP resins some of the application processes have to be modified.

[1] Horn, Nicole, Modellversuche zur Aufklärung des Härtungsmechanismusses comonomerenfreier ungesättigter Polyester Diplomarbeit im Studiengang Chemieingenieurwesen Fachhochschule Hamburg 1998
[2] Bunardi, Wenli, Wechselwirkungen bei der kombinierten thermischen und UV-Härtung bei monomerenfreien UP-Harzen Diplomarbeit im Studiengang Chemieingenieurwesen Fachhochschule Hamburg 1999

Macromol. Symp. **2003**, *199*, 343-350

New Generation of Fire Retardant Polyester Resins

Ewa Kicko-Walczak

Industrial Chemistry Research Institute, Department for Polyesters and Materials for Medicine, Unsaturated Polyester Resins Laboratory Rydygiera 8, 01-793 Warszawa, Poland
E-mail: ewa.kicko-walczak@ichp.pl

Summary: Smoke evolution [in a smoke chamber (750 x 750 x 1000) ± 5 mm, Polish Standard PN-91/K-02501 equivalent to UIC 561-OR (1991)] was studied in, and the oxygen index flammability test (Polish Standard PN-76/C-89020) was carried out for, glass-reinforced polyester (GRP) laminates obtained with unsaturated polyester (UP) resins containing chlorine and bromine in the chain. In these studies, the effect on these properties of such additives as $ZnSnO_3$ (ZS), $ZnSn(OH)_6$ (ZHS), $Al(OH)_3$ or $Mg(OH)_2$ and Sb_2O_3 in an amount of up to 30 mass-% was determined. The most efficient ignition and smoke-evolution retarder from among the investigated compounds was ZS and ZHS, whereas an essential reduction in smoke evolution was observed also with Sb_2O_3. GRP laminates with these additives meet the fire-safety recommendations concerning smoke evolution from materials used in transportation means and in the building industry.

Keywords: FR additives, oxygen index, polyester resins, smoke evolution

Introduction

Smoke - forming tendency of plastics has been systematically studied, particularly for poly(vinyl chloride) presumably because this polymer has been used most extensively in appliance manufacture.[1]

It appears that, regardless of the nature of a halogen - containing polymer, methods for reducing the optical density of smoke and the retardancy mechanisms involved may be assumed to exhibit close similarities. In the search for similarities in other polymers to unsaturated polyester resins (UPR) smoke suppression studies on thermoplastic polymers, e.g., ABS, compounded with decabromobiphenyl were taken into consideration.[2]

Burning UPR and polyester - glass laminates (GPR) produce black smoke in considerable amounts. The accompanying emission of toxic fumes (HCl, HBr, CO), especially noxious with

 DOI: 10.1002/masy.200350929

slow - burning halogen - containing UPR, makes fire - fighting difficult or even impossible. Therefore, preparation of flame - retarded and simultaneously smoke - suppressed polyester resins is of considerable importance.

This goal was deemed achievable by compounding UPR with additive flame retardants. Most efficient retardants should modify UPR and GRP pyrolysis reactions toward suppressing the formation of fuliginous aromatic pyrolyzates (benzene, toluene) and increasing that of charring aliphatics. Production of smoke and emission of toxic gases from burning UPR are related to the nature and formulation of the resin. By way of illustration, styrene is more advisable at lower proportions; 1,2-propylene glycol is a more smoky component than other glycols; so is phthalic acid compared with isophthalic acid.[3] Admittedly, a "specially" selected structure of the polyester resin chain does not guarantee flame retardancy in the product.

To impart flame retardancy and smoke suppression to UPR and GRP, the resins have been compounded with aluminium hydroxide [Al(OH)3] which, on burning, decomposes at a temperature of 230°C to yield 34% of water: this cools the plastic and extinguishes the fire.[4] Again, in the pyrolysis of UPR, the hydroxide reduces smoke production and emission of HCl and HBr, but only if added in abundant amounts (up to 200 phr), prohibitively high in industrial practice. Smoke suppressants are therefore desired which are effective when incorporated into the formulation in relatively small proportions.

Experimental

Materials

- Polimal 120 (Chemical Works "Organika - Sarzyna", Sarzyna, PL), an isophthalic - maleic - propylene glycol UPR containing 35% of styrene and no halogens;
- Polimal 161 (same manufacturer), a maleic - phthalic - epichlorohydrin UPR containing an incorporated bromine compound, styrene 35%, chlorine (Cl) 9.3%, and bromine (Br) 6.7%;
- Polimal (same manufacturer) based on the HET acid, i.e., a UPR prepared from maleic anhydride, HET acid and diethylene glycol, containing styrene 33% and chlorine (Cl) 21%;
- A glass mat, surface density 450 g/m^2, prepared from a borosilicate glass;
- Aluminium hydroxide, $Al(OH)_3$ (Consulting Enterprise ADOB, Poznań, PL);

- Antimony trioxide (PPOCh, Gliwice, PL);
- Magnesium hydroxide, $Mg(OH)_2$ (Merck - Schuchardt, Hohenbrunn, Germany);
- Molybdenum trioxide (PPOCh, Gliwice, PL);
- Calcium sulfate (PPOCh, Gliwice, PL);
- Zinc hydroxystannate, $ZnSn(OH)_6$ (ALCAN Chemical Europe, Buckinghamshire, GB);
- Zinc stannate, $ZnSnO_3$ (same manufacturer);
- Zinc borate, $2ZnO.3B_2O_3 \cdot 3H_2O$ (same manufacturer);
- Methyl ethyl ketone peroxide (Ketonox), a 40% solution in dimethyl phthalate;
- Cobalt naphthenate, a 1% Co solution in styrene.

Scope of Study

An overview of the literature and this Institute's past research works made the following resins advisable for study:

(i) an isophthalic UPR;

(ii) halogen - containing resins, viz., a maleic - phthalic - epichlorohydrin resin, a maleic - phthalic - epichlorohydrin resin compounded with a boron compound, and a UPR prepared from hexachloroendomethylenetetrahydrophthalic acid (HET acid).

The resins were compounded with selected metal oxides, metal hydroxides and mixtures thereof to benefit by synergism. Synergism occurs between tin, zinc, antimony, molybdenum, calcium and magnesium compounds. For halogen - containing UPR, tin and zinc compounds, viz., $ZnSnO_3$ and $ZnSn(OH)_6$, appear to be particularly promising.

Indigenous UPR and selected smoke suppressants were used for investigations.

Each composition examined included 100 weight parts of resin and 5-30 weight parts of additive(s). The curing compound was a mixture of methyl ethyl ketone peroxide (3 phr) and the (1% Co) cobalt naphthenate (0.4 phr). From each composition a laminate was prepared involving three layers of a ~30% glass mat (surface density, 450 g/m^2). The laminate was conditioned at room temperature, relative air humidity 50%, for 24h and then cured up at 353K (80°C). The laminate was cut into strips, 120 x 100 mm in size. Samples identical in composition, 2-3 mm thick, were examined in triplicate.

Smoke tests were carried out at the Polish State Railway Research Institute's Laboratory for Plastics, Warsaw.

Smoking Intensity

Smoking intensity was studied in conformity with the Polish Standard PN-91 K-02501, Rolling Stock, "Intensity of Smoking of Materials on Burning; Requirements and Tests". The method is to measure the quantity of light passing through the smoke and falling upon a surface during the initial period of the 4-min period of burning time. A special smoke chamber made of heat - insulated material, (750 x 750 x 1000 mm) ±5 mm in internal size, was used. The chamber was equipped with an ignition system, an illuminance measuring system, and a ventilation system.

Results

Illuminance (E, lx) averages of the triplicate measurements were plotted (Figs. 1, 2) against the experimental burning time (t, s) and the luminous energy (S, lx•s) transferred during the initial period, t = 4 min, of experiment duration was evaluated as the surface beneath the curve E=f(t), viz.,

$$S=\int_0^t E\,dt$$

where: S - exposure, E - illuminance (exposure intensity), t - exposure time.

The above terminology is consistent with that of the PN-91-/K-02501.

Smoking intensity was thus expressed in terms of the luminous energy that was transferred throughout the initial burning time (t = 4 min) of the material examined.

Flammability

Flammability tests were carried out in conformity with the Polish Standard PN-76/C-89020 "Flammability Tests by the Oxygen Index Method". The method is to establish the minimum oxygen concentration in a nitrogen - oxygen mixture that sustains combustion of a material sample fixed vertically inside the measuring chamber.

The relative efficiency, e_r, of a flame retardant was expressed as

$e_r=(OI-OI_O)/C$

where: OI, OI_O = the oxygen indices (% by vol.) of respectively flame - retarded and unretarded UPR laminates, C = the retardant content, %, in a UPR resin.

Discussion and Conclusion

Tables 1 and 2 show respectively the OI indices determined for the laminates examined and the relative efficiencies evaluated for the retardants studied. In the series of the flame - retarded halogenated resins, flammability is seen (Table 1) to decrease almost rectilinearly. The OI of the HET acid - based resin is higher than could be expected from the amount of incorporated chlorine. This fact appears to be associated[5, 6] with the HET acid residues which decompose in the course of polyester pyrolysis to revert to starting diene synthesis reactants. The resulting hexachlorocyclopentadiene (HEX) provides a protective barrier against the access of oxygen. The other chlorine - containing UPR released chloride or chlorine. HEX can affect thermal degradation of the original polyester chain by favouring the formation of char rather than of volatile aromatic pyrolyzates.[13]

Table 1. Smoking intensities and oxygen indices determined for GRP laminates.

Sample No.	UPR used in GRP laminate	Smoke suppressant	Smoking intensity, lx•s	OI, vol.%
1	[1)] X-free	-	9100	19.1
2	Cl	-	4706	25.8
3	Cl/Br	-	5100	26.1
4	Cl(HET acid)	-	6363	30.4
5	Cl/X-free	-	8700	24.1
6	Cl/X-free	$Al(OH)_3$, $CaSO_4$, MoO_3	8247	26.2
7	Cl/Br/X-free	$ZnSn(OH)_6$, $Al(OH)_3$	9204	28.5
8	Cl/Br/X-free	$ZnSnO_3$, Sb_2O_3	9523	26.1
9	Cl(HET acid)/ X-free	$ZnSnO_3$, $Mg(OH)_2$	12159	38.7
10	Cl(HET acid)/ X-free	$ZnSn(OH)_6$, $2ZnO\ 3B_2O_3 \circ 3H_2O$	17500	44.4

[1)] X = halogen, Cl = epichlorohydrin`s chlorine incorporated into resin formulation, Br = bromine incorporated into resin formulation.

The OI data (Table 1) for the GRP laminates examined allow simple flammability class assignments. According to the Polish Standard PN-84/K-02500 the laminate characterized by an OI value of 21 to 28% (determined in conformity with PN-76/C-89020) belongs to the slow - burning material class and the one with OI>28% belongs to the non-flammable material class. By this definition, the laminates Nos. 4, 7, 9 and 10 are non-flammable, whereas Nos. 2, 3, 5, 6 and 8 fall in slow - burning class (Table1). Smoke intensity data are given in Table 1, Column 4. They confirm the fact that any halogen - free resin produces less smoke than does a halogen - containing resin.

A desirable smoking intensity (as determined in conformity with PN-91/K-01501) could be achieved by blending a halogen - free resin with a chlorine - containing resin. An optimum OI was obtained. The observed smoking intensities (Table 1, Figs. 1, 2) show that a suitable combination of smoke suppressants results in a smoking intensity successfully lowered.

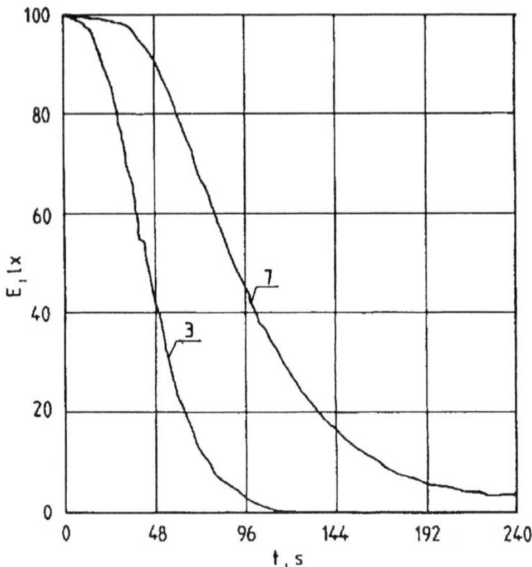

Fig. 1. Illuminance (E,lx) vs. burning time (t,s) for laminates Nos.3 and 7 Table 1.

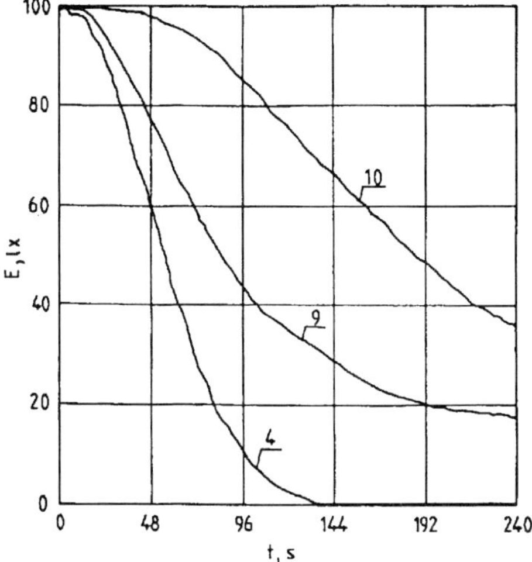

Fig. 2. Illuminance (E,lx) vs. burning time (t,s) for laminates Nos.4, 9, 10 of Table 1.

Zinc hydroxystannate [$ZnSn(OH)_6$] and zinc stannate [$ZnSnO_3$] have been found to be particularly effective as smoke suppressants especially when used in chlorinated and brominated resins.[12] The present data confirm the synergism between these suppressants and the halogens incorporated into the polyester chain in the Polish UPR examined. The observed smoking intensities (Table 1) show that some laminates fall in the medium intensity class, D2 according to PN-91/K-02501. These are the laminates Nos. 1, 7, 8, 9, 10 with luminous energies ranging from $9 \cdot 10^3$ to $1.35 \cdot 10^4$ lx •s (Table 1).

Table 2. Relative efficiencies, e_r, of flame retardants in GRP Laminates.

	$ZnSn(OH)_6$, $2ZnO\ 3B_2O_3 \circ 3H_2O$	$ZnSnO_3$, Sb_2O_3	$Al(OH)_3$, $CaSO_4$
Cl	0.12	0.15	0.14
Cl/Br	0.28	0.21	0.18
Cl/(HET acid)/[1]X-free	0.72	0.58	0.24

[1] X = halogen

Such laminates are suitable for rolling stock construction and/or repairing applications including floors, walls, ceilings and toilets. Plywood and hardboard can thus be eliminated. This replacement would result in prolonged service life, improved external appearance and, most of all, enhanced fire safety of railway carriages.

[1] R.P. Lattimer, W.J. Kroenke.: *J. Appl. Polym. Sci.* **1981**, 26, 1191.
[2] M. M. Hirschler, O. Tsika.: *Eur. Polym. J.* **1983**, 10, 375.
[3] D. P. Miller, R. V. Petrella, A. Manca.: *Modern Plastics* **1978**, 53, nr 9, 95.
[4] D. Scholz: *Off Jahrestagung AVK*, Freudenstadt **1981**.
[5] J. E. Selley, P. W. Vaccarella: *Plast. Engn.* **1979**, 35, 2, 43.
[6] G. A. Skinner: *Fire and Materials* **1976**, 1, 154.
[7] H.P. Dorge, J. Wismer: *Cell Plast.* **1972**, 8, 311.
[8] D. A. Church, F. W. Moore: *Plast. Engn.* **1975**, 31, 12, 36.
[9] G. A. Skinner, P. J. Haines: *Fire and Materials* **1986**, 10, 63.
[10] R. W. Adams: *Plast. Engn.* **1988**, 44, 59.
[11] E. Kicko – Walczak, A. Kamiński, P. Penczek: *Polimery* **1990**, 38, 33.
[12] P. A. Cusack., M. S. Heer, A. W. Monk: *Poly,. Degrad. and Stab.* **1991**, 32, 17.
[13] E. D. Weil, M. M. Hirschler, N. G. Patel, M. M. Said, S. Shaikr: *Fire and Materials* **1992**, 16, 159.

New Organosoluble Polyimides with Low Dielectric Constants Derived from Bis[4-(2-Trifluoromethyl-4-aminophenoxy)phenyl] Diphenylmethylene

Der-Jang Liaw, Wen-Tsung Tseng*

Department of Chemical Engineering, National Taiwan University of Science and Technology, Taipei, Taiwan
E-mail: liaw@ch.ntust.edu.tw

Summary: A new kink diamine with trifluoromethyl group on either side, bis[4-(2-trifluoromethyl-4-aminophenoxy)phenyl]diphenylmethane(**BTFAPDM**), was reacted with various aromatic dianhydrides to prepare polyimides via poly (amic acid) precursors followed by thermal or chemical imidization. Polyimides were prepared using 3,3', 4,4'-biphenyltetracarboxylic dianhydride(1), 4,4'-oxydiphthalic anhydride(2), 3,3',4,4'-benzophenonetetracarboxylic dianhydride (3), 4,4'-sulfonyldiphthalic anhydride(4), and 4,4'-hexafluoroisopropylidene-diphathalic anhydride(5). The fluoro-polyimides exhibited low dielectric constants between 2.46 and 2.98, light color, and excellent high solubility. They exhibited glass transition temperatures between 227 and 253°C, and possessed a coefficient of thermal expansion (CTE) of 60-88 ppm/°C. Polymers **PI-2**, **PI-3**, **PI-4**, **PI-5** showed excellent solubility in the organic solvents: *N*-methyl-2-pyrrolidinone (NMP), *N,N*-dimethylacetamide (DMAc), *N,N*-dimethylformamide (DMF), dimethyl sulfoxide (DMSO), pyridine and tetrahydrofuran (THF). Inherent viscosity of the polyimides were found to range between 0.58 and 0.72 dLg-1. Thermogravimetric analysis of the polyimides revealed a high thermal stability decomposition temperature in excess of 500°C in nitrogen. Temperature at 10 % weight loss was found to be in the range 506-563°C and 498-557°C in nitrogen and air, respectively. The polyimide films had a tensile strength in the range 75-87 MPa; tensile modulus, 1.5-2.2 GPa; and elongation at break, 6-7%.

Keywords: bis[4-(2-trifluoromethyl-4-aminophenoxy)phenyl]diphenylmethane-(BTFAPDM), dielectric constants, kink, polyimide, trifluoromethyl

Introduction

Aromatic polyimides are thermally stable polymers that exhibit excellent mechanical strength and stability. During the past decade, interests in these polymers have risen in response to increasing technological applications in a variety of fields such as aerospace,

 DOI: 10.1002/masy.200350930

automobile, and microelectronics.[1] Generally, aromatic polyimides are insoluble in organic solvents, and have extremely high glass transition or melt-temperatures, which preclude melt processing. Hence, a great deal of efforts have been made to improve the processing characteristics of these intractable polymers. One of the successful approaches to improve solubility and processability of polyimides with minimal detrimental effect on their high thermal stability is the introduction of bulky substituents[2-4] or pendent groups[5,6] along the polymer backbone.

A successful approach that has improved the solubility and processability of polyimides is the introduction of kink linkages such as ether, sulfone, ketone, and methylene groups.[7-9] The most effective approach to obtain organo–soluble polyimides is the incorporation of substituted methylene linkages, such as isopropylidene[10-13] and hexafluoroiso-propylidene[14,15] groups, which provides kinks in the backbone leading to an increased solubility of the polymer. Recently, we have reported polyimides with fluorine-containing kinks along the backbone [15]. It was observed that polyimides containing kink unit exhibit good solubility and mechanical property, and higher thermal stability.

In general, the solubility of polyimides can be improved by introducing flexible or bulky pendent groups such as phenyl,[16] cardo,[17] and fluorine–containing groups[18,19] on polymer backbone. Trifluoromethyl is an effective substituent that can improve the solubility[20] of polyimides without sacrificing the thermal, mechanical, and dielectric properties. Other special features of trifluoromethyl–containing polyimides are low dielectric constants and colorlessness. Hence, considerable attention has been devoted to the trifluoromethyl–containing polyimides in recent years. The reactivity of aromatic diamine, the monomer of polyimide, is greatly dependent on the chemical feature and position of the fluorine–containing substituents in the monomer because of the low polarizability and high electronegativity of fluorine atoms. Usually, aromatic diamines with fluorine or fluorinated moieties ortho to the amino groups yield low molecular weight polyimide by the conventional polycondensation procedure. Therefore, high reactive fluorine–containing aromatic diamines are required.

Experimental

Materials

Bis(4-hydroxyphenyl)diphenylmethane (**BHPP**) was successfully prepared by refluxing a mixture of dichlorodiphenylmethane and phenol (molar ratio 1:2) in xylene. Reagent grade aromatic dianhydrides such as 3,3',4,4'-biphenyltetracarboxylic dianhydride (**DA1**, from CHRISKEV), 4,4'-oxydiphthalic anhydride (**DA2**, from TCI), 3,3', 4,4'-benzophenonetetracarboxylic dianhydride (**DA3**, from CHRISKEV), 4,4'-sulfonyldiphthalic anhydride (**DA4**, from New Japan Chemical Co.), and 4,4'-hexafluoroisopropylidenediphathalic anhydride (**DA5**, from CHRISKEV) were recrystallized from acetic anhydride prior to use. *N,N*-Dimethylacetamide (DMAc), *N,N*-dimethylformamide (DMF) and pyridine were vacuum distilled over calcium hydride prior to use.

Monomers Synthesis

Scheme 1 illustrates the synthesis route to diamine bis[4-(2-trifluoromethyl–4-aminophenoxy) phenyl] diphenylmethane (**BTFAPDM**).

$$\textbf{BHPP} + 2\,O_2N\text{–}C_6H_3(CF_3)\text{–}Cl \xrightarrow[\text{DMF}]{K_2CO_3} \textbf{BTFNPDM}$$

$$\textbf{BTFNPDM} \xrightarrow[\text{Pd/C}]{H_2NNH_2\cdot H_2O} \textbf{BTFAPDM}$$

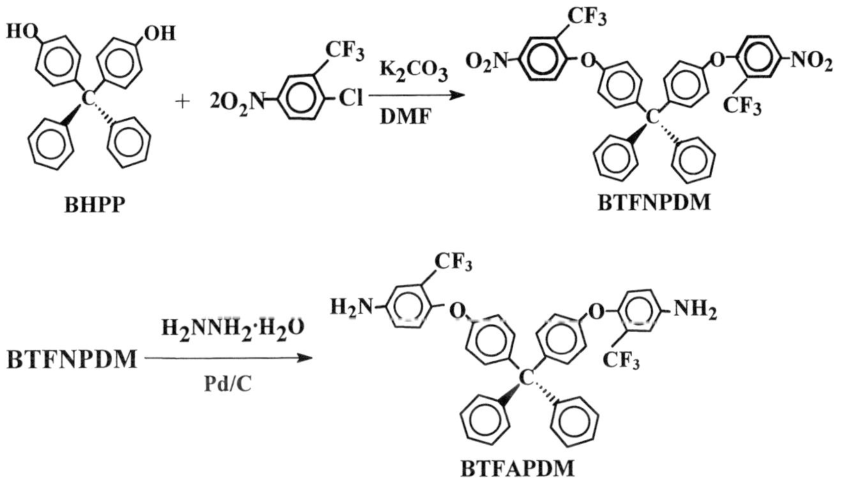

Scheme 1. Synthesis route to diamine bis[4-(2-trifluoromethyl–4-aminophenoxy) phenyl] diphenylmethane (**BTFAPDM**).

Bis[4-(2-trifluoromethyl–4-nitrophenoxy)phenyl] diphenylmethane (BTFNPDM). A mixture of **BHPP** (3.52 g, 0.01 mol), 2-chloro-5-nitrobenzotrifluoride (6.12 g, 0.02 mol), potassium carbonate (2.78 g, 0.02 mol) and *N,N*-dimethylformamide (DMF, 50 mL) was refluxed for 8 h. The mixture was cooled and poured into methanol/water (v/v=1:1) to precipitate the product. The crude product was recrystallized from DMF to provide a yellow product in 77 % yield (m.p. 227°C). The IR spectrum (KBr) exhibited absorptions at 1587 and 1335 cm^{-1} (NO_2), 1270 cm^{-1} (C-O-C). ^{1}H-NMR ($CDCl_3$): δ(ppm)= 7.00-7.04 (m, 6H); 7.22-7.24 (t, 6H); 7.27-7.29 (t, 4H); 7.32-7.35 (d, 4H); 8.28-8.31(d, 2H); 8.53-8.54 (s, 2H). ^{13}C-NMR ($CDCl_3$): δ(ppm)=64.5; 117.48; 120.93; 121.19; 121.34; 123.51; 124.15; 128.15; 129.07; 131.13; 133.36; 142.05; 144.82; 146.24; 152.44; 161.06. ANAL. calcd. for $C_{39}H_{24}O_6F_6N_2$: C, 64.11 %; H, 3.31 %; N, 3.83 %; found : C, 64.25 %; H, 3.01 %; N, 3.73 %.

Bis[4-(2-trifluoromethyl–4-aminophenoxy)phenyl] diphenylmethane (BTFAPDM). Hydrazine monohydrate (10 mL) was added dropwise to a mixture of **BTFNPDM** (5.4 g, 7 mmol), ethanol (40 mL), and a catalytic amount of Pd/C (0.054 g) at reflux temperature. After the addition was complete, the reaction was continued at reflux temperature for another 24 h. The mixture was then filtered to remove Pd/C. After cooling, the precipitated crystals were isolated by filtration and recrystallized from ethanol twice in 45 % yield (m.p. 198°C). The IR spectrum (KBr) exhibited absorptions at 3410 and 3332 cm^{-1} (N-H), 1230 cm^{-1} (C-O-C). ^{1}H-NMR ($CDCl_3$): δ(ppm)=3.71 (s, 4H); 6.73-6.75 (d, 2H); 6.82,6.84 (d, 2H); 6.87,6.89 (d, 2H); 6.93 (d, 4H); 7.13-7.28 (m, 14H). ^{13}C-NMR ($CDCl_3$): δ(ppm)=62.9; 112.95; 116.49; 119.35; 121.97; 122.75 ; 123.10; 124.69; 127.60; 131.11; 132.35; 142.57; 146.0; 146.93; 159.35. ANAL. calcd. for $C_{39}H_{28}O_2F_6N_2$: C, 69.85 %; H, 4.21 %; N, 4.18 %; found: C, 69.14 %; H, 4.40 %; N, 4.72 %.

Polymer Synthesis

To a stirred solution of **BTFAPDM** (0.6707 g, 1 mmol) in DMAc (5 mL), 4,4'-hexafluoroisopropylidenediphathalic anhydride (**DA5**) (0.4442 g, 1 mmol) was gradually added. The mixture was stirred at room temperature for 2 h under argon atmosphere to form the poly (amic acid) (**PAA5**). Chemical imidization was also carried out by adding an extra DMAc, and an equimolar mixture of acetic anhydride and pyridine into the above-mentioned poly(amic acid) solution with stirring at room temperature for 1 h, and then heating at 100°C for 3 h. It was subsequently poured into methanol and the yellow solid precipitate was filtered off, washed with methanol and hot water, and then dried at 100°C for 24 h to afford polymer **PI-5**. All other polyimides were prepared using the similar procedure.

Measurements

Melting points were measured in capillaries on a BÜCHI apparatus (model BÜCHI 535). IR spectra were recorded in the range 4000-400 cm^{-1} for the synthesized monomers and polymers in KBr disks (JASCO IR-700 spectrometer). The inherent viscosities of all polyimides were measured using Ubbelohde viscometer. NMR spectra were recorded using a Varian VXR400S (^{1}H at 399.96 MHz and ^{13}C at 100.58 MHz). Thermogravimetric data were obtained on a Du Pont 2200 in flowing nitrogen (60 cm^3 min^{-1}) at a heating rate of 20°C*min^{-1}. Differential scanning calorimetry analysis was performed on differential scanning calorimeter (Du Pont 2000) at a heating rate of 20 °C*min^{-1}. Tensile properties were determined from stress-strain curves obtained with a Orientec Tensilon with a load cell of 10 kg. A gauge of 3 cm and a strain rate of 2 cm min^{-1} were used for this study. Measurements were performed at room temperature with film specimens (0.5 cm wide, 6 cm long, and ca. 0.5 mm thick). The in-plane linear coefficient of thermal expansion (CTE) was obtained from a TA TMA-2940 thermomechanical analyzer (10°C*min^{-1} from 30 to 300°C, 0.05N). The CTE value was measured on the temperature scale between 100 to 150°C. Dielectric constants were measured by the parallel–plate capacitor method using a dielectric analyzer (TA Instrumenents DEA 2970) in the frequency 1KHz on thin films. Gold electrodes were vacuum deposited on both surfaces of dried films, followed by measuring at 25°C in a sealed chamber at 0% relative humidity.

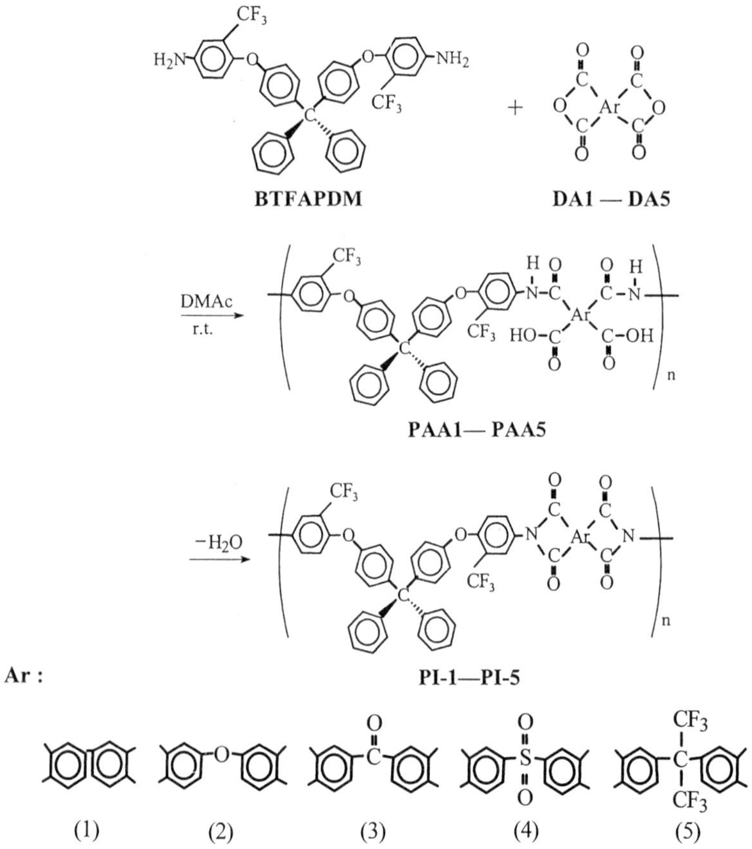

BTFAPDM DA1 — DA5

DMAc r.t.

PAA1— PAA5

$-H_2O$

PI-1—PI-5

Ar :

(1) (2) (3) (4) (5)

Scheme 2. Synthesis of trifluoromethyl-containing polyimides.

Results and Discussion

Preparation of Polyimides

Polyimides were prepared by the conventional two-step polymerization method, as shown in Scheme 2, involving ring-opening polyaddition forming poly (amic acid) and subsequent thermal or chemical imidization. Adding the dianhydride to the diamine solution at room temperature gave viscous poly (amic acid) solutions. The poly (amic acid) films did not break on folding/ slight manual stretching; they were thermally imidized at 300°C under vacuum to produce polyimide films. Alternatively, chemical imidization of poly (amic acid)s with a dehydrating agent such as a mixture of acetic anhydride and pyridine was also effective in obtaining polyimides. Before adding dehydrating agents, extra DMAc must be added in the poly(amic acid) solutions to prevent gelation while imidization. However, polymer **PI-1** underwent gelation on adding dehydrating agents due to their poor solubility, which is listed in Table 1. The IR spectra

Table 1. Inherent viscosity, solubility and film properties of polyimides.

Polymer Code	ηinh [a] ($dL \cdot g^{-1}$)	Solubility of polyimides [c]						Film Characteristics
		NMP	DMAc	DMF	DMSO	Pyridine	THF	
PI-1	— [b]	-	-	-	-	-	-	— [d]
PI-2	0.65	++	++	++	+-	++	++	Very light brown to colorless, Transparent, flexible
PI-3	0.72	++	++	++	-	+	++	Very light brown to colorless Transparent, flexible
PI-4	0.58	++	++	++	+	++	++	Very light yellow to colorless, Transparent, flexible
PI-5	0.70	++	++	++	++	++	++	Very light yellow to colorless, Transparent, flexible
Ref-1	—	-	-	-	-	-	-	Light amber, flexible

a. Measured in DMAc at a concentration of $0.5\ g \cdot dL^{-1}$ at 30 °C.

b. Polymer not soluble in DMAc.

c. Solubility: ++, soluble at room temperature; +, soluble on heating at 70°C; +-, partially soluble at 70°C; -, insoluble.

Abbreviations: NMP, *N*-methyl-2-pyrrolidinone; DMAc, *N,N*-dimethylacetamide; DMF, *N,N*-dimethylformamide; DMSO, dimethylsulfoxide; THF, tetrahydrofuran.

d. Not available.

Ref-1:

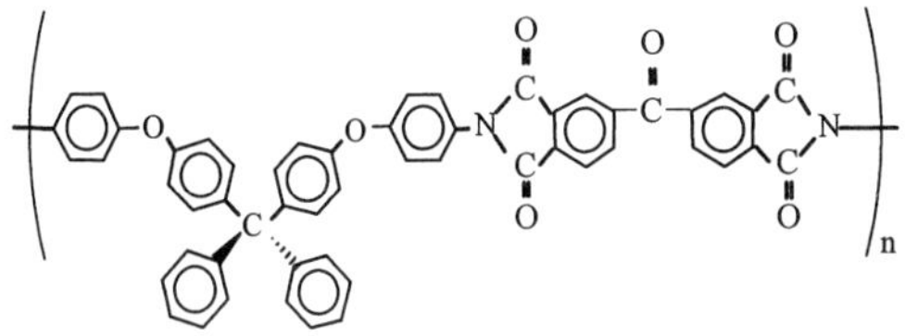

Ref-2:

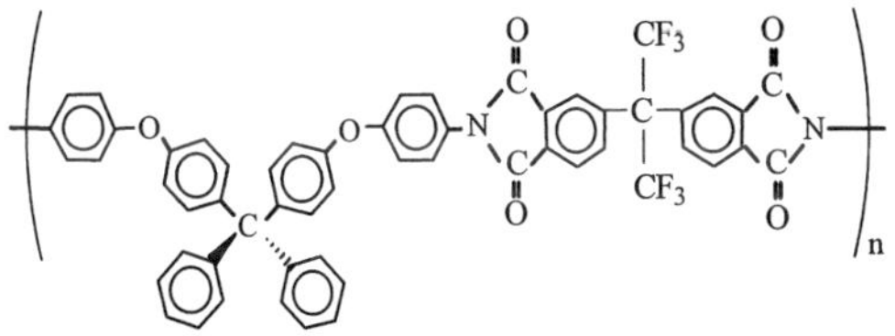

of polymers confirmed the formation of polymers. The IR spectrum of polyimide **PI-5,** obtained by chemical imidization, is shown in Figure 1. The characteristic bands at 1786

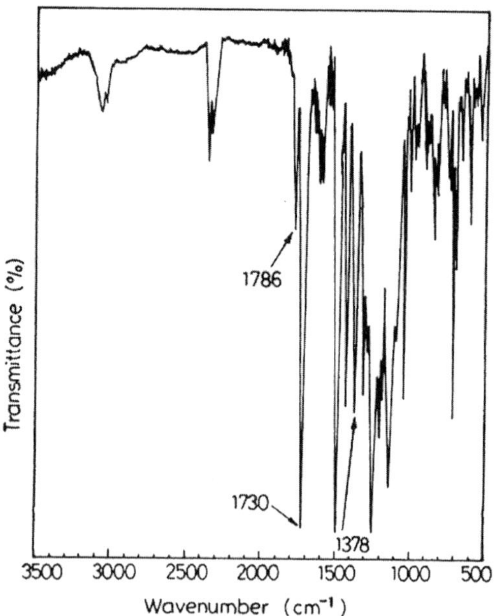

Fig. 1. IR spectrum of **PI-5**.

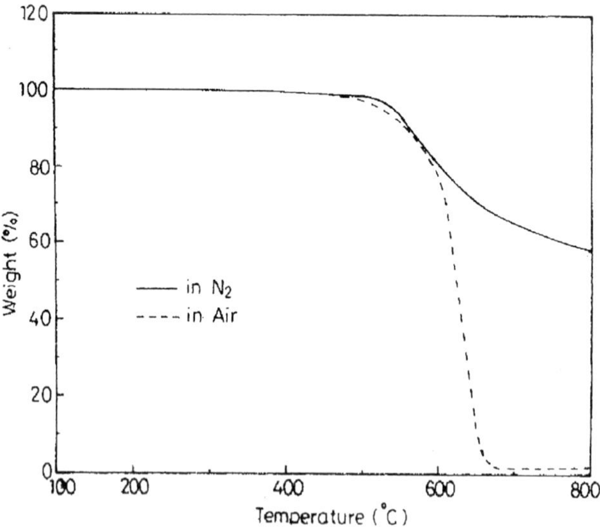

Fig. 2. TG curves of **PI-5** in nitrogen and air atmosphere at a heating rate of 20°C*min^{-1}.

and 1730 cm^{-1} are attributed to the asymmetric and symmetric stretches of imide carbonyl groups, respectively. The C-N stretching absorption at 1378 cm^{-1} confirmed the imide formation.

Polymer Properties

Solubility and film characteristics of the polyimides are listed in Table 1. The fluoro–polyimide films were very colored to almost colorless. Many non-fluorinated polyimide films are known to be yellow to dark amber on color, whereas the fluorinated polyimide films are almost colorless. Incorporation of fluorine–containing groups in the polyimide structure would reduce the refractive index and optical less. Polymers **PI-2**–**PI-5** exhibited a good solubility in a variety of solvents such as *N*-methyl-2-pyrrolidinone (NMP), *N,N*-dimethylacetamide (DMAc), *N, N*-dimethylformamide (DMF), dimethyl sulfoxide, pyridine, and tetrahydrofuran (THF) at room temperature or upon heating at 70°C. For comparison, polyimide **Ref-1** and **Ref-2** containing diphenylmethane without trifluoromethyl were prepared by polymerizing bis[4-(4-aminophenoxy)phenyl] diphenylmethane with dianhydride **DA3** and dianhydride **DA5**. It was observed that polyimide **Ref-1** was nearly insoluble in test solvents and showed poorer solubility than its analogous polymer **PI-3**. It may be attributed to the presence of the fluorine.

Thermal properties of the polyimides are tabulated in Table 2. Glass transition temperatures of the polymers, determined by means of differential scanning calorimeter (DSC), were found to be in the range of 227-253°C. As expected the oxy-bridge containing polyimide **PI-2** had the lowest T_g.[10] The glass transition temperature of **PI-5** was higher compared to **Ref-2**. This is reasonable because fluorine atoms of the trifluoromethyl groups are bulky. The bulky trifluoromethyl groups restrict the free rotation of the polymer chains.[12] Our previous studies demonstrated a similar tendency.[11] The thermal stability of these polyimides was evaluated by thermo-gravimetric (TG) analysis. Representative TG curves of the polymer **PI-5** in nitrogen and air atmosphere are shown in Figure 2. On examining the figure the excellent stability of the polymer is apparent with a decomposition temperature in excess of 500°C in nitrogen. The decomposition temperature corresponding to 10% weight loss of these polyimides are summarized in Table 2. They reach 506-563°C and 498-557°C in nitrogen and air, respectively. Most of the polyimides suffered more weight loss in nitrogen than in air atmosphere below 600°C, indicating the high thermo-oxidative stability of the polymers. A comparison of the Td_{10} values of polymer **PI-5** and **Ref-2** revealed the higher thermal stability of the polymer with the fluorine-containing group than the polymer having no fluorine–containing unit. The residual weights in nitrogen ranged from 48-60 % at 800°C. As can be seen from Table 3 the dielectric constants of polyimides are low in general and in particular polymer **PI-5**, containing trifluoromethyl and hexafluoropropane (6F) groups was quite low. Tensile properties of the polyimide films are summarized in Table 3. The polymer films had a tensile strength of 75-87 MPa, elongation at break of 6-7 %, and tensile modulus of 1.5-2.2 GPa. Among the polymers, **PI-3** showed better tensile properties. Table 3 also shows that the CTEs of polyimides are moderate, ranging from 60-88 ppm/°C. The release of H_2O molecules during imidization process has been reported to affect the CTE of fluoropolymers.[21] In addition, we believe that the film shrinkage during imidization and the resulting possible disturbances in the molecular orientation also serious effect the CTE.

Table 2. Thermal properties of polyimides.

Polymer Code	T_g *a* (°C)	Decomposition Temperature (°C) *b*		R_{800} *c* (%)
		In N_2	In Air	
PI-1	248	553	540	59
PI-2	227	506	498	48
PI-3	247	553	537	54
PI-4	237	518	517	60
PI-5	253	563	557	60
Ref-2	239	556	552	66

a. Glass transition temperature (T_g) was measured by DSC at a heating rate of 10 °C· min^{-1}.

b. Temperature at which 10 % weight loss occurred, as recorded on TGA at a heating rate of 20 °C· min^{-1}.

c. Residual weight % at 800 °C in nitrogen.

Table 3. Physical properties of polyimide films.

Polymer Code	Tensile Strength (MPa)	Elongation at Break (%)	Tensile Modulus (GPa)	Dielectric Constant	CTE [b] (ppm/°C)
PI-1	—[a]	—[a]	—[a]	—[a]	—[a]
PI-2	75	7	1.5	2.78	70
PI-3	87	6	2.2	2.98	60
PI-4	86	6	1.7	2.90	63
PI-5	84	7	1.5	2.46	88

a. Polymer was too brittle to be measured.

b. The CTE value was measured on the temperature scale between 100 and 150°C.

Conclusion

In this study a series of new polyimides containing diphenylmethylene unit were prepared from a new diamine bis[4-(2-trifluoromethyl-4-aminophenoxy)phenyl] diphenylmethane. Some of the polymers exhibited high solubility in common organic solvents. The polyimides showed good mechanical properties and high thermal decomposition temperature. The fluorine–containing polyimide showed a better thermal stability. Thus, the polyimides seem attractive candidates for practical applications such as processable high–temperature engineering plastics.

Acknowledgments

The authors thank National Science Council of the Republic of China for the financial support of this work under grant NSC 90-2216-E-011-027.

[1] M. K. Ghosh and K. L. Mittal., "*Polyimide: Fundamentals and Applications* ", Marcel Dekker, New York, 1996.
[2] H. J. Jeong, Y. Oishi, M. Kakimoto, and Y. Imai, *J. Polym. Sci. Part A : Polym. Chem.*, **1991**, *29*, 39.
[3] D. J. Liaw and B. Y. Liaw, *Polym. J.*, **1996**, *28*, 970.
[4] D. J. Liaw and B. Y. Liaw, *J. Polym. Sci., Part A: Polym. Chem.*, **1997**, *35*, 1527.
[5] J. A. Mikroyannidis, *Macromolecules*, **1995**, *28*, 5177.
[6] K. H. Park, T. Tani, M. A. Kakimoto and Y. Imai, *J. Polym. Sci., Part A: Polym. Chem.*, **1998**, *36*, 1767.
[7] M. I. Bessonov and N. P. Kuznetsov, " *Polyimides: Synthesis, Characterization and Applications* ", K. L. Mittal, Ed., Plenum Press, NY, **1984**, p. 385.
[8] F. P. Glatz and R. Mulhaupt, *Polym. Bull.*, **1993**, *31*, 137.
[9] D. J. Liaw and B. Y. Liaw, *Macromo. Chem. Phys.*, **1998**, *199*, 1473.
[10] D. J. Liaw, B. Y. Liaw and C. M. Yang , *J. Polym. Sci., Part A : Polym. Chem.*, **1999**, *50*, 332.
[11] D. J.Liaw, B. Y. Liaw, P. N. Hsu and C. Y. Hwang, *Chem. Mater.*, **2001**, *13*,1811.
[12] D. J. Liaw, B. Y. Liaw and Y. C. Cheng, *Macromo. Chem. Phys.*, **1999**, *202*, 1625.

[13] Y. S. Negi, Y. Suzui, I. Kawamura, T. Hagiwara, Y. Takahashi, M. Iijima, M. Kakimoyo and Y. Imai, *J. Polym. Sci., Part A : Polym. Chem.*, **1992**, *30*, 2281.
[14] A. C. Misra, G. Tesoro, G. Hougham and S. M. Pendharkar, *Polymer*, **1992**, *33*, 1078.
[15] D. J. Liaw, B. Y. Liaw and J. M. Tseng, *J. Polym. Sci., Part A : Polym. Chem.*, **1999**, *37*, 2629.
[16] D. J. Liaw and B. Y. Liaw, *Polym. J.,* **1999**, *31*, 1270.
[17] K. Xie, J. G. Liu, H. W. Zhou, S. Y. Zhang, M. H. He and S. Y. Yang, *Polymer*, **2001**, *42*, 7267.
[18] K. U. Jeong, Y. J. Jo and T. H. Yoon, *J. Polym. Sci., Part A: Polym Chem.,* **2001**, *39*, 3335.
[19] K. Liu, S. Y. Zhang, J. G. Liu, M. H. He and S. Y. Yang, *J. Polym. Sci. Part A: Polym. Chem.,* **2001**, *39*, 2581.
[20] H. S. Lee and S. Y. Kin, *Macromo. Rapid Commun.*, **2002**, *23*, 665.
[21] A. E. Feiring, B. C. Auman and E. R. Wonchoba, *Macromolecules,* **1993**, *26*, 2779.

Macromol. Symp. **2003**, *199*, 363-373

HALS in Polyamide 6 Polymerization

Roberto Filippini Fantoni,[*1,2] *Giorgio Sanfilippo*[1]

[1]Mazzaferro Tecnopolimeros Ltda., via Anchieta, km 18, CEP 09893-000 São Bernardo do Campo (SP), Brazil
Fax: **55-11-43415298, e-mail: filippini@grupomazzaferro.com.br
[2]Via Corridoni, 68, 24124 Bergamo, Italy
Fax: 035-360437, e-mail: r.filippini@cyberg.it

Summary: The use of Hindered Amine Light Stabilizers (HALS) directly in polyamide 6 polymerization can cause some problems. The following two problems were the focus of our project:

1) We investigated, from a theoretical point of view, the results of introducing directly one of the precursors of HALS into polyamide 6 polymerization. For the investigation, 4 amino-2,2,6,6-tetramethylpiperidine (triaceton-diamine or TAD) was chosen. We considered the TAD chain-ending effects and their influence on the total amount of amino chain-endings that can be titrated, a parameter of primary importance in the fibre field.
2) We examined the amide interchange reaction in the case of an HALS containing two amide groups, using a product available on the market, N,N'-bis(tetramethyl-4-piperidyl)isophtalamide. In this case we were able to generate a couple of equations that allows one to calculate the quantity of amide interchanged HALS. This was done by comparing the results (molecular weight and chain-endings analysis) of polymerization with and without HALS.

Keywords: HALS, light stabilizers, polyamide, polymerization, transamidation

Introduction

In the last decades, the spinning technology has required a considerable increase of the spinning speed both in the POY (pre-oriented yarn) and in the FDY (full draw yarn) technologies. This evolution has forced industries to be more selective in the choice of the polymers. An excellent polymer for high spinning speed is a polyamide 6 that contains a suitable quantity of thermal and light stabilizers.

The possibility to add amine or amide light stabilizers (HALS) during polyamide 6 polymerization should be investigated from the producers of these polymers.

The chain-ending reactions of the HALS containing amine groups and the amide interchange reactions of the HALS containing amide groups are very important from both theoretical and technological point of view.

 DOI: 10.1002/masy.200350931

Experimental Part

Relative Viscosity Determination

Relative viscosity is determined by dissolving the polyamide samples in a solution of sulfuric acid 95.7%±0.2% (w/w) at a temperature of 50÷55°C, at a polyamide concentration of 0.01 g/mL. When the polyamide is completely dissolved, let the solution cool down to 20°C and measure the flow-time in a suitable Ubbelohde Viscometer, placed in a thermostatic bath regulated at 20.0±0.1°C. Relative viscosity is calculated by dividing this flow time by that of the solvent measured in the same viscometer.

Number-Average Molecular Weights

The number-average molecular weights we used to determine the percentage of amide interchange through the equation (12), are calculated from the relative viscosity (η_{rel}) through the equations (1) and (2):

$$\overline{M}_n = 11500\left(\eta_{rel} - 1\right) \tag{1}$$

$$\overline{M}_n = F\left(\eta_{rel} - 1\right) \tag{2}$$

These equations have been used for a long time and have been verified many times.

Equation (1) must be used with a non-chain-ended polyamide 6 or with a monofunctional chain-ending.[1] Equation (2) must be used with a polyfunctional chain-ending, when the molecular distribution index is other than D=2.

Factor F depends on the molecular weight distribution,[1][2] following equation (3):

$$F = \frac{4.044 - D}{1.777 \text{x} 10^{-4}} \tag{3}$$

In case of bi-functional chain-endings index D will be obtained from the ratio R between non-terminated and terminated chains[3] as showed in equation (4):

$$D = 2 - 2\left[1/(2 + R)\right]^2 \tag{4}$$

Titration of Amino and Carboxyl End-Groups

We dissolved the polyamide in 2,2,2 trifluoroethanol and we titrated the amino end-groups with 0.02N aqueous hydrochloric acid and then we back-titrated with 0.02N aqueous sodium hydroxide to detect the carboxyl end-groups.[4]

Industrial Polymerization

To carry out our experiment, we generated two series of three polymerization batches.

All the polymerizations were carried out in industrial batch plant.

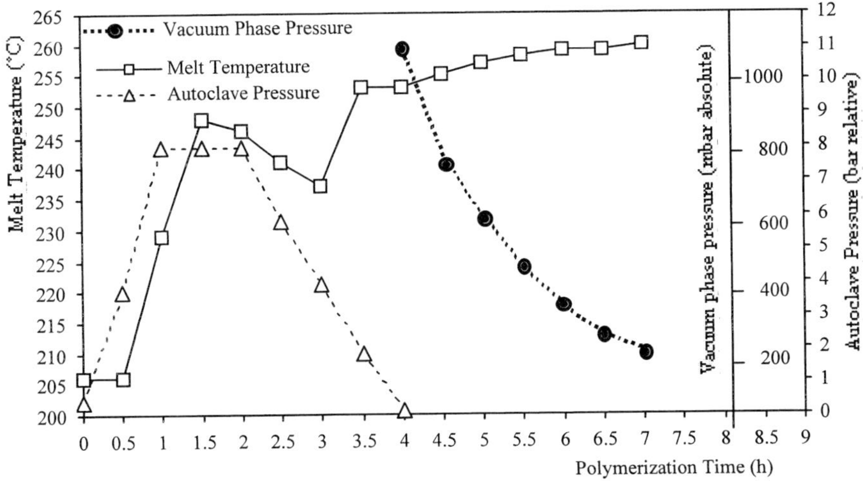

Fig. 1. Melt temperature, autoclave pressure in the pressure and vacuum phases as function of polymerization time.

The conditions of batch polymerization were identical to those used in industrial plants and were kept the same for all the polymerizations. In Figure 1 we graphed the variables relevant to these conditions such as the polymer temperatures and the pressures in the pressure and vacuum phases.

Table 1. Recipes and final compositions relevant to the two series of batches polymerized with and without TAD-IPA.

	Blank polymerizations without TAD-IPA				Polymerization with TAD-IPA			
	Autocl. charge	Extract. dried PA6	Percent in final PA6	Chain-endings conc.	Autocl. charge	Extract. dried PA6	Percent in final PA6	Chain-endings conc.
Components	*Kg*	*Kg*	*wt.-%*	*mmoles/ /kg*	*Kg*	*Kg*	*wt.-%*	*mmoles/ /kg*
Caprolactam	1400.00	1225.00	98.986	-----	1400.00	1225.00	98.599	-----
Benzylamine	2.80	2.27	0.184	17.20	2.80	2.27	0.183	17.10
Acetic acid	1.40	1.21	0.098	16.33	1.40	1.21	0.098	16.33
TAD-IPA	-----	-----	-----	-----	4.20	4.62	0.372	8.40
Antifoam	0.88	0.40	0.032	-----	0.88	0.40	0.032	-----
CL+Oligom.	-----	8.04	0.650	-----	-----	8.28	0.666	-----
Water	50.00	0.62	0.050	-----	50.00	0.62	0.050	-----
TOTAL	1455.08	1237.54	100.000	-----	1459.28	1242.40	100.000	-----

In Table 1 we reported the recipes for three batches without HALS and for three batches with the presence of 0.3 wt.-% (weight percent) on caprolactam of the HALS.

In Table 2 we reported the results of the analysis on the extracted and dried polyamide, for the six batches and the average values.

Table 2. Analysis relevant to the final polyamides 6 obtained in the two series of batches polymerized with and without TAD-IPA.

		Blank polymerizations without TAD-IPA				Polymerization with TAD-IPA			
		Batches				Batches			
Properties	Unities	1st	2nd	3rd	Average	1st	2nd	3rd	Average
η_{rel}	------	2.674	2.665	2.679	2.673	2.606	2.624	2.617	2.616
[COOH]	*meq/kg*	34.4	35.7	34.6	34.9	35.3	33.9	33.4	34.2
$[NH_2]$	*meq/kg*	35.7	36.6	36.3	36.2	54.6	51.9	53.7	53.4
CL+Olig.	*wt.-%*	0.60	0.67	0.56	0.61	0.72	0.59	0.65	0.65
Water	*wt.-%*	0.048	0.054	0.053	0.052	0.054	0.061	0.064	0.060

Results and Discussion

HALS as Chain-Ending

Because the use of additives containing HALS is ideal to provide thermal and light resistance, different producers of these chemicals (generally HALS are obtained from tetramethylpiperidine) have tried to make appropriate derivatives and relevant polyamide based master-batches. Among these producers, BASF[5][6] and Allied[7] have patented the use of triacetondiamine (TAD - 4amino2,2,6,6 tetramethyl-piperidine) and relevant derivatives. TAD contains in its molecule a perfectly polycondensable primary amine and a sterically hindered secondary amine which does not polycondensate under usual melt-polycondensation conditions but which is perfectly dyeable as chain-endings. Moreover the molecule is thermically stable at the polymerization conditions.

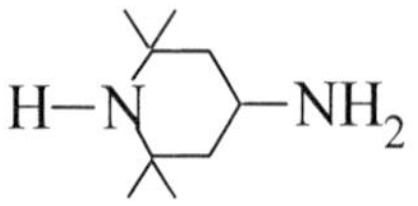

The dosage of TAD is critical when adding a small quantity of this product to the polymerization process. This is due to its strong chain-ending effect, which is similar to that of a monofunctional amine, and to its strong repercussion on the total number of amino chain-endings. The effect is so strong that in order to reduce the number of these total amino groups (primary of the macromolecule chains and secondary of TAD), it is almost always necessary to add a bicarboxylic acid and consequently another chain-ending (in addition to TAD). The following example is given to illustrate this point.

Taking into consideration a polyamide 6 with a molecular weight of 19000 (52.6 mmoles/kg) chain-ended with 0.2 wt.-% of TAD (12.8 mmoles/kg) and a sufficient quantity of adipic acid to obtain 42 meq/kg of titratable amino group, the chain situation will be the following (all the concentrations are relevant to the final polyamide 6, extracted and dried):

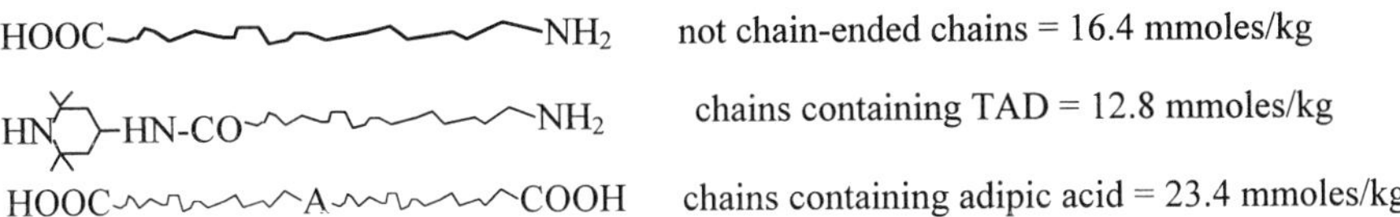

Thus, the use of two chain-endings, one of them being TAD, allow us to reach a good molecular weight stability during spinning (by the use of large quantity of chain-endings) and a good thermal stability during high speed spinning (by the presence of TAD).

In addition the dyeability can be predicted because of the presence of the 42 meq/kg of dyeable groups: 29.2 meq/kg supplied by the primary amino group and 12.8 meq/kg supplied by the secondary amino group of TAD.

Effectiveness as Light Stabilizer

When we started this research, there were some questions on the effectiveness of using TAD as a light stabilizer in lieu of low molecular weight HALS. These questions were related to TAD's reduced mobility when linked to a macromolecular chain.

In another paper we showed that the light stabilization is identical for a polyamide 6 with TAD as chain-ending and a polyamide 6 blended with HALS containing the same concentration of sterically hindered amine.[8]

Amide Interchange Determination

When feeding HALS during polymerization, one has to take into account the many possible interferences with the polymerization reactions by thermal and thermo-oxidative reactions. Many degradation by-products of the thermal or thermo-oxidative reactions can interfere during the polymerization process and therefore selecting thermally stable HALS for the polyamide 6 polymerization is critical.

In addition to the chain-ending effect of an HALS having a reactive primary amino group in its structure, the interference created by the amide interchange between the amide group of an amide type HALS and the polyamide chains has to be taken into consideration. This interference is not negligible and can produce unexpected variation in terms of molecular weight.

The HALS we have to take into account must be a primary amide and there are many products available on the market.

We decided to test both experimentally and theoretically one of the products available on the market. N,N'-bis(tetrametyl-4-piperidyl)isophthalamide (TAD-IPA) can be easily purchased and is broadly used in polyamide 6 spinning:

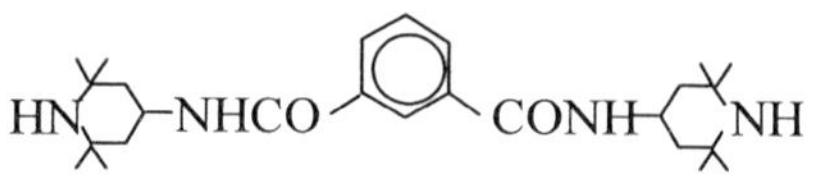

HN–NHCO–(C6H4)–CONH–NH

This HALS is stable under standard polymerization conditions but the amount that will transamidate is strictly dependent on the polymerization conditions (temperature, time, pressure, catalysts) and must be evaluate on a case-by-case basis.

To evaluate this parameter by analysis is quite complicated due to the low solubility of polyamide and, moreover, it requires sophisticated instrumentation such as NMR, GC/MS, etc.

A simpler way to get this result was obtained by comparing the results of two series of three polymerizations carried out under identical conditions. The comparison terms were the number-average molecular weight (or relative viscosity) and the concentrations of different types of terminated chains, values that can be calculated by knowing the end-groups (amino and carboxyl) and the HALS concentrations in the final polyamide.

Simulation

Let us consider a polyamide 6 produced with this recipe:

Caprolactam = 99.7 p. Benzylamine = 0.2 p. Acetic acid = 0.1 p. Water = 3.0 p.

The following leakage was predicted during the polymerization: benzylamine = 18 wt.-% acetic acid = 13 wt.-%. It is assumed that the extraction plant can extract 10 wt.-% of caprolactam and its oligomers. Therefore the chain situation will be illustrated in Figure 2.

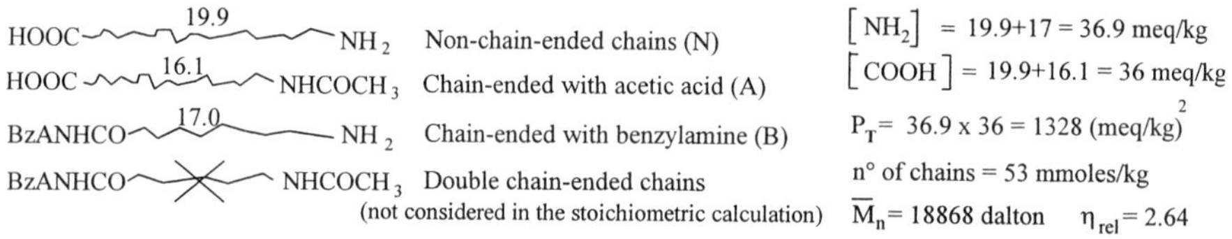

HOOC～～19.9～～NH_2 Non-chain-ended chains (N)

HOOC～～16.1～～$NHCOCH_3$ Chain-ended with acetic acid (A)

BzANHCO～～17.0～～NH_2 Chain-ended with benzylamine (B)

BzANHCO～～✕～～$NHCOCH_3$ Double chain-ended chains (not considered in the stoichiometric calculation)

$[NH_2]$ = 19.9+17 = 36.9 meq/kg

$[COOH]$ = 19.9+16.1 = 36 meq/kg

P_T = 36.9 x 36 = 1328 $(meq/kg)^2$

n° of chains = 53 mmoles/kg

$\overline{M}_n$ = 18868 dalton η_{rel} = 2.64

Fig. 2. Situation of the chains in case of a PA6 terminated with a pair of chain-ending.

If, in the initial recipe, we add 0.25 wt.-% of TAD-IPA (MW=442.6 dalton), this will correspond to 0.278 wt.-% or 6.28 mmoles/kg in the final polyamide 6.

If no amide interchange occurs during the polymerization, the only change would be in the titratable amino group: $[NH_2]_{TIT} = [NH_2]_{PC} + [NH]_{TAD\text{-}IPA}$ = 36.9+6.28x2 = <u>49.46 meq/kg</u>

Molecular weight, relative viscosity, carboxyl end-groups and polycondensable amino end-groups remain unchanged.

If some amide interchange occurs during the polymerization, the chain situation will get more and more complicate because a large number of possible reactions between the TAD-IPA and the different chains have be to considered. Specifically, six types of amide interchange reactions can occur.

Taking into account the whole schema, we can obtain a total of twelve type of chains with different terminations and ten types of non-macromolecular products. The possible cyclic oligomers are not taken into consideration because they do not affect chain-endings.

Fortunately, this complicated schema can be simplified in the calculation. This is because, from a stoichiometrical point of view, the number of chains to be taken into consideration is small. Linear monomeric products and chains with double terminations can all be incorporated in the chains we take into consideration for the calculation.

For this simulation, the following parameters were taken into account:

(1) The chain ending and the dyeable amino-group situation in the polyamide after the possible amide-interchange.

(2) The case when amide-interchange reactions occurred either completely or at only 50%.

(3) The possibility that the transamidation occurred only by one of the two amide groups of TAD-IPA.

It should be noted that in reality the situation is always a mixture of these three possibilities. However, interesting conclusions can be drawn from these simplified assumptions.

Case A – Complete amide interchange

In this situation, in order to perform our calculation, we only have to add chains H and K to the previously considered chains N, A and B. In fact, as already mentioned, from a stoichiometrical point of view, a lot of chains can be merged with the previous types.

In order to keep things as simple as possible, only the amide interchange with non-terminated chains was considered. The same conclusions, however, can be drawn if the amide interchange occurs on the four types of chains considered in the polymerization without TAD-IPA. If H mmoles/kg will transamidate we obtain H mmoles/kg of chains containing isophthalic acid as regulator and K chains terminated with TAD, being K=2H:

OC CO (TAD-IPA: N–H H–N, HN ... NH) + 2 HOOC~~NHCO~~NH_2 → (Total transamidation) HOOC~~OC–C$_6$H$_4$–CO~~COOH (H) 6.28 mmoles/kg + 2 HN–HN–OC~~~~NH_2 (K) 12.56 mmoles/kg

As illustrated in the above reaction, three polymer chains are obtained from two polymer chains. This means a decrease in molecular weight and, at the same time, a variation of the molecular weight distribution index, because the H chains are terminated with a bicarboxylic acid. The final result will be a decrease of molecular weight, while the titratable and polycondensable amino groups, as well as the carboxyl groups, remain constant.

Starting from the initial situation and considering the type of chains we already have described, we can obtain:

$$[NH_2]_{TIT} = N+B+2K = N+17.0+25.12 = \underline{N+45.12} \quad (5)$$

$$[NH_2]_{PC} = N+B+K = N+17.0+12.56 = \underline{N+29.56} \quad (6)$$

$$[COOH] = N+A+2H = N+16.1+12.56 = \underline{N+28.66} \quad (7)$$

The polymerization conditions being the same, the P_T can be considered constant, and thus:

$P_T = (N+29.56)(N+28.66) = 1328$ which yields $N = \underline{7.34\ meq/kg}$

From this value we will obtain: $[NH_2]_{PC} = \underline{36.90\ meq/kg}$ $[COOH] = \underline{36.00\ meq/kg}$ $[NH_2]_{TIT} = \underline{49.46\ meq/kg}$

As predicted, the polycondensable and titratable amino groups as well as the carboxyl groups remained unchanged while the molecular weight and the relative viscosity decreased:

Total number of chains = N+A+B+3H = 59.28 mmoles/kg.

From this value we can calculate the following parameters:

$$\overline{M}_n = 10^6/59.28 = 16869 \text{ dalton} \qquad D = 1.982^{(a)} \qquad \eta_{rel} = \overline{M}_n/F + 1 = 16869/11603 + 1 = 2.454^{(a)}$$

[a] D can be calculate from the chain situation [equation (4)], while F is obtained from D [equation (3)]

Case B – 50% of total amide interchange

With a calculation that is very similar to the one outlined in *Case A* and taking into consideration that 50% of the TAD-IPA remains unreacted, we obtain the following:

N = non terminated chains = 13.62 mmoles/kg

$[NH_2]_{PC} = \underline{36.90\ meq/kg}$ $[COOH] = \underline{36.00\ meq/kg}$ $[NH_2]_{TIT} = \underline{49.46\ meq/kg}$

$$\overline{M}_n = 10^6/56.14 = 17812 \text{ dalton} \qquad D = 1.994 \qquad \eta_{rel} = \overline{M}_n/F + 1 = 17812/11531 + 1 = 2.545$$

Also in this case, the only variables are molecular weight and relative viscosity that are intermediate between the previous two already considered.

Case C – 100% of partial amide interchange

For a more complete evaluation, we have taken into account the possibility of a partial transamidation, when only one of the amide groups of TAD-IPA has undergone an amide interchange:

Using the same calculation, we obtained, as predicted, the same results found in *Case A*. Only the MWD index and, consequently the relative viscosity, are a little higher. The difference in viscosity is so small that it can be considered negligible from a practical standpoint. This means that we are not able to distinguish between partial and total transamidation

$$\overline{M}_n = 10^6/59.28 = 16869 \text{ dalton} \qquad D = 2.00 \qquad \eta_{rel} = \overline{M}_n/F + 1 = 16869/11500 + 1 = 2.467$$

Evaluation of Amide Interchange

To determine the percentage of TAD-IPA that will undergo transamidation, we can find a suitable equation by taking the same type of mathematical approach we used in the previous examples. P_T is obtained from the blank polymerization (without TAD-IPA) shown in the experimental part of this discussion. Using the same symbology for the different type of chains we can set an equation based on P_T [equation (8)]:

$[NH_2]_{PC}$ = (N+B+K) [COOH] = (N+A+2H)

$$P_T = (N+B+K)(N+A+2H) = (N+B+2H)(N+A+2H) \tag{8}$$

Now, if H represents the mmoles/kg of transamidated TAD-IPA, S the total quantity of TAD-IPA in the final polyamide, and T the fraction of transamidated TAD-IPA, we obtain this relation: ST = H.

Therefore we can transform the previous equation (8) in this way:

$$P_T = (N+B+2ST)(N+A+2ST) \tag{9}$$

From this equation it is easy to obtain other interesting relations between these variables:

$$\text{number of chains (mmole/kg)} = \frac{1}{2}\left[\sqrt{(A-B)^2 + 4P_T} + A + B + 2ST\right] \tag{10}$$

$$\overline{M}_n = 2\text{x}10^6 \Big/ \left[\sqrt{(A-B)^2 + 4P_T} + A + B + 2ST\right] \tag{11}$$

From the equation (11) is possible to obtain the equation (12), which is the value of the transamidated TAD-IPA fraction:

$$T = \frac{1}{2S}\left[\left(2\text{x}10^6/\overline{M}_n\right) - (A+B) - \sqrt{(A-B)^2 + 4P_T}\right] \tag{12}$$

By substitution with the equation (2) we can obtain the equation (13):

$$T = \frac{1}{2S}\left\{\left[2x10^6/F(\eta_{rel} - 1)\right] - (A + B) - \sqrt{(A - B)^2 + 4P_T}\right\} \tag{13}$$

F being dependent on T, we need to obtain a much more complicated equation. Usually, due to the small number of chains containing a bifunctional chain-ending (obtained by transamidation), we can use the value 11500 for the factor F. This estimation will produce only a negligible error, quite often lower than the precision of Relative Viscosity measurement itself:

$$T = \frac{1}{2S}\left\{\left[173.913/(\eta_{rel} - 1)\right] - (A + B) - \sqrt{(A - B)^2 + 4P_T}\right\} \tag{14}$$

These equations are very useful for calculating the quantity of transamidated TAD-IPA. These equations can only be used for polyamide 6 chain-ended with a monoamine, a monocarboxylic acid, and polymerized with TAD-IPA. However, this case being one of the more complicated (double chain-endings), using the same logic, it is very easy to find similar (and usually simpler) equations for different chain-ending (single or couple) applications.

How to Apply the Equations in a Real Case

To illustrate how to use this equation in a real situation, we carried out two series of three polymerizations with and without TAD-IPA as explained in the experimental section of this paper. The polymerization recipes (Table 1) and the relevant average results (Table 2) of the series are also shown in the experimental section.

From the values of the first series of polymerizations (Table 2), we can calculate P_T using the average values of the concentrations of carboxyl and amine groups: P_T=1263 (meq/kg)^2.

The values for A, B and S to be used in equation (14) must be obtained from Table 1. The value for viscosity is obtained from Table 2:

η_{rel}=2.616 A=16.3 mmoles/kg B=17.1 mmoles/kg S=8.40 mmoles/kg

Because there is only a very small concentration of chains containing a bicarboxylic acid in our example, F will be close to 11500 (by iteration: D=1.998; F=11513). Using an estimation of 11500 for F and using equation (14), we obtain a value for T of 0.187. This means that only 18.7% of the original TAD-IPA is transamidated.

With different polymerization conditions, specially in continuous plant where the polymerization times are higher, the quantity of TAD-IPA that will transamidate is higher and can reach values around 40%. In this case the molecular weight will be dramatically reduced and these amide interchange reactions must be took into consideration.

Conclusion

The chain ending reaction of TAD, alone or with other bicarboxylic acids, was discussed for polyamide 6 polymerization. A method for calculating the percentage of transamidation of a light stabilizer, like TAD-IPA, was also presented for polyamide polymerization. This calculation uses the polymer's number-average molecular weight (or relative viscosity in sulfuric acid) and the amino and carboxyl end-groups concentrations.

[1] A. Filippi, R.F. Filippini, *"Giornate di studio sulla policondensazione"*, Associazione Italiana di Scienza e Tecnologia (AIM), 1981, p.87

[2] J. Lin, Z.L. Tang, R.F. Filippini, *"The 2nd China-Italy Joint Polymer Symposium"*, Polymer Division of Chinese Chemical Society – Hangzhou, 1995, p.85

[3] R.F. Filippini, M. Fornaroli, M. Farina, *"4th Convegno Italiano di Scienza delle Macromolecole"*, AIM - Acta 1979 - p.107

[4] Z.L. Tang, J. Lin, R.F. Filippini, *Die Angew. Makr. Chemie* , **1997**, *250*, 4321

[5] Ger. 4413177 (1995), BASF AG, invs.: K. Weinerth, K. Mell, P. Matthies, L. Beer; *Chem. Abstr.* **1996**, *124*, 89003c

[6] Ger. 4429089 (1996), BASF AG, invs.; K. Weinerth, K. Mell, P. Matthies, L. Beer; *Chem. Abstr.* **1995**, *124*, 89028q

[7] U.S. 5618909 (1997), AlliedSignal Inc., invs.:R. Lofquist, R.Y. Mohajer; *Chem. Abstr.* **1997**, vol *126*, 226478

[8] L. Battisti, R.F. Filippini, G. Sanfilippo *"15th Convegno Italiano di Scienza e Tecnologia delle Macromolecole"* , AIM, Trieste, Acta 2001 (Only CD Edition)

Synthesis, Characterization and Properties of Polycarbonate Containing Carboxyl Side Groups

*Ruifeng Zhang, J. A. Moore**

Department of Chemistry, Rensselaer Polytechnic Institute, Troy, NY 12180, USA
E-mail: moorej@rpi.edu

Summary: Diphenolic Acid, DPA [bis(4-hydroxyphenyl)pentanoic acid]can be made from cellulose-rich waste. The t-butyl ester was converted to homo- and copolycar- bonates (with bis-phenol-A, BPA). Deblocking the ester yielded polycarbonates with pendent carboxyl groups that exhibit all the properties of polyelectrolytes and retain solubility in aqueous base without degradation for long periods.

Keywords: copolymerization, functionalization of polymers, polycarbonates, renewable resources, water-soluble polymers

Introduction

Recent progress made in the field of green chemistry is the development of diphenolic acid. It is synthesized by the condensation of phenol and levulinic acid, a compound prepared from the recycling of cellulose-rich wasters such as wood, paper, sewage sludge, paper mill sludge and food processing waste.[1] Because the diphenolic acid has a structure similar to bisphenol A (BPA) and a price only one third of the cost of the latter, it is interesting and of potential commercial value to develop polycarbonate derivatives and functional polycarbonates based on the use of diphenolic acid.

In previous work, high molecular weight polycarbonate derivatives have been synthesized by the condensation of diphenolate esters and phosgene.[2,3,4] In the current research, a new kind of polycarbonate derivative containing unprotected carboxyl side groups has been synthesized. The reaction route is shown in Scheme 1.

It has been found that the free carboxyl group on diphenolic acid caused cross-linking and branching when directly polymerized with phosgene. Therefore, to obtain well-defined linear polycarbonates, the carboxyl group was first protected by esterification with *tert*-butanol.[5] After the formation of the corresponding homo-polycarbonate the *t*-butyl group can be quantitatively cleaved with trifluoroacetic acid at room temperature or by heating above 200 °C.

 DOI: 10.1002/masy.200350932

The obtained polycarbonate acid can be dissolved in dilute, aqueous NaOH solution and exhibits properties typical of a polyelectrolyte.[6]

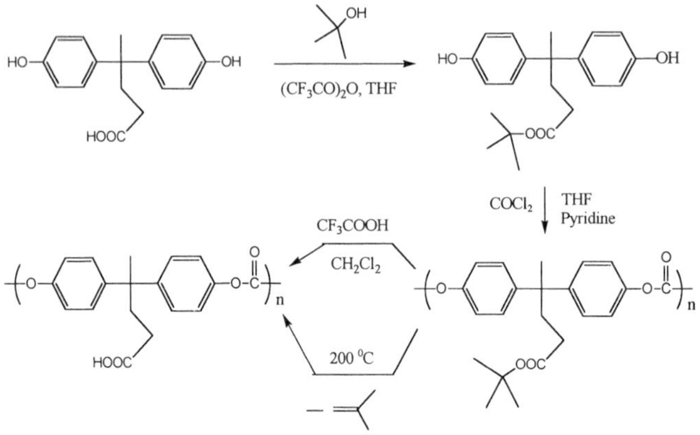

Scheme 1. Synthesis of Polycarbonate Carboxylic Acid.

Carboxyl groups in a polymer may act as cross-linking agents through hydrogen bonding, which influences the physical properties of the polymer. By using small amounts of Zn^{2+} or Al^{3+} ions the carboxyl groups can be cross-linked more strongly by generation of ionic bonds.[7] It has also been found that the presence of carboxyl groups influence the thermo-stability of the polymer. The glass transition temperatures of polymers with different extents of cross-linking have been investigated in this work (vide infra).

Experimental Part

General Methods

Fourier-Transform Infrared (FTIR) spectra were recorded on a Perkin Elmer Paragon 1000 spectrophotometer. Nuclear Magnetic Resonance (NMR) spectra were recorded on a Varian Unity 500 spectrometer. Viscosity measurements were carried out in aqueous solution with a Cannon-Ubbelohde viscometer. All solutions were filtered through Gelman Acrodisc CR PTFE syringe filters with a pore size of 0.8 μm. Differential Scanning Calorimetry (DSC) and Thermogravimetric Analysis (TGA) were carried out on Perkin Elmer Series 7 instruments at a heating rate of 10 °C/min under nitrogen. TGA measurements were carried out at a heating rate of 20 °C/min under nitrogen. Gel Permeation Chromatography (GPC) was performed using a Waters 2410 liquid chromatograph equipped with a refractive index detector. A universal calibration was made with narrowly dispersed polystyrene standards.

Solvents and Reagents

THF was distilled from sodium metal. Methylene chloride was distilled from calcium hydride. Diphenolic acid and Bisphenol-A were purchased from Aldrich and used without further purification. Pyridine was distilled from calcium hydride and stored over 4A molecular sieves. *Tert*-butanol was distilled from calcium hydride before use.

Synthesis

1. *tert*-butyl 2,2-bis(4-hydroxyphenyl)valerate

In a 500-ml, 3 neck round bottom flask diphenolic acid (30.1g, 0.1 mol) was dissolved in 100 ml of THF. An ice bath was used to cool the solution to which trifluoroacetic anhydride (50 ml) was added dropwise through an addition funnel over a period of 10 minutes. After The ice bath was then removed and the reaction mixture was stirred at room temperature for 3 hours. Under cooling in an ice bath 80 ml of tert-butanol was added through an addition funnel to the reaction mixture over 5 minutes. The ice bath was then removed and the reaction mixture was stirred overnight at room temperature. A 10% aqueous potassium carbonate solution was used to neutralize the reaction mixture. The oil-like product was extracted with methylene chloride in a separatory funnel. The methylene chloride was removed on a rotary evaporator and the residue was poured into a large excess of hexane. The precipitated product was filtered and dried. The crude product was dissolved in aqueous KOH solution. Dry ice was added and the product gradually precipitated. The solid was filtered and washed with pure water. To remove unreacted diphenolic acid completely, this procedure should be repeated twice. Finally, the obtained white solid was dried at 60 °C for 12 hours. The pure product should be white crystals (m.p. 131-132 °C, 20.4g, yield: 58.9%).

^{1}HNMR(acetone-d_6) ppm 8.18(s, 1H, O**H**), 7.06(d, 4H, aromatic ring), 6.78(d, 4H, aromatic ring), 2.34(t, 2H, CH_2-C**H**$_2$-COO), 2.03(t, 2H, C**H**$_2$-CH_2-COO), 1.57(s, 3H, C-C**H**$_3$), 1.42(s, 9H, O-C(C**H**$_3$)$_3$). ^{13}C NMR (acetone-d_6) 177.44 (C=O), [155.45, 140.46, 128.38, 114.96] (aromatic carbon atoms), 79.52 (O-**C**-(CH_3)$_3$), 44.40 (O-C-(**CH**$_3$)$_3$), 33.85(O=C-**C**H_2-CH_2),

27.63 (O=C-CH_2-**CH_2**), 13.47(C-**CH_3**). FTIR (KBr) 3413 cm^{-1}(-OH), 2972 cm^{-1}, 1732 cm^{-1} (COO-) 1612 cm^{-1}, 1511 cm^{-1}, 1368 cm^{-1}, 1225 cm^{-1}(C-O), 1190 cm^{-1}, 1153 cm^{-1}, 1012 cm^{-1}, 834 cm^{-1}(ϕ-H).

2.Poly[oxycarbonyloxy-1,4-phenylene-4-tert-butyloxycarbonyl-1-methyl-propylidene-1,4-phenylene]

A 3-neck 250 ml round bottom flask was equipped with a stir bar, a gas inlet and outlet tubes. The gas inlet tube was connected to a phosgene tank with a flow meter and regulator. The outlet tube was connected to a trap containing ammonium hydroxide. A water bath was placed under the reaction flask. *tert*-Butyl diphenolate (15g, 0.04 mol), pyridine (20 ml) and THF (80 ml) were added to the flask. Phosgene was slowly introduced to the reaction mixture at a flow rate of 200 ml/min. The reaction mixture was kept at 20°C. The phosgene flow was stopped after 45 minutes. The reaction mixture was washed with 10% aqueous HCl solution and a large amount of methanol, sequentially. The precipitate was collected by filtration. The polymer was dissolved in THF and precipitated in methanol. The purified polymer was dried in vacuum overnight at 60 °C (14.7 g, yield: 91%). ^{1}HNMR($CDCl_3$) ppm 7.12(m, 8H, aromatic), 2.33(t, 2H, CH_2**CH_2**COO), 1.95(t, 2H, **CH_2**CH_2COO), 1.55(s, 3H, C**CH_3**), 1.34(s, 9H, O-C(**CH_3**)$_3$). FTIR (KBr) 2974 cm^{-1}, 1774 cm^{-1} O-COO-), 1725 cm^{-1} (COO-), 1504m^{-1}, 1391m^{-1}, 1229m^{-1}(C-O), 1195cm^{-1}, 1163cm^{-1}, 1011cm^{-1}, 834 cm^{-1}(ϕ-H). A similar procedure was used to synthesize polycarbonate containing both *tert*-butyl diphenolate and bisphenol-A units.

3. Poly[oxycarbonyloxy-1,4-phenylene-(4-carboxy-1-methyl-propylidene)-1,4-phenylene]

The above-mentioned polymer (3.0 g) was dissolved in 40 ml of methylene chloride, and 5 ml of trifluoroacetic acid was added. The reaction mixture was kept at room temperature for 3 hours. Finally, methylene chloride was evaporated under vacuum at room temperature. The residual liquid was treated with a large amount of water. The precipitated polymer was washed with distilled water several times and then dried in vacuum at 50 °C for 5 hours (2.5g, yield: 98%). ^{1}HNMR(DMSO-d_6) pm 7.27(m, 8H, aromatic), 2.38(t, 2H, CH_2**CH_2**COO), 1.96(t, 2H, **CH_2**CH_2COO), 1.57(s, 3H, C**CH_3**). FTIR (KBr) 2971 cm^{-1}, 1773 cm^{-1} O-COO-), 1709 cm^{-1} (COOH), 1504m^{-1}, 1229m^{-1}(C-O), 1195cm^{-1}, 1163cm^{-1}, 1014cm^{-1}, 831 cm^{-1}(ϕ-H).

4. The introduction of metal ions into the polymer matrix

Zn^{2+} and Al^{3+} ions can be quantitatively introduced into the polycarbonate containing free carboxyl groups as shown in Scheme 2. Polymer acid, 0.5 g, was dissolved in 20 ml of completely dried THF. Solid aluminum isopropoxide was added to the solution, followed by heating to reflux for 20 minutes. An insoluble and transparent gel was obtained. The solvent was removed in vacuum at 50 °C for 10 hours. The preparation of the Zn^{2+}-containing polymer was carried out at room temperature. When adding a 1.0 M hexane solution of diethylzinc, a similar gel was obtained and dried under vacuum under the same condition as for aluminum ions.

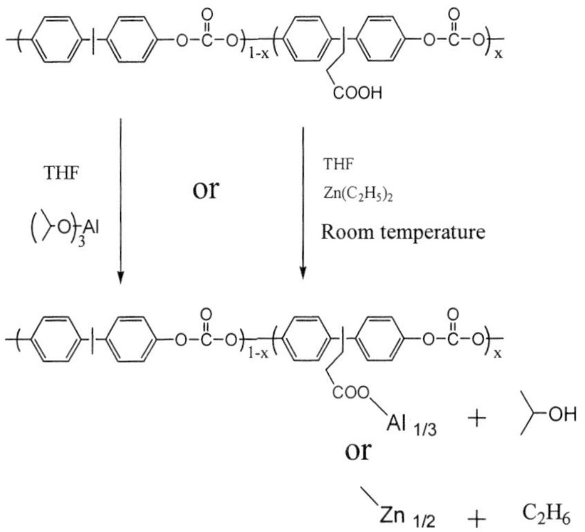

THF or THF $Zn(C_2H_5)_2$ Room temperature

Scheme 2. Introduction of metal ions into the polymer matrix.

Results and Discussion

1. Polymer Characterization

The ^{1}HNMR and FTIR spectra of *tert*-butyl diphenolate polycarbonate and free diphenolic acid polycarbonate are shown in Figurel and Figure 2, respectively. The spectra are in accord with the expected structure of the two polymers. A series of copolymers with the following structure have been synthesized and their molecular weights were measured by GPC. The data are collected in Table 1.

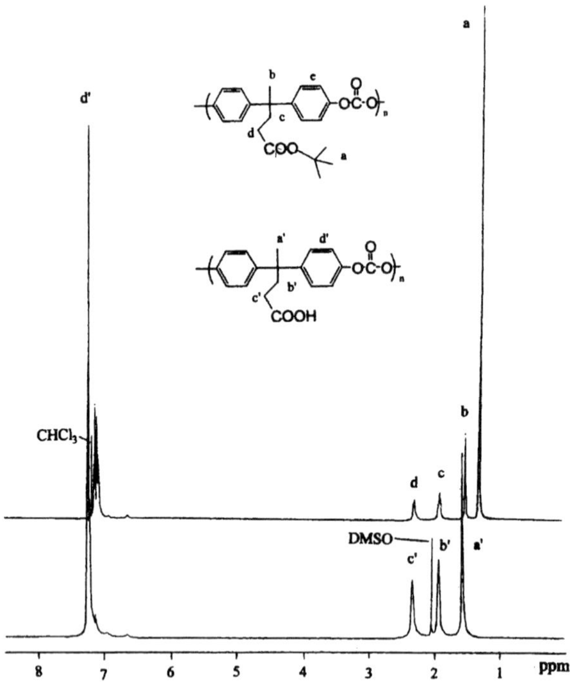

Fig. 1. ^{1}H NMR spectra of tert-butyl diphenolate polycarbonate (top) and diphenolic acid polycarbonate (bottom).

Table 1. Molecular weight of co-polycarbonate esters.

X	$M_w \times 10^{-4}$	$M_n \times 10^{-4}$	PDI
0.05	3.4	2.1	1.62
0.10	2.8	1.3	2.15
0.20	1.5	0.8	1.87
0.50	1.9	1.1	1.73
0.75	2.0	1.2	1.67
1.00	2.3	1.5	1.53

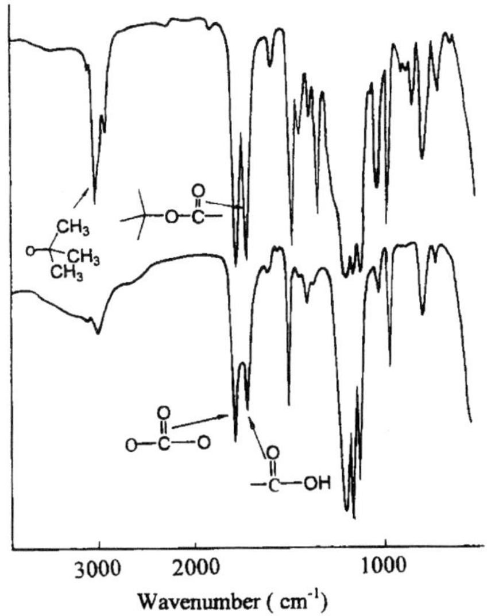

Fig. 2. FTIR spectra of tert-butyl diphenolate polycarbonate and diphenolic acid polycarbonate.

The glass transition temperatures of polycarbonate ester and polycarbonate acid have also been investigated by DSC. The results are shown in Figure 3. For the polyester the lowest T_g can be found at a composition of x = 0.5. The introduction of the *tert*-butyl side group led to a decrease of T_g as compared with the BPA-based homopolycarbonate. After the *tert*-butyl side group was cleaved with trifluoroacetic acid in methylene chloride the obtained polymer has the following structure:

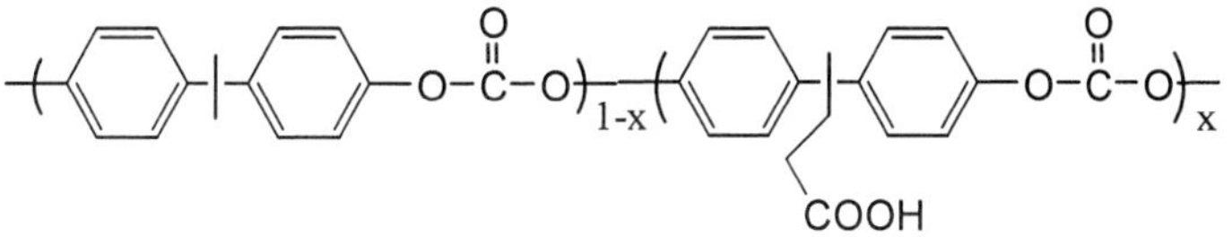

The dependence of T_g on polymer composition can be seen in Figure 3. T_g first decreases from 148 °C (x = 0) to 131 °C (x = 0.10) and then increases from 131 °C to 147 °C with increasing content of carboxyl groups. The minimum value is observed at x = 0.10. Clearly the influence of hydrogen bonding can be seen to retard local chain motion compared to the ester-containing polymer.

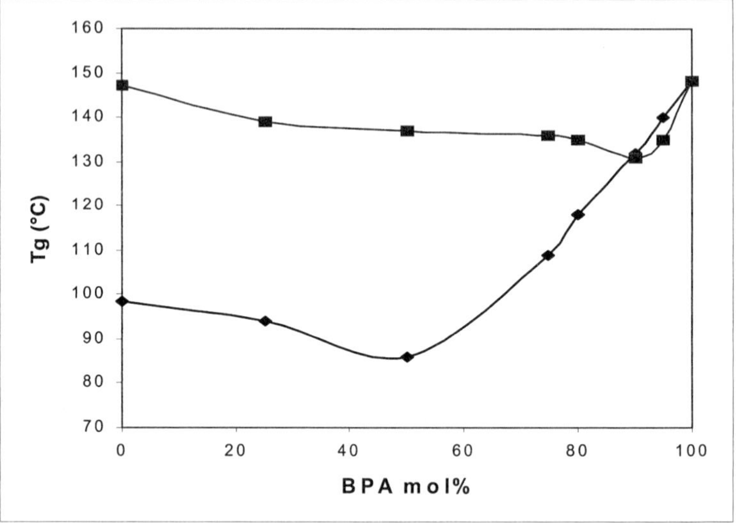

Fig. 3. The T_g of polycarbonate ester (bottom) and polycarbonate acid (top) of different compositions.

The viscosity of the polycarbonate containing free carboxylate groups (X = 0.75 and 1.0) in aqueous base solution has also been investigated. The results are shown in Figure 4. It is clear that the two polymers show typical behavior of polyelectrolytes because viscosity increases markedly as concentration decreases as a result of coil expansion.

2. The Thermal Transformation of *tert*-butyl diphenolate Polycarbonate

Figure 5 shows the TGA data of both *tert*-butyl diphenolate polycarbonate(solid line) and free diphenolic acid polycarbonate (broken line). The latter was prepared by chemical de-protection as mentioned above. It is clear that the former loses about 15% of its weight at around 200 °C, which means that the *tert*-butyl ester can be transformed into the free acid accompanied by the generation of one molecule of isobutene. In the temperature range after liberation of the protecting group, the two curves are very similar, which indicates that they correspond to the same material (free diphenolic acid polycarbonate).

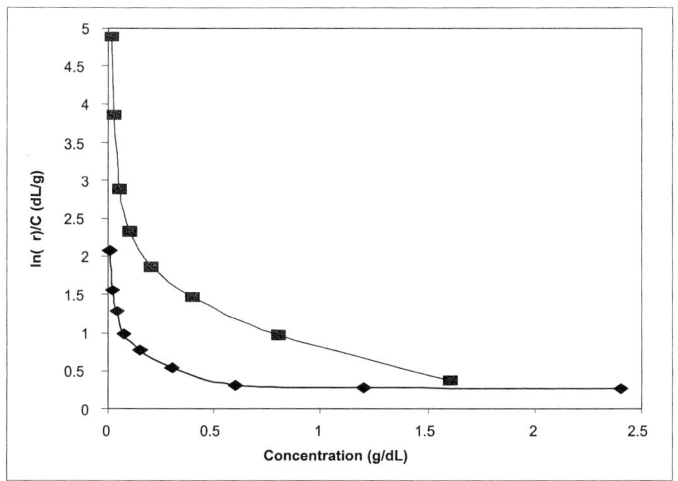

Fig. 4. The concentration-dependent viscosity of polyacids in aqueous solution at 20 °C: x = 1.0, pH = 9 (top); x = 0.75, pH = 10 (bottom).

3. The Formation of Hyperbranched Polymer[5]

As seen in Figure 5, when heating the free acid polycarbonate in the region from 250 °C to 300 °C, there is a weight loss of 12%. To know what happened in this region, IR spectra have been recorded when the sample was heated at 260 °C for different times (see Figure 6). Two significant changes appeared in the IR spectra as the heating time increased. The peak at 3380 cm^{-1} appeared as a result of the generation of hydroxyl groups during the heating process. The peak at 1773 cm^{-1} and 1704 cm^{-1} in the spectrum of the sample before heating should be ascribed to the carbonate bonds and carboxyl groups. Both of them gradually disappeared and a new peak at 1753 cm^{-1} appeared. The above results provide evidence for the formation of hyperbranched polymer through the transesterification reaction of carboxyl groups and carbonate bonds accompanied by the elimination of one molecule of carbon dioxide. This elimination corresponds to the weight loss of 12% observed in Figure 5. In the meantime, new ester bonds and phenol groups are produced in a reaction that might be described as shown below in Scheme 3.

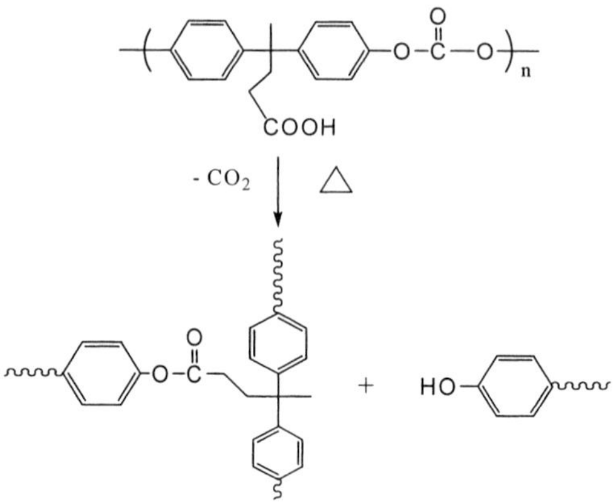

Scheme 3. Transesterification of polycarbonate acid to hyperbranched polyester.

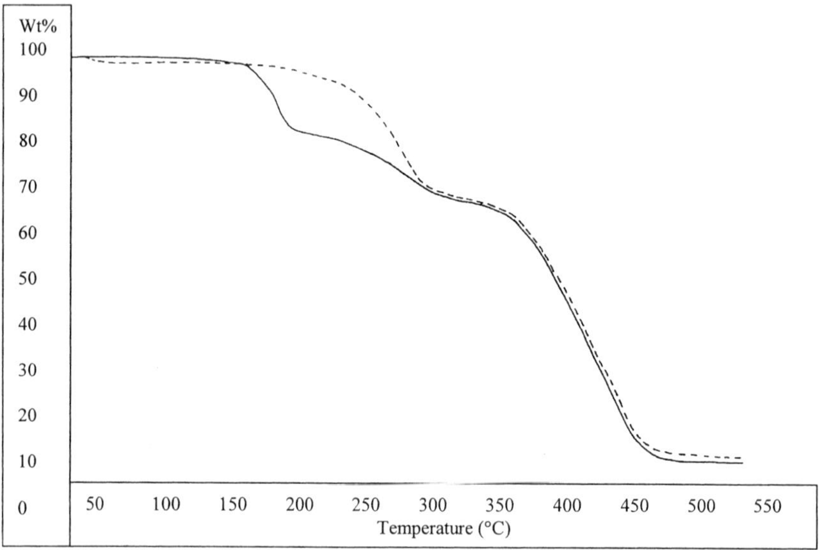

Fig. 5. TGA trace of tert-butyl diphenolate polycarbonate (solid line) and free diphenolic acid polycarbonate (broken line).

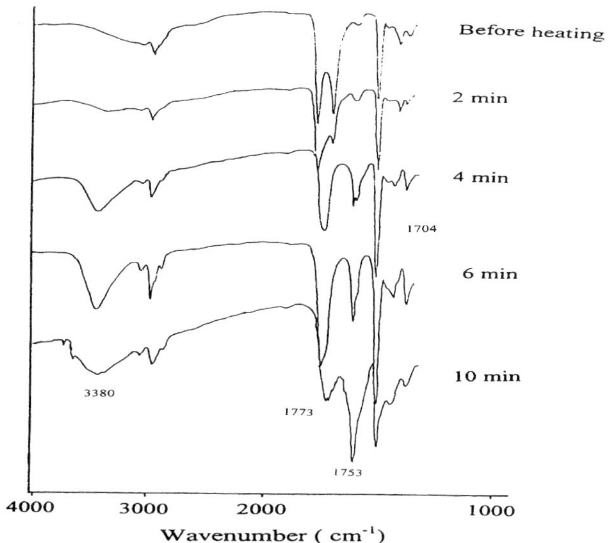

Fig. 6. IR spectra of free acid polycarbonate after heating at 260°C for different times.

4. Properties of Metal Ion Cross-linked Polycarbonate

Polycarbonate acid can react with diethyl zinc very quickly to produce a gel. The DSC measurement of these gels showed that the T_g of the materials increased as the wt-% of Zn increased (see Fig. 7). However, as the Zn wt-% reaches as high as 5 % the glass transition can no longer be observed from DSC measurements. A similar phenomenon has also been observed from the DSC results of Al cross-linked polycarbonate acid (see Fig. 8). Although Zn-complexed polymer showed relatively lower T_g values as compared with Al-complexed polymer the glass transition also disappeared as the Zn wt % reached 5 %. It has also been found that in the case of polymers with a low content of Zn (for example 1% or 2%) the formed gel can be made soluble again by heating at around 60 °C. However, the obtained solution can not be reversibly converted into a gel by cooling. This phenomenon has not been found in the Al cross-linked system and is currently being studied further.

5. The Titration of Polycarbonate Acid

The polycarbonate acid can be dissolved in dilute aqueous NaOH solution. The formed carboxlate groups will act as weak bases. In fact, every carboxlate group on the polymer has a different ionization constant, depending on its local environment. Therefore when this polycarboxylate is neutralized with HCl the titration curve should be quite different from that of a low molecular weight base. Fig. 9 shows the titration curve of sodium polycarbonate carboxylate with aqueous HCl solution. For the sake of comparision a curve of the same NaOH titrated with the same HCl solution is also included. As more HCl is added to the basic polymer solution the pH value gradually decreases from pH = 12 to pH = 3 without any sharp transition. Once the pH reaches 8 the polymer preciptates rapidly from the solution as the polycarbonate acid. This titration behavior is typical of a polyacid, which is quite different from low molecular weight analogs.

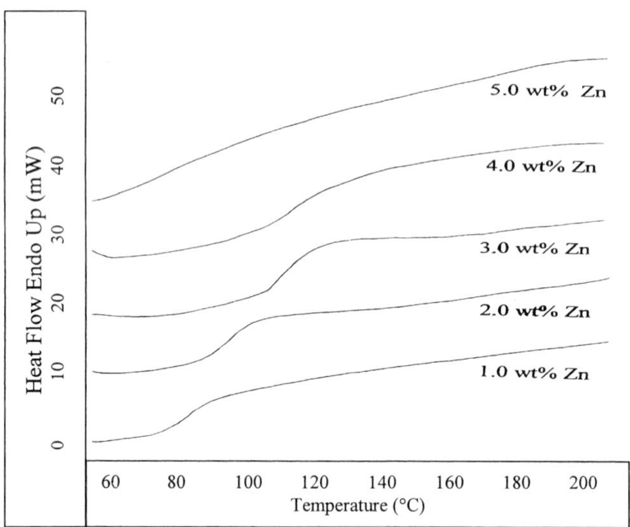

Fig. 7. DSC results for polycarbonate containing different amount of Zn^{2+}.

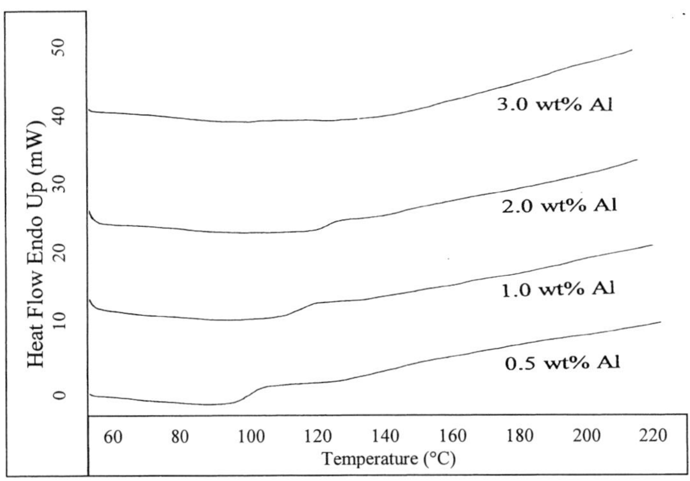

Fig. 8. DSC results for Polycarbonate containing different amount of Al^{3+}.

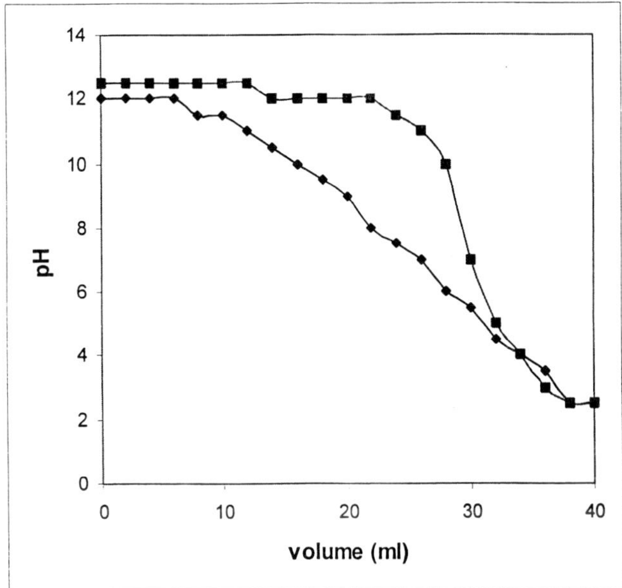

Fig. 9. Titration curve of NaOH by HCl without (top) and with 0.5g of polycarbonate acid (bottom). The concentration of NaOH and HCl is 0.01M. The volume of NaOH solution is 30 ml.

6. Stability of Polycarbonate Acid in Aqueous Basic Solution

Usually NaOH or KOH can catalyze the hydrolytic degradation of polycarbonate in a medium rich in water.[8] The carbonate bond is sensitive to inorganic or organic bases. It is interesting and important to investigate the stability of the polycarbonate acid because the polymer is soluble in aqueous base solution. Figures 10 to 12 contain data collected to measure the change in viscosity of the polymer solution as a function of time at particular concentrations and temperatures. The results show that the polycarbonate carboxylate is stable for some time in basic conditions at relatively lower temperatures. In strongly basic conditions and higher temperatures the polymer will degrade. It is proposed that the negative charge of the carboxylated backbone repels OH^- ion and prevents it from attacking the backbone carbonate bonds. This rationalization of why the polycarbonate acid is not as unstable as might be expected is under further investigation.

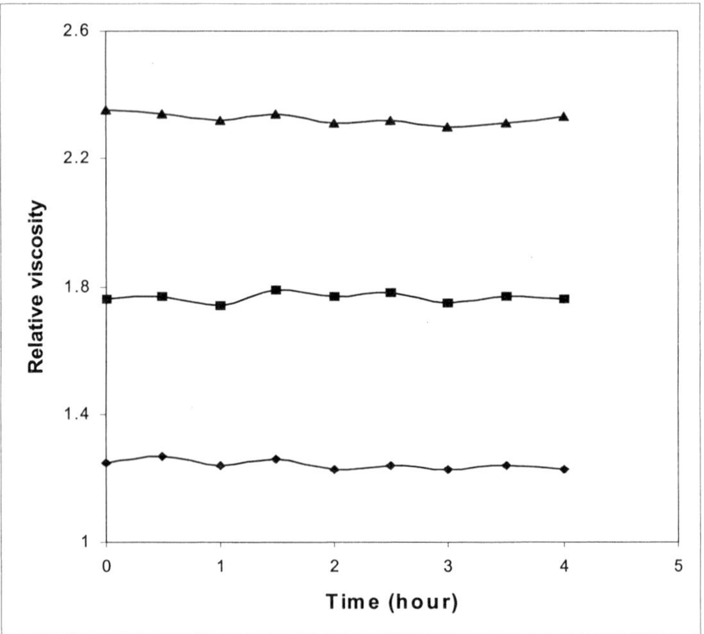

Fig. 10. Stability of homopolycarbonate acid at pH = 8 and different temperatures from top to bottom (2.0 g/dl, 75 °C; 1.0 g/dl, 50 °C; 0.5 g/d, 20 °C).

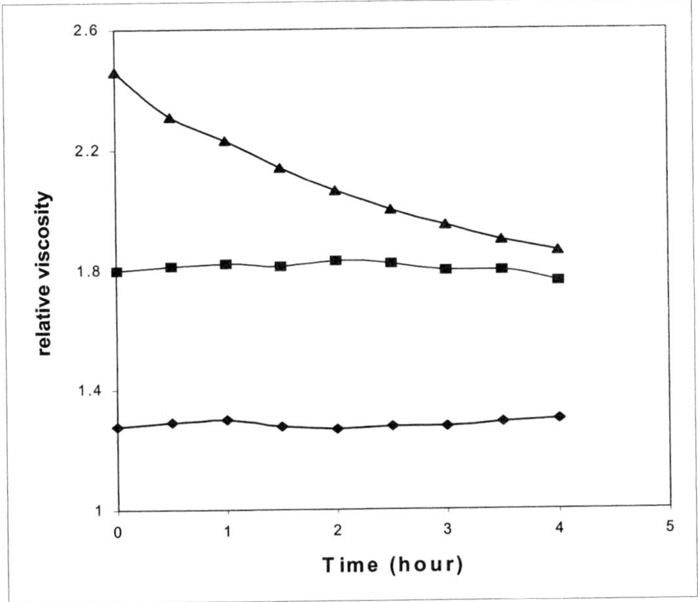

Fig. 11. Stability of homopolycarbonate acid at pH = 10 and different temperatures from top to bottom (2.0 g/dl, 75 °C; 1.0 g/dl, 50 °C; 0.5 g/d, 20 °C).

Conclusion:

Polycarbonate derivatives containing diphenolic acid as a repeat unit have been synthesized in three steps --- esterification of diphenolic acid with *tert*-butanol, condensation with phosgene, and cleavage of the tert-butyl group with trifluoroacetic acid or by thermolysis. Heating the polycarbonate acid at higher temperature leads to the formation of hyperbranched polymer. The structures and properties of obtained polymers have been characterized and investigated. The free diphenolic acid polycarbonate showed polymeric electrolyte behavior and good stability in aqueous base solution. The free carboxyl groups on the polycarbonate acid derivative can be cross-linked by using metal ions such as Al^{3+} and Zn^{2+}. The cross-linked polymer showed increasing T_g values with increasing amounts of metal ion.

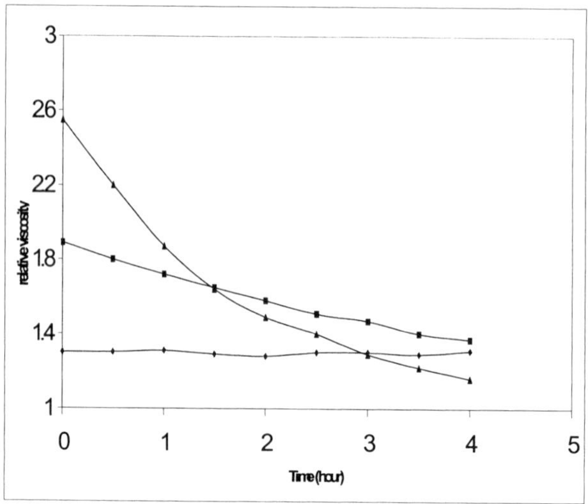

Fig. 12. Stability of homopolycarbonate acid at pH = 12 and different temperatures from top to bottom (2.0 g/dl, 75 °C; 1.0 g/dl, 50 °C; 0.5 g/dl, 20 °C).

[1] J. J. Bozell, L. Moens, D. C. Elliot, Y. Wang,G. G. NeuenSchwander, S. W. Fitzpatrick, R. J. Bilski, J. Jarnefeld, *Resources, Conservation and Recycling*, **2000**, *28*, 227.

[2] R. P. Fisher, G. R. Hartranft, J. S. Heckles, *J.Appl. Polym. Sci.*, **1986**, *32*, 6191.

[3] L. H. Tagle, F. R. Diaz, A. Donoso, *J.M.S.-Pure Appl. Chem.*, **1996**, *A33(11),* 1643.

[4] J. A. Moore, T. Tannahill, *High Performance Polymers,* **2001**, *13(2),* 5305.

[5] Q. Hua, M. S. Thesis, Rensselaer Polytechnic Institute, Troy, NY, 2002.

[6] H.P. Gregor, L. B. Luttinger, E. M. Loebl, *J. Am. Chem. Soc.,* **1954**, *76*, 5879.

[7] J.-S. Kim, A. Eisenberg, *"Introduction to Ionomers"*, Wiley-Interscience, New, York, 1998.

[8] C. A. Pryde, M. Y. Hellman, *J. Appl. Polym. Sci.*, **1980**, *25*, 2573.

New Silicon-Containing Heterocyclic Polyimides

Burkhard Schulz,[1] *Elena Hamciuc,*[2*] *Thomas Köpnick,*[3] *Yvette Kaminorz,*[4] *Maria Bruma*[2]

[1] University of Potsdam, FZDOBS, Am Neuen Palais 10, 14469 Potsdam, Germany
[2] Institute of Macromolecular Chemistry, Aleea Ghica Voda 41A, 6600 Iasi, Romania
Email: ehamciuc@icmpp.tuiasi.ro mbruma@icmpp.tuiasi.ro
[3] Institute of Thin Film Technology and Microsensors, Kanntstr 55, 14513 Teltow, Germany
[4] University of Potsdam, Institute of Physics, Am Neuen Palais 10, 14415 Potsdam, Germany

Summary: New aromatic polyimides and polyamide-imides with phenylquinoxaline rings and dimethylsilane units have been synthesized by solution polycondensation reaction of aromatic diamines containing phenylquinoxaline units with bis(3,4-dicarboxyphenyl)-dimethylsilane dianhydride, or with a diacid chloride resulting from the reaction of this dianhydride with *p*-aminobenzoic acid. These polymers were easily soluble in organic solvents, such as N-methylpyrrolidinone and dimethylacetamide, and showed high thermal stability with decomposition temperature being above 440°C and glass transition temperature in the range of 245-285°C. Very thin coatings were deposited from polymer solutions onto silicon wafers and exhibited smooth, pinhole-free surface in atomic force microscopy investigations. Some of these polymers showed blue fluorescence in solution and films, with a maximum in the range of 415-425 nm.

Keywords: dimethylsilane, phenylquinoxaline, photoluminescence, polyimides, thin films

Introduction

High performance polymers are used in applications demanding service at enhanced temperatures while maintaining their structural integrity and an excellent combination of chemical, physical and mechanical properties. Wholly aromatic polyimides are generally the polymers of choice for these applications due to their many desirable characteristics including good thermooxidative stability and excellent mechanical properties.[1,2] However, these polymers are processed with great difficulty because they are insoluble and infusible. Being known that the introduction of flexible groups such as dimethylsilane or ether linkages into the backbone of a polyimide leads to soluble products having a high thermal stability, the synthesis of copolymers containing such groups is a promising way to easy processable compounds. On the other hand, it was shown that poly(phenylquinoxaline)s are a family of aromatic polymers known for their excellent thermal and chemical stability.[3,4] The pendent

 DOI: 10.1002/masy.200350933

phenyl substituents improve the solubility and processing characteristics of these polymers as well as the thermooxidative stability over the unsubstituted polymers. Therefore, we considered that by introducing silicon together with phenylquinoxaline rings into the repeating unit of aromatic polyimides, polymers with substantialy improved solubility and processability could be obtained, having a better combination of properties. Thus, we have synthesized polyphenylquinoxaline-imides and polyphenylquinoxaline-imide-amides by the condensation reaction of certain diamines containing phenylquinoxaline rings with bis(3,4-dicarboxyphenyl)-dimethylsilane dianhydride, or with a diacid chloride resulting from the reaction of this dianhydride with *p*-aminobenzoic acid followed by treating with thionyl chloride. The solubility, inherent viscosity, molecular weight, thermal stability, glass transition temperature, fluorescence properties and the quality of the thin films made from these polymers have been investigated and compared with those of related compounds.

Experimental

Synthesis of the Monomers

Four aromatic diamines containing phenylquinoxalines units (**1**), one silicon-containing dianhydride namely bis(3,4-dicarboxyphenyl)-dimethylsilane dianhydride (**2**) and one silicon-containing diacid chloride (**3**) have been prepared and their structures are shown in scheme 1. The diamines **1** were synthesized by the condensation reaction of 3,4,4'-triamino-diphenylether (2 mol) with bis(α-diketone)s (1 mol), in ethanol at reflux temperature, according to the literature.[5]

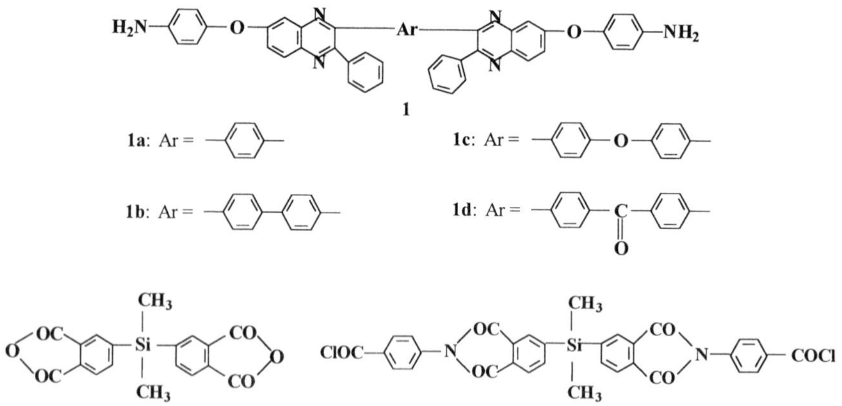

H_2N O N N Ar N N O NH_2

1

1a: Ar =

1c: Ar = O

1b: Ar =

1d: Ar = C O

CH_3 OC Si CO O OC CO CH_3 O

ClOC N OC CO CH_3 Si CH_3 CO N CO COCl

2

3

Scheme 1. Structures of the monomers.

Bis(3,4-dicarboxyphenyl)-dimethylsilane dianhydride **2** was prepared by a multistep reaction in which 4-bromo-*o*-xylene reacted with dimethyldichlorosilane to produce the bis(3,4-dimethylphenyl)-dimethylsilane that underwent the oxidation with potassium permanganate resulting in bis(3,4-dicarboxyphenyl)-dimethylsilane; the dehydration in refluxing acetic anhydride gave the corresponding dianhydride.[6]

The silicon-containing diacid chloride **3** has been synthesized in two steps by following a procedure described for similar diacid chloride[7]: the reaction of dianhydride **2** (1 mol) with *p*-aminobenzoic acid (2 mol) in glacial acetic acid at reflux afforded a silicon-containing dicarboxylic acid which by further reaction with thionyl chloride gave the corresponding diacid chloride **3**.

Synthesis of the Polymers

Silicon-containing polyphenylquinoxaline-imides **4** were prepared by solution polycondensation of equimolar amounts of diaminophenylquinoxalines **1** with bis(3,4-dicarboxyphenyl)-dimethylsilane dianhydride **2** in N-methylpyrrolidinone (NMP) as a solvent at a total concentration of 12-15% solids. The reaction mixture was stirred at room temperature for 6-8 h and at 180-190°C for another 3-4 h The resulting polyimide **4**was isolated by precipitation in water. The structures of these polymers are shown in scheme 2.

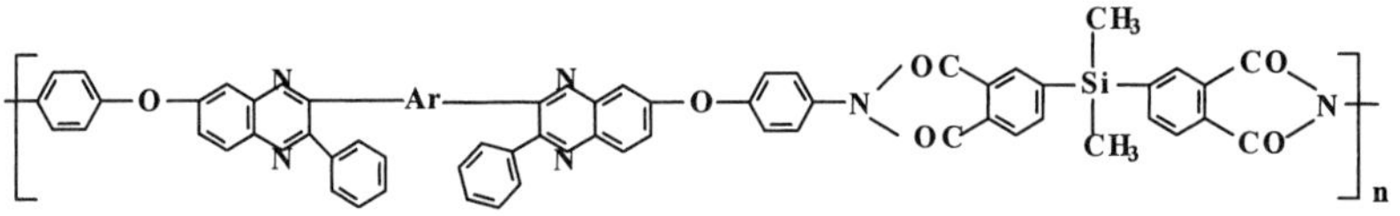

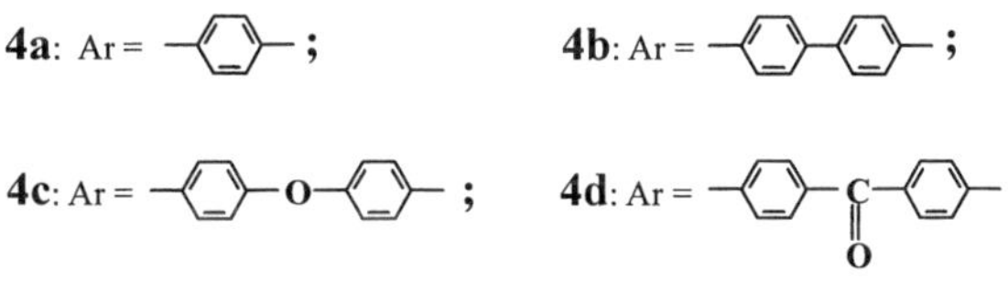

Scheme 2. Structures of silicon-containing polyphenylquinoxaline-imides **4**.

Silicon-containing polyphenylquinoxaline-imide-amides have been prepared by polycondensation of equimolar amounts of a diamine **1** with silicon-containing diacid chloride **3** in NMP as a solvent and with pyridine as an acid acceptor at 10-20°C. The total concentration of the monomers in the reaction mixture was 8-10% and the reaction time was

3-4 h. The resulting viscous solution was slowly poured into water to precipitate the fibrous polymer. The structures of these polymers are shown in scheme 3.

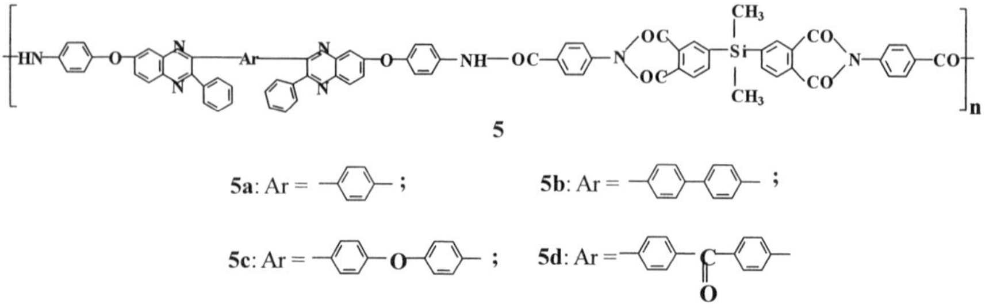

Scheme 3. Structures of silicon-containing polyphenylquinoxaline-imide-amides **V**.

Preparation of Polymer Films

Films of silicon-containing polyphenylquinoxaline-imides **4** were prepared by casting a solution of 5% concentration of polymer in chloroform onto glass plates, followed by drying at room temperature for 24 h under a Petri dish and for another 2 h at 130°C. Films of silicon-containing polyphenylquinoxaline-imide-amides **5** were similarly prepared by using 10% polymer solution in NMP which was cast onto glass plates and dried gradually from room temperature to 210 °C during 2 h.

Measurements

Infrared spectra were recorded with a Nicolet Magna FTIR spectrometer in transmission mode by using monomers or precipitated polymers ground in potassium bromide pellets. The inherent viscosities (η_{inh}) of the polymers were determined with an Ubbelohde viscometer, by using polymer solutions in NMP, at 20°C, at a concentration of 0.5 g/dL. Weight average molecular weights (Mw) and number-average molecular weights (Mn) were determined by means of gel permeation chromatography (GPC) using a Waters GPC apparatus, provided with Refraction and Photodiodearray Detectors and Phenomenex-Phenogel MXN column. Measurements were carried out with polymer solutions having 0.2% concentration in chloroform, and by using chloroform as eluent. Standard polystyrene of known molecular weight was used for calibration. The thermogravimetric analysis (TGA) of the precipitated polymers was performed in air by using a Perkin-Elmer TGA-7 equipment at a heating rate of 10°C/min. The initial decomposition temperature (IDT) is characterized as the temperature at which the sample achieves a 5% weight loss. The temperature of 10% weight loss (T_{10}) in the

TGA curve was also recorded. The glass transition temperature (T_g) of the precipitated polymers was determined with a Mettler differential scanning calorimeter DSC 12 E using a heating rate of 20°C/min under nitrogen. The mid point of the inflection in DSC curve resulting from the second heating was regarded as T_g temperature. Model molecules for a polymer fragment were obtained by molecular mechanics (MM^+) by means of the Hyperchem program, version 4.0.[8] The surfaces of the very thin films as-deposited on silicon wafers were studied by atomic force microscopy (AFM) with a SA1/BD2 apparatus (Park Scientific Instruments) in the contact mode. For absorption and fluorescence measurements the polymers **4** were dissolved in NMP at a concentration of 5mg/mL in NMP and the solutions were spin-coated onto a quartz substrate, followed by heating to remove the solvent. The absorption of the polymer solution as well as of the films was measured in transmission geometry using a Perkin Elmer Lambda 16 spectrophotometer. The measurements of fluorescence were performed in a front-face arrangement using a modular fluorescence spectrometer Alphascan from PTI with a special fiber goniometer. The excitation light was transferred by an optical fiber (400 μm diameter) aligned normal to the surface. The emitted light was collected by an optical fiber (600 μm diameter) inclined by 25 degrees with respect to the surface normal. All spectroscopic measurements were carried out under ambient conditions.

Results and Discussion

The structure of the polymers **4** and **5** was identified by IR spectra and elemental analyses. The strong absorption bands at 1780, 1730 and 730 cm^{-1} confirmed the presence of the imide rings. Absorption bands at 1240 cm^{-1} and 810 cm^{-1} are characteristic for methyl-silane bonds and are present in all the spectra. Aromatic C-H absorptions were found at 3070 cm^{-1}, while aromatic C=C bonds were found at 1600 cm^{-1} and 1500 cm^{-1}. The data obtained for C, H and N content, in elemental analysis, are in good agreement with the calculated values. In the spectra of polymers V strong and wide absorptions appeared at 3450 cm^{-1} due to NH in amide groups.

All these polymers are easily soluble in organic solvents such as N-methylpyrrolidinone, dimethylacetamide (DMA), and the polymers **4** are also soluble in pyridine (Py) and chloroform, which is very useful from a practical point of view. The improved solubility of these polymers can be explained by the presence of dimethyl-silane groups and of large-volume phenylquinoxaline rings which contribute to create a distance between the macromolecular chains and facilitates the diffusion of small molecules of solvent. The shape of a macromolecule fragment is far from a linear rigid rod shape of a traditional aromatic polyimide, as shown in Figure 1. The good solubility makes the present polymers potential candidates for practical applications in spin coating and casting processes.

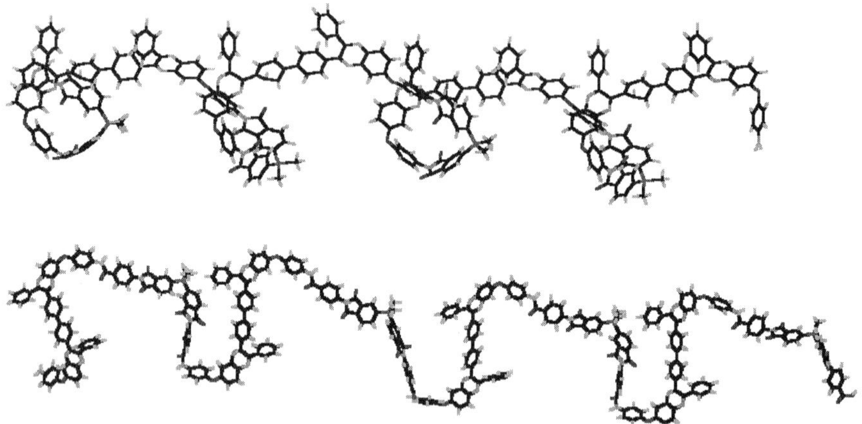

Fig. 1. Model molecules (four repeating units) of polymers **4b** (top) and **5b** (bottom).

The inherent viscosities, measured in NMP solution, were in the range of 0.4-0.8 dL/g (Table 1). The molecular weight of polymers was determined by GPC. The molecular weight Mw values of **4** are in the range of 30000-72000 g/mol, Mn in the range of 10000-16000 g/mol and Mw/Mn in the range of 2.6-4.3 (Table 1). The molecular weight of **5** are: Mw in the range of 37000 – 74000, Mn in the range of 20000-39000, and polydispersity Mw/Mn in the range of 1.8-2.0.

Table 1. GPC data of the polymers **4** and **5**.

Polymer	η_{inh} [a)] (dL/g)	Mw (g/mol)	Mn (g/mol)	Mw/Mn
4a	0.46	35000	11000	3.1
4b	0.43	42000	10000	4.1
4c	0.42	30000	11000	2.6
4d	0.58	72000	16000	4.3
5a	0.40	37000	20000	1.8
5b	0.38	54000	30000	1.8
5c	0.88	74000	39000	1.9
5d	0.58	51000	25000	2.0

[a)] Determined in NMP at 20°C, at a concentration of 0.5 g/dL.

All these polymers possess remarkable film-forming ability. The films which were obtained by casting technique, having a thickness of tens of micrometres were tough, flexible and creasable. Very thin coatings having a thickness in the range of tens of nanometres which have been prepared by spin coating technique onto silicon wafers showed a very strong adhesion to the silicon support and high quality when studied by atomic force microscopy (AFM). The as-deposited films exhibited very smooth and homogeneous surfaces over large scanning ranges (1-100 μm). They did not show any pinholes or cracks and were practically deffectless. These qualities are very important when such films are used in microelectronic devices.[9] A typical AFM image for polymers **5** is shown in Figure 2.

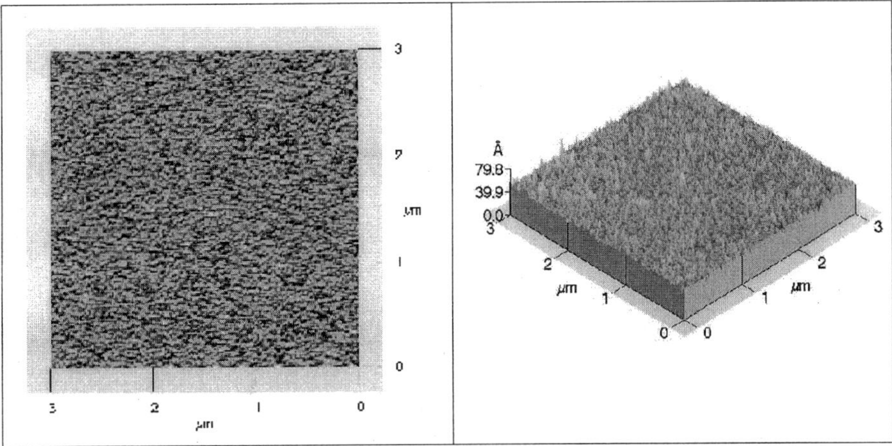

Fig. 2. AFM image of a film made from polymer **5d** (left: top view; right: side view).

The thermal stability of the polymers was evaluated by thermogravimetric analysis (TGA), which was performed in air. All polymers exhibited high thermal stability. The polyimides **4** begin to decompose in the range of 485-505°C, while the polyphenylquinoxaline-imide-amides **5** begin to decompose at a slightly lower temperature, in the range of 446-466°C. (Table 2).

The temperatures of 10% weight loss were in the range of 519-545°C for polyphenylquinoxaline-imides **4** and in the range of 490-500°C for polyphenylquinoxaline-imide-amides **5**. By comparing the present aromatic polyimides and polyimide-amides containing phenylquinoxaline rings and dimethylsilane linkages with related polyimide-phenylquinoxalines[10] and polyimide-amide-phenylquinoxalines[11] that have been synthesized from the same diamines but with a dianhydride or with a diacid chloride which do not contain silicon, it can be seen that their decomposition temperatures are in the same range. This shows that the presence of dimethylsilane groups preserves the high thermal stability while improving the solubility. Also, comparison of these polymers with related silicon-containing polyimides[12] that were obtained from bis(3,4-dicarboxyphenyl)-dimethylsilane dianhydride and conventional aromatic diamines, without heterocyclic units, shows that their thermal stabilities are similar to each other, which means that the introduction of phenyl-quinoxaline rings into their chains did not significantly influence the thermal resistance.

Table 2. Thermal properties of the polymers **4** and **5**.

Polymer	IDT[a)] (°C)	T_{10} (°C)	T_g (°C)	Polymer	IDT[a)] (°C)	T_{10}[b)] (°C)	T_g[c)] (°C)
4a	485	519	265	**5a**	452	495	280
4b	492	532	266	**5b**	457	500	284
4c	505	545	245	**5c**	466	500	278
4d	498	532	255	**5e**	446	490	282

[a)] Initial Decomposition Temperature = Temperature of 5% weight loss.
[b)] Temperature of 10% weight loss.
[c)] Glass transition temperature.

The glass transition temperatures (T_g) of the present polymers polyphenylquinoxaline-imides **4**, evaluated from DSC curves, are in the range of 245-266°C, while those of polyphenylquinoxaline-imide-amides **5** are in the range of 278-284°C (Table 2). There is a large interval between T_g and the decomposition temperature of each of these polymers which makes them attractive for thermoforming processing.

The light-emitting ability of these polymers was evaluated by recording the absorption and fluorescence spectra. The absorption of the polymers **4** in solution showed a maximum in the range of 363 nm - 374 nm (table 3). The absorption behavior of these polymers does not differ very much from each other. Therefore the absorption properties are mainly determined by phenylquinoxaline units. The absorption maximum of the films was 372 nm - 380 nm.

All polymers **4** showed blue fluorescence in solution and films, with a maximum in the range of 415 to 425 nm (table 3 and figure 3). The slightly blue shift of polymer **4a** is probably not caused by the chemical structure of the polymer but rather by re-absorption effects. The intensities of the fluorescence of the polymers **4a**, **4b** and **4c** are comparable to each other (in the range of 4000 cps).

Table 3. UV absorption and fluorescence properties of the polymers **4**.

Polymer	Ar	Absorption (nm)		Fluorescence (nm)	
		Solution	Film	Solution	Film
4a		374	374	415	425
4b		374	380	420	425
4c	O	369	374	420	425
4d	C O	363	372	420	425

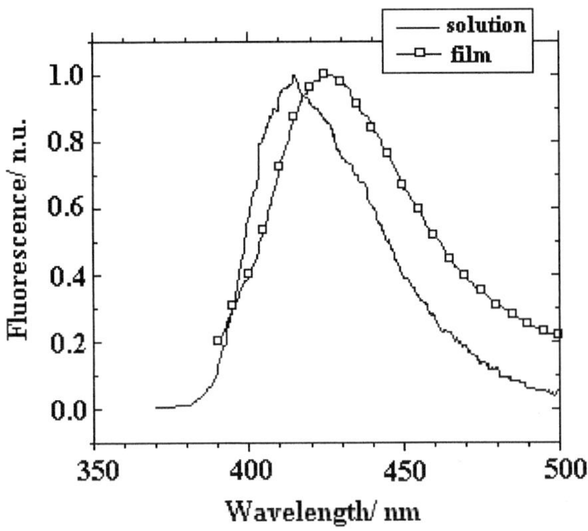

Fig. 3. The fluorescence spectra of the polymer **4a** in solution and as spin-coated film.

Conclusions

New silicon-containing heterocyclic polyimides and polyimide-amides have been synthesized by solution polycondensation of aromatic diamines containing preformed phenylquinoxaline groups with a dianhydride or with a diacid chloride, both incorporating dimethylsilane linkages. They show high thermal stability, with decomposition starting above 480°C for polyimides, and above 440°C for polyimide-amides, in air atmosphere, and a glass transition in the range of 245-266°C for polyimides and 278-284°C for polyimide-amides. The large interval between glass transition and decomposition temperature may be advantageous for their processing by a thermoforming technique. The highest quality of these silicon-containing heterocyclic polyimides and polyimide-amides is their remarkable solubility in organic solvents such as NMP and DMA, which associated with their high thermal stability is very important from practical point of view. The present polymers can also be processed from solutions into flexible thin and ultrathin, pinhole-free films having strong adhesion to silicon wafers. Some of them showed blue fluorescence in solution and films, with a maximum in the range of 415-425 nm. Potential applications in microelectronics, optoelectronics or other related fields are foreseen.

Acknowledgements

We acknowledge with great pleasure the financial support provided to EH by DAAD (Deutscher Akademischer Austauschdienst) and to MB by AiF (Arbeitsgemeinschaft industrieller Forschungsvereinigungen) in Germany. Our warm thanks also go to B. Stiller and M. Ehlert at Potsdam University for AFM and TGA measurements and M. Bierbrauer at the Institute of Thin Film Technology and Microsensors, Teltow, Germany, for GPC analyses.

[1] J. E. McGrath, D. L. Dunson, S. J. Mecham, J. L. Hedrick, *Adv. Polym. Sci.* **1999**, *140*, 61 .
[2] M. Hasegawa, K. Horie, *Prog. Polym. Sci.* **2001**, *26*, 259.
[3] P. Hergenrother, in "*Encyclopedia of Polymer Science and Engineering*", 2nd ed., vol 13, Wiley & Sons, New York, 1988, p. 55.
[4] M. Bruma, in "*Handbook of Thermoplastics"* O. Olabisi, Ed. (Marcel Dekker, New York 1997, p. 771.
[5] V. V. Korshak, E. S. Krongauz, N. M. Belomoina, H. Raubach, D. Hein, *Acta Polym.* **1983**, *34,* 213.
[6] J. R. Pratt, S. F. Thames, *J. Org. Chem.* **1973**, *38,* 4271.
[7] M. Bruma, F. Mercer, J. Fitch, P. Cassidy, *J. Appl. Polym. Sci.*, **1995**, *56,* 527.
[8] Hypercube Inc. (Ontario), *Hyperchem* version 4 (1994).
[9] N. Xu N, M. R. Coleman, *J. Appl. Polym. Sci.* **1997**, *66,* 459.
[10] I. Sava I, M. Bruma, N. Belomoina, F. W. Mercer, *Angew. Makromol. Chem.* **1993**, *211,* 113.
[11] E. Hamciuc, M. Bruma, F. W. Mercer, N. M. Belomoina, C. I. Simionescu, *Angew. Makromol. Chem.* **1995**, *227,* 19.
[12] M. M. Koton, T. I. Zhukova, F. S. Florinskii, T. M. Kiseleva, L. A. Laius, Y. N. Sazanov, *Vysokomol. Soedin., B,* **1980**, *22,* 4.

Transfer Molding Imide Resins Based on 2,3,3′,4′-Biphenyltetracarboxylic Dianhydride

J. G. Smith Jr.,[*1] J. W. Connell,[1] P. M. Hergenrother,[1] L. A. Ford,[2] J. M. Criss[3]*

[1]National Aeronautics and Space Administration, Langley Research Center, Hampton, VA 23681-2199, USA
E-mail: joseph.g.smith@larc.nasa.gov.
[2]National Research Council Research Associate at NASA LaRC
[3]M and P Technologies, Inc., Marietta, GA 30067, USA

Summary: Phenylethynyl containing imide oligomers have been under investigation as part of an effort to develop resins for non-autoclave composite fabrication processes such as resin transfer molding (RTM). These high performance/high temperature composites are potentially useful on advanced aerospace vehicles such as reusable launch vehicles (RLVs). New phenylethynyl terminated imide oligomers (PETI) based upon 2,3,3',4'-biphenyltetracarboxylic dianhydride (a-BPDA) were prepared and characterized primarily by rheological behavior and cured glass transition temperature (Tg). In comparison to resins from the symmetrical isomer (3,3',4,4'-biphenyltetracarboxylic dianhydride, s-BPDA), a-BPDA afforded corresponding resins with lower melt viscosities and upon curing, higher Tgs. Several resins exhibited an attractive combination of properties such as low and stable melt viscosities required for RTM composite fabrication, high cured Tgs, and moderate toughness. One resin (P10) was used to fabricate flat, void free laminates by RTM. The laminates exhibited high mechanical properties at temperatures to 288°C. The chemistry and physical properties of these new PETIs and the laminate properties of one composition are discussed.

Keywords: a-BPDA, high temperature polymers, PETI-330, polyimides, resin transfer molding

1 Introduction

Advanced aerospace vehicles such as high speed aircraft and reusable launch vehicles (RLV) require high performance/high temperature resins for use as adhesives and composite matrix resins for structural applications. Due to the excellent physical and mechanical properties of

 DOI: 10.1002/masy.200350934

aromatic polyimides,[1,2] these materials are under investigation for these applications. However, even moderate molecular weight polyimides are difficult to process into large bonded panels and composites. The monomeric reaction approach, such as used with PMR-15, has been practiced for many years as a means of obtaining processability and accordingly, good quality composites. An alternative approach that is essentially free of toxicity and reduces the problems associated with residual solvent is through the use of controlled molecular weight oligomers containing phenylethynyl groups.

Controlled molecular weight imide oligomers containing phenylethynyl groups have been under investigation for a variety of electronic and structural applications for more than 20 years.[3-28] Phenylethynyl terminated imide oligomers were extensively evaluated as adhesives and composite matrices in the high speed civil transport (HSCT) program.[29-37] To be considered for HSCT applications, the resins had to: (1) be non-toxic, (2) provide bonded panels and composites using conventional fabrication processes, (3) exhibit performance as required by the vehicle, and (4) be cost-effective. The main operational issues for the HSCT included high temperature, high stress, object impact, thermal cycling, and exposure to moisture and aircraft fluids. In addition, components were required to retain acceptable mechanical properties at 177°C for 60,000 hours.

In the HSCT program, many phenylethynyl containing imide oligomers were evaluated. One phenylethynyl terminated imide (PETI), designated PETI–5, was selected as the adhesive and composite matrix resin. The material was prepared from 3,3',4',4-biphenyltetracarboxylic dianhydride (s-BPDA) and an 85:15 ratio of 3,4'-oxydianiline (3,4'-ODA) and 1,3-bis(3-aminophenoxy)benzene (1,3,3-APB) at a calculated number average molecular weight ($\overline{M}_n$) of 5000 g/mol and endcapped with 4-phenylethynylphthalic anhydride (PEPA). Due to its oligomeric nature, PETI-5 exhibited good autoclave processability during the fabrication of large bonded and composite structures at 350-371°C for 1 hour under 1.4 MPa (200 psi) or less.[37] The thermal cure of the phenylethynyl group involved chain extension, branching, and light crosslinking without the evolution of volatile by-products. The cured resin with a glass transition temperature (Tg) of ~270°C provided an excellent combination of properties including high toughness, high strength, moderate modulus, good moisture and solvent resistance.

However, the fabrication of support structures (e.g. frames, ribs, stringers) in the HSCT program was preferably performed by resin infusion (RI) and resin transfer molding (RTM) processes due to their low manufacturing costs. These methods required a resin with low (preferably <3 Pa•s) and stable (>2 hr at the injection or infiltration temperature) melt viscosity during part fabrication. Even though PETI-5 exhibited good autoclave processability, the melt viscosity was too high (~6000 Pa*s at 370°C) to fabricate composites by RI and RTM. Initial attempts at lowering the melt viscosity of PETI-5 to make it amenable to RTM through molecular weight reduction[29] and the incorporation of a reactive additive[30] met with limited success. In the first approach, melt viscosities of lower molecular weight versions of PETI-5 decreased relative to that of the $\overline{M}n$= 5000g/mol version, but were still too high for RTM. The second approach reduced the melt viscosity of PETI-5 to acceptable levels, unfortunately, >80 % of the additive was required to be effective. Additionally, the minimum melt viscosities under both approaches occurred at temperatures where the phenylethynyl group reacted appreciably, leading to an unstable melt.

Work continued on the modification of the oligomer composition with the same monomers used to make PETI-5. This was partly due to the desire of the HSCT program which preferred a PETI-5 like material for RTM and RI. The work primarily involved adjustments to the diamine ratio as well as the molecular weight of the oligomer. Results from this effort afforded resins that were processable by RI and RTM.[31-34] These volatile-free oligomers exhibited stable, low melt viscosities at temperatures <290°C where the phenylethynyl group was slow to react. One composition, designated PETI-RTM, was scaled-up to 30 kilogram quantities and successfully used in the fabrication of high quality 2.4 m long curved F-frames by RTM. The cured Tg of PETI-RTM was 256°C, making it an attractive material for the HSCT.

With the closing of the HSCT program in 1999, work continued to develop RI/RTM resins with higher cured Tgs and better mechanical properties than PETI-RTM for future use in composite structures for RLVs. The operational environment for RLV composites differs significantly from that of the HSCT in that higher temperature performance is needed for shorter periods of time. Numerous compositions were evaluated which looked at other diamines than those used in the PETI-5. From this work, RTM resins emerged with higher cured Tgs.[35-36] One composition designated PETI-298 exhibited a cured Tg of ~300°C while retaining excellent RTM processability. It differed from PETI-RTM in that 1,3,3-APB was replaced with 1,3,4-APB. This change provided an improvement over PETI-RTM with better mechanical properties at elevated temperatures due to a higher Tg. Good initial mechanical properties were obtained on flat laminates and excellent retention of room temperature properties were observed after aging at 288°C in air for 1000 hrs.[36] Flat laminates fabricated with this resin by vacuum assisted resin transfer molding (VARTM) gave properties comparable to the RTM laminates, however the void content was ~4%.[38] Work is underway to improve the quality of VARTM/PETI-298 laminates.

Over the past 15 years, polyimides from 2,3',3',4-biphenyltetracarboxylic dianhydride (asymmetric isomer, a-BPDA) were shown to have lower melt viscosities and higher Tgs than the corresponding polymers from s-BPDA.[39-44] This is presumably due to the highly irregular structure of the polyimide emanating from a-BPDA. In addition, a-BPDA based PETIs were shown to have lower melt viscosities and when cured, higher Tgs than the same PETI based upon s-BPDA. [45-47] However, none of these a-BPDA derived PETIs had a melt viscosity low enough for RTM. Thus, a series of low molecular weight PETIs derived from a-BPDA were prepared in an attempt to obtain RTM/VARTM resins with improved processability and when cured, higher Tgs and higher laminate properties.[48] The chemistry, physical and mechanical properties of these new resins and their RTM laminates are described herein.

2 Experimental

2.1 Starting Materials

The following chemicals were obtained from the indicated sources and used without further purification: 3,4'-oxydianiline [3,4'-ODA, Mitsui Petrochemical Ind., Ltd., melting point (m.p.) 84°C]; 1,3-bis(4-aminophenoxy)benzene (1,3,4-APB, Chriskev, m.p. 115°C); 1,4-bis(4-aminophenoxy)benzene (1,4,4-APB, Chriskev Co., m.p. 173°C); 1,3-diaminobenzene (m-PDA, Aldrich Chemical Co., m.p. 66°C); 2,2'-bis(trifluoromethyl)benzidine (TFMBZ, Chriskev Co., m.p. 182°C); 3,3',4,4'-biphenyltetracarboxylic dianhydride (s-BPDA, Allco Chemical Co., m.p. 227°C); 4-phenylethynylphthalic anhydride (PEPA, Imitec, Inc. or Daychem Laboratories, Inc., m.p. 152°C) and N-methyl-2-pyrrolidinone (NMP, Fluka Chemical Co.). 2,3,3',4'-Biphenyltetracarboxylic dianhydride (a-BPDA, Ube Industries Inc., m.p. 197°C) was recrystallized from toluene and acetic anhydride. 2,2'-Dichlorobenzidine (CLBZ, m.p. 166°C) [49] was used as-received from Professor Frank W. Harris, The University of Akron. All other chemicals were used as-received without further purification.

2.2 Synthesis of Phenylethynyl Terminated Imide (PETI) Oligomers

PETI oligomers were prepared at a calculated $\overline{M}n$ of 750 g/mol by the reaction of the appropriate amount of aromatic dianhydride(s) with the appropriate amount of aromatic diamines and endcapped with PEPA. The oligomers were prepared by initially dissolving the aromatic diamines in NMP at room temperature under nitrogen. The appropriate quantities of dianhydride(s) and endcapper (PEPA) were subsequently added in one portion as a slurry in NMP and rinsed into the flask with additional NMP. The reactions were allowed to stir for ~24 hrs at ambient temperature under nitrogen prior to thermal imidization. Imide oligomer was prepared directly from the amide acid solution by azeotropic distillation with toluene under a Dean Stark trap to effect cyclodehydration. In general, the imide oligomers remained soluble during the imidization process. Imide powders were obtained by precipitating the reaction mixtures in water and washing in warm water. The yellow powders were dried at ~200°C to constant weight with yields >95%.

2.3 Composite Laminates

Laminates of PETI oligomer were made by RTM using AS-4 8 Harness Satin (AS-4 8HS, Hexcel) carbon fiber fabric. The fabric sizing was removed by bagging the fabric in Kapton™, sealing with a high temperature sealant, and heating at 400°C for 2 hr under vacuum. PETI oligomer was injected into the unsized AS-4 8HS carbon fiber fabric positioned on an Invar tool using a high temperature injector. The maximum processing parameters of the injector, designed and built by Radius Engineering according to Lockheed Martin specifications, are a temperature of 288°C, flow rate of 500 cc/min, and pressure of 2.75 MPa. The tool containing the fabric was loaded into a press, heated to 316°C, and held at 316°C for 1.5 hrs prior to resin injection. PETI oligomer was degassed in the injector by heating to 288°C and holding for 1 hr prior to injection. The degassing step was required primarily to remove moisture, residual solvent, and air from the resin. The molten resin was used to infiltrate 8-ply stacks of unsized AS-4 8HS fabric [equivalent to 16 plies of unitape, lay-up of $(0/90)_{4S}$] in an Invar tool under 1.34 MPa hydrostatic pressure during the entire process cycle. After the resin was injected at 288°C at a rate of 200 cc/min, the part was held under 1.34 MPa of hydrostatic pressure while heating to 371°C and holding at 371°C for 1 hr. The 33 cm x 36 cm laminates were cooled under pressure. The laminates were ultrasonically scanned (C-scanned, pulse echo), cut into specimens, and tested for mechanical properties. The panels were examined for microcracks by a microscope up to 400X magnification. Resin content, fiber volume, and void content were determined by acid digestion using a 1:1 (w/w) solution of concentrated sulfuric acid and 30% hydrogen peroxide. Calculations were based on a density of 1.79 X 10^3 Kg/m^3 (1.79 g/cc) for AS-4 8HS fabric and 1.37 X 10^3 Kg/m^3 (1.37 g/cc) for the cured PETI resin. Open hole compression (OHC) properties (Northrup Grumman Test[25]) were determined on specimens 25.4 cm by 3.81 cm with a 0.64 cm hole in the center. Un-notched compression properties (Boeing Test, BSS 7260) were determined on specimens 8.08 cm by 1.27 cm. Short beam shear (SBS) strength (ASTM D2344-84) was determined on specimens 0.64 cm by 1.91 cm. Unstressed specimens were aged at 288°C in a forced air oven. Five specimens were tested under each condition.

2.4 Other Characterization

Differential scanning calorimetry (DSC) was performed on powdered samples using a Shimadzu DSC-50 thermal analyzer at a heating rate of 20°C/min with the Tg taken at the inflection point of the differential heat flow (ΔH) versus temperature curve. PETI oligomers were cured at 371°C for 1 hr in a sealed aluminum pan, quenched, and the samples were rerun to obtain the Tgs of the cured resins. Dynamic mechanical thermal analysis (DMTA) was conducted on a Bohlin VOR Rheometer equipped with a solids fixture for measuring rectangular specimens. An oscillatory deformation of ~1% strain, a frequency of 1 Hz which translates to an angular velocity of 6.28 rad/sec, and a stepwise temperature test mode sweep at a heating rate of 10°C/min were used in the analysis. Rheological measurements were conducted on a Rheometrics System 4 rheometer at a heating rate of 4°C/min. Specimen disks (2.54 cm in diameter and 1.5 mm thick) were prepared by compression molding imide powder at room temperature. The compacted resin disk was subsequently loaded in the rheometer fixture with 2.54 cm diameter parallel plates. The top plate was oscillated at a variable strain and a fixed angular frequency of 100 rad/sec while the lower plate was attached to a transducer, which recorded the resultant torque. Complex melt viscosity (η^*) as a function of time (t) was measured at several temperatures.

3 Results and Discussion

3.1 Synthesis of Phenylethynyl Terminated Amide Acid and Imide Oligomers

The work reported herein is a continuation of an effort originally begun under the HSCT program to develop PETI resins that could be used to fabricate carbon fiber reinforced composites by low viscosity techniques such as RTM. In addition to being fabricated by RTM, the composites needed to display high cured Tgs and high mechanical properties. The early RTM PETIs were based upon PETI-5 monomers at different calculated $\overline{M}n$ s and diamine ratios.[31-34] The work resulted in the successful development and article demonstration of PETI-RTM, having a Tg of 256°C when cured.

At the conclusion of the HSCT program, this technology was transitioned to a space transportation program to develop RTM processable resins with high cured Tgs and good mechanical properties for future composite applications on RLVs. This work was still based on s-BPDA, but utilized other diamines than those in PETI-5 to increase the cured Tg while maintaining RTM processability.[35-36] Several attempts at increasing the cured resin Tg through incorporation of a pendent phenylethynyl group (3,5-diamino-4'-phenylethynyl benzophenone) met with limited success. These resins exhibited higher cured Tgs than the early RTM resins but the as-processed laminates exhibited microcracks.

One monomer used in the early RTM PETI resin synthesis was 1,3,3-APB. Although this monomer was known to provide improvement in processability,[50-51] it also contributed to a low Tg. Consequently work was initiated at incorporating the other isomers of APB (1,3,4-APB and 1,4,4-APB) into the oligomer and determining the effects upon RTM processability, cured resin Tg, and laminate properties.[35-36] Compositions containing the 1,4,4-APB monomer increased the cured Tg as expected but the melt viscosity of the resin also increased by several orders of magnitude. Compositions prepared with 1,3,4-APB provided resins with low and stable melt viscosities amenable to RTM processability and high cured Tgs. One composition containing 1,3,4-APB, designated PETI-298, exhibited a cured Tg of ~300°C and a stable melt viscosity of 0.6 Pa•s at 280°C.[35-36] Flat laminates fabricated by RTM from this resin exhibited good mechanical properties, no microcracking, and no detectable void content.

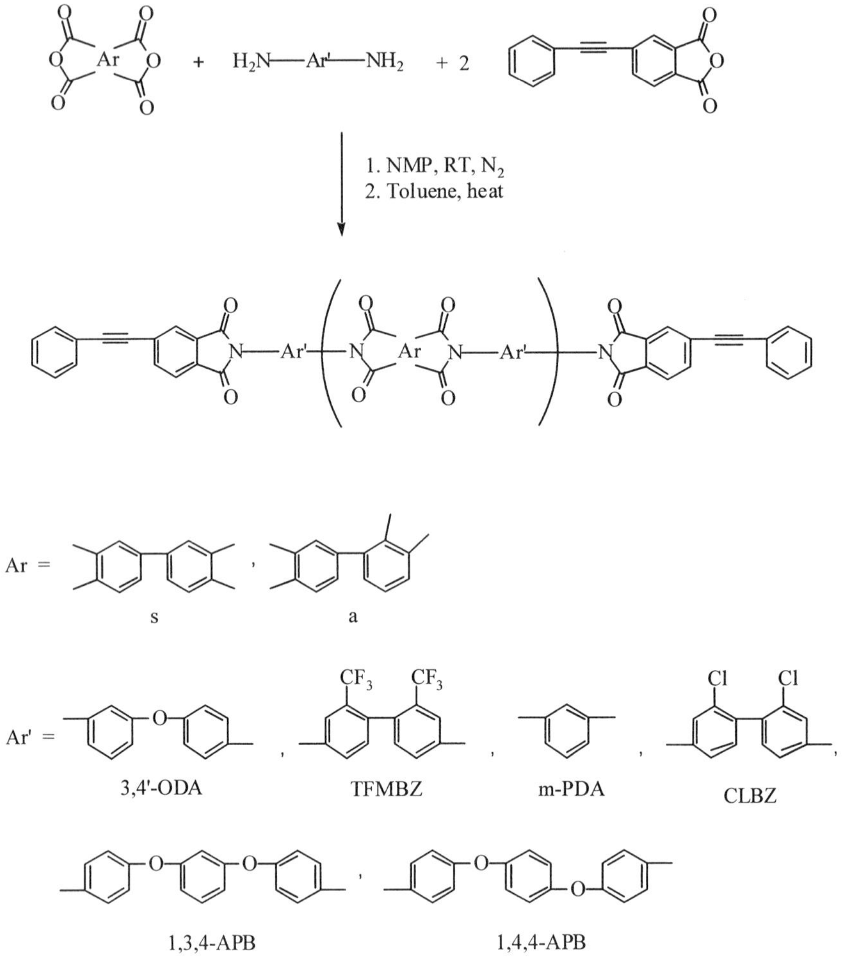
Ar
H2N—Ar'—NH2
+ 2
1. NMP, RT, N2
2. Toluene, heat
N—Ar'
Ar
N—Ar'
N
Ar =
s
a
CF3
CF3
Cl
Cl
Ar' =
3,4'-ODA
TFMBZ
m-PDA
CLBZ
1,3,4-APB
1,4,4-APB

Fig. 1. Synthesis of phenylethynyl containing imide oligomers.

Since a-BPDA derived PETIs were shown to exhibit low melt viscosity and when cured, high Tgs, a series of new PETIs based upon a-BPDA were prepared and evaluated for RTM processability. The PETI oligomers were prepared by the classic amide acid route at a calculated $\overline{Mn}$ of 750 g/mol (Fig.1). Initially, the diamine(s) were dissolved in NMP with subsequent addition of the dianhydride(s) (a and/or s-BPDA) and PEPA endcapper as a slurry. The reaction mixture was stirred at ambient conditions for ~24 hrs prior to conversion of the phenylethynyl terminated amide acid oligomer to the corresponding PETI oligomer. Cyclodehydration was accomplished by azeotropic distillation with toluene. The PETI oligomers remained soluble even upon cooling to room temperature. Isolation was performed by precipitation in water. Previous work has shown that the high stoichiometric offset employed in the preparation of these oligomers resulted in the formation of a complex mixture consisting of various molecular weight oligomers and simple compounds.[32-33] The simple compounds in the products were determined to be crucial in providing resins with the appropriate melt viscosity required for RTM.

3.2 Characterization of PETI Resins

The thermal and rheological properties for the PETI oligomers are discussed with respect to the mole % of 1,3,4-APB used. These mole percentages were 50, 75, and 100. Some miscellaneous compositions which did not fit into this delineation are discussed separately.
The initial Tg and crystalline melt transition (Tm) were obtained by DSC. Powdered samples were cured at 371°C for 1 hr in DSC pans and the samples rescanned to determine the cured resin Tg. The complex melt viscosity (η^*) was measured using PETI discs compression molded at room temperature. The test chamber of the rheometer was at room temperature for specimen insertion. The specimen was heated from 23 to 280°C at a heating rate of 4°C/min and held for 2 hrs at 280°C to assess melt stability and then heating was continued to 371°C at the same heating rate and held for 0.5 hr to cure the material. The η^* is presented initially and after the 2 hr hold at 280°C.

3.2.1 Compositions Based on 100 mole % 1,3,4-APB

To study the effect of a-BPDA content on the cured resin Tg and the melt viscosity of s-BPDA based transfer molding resins, PETIs were prepared using only 1,3,4-APB as the diamine. The a-BPDA was used at 25, 50, and 100 mole % with the difference being s-BPDA. The results are presented in Table 1.

Table 1. PETI Oligomers Based on 100 mole % 1,3,4-APB.

Oligomer	BPDA	Tg (Tm), °C Initial[1]	Cured[2]	η* @ 280°C for 2 hr, Pa·s
P1	s	123 (246)	298	13.5-26.0
P2	75% s, 25% a	149 (239)	301	0.4-0.7
P3	50% s, 50% a	168 (222)	307	0.8-1.0
P4	a	--- (200)	296	0.1-0.4

1. Initial Tg determined on powdered samples by DSC at a heating rate of 20°C/min.
2. Cured Tg determined on samples held in the DSC pan at 371°C for 1hr.

The cured resin Tg increased as the amount of a-BPDA went from 0 to 50 mole % (P1 – P3). However, the cured Tg of P4 did not increase although it used 100 mole % a-BPDA. This was unexpected, since the a-BPDA based materials have typically exhibited increased Tgs relative to similar resins prepared from the symmetric isomer. The reason for this is unclear, but may be associated with the large stoichiometric offset and the relatively small amount of dianhydride used. The effect of a-BPDA upon the melt viscosity was as expected. For a loading level of 25 mole % a-BPDA (P2), the melt viscosity decreased by 2 orders of magnitude with respect to P1, based solely on s-BPDA. Further reduction in the melt viscosity with increasing a-BPDA was not observed.

3.2.2 Compositions Based on 75 mole % 1,3,4-APB

Based on the results of oligomers from 100 mole % of 1,3,4-APB where a profound effect upon the melt viscosity and cured resin Tg was obtained by the incorporation of a-BPDA, it was of interest to see the effect of replacing s-BPDA in PETI-298 with a-BPDA. PETI-298 is prepared from a 75:25 ratio of 1,3,4-APB and 3,4'-ODA, s-BPDA and endcapped with PEPA. It was anticipated that with a higher cured resin Tg, the melt viscosity would likewise decrease by an order or two in magnitude. The results are presented in Table 2.

The direct replacement of s-BPDA in PETI-298 with a-BPDA to afford P5 resulted in an increase in the cured resin Tg of 14°C. However the melt viscosity remained essentially the same. Similar results to that of PETI-298 and P5 were obtained when m-PDA was substituted for 3,4'-ODA in the PETI-298 and P5 formulations to afford P6 and P7, respectively. The use of a-BPDA appears to have a more pronounced effect upon the melt viscosity in compositions where the melt viscosity is high.

Table 2. PETI Oligomers Based on 75 mole % 1,3,4-APB.

Oligomer	BPDA	25 mole% Diamine	Tg (Tm), °C Initial	Tg (Tm), °C Cured	η* @ 280°C for 2 hr, Pa·s
PETI-298[1]	s	3,4'-ODA	139	298	0.6-1.4
P 5	a	3,4'-ODA	147	312	0.4-3.0
P6	s	m-PDA	148 (174, 226, 272)	309	1.9-4.1
P7	a	m-PDA	151	318	1.2-18
P8	a	TFMBZ	ND[2] (179)	320	0.3-1.4
P9	a	CLBZ	193	328	0.2-0.7

1. Data from references 35 and 36.
2. ND = not detected.

As discussed in previous work, the incorporation of 25 mole% of 3,4'-ODA in P1 to afford PETI-298 did not affect the cured resin Tg but did reduce the melt viscosity by two orders of magnitude[35]. Based on these results, it was of interest to evaluate the effect of a-BPDA upon the material properties by replacing 3,4'-ODA in PETI-298 with rigid diamines. Substitution of 3,4'-ODA in PETI-298 and P5 with m-PDA to afford P6 and P7, respectively, increased the cured resin Tg by 6-11°C with the greater effect seen with the s-BPDA based material. As mentioned above, a negligible effect upon the melt viscosity was observed using the more rigid diamine, m-PDA, as compared to the more flexible 3,4'-ODA regardless of BPDA isomer used. Further increases in the cured resin Tgs were observed using TFMBZ (P8) and CLBZ (P9) in the a-BPDA based oligomers with a negligible effect upon the melt viscosity as compared to P5 and P7. The order of increasing cured resin Tgs for the a-BPDA oligomers was: 3,4'-ODA (P5) < m-PDA (P7) < TFMBZ (P8) < CLBZ (P9). The use of TFMBZ and CLBZ was for academic purposes since their high cost prohibits their use in commercial RTM resins.

3.2.3 Compositions Based on 50 mole % 1,3,4-APB

Since the more rigid diamines in P7, P8, and P9 increased the cured resin Tgs without effecting the melt viscosities as compared to PETI-298; it was of interest to further increase the concentration of these diamine components from 25 to 50 mole %. It was anticipated that further increases in the cured resin Tg could be achieved without detrimentally affecting the melt viscosity. The results are presented in Table 3.

Table 3. PETI Oligomers Based on 50 mole % 1,3,4-APB.

	50 mole%	Tg (Tm), °C		η* @ 280°C
Oligomer	Diamine	Initial	Cured	for 2 hr, Pa·s
P10	m-PDA	ND (182)	330	0.06-0.09
P11	TFMBZ	ND (164)	345	0.6-2.0
P12	CLBZ	148	349	7-21

Increasing the mole percentage of m-PDA to 50% in P10 from 25 mole % in P7 resulted in an additional 12°C increase in the cured resin Tg as compared to P7. The melt viscosity decreased by two orders of magnitude in P10 as compared to P7. Similar effects upon the cured resin Tg were observed for increasing the amount of TFMBZ and CLZB from 25 to 50 mole % (P11 as compared to P8 and P12 as compared to P9, respectively). The cured resin Tgs increased ~20-25°C for these polymers as compared to the cured polymers from oligomers containing only 25 mole % of TFMBZ and CLBZ. No effect upon the melt viscosity was observed for increasing the mole % of TFMBZ to 50 (P11 as compared to P8). For compositions containing CLBZ as the diamine, the melt viscosity did increase by one to two orders of magnitude when the mole % was increased to 50 (P12 as compared to P9).

3.2.4 Compositions Based on 50 mole % 1,3,4-APB and 50 mole% m-PDA

Because of the results for oligomers from 100 mole % 1,3,4-APB where the partial replacement of s-BPDA with a-BPDA was evaluated, it was of interest to investigate this with respect to P10. Oligomer P10 was selected due to the combination of low melt viscosity and high cured resin Tg. The a-BPDA in P10 was replaced with 50, 75, and 100 mole % s-BPDA and the effects on the properties are presented in Table 4.

Table 4. PETI Oligomers Based on 50 mole % 1,3,4-APB and 50 mole% m-PDA.

Oligomer	BPDA	Tg (Tm), °C Initial	Cured	η* @ 280°C for 2 hr, Pa·s
P10	a	ND (182)	330	0.06-0.09
P13	50% s, 50% a	139 (173)	332	1.4-4.3
P14	75% s, 25% a	145 (168)	333	1.1-3.1
P15	s	-- (169, 236)	325	$>10^4$

Upon replacement of a-BPDA with 50 and 75 mole % of s-BPDA to afford P13 and P14, respectively, the cured resin Tg remained relatively unchanged with respect to P10. The melt viscosity increased by 2 orders of magnitude for both oligomers as compared to P10. The values for P13 and P14, however, were still within the range useful for resin transfer molding. When a-BPDA in P10 was totally replaced with s-BPDA to afford P15, the melt viscosity increased by 5 orders of magnitude with respect to P10 with a concurrent decrease in the cured resin Tg. This was expected since a-BPDA increases the cured resin Tg and reduces the melt viscosity as reported for other PETIs.[45-47] These results suggest that little a-BPDA is required to have a dramatic impact upon the melt viscosity of high viscosity s-BPDA oligomers.

3.2.5 Miscellaneous Compositions

The Tgs and melt viscosities of other oligomers are presented in Table 5. Oligomers P5, P9, P10, and P12 are included for comparison. Oligomer P16, made with the more rigid 1,4,4-APB as compared to P5 prepared with 1,3,4-APB, exhibited a 27°C increase in the cured resin Tg. However, the melt viscosity increased by 2 to 3 orders of magnitude. Although the melt viscosity of oligomer P5 was out of the range required for RTM, it illustrated that the use of a-BPDA in place of s-BPDA in this composition had the effect of reducing the melt viscosity by 2 to 3 orders of magnitude.[35,36] By reducing the CLBZ content to 40 mole% to afford P17, the melt viscosity was reduced to a comparable level to that of P9 with a minimal reduction in the cured resin Tg as compared to P12. The replacement of 1,3,4-APB in P10 with 3,4'-ODA to afford P18 provided an increase of 12°C in the cured resin Tg. The melt viscosity increased by 2 orders of magnitude using the more rigid 3,4'-ODA and was comparable to that of P13 and P14 where lower amounts of a-BPDA were used.

Table 5. Other PETI Oligomers.

	Diamine	Tg (Tm), °C		η* @ 280°C
Oligomer	Composition, (%)	Initial	Cured	for 2 hr, Pa·s
P5	1,3,4-APB (75), 3,4'-ODA (25)	147	312	0.4-3.0
P16	1,4,4-APB (75), 3,4'-ODA (25)	ND[4]	339	41-480
P9	1,3,4-APB (75), CLBZ (25)	193	328	0.2-0.7
P12	1,3,4-APB (50), CLBZ (50)	148	349	7-21
P17	1,3,4-APB (60), CLBZ (40)	145	345	0.3-1.1
P10	1,3,4-APB (50), m-PDA (50)	ND (182)	330	0.06-0.09
P18	3,4'-ODA (50), m-PDA (50)	ND	342	1.4-6.0

3.3 Composites

Based on the combination of cured resin Tg and melt viscosity, oligomer P10 (designated as PETI-330), was selected for evaluation as flat laminates fabricated by RTM. A typical melt viscosity curve for PETI-330 is shown in Fig. 2. At 280°C, the phenylethynyl groups do not react or react very slowly. Thus, the melt viscosities of the material is stable for >2 hrs at this temperature.

The laminates were made with 8 plies of unsized AS-4 8HS carbon fiber fabric with a $(0/90)_{4S}$ lay-up. Fabric sizing was removed by heating at 400°C for 2 hrs under vacuum prior to insertion in the tool. To assess volatile content of the powder, PETI-330 was analyzed by dynamic TGA. At 350°C, <1% mass loss was observed. The PETI-330 powder was charged to the resin chamber and degassed prior to injection into the tool. Laminate fabrication involved injecting the molten oligomer at ~280°C into the preheated tool followed by a cure at 371°C for 1 hr under ~1.4 MPa hydrostatic pressure. All of the laminates were of high quality as determined by C-scan and photomicroscopy.

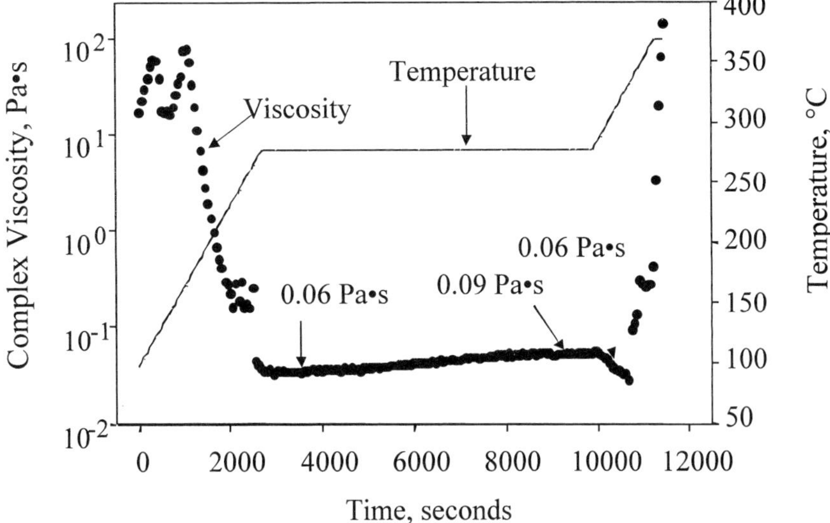

Fig. 2. Melt viscosity vs. temperature curve of PETI-330 (oligomer P10).

The PETI-330 laminate properties are summarized in Table 6 with those of PETI-298 included for comparison. Laminate Tgs were lower for both materials than those obtained for the cured polymers by DSC. Past work has shown that PETIs cured in air exhibited higher Tgs than those cured in a low air environment,[28] as found in the RTM tool. The laminates had ≤2% porosity with fiber volumes of ~57 % as determined by acid digestion. The close fiber volumes allowed for a direct comparison of properties without normalization.

The room temperature (RT) open hole compressive (OHC) strength and short beam shear (SBS) strength of PETI-330 were comparable to those of PETI-298. The retention of OHC and SBS strengths at 288°C was significantly better for PETI-330 than for PETI-298 due to its higher Tg. The OHC modulus was comparable for both materials regardless of test temperature. The RT un-notched compressive strength was lower for PETI-330 as compared to PETI-298. Overall, the mechanical properties of the unoptimized PETI-330 laminates were very good.

Table 6. Laminate Physical and Mechanical Properties.[1]

Property	Test Temp., °C	P10[2] (PETI-330)	PETI-298[3,4]
Cured Tg, °C (DSC of resin)	----	330	298
Cured Tg, °C (TMA of composite)	----	313	289
Fiber Volume, %	----	57	57.3
Un-notched Comp. Str., MPa	23	442	503
OHC Str., MPa	23 288	250 218	244 161
OHC Mod., GPA	23 288	42 40	40 38
SBS Str., MPA	23 232 288	38 37 34	39 32 23

1. 8 plies, quasi-isotropic lay-up.
2. P10 (PETI-330) composition: 50:50 mixture of 1,3,4-APB and m-PDA with a-BPDA and endcapped with PEPA.
3. PETI-298 composition: 75:25 mixture of 1,3,4-APB and 3,4'-ODA with s-BPDA and endcapped with PEPA.
4. Data from references 35 and 36.

4 Conclusion

New PETI oligomers based on a-BPDA were prepared and evaluated. Oligomers with low and stable melt viscosities amenable to RTM processing were obtained which exhibited higher cured Tgs as compared to similar compositions based on s-BPDA. One composition (oligomer P10) displayed excellent processability by RTM and provided flat laminates with no voids and free of microcracks that exhibited good mechanical properties at RT with good retention of properties at 288°C. Further development of this composition, designated as PETI-330, is underway.

Acknowledgments

The authors would like to acknowledge the support of Professors Eric A. Mintz, David R. Veazie and Mr. Brian Shonkwiler in the High Performance Polymer and Composite Center (HiPPAC) at Clark Atlanta University. The authors would like to acknowledge Dr. Michael

Meador of NASA Glenn Research Center for partial financial support of this work. The authors wish to thank Dr. Frank W. Harris of the University of Akron for the kind donation of 2,2'-dichlorobenzidine (CLBZ).

The use of trade names of manufacturers does not constitute an official endorsement of such products or manufacturers, either expressed or implied, by the National Aeronautics and Space Administration.

[1] D. Wilson, H. D. Stenzenberger, and P. M. Hergenrother, "*Polyimides*", Blackie and Sons Ltd., Glasgow, United Kingdom, 1990.
[2] C. E. Sroog, *Prog. Polym. Sci.* **1991**, 16, 561.
[3] F. W. Harris, S. M. Padaki and S. Vavaprath, *Polym. Prepr.* **1980**, 21(1), 3.
[4] F. W. Harris, A. Pamidimukkala, R. Gupta, S. Das, T. Wu and G. Mock, *Ibid* **1983**, 24 (2), 324.
[5] F. W. Harris, K. Sridhar and S. Das, *Ibid* **1984**, 25 (1), 110.
[6] F. W. Harris, A. Pamidimukkala, R. Gupta, S. Das, T. Wu and G. Mock, *J. Macromol. Sci.-Chem. A* **1984**, 24 (8/9), 1117.
[7] S. Hino, S. Sato and O. Suzski, Jpn. Kokai Tokyo Koho JP (1988), 63, (196), 564. *Chem. Abstr.* **1989**, 110, 115573w. U. S. Patent 5,066,771 (1991) to Agency of Industrial Science and Technology, Japan.
[8] C. W. Paul, R. A. Schultz and S. P. Fenelli, in "*Advances in Polyimide Science and Technology*", C. Feger, M. M. Khoyasteh and M. S. Htoo Eds., Technomic, Lancaster, PA 1993, pp 220.
[9] R. G. Bryant, B. J. Jensen and P. M. Hergenrother, *Polym. Prepr.* **1993**, 34 (1), 566.
[10] B. J. Jensen, P. M. Hergenrother and G. Nwokogu, *Polymer* **1993**, 34 (3), 630.
[11] G. W. Meyer, S. Jayaraman and J. E. McGrath, *Polym. Prepr.* **1993**, 34 (2), 540.
[12] S. J. Havens, R. G. Bryant, B. J. Jensen and P. M. Hergenrother, *Ibid* **1994**, 35 (1), 553.
[13] P. M. Hergenrother and J. G. Smith, Jr., *Ibid* **1994**, 35 (1), 353. *Polymer* **1994**, 35 (22), 4857.
[14] G. W. Meyer, T. E. Glass, H. J. Grubbs and J. E. McGrath, *Ibid* **1994**, 35 (1), 549.
[15] J. A. Johnston, F. M. Li, F. W. Harris and T. Takekoshi, *Polymer* **1994**, 35 (22), 4865.
[16] T. Takekoshi and J. M. Terry, *Ibid* **1994**, 4874.
[17] J. W. Connell, J. G. Smith, Jr., R. J. Cano and P. M. Hergenrother, *Soc. Adv. Mat. Proc. Eng. Ser.* **1996**, 41, 1102. *High Perform. Polym.* **1997**, 9, 309.
[18] J.A. Hinkley and B.J. Jensen, *High Perform. Polym.* **1996**, 8, 599.
[19] B. Tan, V. Vasudevan, Y.J. Lee, S. Gadner, R.M. Davis, T. Bullions, A.C. Loos, H. Parvatareddy, D.A. Dillard, J.E. McGrath and J. Cella, *J. Polym. Sci.: Pt. A: Polym. Chem.* **1997**, 35, 2943.
[20] J. G. Smith, Jr., J. W. Connell and P. M. Hergenrother, *Polymer***1997**, 38 (18), 4657.
[21] J. W. Connell, J. G. Smith, Jr. and P. M. Hergenrother, *Intl. SAMPE Tech. Conf. Series* **1997**, 29, 317.
[22] R. G. Bryant, B. J. Jensen and P. M. Hergenrother, *Soc. Adv. Mat. Proc. Eng. Ser.* **1994**, 39, 273 (closed papers volume).
[23] B. J. Jensen, R. G. Bryant, J. G. Smith, Jr,. and P. M. Hergenrother, *J. Adhesion* **1995**, 54, 57.
[24] R. J. Cano and B. J. Jensen, *J. Adhesion* **1997**, 60, 113.
[25] T. Hou, B. J. Jensen and P. M. Hergenrother, *Composite Materials* **1996**, 30 (1), 109.
[26] P. M. Hergenrother and M. Rommel, *Soc. Adv. Mat. Proc. Eng. Series.* **1996**, 41, 1061.
[27] M. Rommel, L. Konopka and P. M. Hergenrother, *Intl. SAMPE Tech. Conf. Series* **1996**, 28, 14.
[28] J W. Connell, J. G. Smith, Jr., and P. M. Hergenrother, *J. Macromol. Sci.-Rev. Macromol. Chem. Phys.* **2000**, C40 (2&3), 207.
[29] J. G. Smith, Jr., J. W. Connell and P. M. Hergenrother, *Soc. Adv. Mat. Proc. Eng. Ser.* **1998**, 43, 93. *J. Comp. Matls.* **2000**., 34 (7), 614.
[30] J. W. Connell, J. G. Smith, Jr., P. M. Hergenrother and M. L. Rommel, *Intl. SAMPE Tech. Conf. Series* **1998**, 30, 545.
[31] J. M. Criss, J. W. Connell and J. G. Smith, Jr., *Intl. SAMPE Tech. Conf. Series* **1998**, 30, 341.

[32] J. M. Criss, C. P. Arendt, J. W. Connell, J. G. Smith, Jr., and P. M. Hergenrother, *SAMPE J.* **2000**, 36 (3), 32.

[33] J. G. Smith, Jr., J. W. Connell, P. M. Hergenrother, and J.M. Criss, *Soc. Adv. Mat. Proc. Eng. Ser.* **2000**, 45, 1584.

[34] J.W. Connell, J.G. Smith Jr., and P. M. Hergenrother, U.S. Patent 6,359,107 B1 (2002) to NASA.

[35] J. G. Smith, Jr., J. W. Connell, P. M. Hergenrother, and J.M. Criss, *Soc. Adv. Mat. Proc. Eng. Ser.* **2001** 46, 510.

[36] J. G. Smith, Jr., J. W. Connell, P. M. Hergenrother, and J. M. Criss, *J. Comp. Matls* **2002**, 36 (19), 2255.

[37] P. M. Hergenrother, *SAMPE J.* **2002**, 36 (1), 330.

[38] J. M. Criss, R. W. Koon, P. M. Hergenrother, J. W. Connell, and J. G. Smith, Jr., *Intl. 51SAMPE Tech. Conf. Series* **2001**, 33, 1009.

[39] H. Inoue , H. Okamoto , Y. Hiraoka, *Radiat Phys Chem* **1987**, 29, 283.

[40] H. Yamaguchi in: R. Yokota , M. Hasegawa , editors. *Recent Advances in Polyimides*. Tokyo, Japan: Raytech Co., 1997, p. 5.

[41] M. Hasegawa , N. Sensui , Y. Shindo , R. Yokota (a) *J. Photopolym. Sci.* **1996**, 9, 367. (b) *Macromolecules* **1999**, 32, 387. (c) *J. Polym. Sci.: Pt. B: Polym. Phys.***1999**, 37, 2499.

[42] T. Takahashi , S. Takabayashi , H. Inoue, *High Perf. Polym.* **1998**, 10, 33.

[43] R. Yokota , S. Yamamoto, S. Yano, T. Sawaguchi, M. Hasegawa, H. Yamaguchi, H. Ozawa, R. Sato, in: K. I. Mittal, editor. *Polyimides and Other High temperature Polymers*, Zeist, The Netherlands,VSP, 2001, p. 101.

[44] M. Hasegawa, Z. Shi, R. Yokota, F. He, H. Ozawa, *High Perf. Polym* **2001**., 13, 355.

[45] R. Yokota in: T. Takeichi, M. Kochi, editors. *Proc 7th Japan Polyimide Conf* 1998, p. 21.

[46] R. Yokota in: R. Yokota, editor. *Proc. 9th Japan Polyimide Conf* 2000, p. 12.

[47] R. Yokota, S. Yamamoto, S. Yano, T. Sawaguchi, M. Hasegawa, H. Yamaguchi, H. Ozawa, R. Sato, *High Perf. Polym.* **2001**, 13, S61.

[48] J. G. Smith, Jr., J. W. Connell, P. M. Hergenrother, R. Yokota and J. M. Criss, *Soc. Adv. Mat. Proc. Eng. Ser.* **2002**, 47, 316.

[49] A. J. Bilbo and G. M. Wyman, *J. Am. Chem. Soc.* **1953**, 75, 5312.

[50] A. L. Landis, N. Bilow, R. H. Boschan, and R. E. Lawrence, *Poly. Prepr.* **1974**, 15 (2), 537.

[51] N. Bilow and A. L. Landis, U.S. Patent 4,276,407 (1981) to Hughes Aircraft company.

Active Polycondensation: From Peptide Chemistry to Amino Acid Based Biodegradable Polymers

Ramaz Katsarava

Center for Medical Polymers & Biomaterials, and Department of Chemical Technology of the Georgian Technical University, P.O.Box 24, 380079 Tbilisi, Georgia
Email: kats@gol.ge

Summary: New polycondensation (PC) methods of polymer synthesis using non-traditional active derivative of dicarboxylic acids are reviewed. The new PC methods are named by general name "Active Polycondensation" (APC) to tell them from traditional low-temperature PC. The most of these methods are based on well known in peptide chemistry approaches to the activation of carboxylic groups. In the present paper the syntheses of heterochain polymers of basic classes – polyamides, polyesters, polyurethanes, polyureas, and polybenzazoles by interaction of various active diesters with di- and polyfunctional nucleophiles are discussed in brief. Special attention is given to the synthesis of non-conventional heterochain macromolecular systems, in particular poly(ester amide)s (PEAs), composed of naturally occurring α–amino acids and other non-toxic building blocks like fatty diacids and diols - synthetic analogues of naturally occurring amino acid based polymers – peptides and proteins. The synthesis and properties, biodegradation, and some practical applications of PEAs are discussed in brief.

Keywords: α–amino acids, active polycondensation, biodegradable polymers, biomedical applications, heterochain polymers, poly(ester amide)s

Introduction

Polycondensation (PC) is traditionally subdivided into two types: high-temperature PC (HTPC) and low-temperature PC (LTPC).[1-3] In this paper the overview of the PC methods is chiefly confined by the PC reactions based on aminolysis of derivatives of dicarboxylic acids (polyamidations).

 DOI: 10.1002/masy.200350935

The HTPC was developed mostly in 20-30s of the last century. Under the conditions of HTPC (commonly bring about at 200-250°C in a melt or in a high-boiling solvent) monomers with low reactivity (e.g. dicarboxylic acids or their alkyl esters) are used:

$$\sim\sim\sim\text{-R-CO-OR}_1 + \text{H}_2\text{N-R}^2\text{-}\sim\sim\sim \xrightarrow[\text{-HOR}_1]{200\text{-}250^\circ\text{C}} \sim\sim\sim\text{-R-CO-HN-R}^2\text{-}\sim\sim\sim$$

where: $R_1 = H, CH_3$, etc.

Scheme 1.

These reactions, however, in most cases proceed with low rates and are accompanied by numerous side reactions which lead to decomposition of the functional groups (and hence lead to chain-termination) as well as to the formation of anomalous units in the polymeric backbones ("unit heterogeneity").[4] In case of low-activity aromatic diamines high-molecular-weight polyamides under the conditions of HTPC are not formed at all.[5]

In such cases LTPC, developed mainly in 50-60s of the last century and commonly brought about in temperature range *c.a.* from –30 to 50°C in organic solvents or interfacially, is effective by far. This method is based on the use of chemically activated monomers like diacid chlorides (**Y=none**) and *bis*-chloroformates (**Y=O**):[2,3]

$$\sim\sim\sim\text{-R-Y-CO-Cl} + \text{H}_2\text{N-R}^1\text{-}\sim\sim\sim \xrightarrow[\text{-HCl}]{\text{Acid acceptor}} \sim\sim\sim\text{-R-Y-CO-HN-R}^1\text{-}\sim\sim\sim$$

Scheme 2.

The low-temperature PC promoted further advancement of macromolecular chemistry and helped to solve many practical problems, however it is characterized by a number of drawbacks, among which numerous side reactions ("multi-channeling" of the process) leading to chain-termination and unit-heterogeneity,[2-4] as well as to synthetic limitations, should be noted. Among these side reactions undesirable interactions of aliphatic diacid chlorides with tertiary amines should be specially emphasized. These parasitic reactions prevent the formation of high-molecular-weight polyamides by solution PC.[3]

All the drawbacks of traditional PC methods mentioned above, determined in a significant extent the development in 70-80s of the last century of new PC methods based on non-traditional ways of activating monomers.[6] Among the five approaches developed the method called as "Leaving groups" method looks the most universal and promising. Came from peptide chemistry this method is based on the activation of that part of derivatives of carboxylic acids – esters or amides – which are considered as leaving groups (**X**), i.e. are liberated as low-molecular-weight by-products **HX** after PC, as is shown in Scheme 3.

~~~-R-CO-X + H_2N-R[1]-~~~ ⟶ ~~~-R-CO-NH-R[1]-~~~ + HX

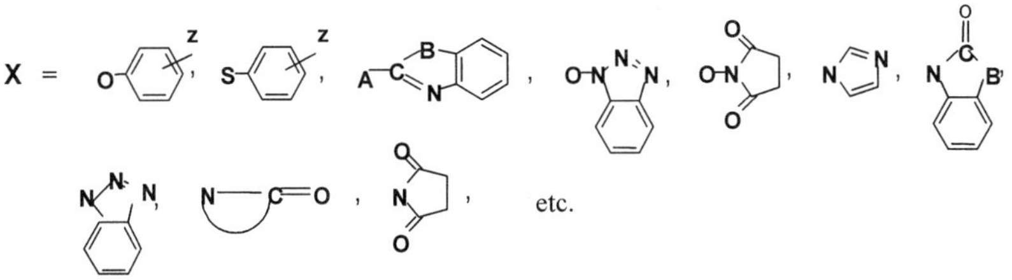

X = O–C₆H₄–Z, S–C₆H₄–Z, A–C(=N)–B (benzo-fused), O–N (benzotriazole), O–N (succinimide), N (imidazole), N–C(=O)–B' (benzo-fused), N (benzotriazole), N–C=O (lactam), N (succinimide), etc.

A = O, S; B = O, S, NH; Z = electron-withdrawing substituents: **F, Cl, Br, NO_2, CN,** etc.;

Scheme 3.

In details active diesters (O- and rarely S-containing leaving groups) and active diamides (N-containing leaving groups) as new polycondensation type monomers, their synthesis and structure, the nature of activating effects, reactivity, etc. are analyzed in ref.[6]

Heterochain Polymers Synthesized *via* Active Diesters Aminolysis Reactions

Polyamides and Polybenzazoles

Active polyamidation proceeds according to Scheme 3. In this scheme R and R^1 can be Alkylene or Arylene. This means four different pairs of interacting monomers followed by the synthesis of all four possible classes of high molecular weight polyamides by APC.

Active polyamidation is especially promising for the synthesis of high-molecular-weight aliphatic polyamides (R and R^1 = Alkylene) by *solution* PC in contrast to traditional LTPC based on the interaction of alkylenediamines with aliphatic diacid chlorides that leads to the synthesis of low-molecular-weight or branched/cured polyamides under the same conditions.[2-4] A wide range of organic solvents both with or without functional groups – from alcohols (like ethanol, 2-propanol, etc.) to common organic (chloroform, methylene chloride, dioxane, THF, etc) and amide type solvents (like DMF, DMA, NMP, etc.) were successfully used as a medium of APC that is allowed by high inertness of active diesters towards these solvents.[7] The *p*-nitrophenoxide leaving group was found to be the best one for synthesizing polyamides from fatty diamines taking into account both availability of *p*-nirtophenol and high-molecular-weights of PEAs obtained. In case of aromatic diamines diesters of higher activity - derivatives of dinitrophenols or pentafluorophenol had to be used. A high stability of aliphatic active diesters towards tertiary amines[7] allowed to use the salts of diamines instead of free diamines in APC (Tertiary amines in this case play a role of acid acceptor transforming *in situ* salts of diamines into free bases). This approach is promising for synthesizing polymers on the basis of diamines unstable as free bases, e.g. diamines containing ester linkages in the molecules, etc. We successfully used this new schemes of polyamidation for the synthesis of biodegradable poly(ester amide)s, discussed below.

A high selectivity of active diesters towards nucleophilic functional groups of close reactivity like *ortho*-amino groups in aromatic *tetra*-amines, turned out suitable for synthesizing high-molecular-weight (η_{red} up to 2.4 dL/g) soluble polybenzimidazoles.[6] The first stage of the synthesis of polybenzimidazoles represents active polyamidation reaction with participating one, the most active *ortho*-amino group with subsequent involvement of the next *ortho*-amino group in benzimidazole ring formation.

Polyurethanes and Polyureas

The synthesis of polyurethanes by APC is based on the aminolysis of active *bis*-carbonates of diols, as is shown in Scheme 4.

~~~-R-O-CO-X + H_2N-R^1-~~~ ⟶ ~~~-R-O-CO-NH-R^1-~~~ + HX

Scheme 4.

This reaction was found to be the most suitable for synthesizing linear aliphatic polyurethanes (i.e. when R and R^1 = Alkylene) revealing excellent film and fiber-forming properties.[8,9] Like the synthesis of aliphatic polyamides above the *p*-nitrophenoxide leaving group is found to be the best one for synthesizing polyurethanes by APC.
The synthesis of polyureas by APC *via* active carbonates follows Scheme 5.

X-CO-X + 2 H_2N-R^1-~~~ ⟶ ~~~-R^1-NH-CO-NH-R^1-~~~ + 2 HX

Scheme 5.

This scheme is suitable for synthesizing unbranched and soluble high-molecular-weight polyureas from both aliphatic (η_{red} up to 0.75 dL/g) and aromatic diamines (η_{red} up to 1.34 dL/g) .[10]

Amino Acid Derived Polymers

Synthesis

It is not surprising that the methods of APC, based on the achievements of peptide chemistry, turned out especially suitable for constructing non-conventional hetero-chain macromolecular systems composed of α–amino acids. These polymers can be considered as synthetic analogues of naturally occurring amino acid derived polymers – proteins, and are promising for numerous biomedical applications. Using the methods of APC developed, we have synthesized a large variety of these polymers which we called as Amino Acid Based Bioanalogous Polymers (AABBPs) (See ref.[11] and refs. cited therein).

Four types of polycondensation monomers based on polyfunctional α–amino acids and dimeric forms of hydrophobic α–amino acids were used for constructing AABBPs:

Polyfunctional α–Diaminocarboxylic acids (mostly lysine and cystine) with protected C-terminus are used as diamines (in N,N'-*bis*-trimethylsililated or salt forms), and α–Amino dicarboxylic acids (L-aspartic and L-glutamic acids) with protected N-terminus are used as dicarboxylic acids (in active esters forms). These compounds are useful for incorporating lateral functionalal groups into AABBPs[10,12-15], and synthesizing non-canonical poly-(dipeptides).[13]

Dimeric forms of hydrophobic α–amino acids - N,N'-diacyl-*bis*-α-amino acids and *bis*-α-(L-amino acid) α,ω-alkylene diesters were also used as monomers for constructing AABBPs. N,N'-diacyl-*bis*-α-amino acids – both diamide diacids[16-19] and diurethane diacids[20] (in forms of active diesters[18,20] or *bis*-azlactones[16,17,19]) are suitable monomers for incorporating various hetero-linkages (amide, urethane) into polymeric backbones as well as for synthesizing AABBPs with di-[18] and tripeptide fragments.[17]

For constructing biodegradable polymeric backbones, however, the most promising as building blocks are *bis*-α-(L-amino acid) α,ω-alkylene diesters (AAADs) containing two ester linkages per molecule which can undergo either nonspecific (chemical) or specific (enzymatic) hydrolysis. As alkyl esters of α-amino acid,[21] AAADs are unstable as free bases and enter into undesirable side reactions. Therefore, they are prepared as stable salts of general formula (Scheme 6).

Di-*p*-toluenesulfonic acid salts were found as the most suitable monomeric forms of AAADs. They were synthesized in a nearly quantitative yields according to very simple procedure – by direct condensation of hydrophobic α-amino acids (2 moles) with fatty diols (1 mole) in refluxed benzene (or toluene)[11,22-25] in the presence of *p*-toluenesulfonic acid monohydrate. Various diols – widely available polymethylene diols[11,22] along with diols from renewable resources – dianhydrohexitols[23,25] were used for preparing AAADs.

$$AH \cdot H_2N\text{-}CH(R^3)\text{-}CO\text{-}O\text{-}R^2\text{-}O\text{-}CO\text{-}CH(R^3)\text{-}NH_2 \cdot HA \qquad \text{(AAAD)}$$

R^2 = $(CH_2)_x$ *with* x = 2-4, 6, 8, 12, [two isosorbide-type bicyclic diol structures]

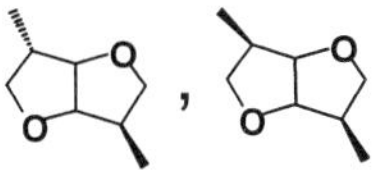

R^3 = CH_3, $CH(CH_3)_2$, $CH_2CH(CH_3)_2$, $CH(CH_3)CH_2CH_3$, CH_2Ph, $(CH_2)_2\text{-}S\text{-}CH_3$.

HA = HCl, HBr, *p*-Toluenesulfonic acid.

Scheme 6.

Neither traditional PC methods could be applied to AAADs due to instability of these compounds (exactly free bases) at elevated temperatures (HTPC) and necessity to use tertiary amines as *p*-toluenesulfonic acid acceptor (LTPC) which, as noted above, cause parasitic reactions with aliphatic diacid chlorides. And only APC proved to be a suitable method for preparing AABBPs from AAADs: poly(ester amide)s (PEAs) by PC with active diesters of dicarboxylic acids,[11,22-25] poly(ester urethane)s (PEURs) and poly(ester urea)s (PEUs) by PC with active *bis*-carbonates of diols and active diphenyl carbonates, accordingly.[26] Due to page limitations this paper deals with the synthesis, properties and some practical applications of PEAs only.

Bis-electrophilic partners of AAADs we used for preparing PEAs – di-*p*-nitrophenyl esters of dicarboxylic acids were synthesized by interaction either a) traditional diacid chlorides with *p*-nitrophenol in the presence of tertiary amines or b) free diacids with *p*-nitrophenol in the presence of condensing agent.[6] The latter allows to realize one of the substantial advantages of APC – wider synthetic possibilities - and to synthesize PEAs based on higher homologues of fatty diacids, dichlorides of which are problematic to be purified up to polycondensation grade. At the same time polymers composed of higher homologues of diacids could be of interest for various biomedical applications due to their anticipated mechanical, physico-chemical and biochemical properties.

The synthesis of AAAD based PEAs was carried out according to Scheme 7.

The APC proceeded smoothly under mild conditions in common organic solvents ($CHCl_3$, DMF, DMA) resulting in high molecular weight (M_w ranged from 24,000 to 167,000 depending on the AAAD and active diester used) PEAs with narrow polydispersity (Mw/Mn ranged from 1.20 to 1,81). The obtained PEAs, having regular but adirectional structure, in most cases were amorphous materials with T_g ranged from 11 to 102°C,[11,23] and only several samples (mostly composed of L-phenylalanine) showed semicrystallinity with T_m in the range 104-124°C.[11]

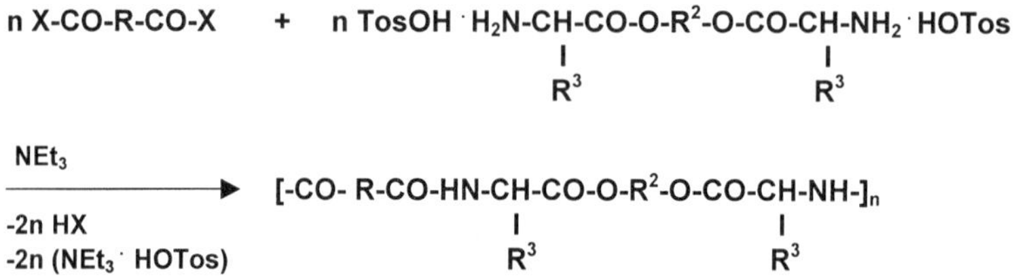

n X-CO-R-CO-X + n TosOH · H_2N-CH(R^3)-CO-O-R^2-O-CO-CH(R^3)-NH_2 · HOTos

$\xrightarrow[\text{-2n HX, -2n (NEt}_3\text{ · HOTos)}]{\text{NEt}_3}$ [-CO- R-CO-HN-CH(R^3)-CO-O-R^2-O-CO-CH(R^3)-NH-]$_n$

Scheme 7.

The PEAs obtained had a wide range of mechanical, physico-chemical and biodegradation properties, however, they did not contain any functional groups (except for terminal ones) which could be used for the attachment of active molecules (drug, bioactive compounds, recognition agents, adhesion promoter, probe, etc.). At the same time, one of the important criteria for the use of synthetic polymers in biomedical applications seems to be that polymers should be not only biodegradable but also have functional groups to which active molecules can be attached covalently or non-covalently. Therefore, very recently[27,28] we obtained functional co-poly(ester amide)s (F-co-PEAs) with lateral COOH groups, using di-*p*-toluenesulfonic acid salt of L-lysine benzyl ester (i.e. α-diaminocarboxylic acid derivative) as a co-monomer with AAAD in APC, followed by debenzylation of the lateral benzyl ester groups of PEAs by catalytic (Pd black) hydrogenolysis.

Biodegradation

To examine the biodegradation property of these regular PEAs a systematic *in vitro* biodegradation study in the presence of hydrolases like trypsin, α-chymotrypsin, lipase and a complex of proteases of Papaya (the last enzyme was used for modeling the catalytic action of nonspecific proteases) were carried out using both automatic potentiometric titration[29] and gravimetric (weight loss) methods.[30] It was found that, in the most cases studied, the PEAs were biodegraded by erosive mechanism, according to the first order kinetics. Spontaneous immobilization (absorption) of the enzymes onto the PEAs surfaces was observed. The surface immobilized enzyme not only accelerated the erosion of the PEAs but also was able to catalyze the hydrolysis of both low-molecular-weight (ATEE) and high-molecular-weight (protein) external substrates. The enzymes could also be impregnated into the PEAs to make them «self-destructive» at a target rate. A comparison of the PEAs' *in vitro* biodegradation data with polylactide (PDLLA) showed that PEAs exhibited a far more tendency towards enzyme catalyzed biodegradation than PDLLA, and PEAs' erosion rates (10^{-1}-10^{-3} mg/cm^2*h) were comparable with erosion rates of polyanhydrides[31] – the fastest biodegradable polymers to date.

A preliminary *in vivo* biodegradation study of the selected PEA sample (implanted as films subcutaneously to rats)[30] with and without lipase-impregnation showed that that PEA was completely absorbed within 1-2 months postimplantation (for the lipase-impregnated ones), and 3-6 months (for the lipase free ones) without any trace of tissue reaction. These findings prompt us to suggest that these new PEAs may have a great potential for designing drug sustained/controlled release devices as well as implantable surgical devices.

Applications

An artificial skin "PhagoBioDerm": this preparation represents a novel wound-healing device consisting of biodegradable PEA impregnated with an antibiotic and lytic bacteriophages. Licensed recently for sale in Georgia, PhagoBioDerm showed an excellent therapeutic effect in the management of infected wounds and ulcers (of both trophic and diabetic origin).[32]

Selected representative of the F-co-PEAs, revealing high-elastic properties along with enzyme catalyzed biodegradation, was used for stent coating.[33] The *in vivo* biocompatibility was tested in porcine coronary arteries, comparing the polymer-coated stents with bare metal stents in 10 pigs. All animals survived till sacrifice 28 days later and follow-up angiography prior to sacrifice revealed identical diameter stenosis in both groups. Histology confirmed similar injury scores, inflammatory reaction, and area stenosis. The results of this study suggest that the polymer is biocompatible and should not elicit an inflammatory reaction, an important pre-requisite for a drug-carrying polymer.

Conclusion

It has been shown that new Active Polycondensation method could be applied for synthesizing hetero-chain polymers of various classes. This method proved to be especially useful for preparing biodegradable poly(ester amide)s composed of nontoxic building blocks like naturally occurring α-amino acids, fatty dicarboxylic acids and diols. Some representatives of the PEAs showed high biodegradation rates along with good biocompatibility and are of interest for various biomedical applications, e.g. as medicated wound dressing/healing materials, stent coatings, drug carriers, etc.

Acknowledgements

We thank: Intern. Sci. (G.Soros) Foundation (USA), JSPS (Japan), DAAD (Germany), CRDF (grants ## GB2-116, GR2-997), ISTC (Gtants ## G-446, G-802), MediVas, LLC (USA), Intralytix, Inc. (USA) for supporting this research.

1) V. V. Korshak, S. V. Vinogradova, "*Ravnovesnaya Polikondensatsia*" (Equilibrium Polycondensation), Nauka, Moscow 1968.
2) V.V.Korshak, S. V. Vinogradova, "*Neravnovesnaya Polikondensatsia*" (Nonquilibrium Polycondensation), Nauka, Moscow 1972.
3) W. P. Morgan, "*Condensation polymers: By Interfacial and Solution Methods*Interscience Publ., New York 1965.
4) V. V. Korshak, "*Raznozvennost' Polimerov*" (Unit Heterogeneity of polymers), Nauka, Moscow 1977.
5) D. A. Holmer, O. A. Pickett, *J. Polym. Sci., Part A-1* **1972**, *10*, 1547.
6) R. Katsarava, *Russian. Chem. Rev.* (British Library) **1991**, *60*, 722.
7) R. Katsarava, D. Kharadze, *Zhurn. Obshch. Khim.* **1991**, *61*, 2413.
8) R. Katsarava, T. Kartvelishvili, D. Kharadze, M. Patsuria, *Vysokomol. Soedin., Ser.A* **1987**, *29*, 2069.
9) R. Katsarava, T. Kartvelishvili, T. Khosruashvili, V. Beridze, *Macromol. Chem. Phys.* **1995**, *196*, 3061.

10) R. Katsarava, T. Kartvelishvili, N. Japaridze, T. Goguadze, T. Khosruashvili, R. P. Tiger, P. A. Berlin, *Makromol. Chem.* **1993**, *194*, 3209.
11) R. Katsarava, N. Arabuli, V. Beridze, D. Kharadze, C.C. Chu, C.Y. Won, *J. Polym Sc. Part A*: *Polym. Chem.* **1999**, *37*, 391.
12) R. Katsarava, D. Kharadze, L. Avalishvili, *Vysokomol. Soedin. Ser. B* **1986**, *28*, 518.
13) R. Katsarava, D. Kharadze, L. Avalishvili, T. Omiadze, *Bull. Georgian Acad. Sci.* **1989**, *134*, 121.
14) R. Katsarava, D. Kharadze, N. Japaridze, L. Avalishvili, T. Omiadze, *Makromol. Chem.* **1985**,*186*, 939.
15) R. Katsarava, T. Kartvelishvili, Y.A. Davidovich, S.V. Rogozhin, *Dokl. Akad. Nauk SSSR* **1982**, *266*, 366.
16) R Katsarava, D. Kharadze, L. Kirmelashvili, *Acta Polymerica* **1985**, *36*, 29.
17) R. Katsarava, D. Kharadze, L. Kirmelashvili, N. Medzmariashvili, T. Goguadze, G. Tsitlanadze, *Makromol. Chem.* **1993**, *194*, 143.
18) R. Katsarava, D. Kharadze, T. Omiadze, G. Tsitlanadze, T. Goguadze, N. Arabuli Z Gomurashvili, *Vysokomol. Soedin. Ser. A* **1994**, *36*, 1462.
19) D. Kharadze, L. Kirmelashvili, N. Medzmariashvili, G. Tsitlanadze, D. Tugushi, C.C. Chu, R. Katsarava, *Polymer Sci.(Russia) Ser. A* **1999**, *41*, 883.
20) R. Katsarava, T. Kartvelishvili, K. Kvintradze, *Macromol. Chem. Phys.* **1996**, *197*, 249.
21) J. P. Greenstein, M. Winitz, "*Chemistry of the Amino Acids*", John Wiley & Sons, Inc., New York-London, 1961.
22) N. Arabuli, G. Tsitlanadze, L. Edilashvili, D. Kharadze, T. Goguadze, V. Beridze, Z. Gomurashvili, R. Katsarava *Macromol. Chem. Phys.* **1994**, *195*, 2279.
23) Z. Gomurashvili, R. Katsarava, H.R. Kricheldorf *J. Macromol. Sci. Pure Appl. Chem.* **2000**, *37*, 215.
24) Y. Fan, M. Kobayashi, H. Kize, *Polym. J.* **2000**, *32*, 817.
25) M. Okada, M. Yamada, M. Yokoe, K. Aoi, *J Appl. Polym. Sci.* **2001**, *81*, 2721.
26) R. Katsarava, T. Kartvelishvili, L. Edilashvili, G. Tsitlanadze, *Macromol. Chem. Phys.* **1997**, *198*, 1921.
27) C.C. Chu, R. Katsarava, Elastomeric functional biodegradable copolyester amides and copolyester urethanes, PCT/US01/27288 WO 02/189477 A2.
28) G. Jokhadze, M. Machaidze, T. Kviria, C. C. Chu, R. Katsarava (*in preparation*)
29) G. Tsitlanadze, T. Kviria, C. C. Chu, R. Katsarava (*pending review*)
30) G. Tsitlanadze, M. Machaidze, T. Kviria, N. Djavakhishvili, C. C. Chu, R. Katsarava (*pending review*)
31) K. W. Leong, B, C. Brott, R. Langer, *J. Biomed. Mater. Res.* **1985**, *9*, 941.
32) K. Markoishvili, G. Tsitlanadze, R. Katsarava, G. J. Morris, A. Sulakvelidze, *Intern. J. Dermatol.* **2002**, *41*, 453.
33) S. H. Lee, I. Szinai, K. Carpenter, R. Katsarava, G. Jokhadze, C. C. Chu, Y. Huang, E. Verbeken, O. Bramwell, I. De Scheerder, M. K. Hong, *Coronary Artery Disease* **2002**, *13*, 237.

Stereochemistry Driven Cocrystallisation Phenomena in Partially Cycloaliphatic Polyamides

Cor Koning,[*1,3] Bert Vanhaecht,[1] Rudolph Willem,[2] Monique Biesemans,[2] Bart Goderis,[4] Bart Rimez[1]*

[1]Department of Physical and Colloidal Chemistry
[2]High Resolution NMR Centre, Free University of Brussels, Pleinlaan 2, 1050 Brussels, Belgium
[3]Laboratory of Polymer Chemistry, Eindhoven University of Technology, P.O. Box 513, 5600 MB Eindhoven, The Netherlands
E-mail: c.e.koning@tue.nl
[4]Chemistry Department, Catholic University of Leuven, Celestijnenlaan 200F, 3001 Heverlee-Leuven, Belgium

Summary: Two series of isomeric copolyamides were synthesised, *viz.* polyamides 12.6 for which the adipic acid residues were partially replaced by *cis/trans*-1,4-cyclohexanedicarboxylic acid (1,4-CHDA), and polyamides 4.14 for which the 1,4-diaminobutane residues were partially replaced by *cis/trans*-1,4-diaminocyclohexane (1,4-DACH). A careful DSC and WAXS analysis learned that only the trans isomers of both 1,4-DACH and 1,4-CHDA are incorporated into the crystalline phase. During DSC analysis, an intitial high trans content is preserved in the case of the non-isomerising 1,4-DACH, whereas the 1,4-CHDA residues gradually isomerise from a high initial trans content to a significanly lower trans content. Since these cis residues are not incorporated into the crystalline domains, the lower second heating melting points of the 1,4-CHDA-based copolyamides in comparison with 1,4-DACH-based copolyamides, having similar cycloaliphatic monomer contents, can be understood.

Keywords: copolymerisation, crystallisation, differential scanning calorimetry (DSC), polyamides, WAXS

Introduction

Although the incorporation of cyclohexyl moieties into the backbone of polyamides and polyesters, and its influence on crystallisation, have been studied quite extensively,[1-10] a question which still remains to be answered is, whether or not both the cis and the trans isomers of either 1,4-cyclohexanedicarboxylic acid (1,4-CHDA) or 1,4-diaminocyclohexane (1,4-DACH) are participating in the formation of crystalline domains in the resulting step-growth

 DOI: 10.1002/masy.200350936

polymers. On forehand one might expect that stable crystals prefer the more 'stretched' trans configuration over the 'kinky' cis configuration, but no scientific evidence is available for such a preference. In our efforts to find this evidence, we decided to study the relation between the stereochemistry of cycloaliphatic residues, and thermal transitions as well as X-ray structures of two series of isomeric copolyamides. In the series based on polyamides 12.6, the adipic acid (AA) residues were partially replaced by *cis/trans*-1,4-cyclohexanedicarboxylic acid (1,4-CHDA), and in the series based on polyamides 4.14, the 1,4-diaminobutane (1,4-DAB) residues were partially replaced by *cis/trans*-1,4-diaminocyclohexane (1,4-DACH). The four types of repeating units, occurring in both isomeric copolyamides 12.6/12.1,4-CHDA and 4.14/1,4-DACH.14 are given in Figure 1.

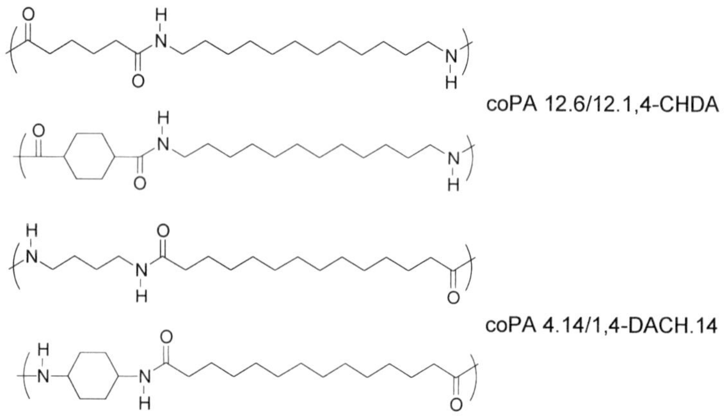

coPA 12.6/12.1,4-CHDA

coPA 4.14/1,4-DACH.14

Fig. 1. Chemical structures of repeating units occurring in copolyamides 12.6/12.1,4-CHDA and 4.14/1,4-DACH.14.

Preliminary DSC results have been reported earlier by us,[11,12] which however did not allow the drawing of hard conclusions regarding the problem we were eager to solve. In the current paper, fine-tuned DSC results are combined with new WAXS data collected at room temperature, after which we were able to get an indisputable answer to the question raised above.

Experimental

Synthesis of the Copolyamides

Equimolar amounts of the desired dicarbonyl dichlorides (a mixture of adipoyl chloride and 1,4-cyclohexanedicarbonyl dichloride with the desired cis/trans ratio for the series 12.6/12.1,4-CHDA, and 1,12-dodecanedicarbonyl dichloride for the series 4.14/1,4-DACH.14) and the desired diamines (1,12-diaminododecane for the series 12.6/12.1,4-CHDA and a mixture of 1,4-diaminobutane and 1,4-DACH with the desired cis/trans ratio for the series 4.14/1,4-DACH.14) were polymerised with stirring in $CHCl_3$ under a nitrogen atmosphere. After polymerisation the solvent was evaporated, the polymers were dissolved in formic acid, followed by two successive precipitations in water, after which they were washed and dried. Details concerning the low temperature polymerisation, during which the initial cis/trans ratio of the cycloaliphatic monomers proved to be preserved, can be found elsewhere.[11,12]

Characterisation of the Copolyamides

Solution ^{1}H-decoupled ^{13}C NMR spectra of the copolyamides dissolved in deuterated trifluoroacetic acid (TFA-d) were obtained in the inverse-gated Fourier transform mode on a Bruker Avance DRX250 instrument equipped with a Quattro probe tuned to 62.93 and 250.13 MHz for ^{13}C NMR and ^{1}H NMR nuclei (details can be found elsewhere[11,12]). Determination of the copolyamide composition and the cis/trans ratio of the cycloaliphatic residues in the polymer main chain was performed from integrated signal areas as obtained by line deconvolution of ^{13}C NMR resonances using the PERCH TLS (Total Line Shape) software.[13]
Polymer solutions (1.0 g/dL in *m*-cresol) were used to determine intrinsic viscosities at 25 °C using a Cannon-Ubbelohde viscometer.

DSC curves were recorded under nitrogen on a Perkin Elmer DSC 7 using a scanning speed of 10 °C/min, both for cooling and heating. After first heating the polyamides were kept at 300 °C for 30 min in order to remove all residual crystallites, capable of acting as nucleating agents during cooling. The T_ms, taken from the first and second heating curves, were defined as the endsets of the endotherms.

WAXS measurements were performed at room temperature using a horizontal Geigerflex diffractometer on a Rigaku Rotaflex RU-200B rotating Cu-anode at a power of 4 kW. Cu K_α

radiation with a wavelength of 1.542 Å was used. Measurements were of the transmission type, and were performed in the diffraction angle range $2 < 2\theta < 60°$. Data were accumulated in steps of $2\theta = 0.05°$, each for a period of 6s.

Results and Discussion

All copolyamides, either belonging to the series 12.6/12.1,4-CHDA or to its isomeric series 4.14/1,4-DACH.14, were synthesised following a recipe during which the initial cis/trans ratio of the cycloaliphatic moieties is totally preserved. This was proven using solution ^{1}H-decoupled ^{13}C NMR spectroscopy, a technique giving both the incorporated comonomer ratio as well as the desired information on the stereochemistry of the built-in 1,4-CHDA and 1,4-DACH residues. The compositions of the copolyamides, as well as their intrinsic viscosities (which are sufficiently high to exclude any possible influence of molar mass on thermal transitions) are given in Tables 1 and 2.

Table 1. Chemical compositions and intrinsic viscosities of synthesized copolyamides 12.6/12.1,4-CHDA.

Entry	Initial molar monomer feed ratio		Molar ratio in final copolyamide		[η](dL/g)
	AA/1,4-CHDA	cis/trans	AA/1,4-CHDA	cis/trans	
1	100/0	-	100/0	-	0.55
2	90/10	2/98	91/9	2/98	0.86
3	85/15	80/20	86/14	80/20	0.75
4	85/15	3/97	84/16	6/94	0.67
5	80/20	80/20	79/21	80/20	0.57
6	75/25	80/20	75/25	80/20	0.54
7	80/20	0/100	74/26	0/100	0.54

Figure 2 shows the first heating melting points of the 12.6/12.1,4-CHDA copolyamides with an initially incorporated 1,4-CHDA cis/trans ratio of either close to 0/100 (◊) or 80/20 (◆). In the same figure, the second heating melting points are shown for both series of copolyamides (■), both obtained after cooling down the melt after storage for 30 min at 300 °C. It is obvious that the incorporation of the cycloaliphatic monomer into the polyamide main chain raises the first heating $T_{m,end}$ significantly with respect to the reference polyamide 12.6, provided that *trans*-1,4-CHDA units are present in the copolyamide. On the other hand, the first heating melting points of copolyamides initially rich in cis moieties correspond remarkably well with the melting point of the homopolyamide 12.6. However, after exposure of both *cis*- and *trans*-1,4-CHDA-rich copolyamides to a temperature of 300 °C for 30 min, the subsequent cooling and reheating yields melting points which show a linear dependency of the mole % 1,4-CHDA, irrespective of the initial cis/trans ratio. It is obvious that the thermal treatment of the initially trans-rich copolyamides results in a significant decrease of the T_m in the second melting. On the other hand, a similar treatment of cis-rich copolyamides results in a significant raise of the T_m in the second melting. It is known that the 1,4-CHDA residues are susceptible to cis/trans isomerisation, for which a mechanism was proposed by Kricheldorf and Schwarz.[1] Solution ^{13}C NMR experiments revealed that, during the thermal treatment for 30 min at 300 °C, isomerisation of the 1,4-CHDA moieties, incorporated in the polyamide main chain, had occurred indeed. As illustrated in Figure 3, for copolyamides based on initially high *trans*-1,4-CHDA contents significant amounts of the trans isomer have been converted into the cis isomer, whereas for copolyamides based on initially high *cis*-1,4-CHDA contents significant amounts of the cis isomer have been converted into the trans isomer. For both copolyamides, after thermal treatment the cis/trans ratio is very similar, indicating that isomerisation equilibrium has been reached, which is in agreement with the straight line in Figure 2 (■).

Table 2. Chemical compositions and intrinsic viscosities of synthesized copolyamides 4.14/1,4-DACH.14.

Entry	Molar monomer ratio in feed		Molar ratio in final copolyamide		[η](dL/g)
	1,4-DAB/1,4-DACH	cis/trans	1,4-DAB/1,4-DACH	cis/trans	
8	100/0	-	100/0	-	0.53
9	95/5	0/100	93/7	0/100	0.59
10	90/10	0/100	86/14	0/100	0.57
11	90/10	0/100	80/20	0/100	0.57
12	90/10	66/34	79/21	57/43	0.48
13	85/15	83/17	80/20	75/25	0.44

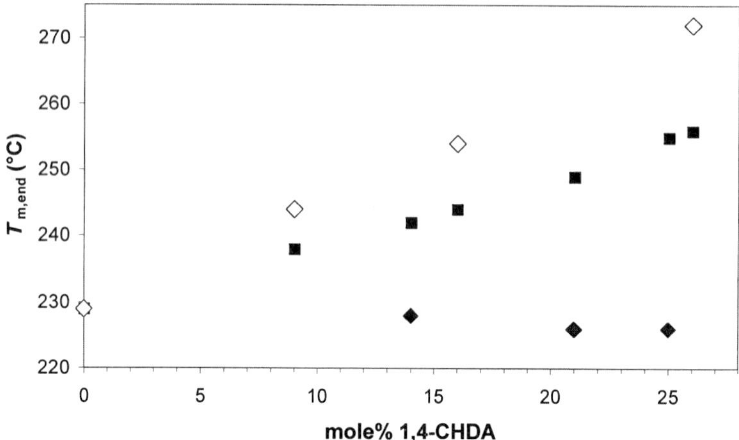

Fig. 2. Comparison of melting points obtained from 1st and 2nd heating of copolyamides 12.6/12.1,4-CHDA with different initial cis/trans ratios. Legend: melting point of originally cis-rich (◆) or trans-rich (◊) copolyamides obtained from 1st heating; (■) melting point after annealing for 30 min at 300 °C.

Combining the results presented in Figure 2 and Figure 3, it is obvious that the melting point only increases with respect to the reference polyamide 12.6 if detectable amounts of the 'stretched' and rigid *trans*-1,4-CHDA isomer are available for cocrystallisation with the adipic acid residues. So, based on the DSC experiments, we are bound to conclude that the *trans*-1,4-CHDA residues are capable of cocrystallising in a common crystal lattice with adipic acid residues. In view of the similar first heating melting points for polyamide 12.6 and the copolyamides rich in *cis*-1,4-CHDA moieties, it is tempting to conclude that the *cis*-1,4-CHDA moieties are excluded from the crystalline phase. However, WAXS diffractograms of both polyamide 12.6 and the copolyamides are required to draw this conclusion, since a possible melting point raising effect of the incorporated rigid *cis*-1,4-CHDA moieties might just be compensated by the inferior quality of the generated crystals, caused by the 'kinky' structure of the cis isomer.

Figure 4 shows WAXS diffractograms of homopolyamide 12.6, of a copolyamide containing some 1,4-CHDA with a high trans content (entry 4 in Table 1), and of a copolyamide containing a similar amount of 1,4-CHDA with a high cis content (entry 3 in Table 1), all recorded at room temperature.

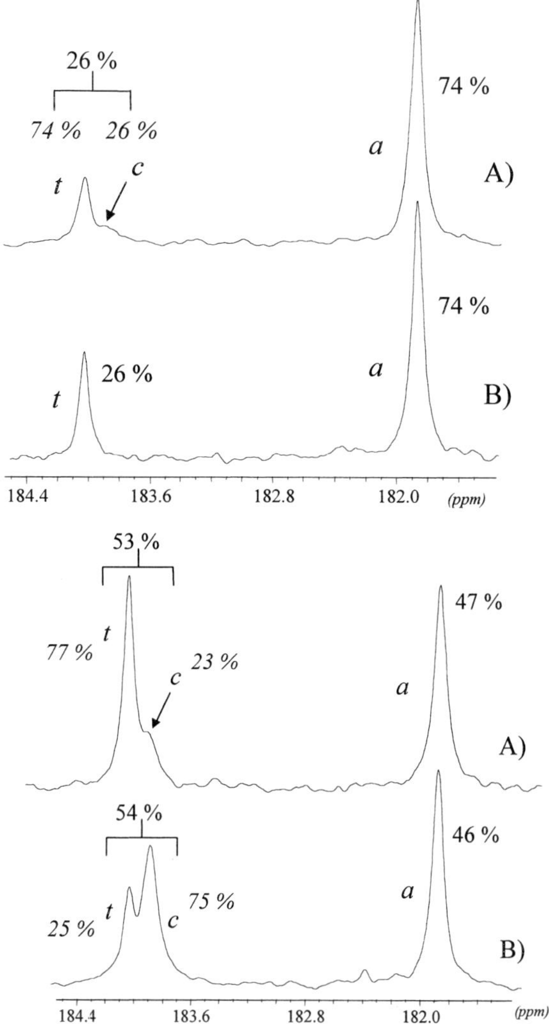

Fig. 3. Solution ^{13}C NMR analysis (TFA-d) of copolyamides 12.6/12.1,4-CHDA, A) after and B) before submission of the samples to 300 °C for 30 min. The partial isomerisation of the 1,4-CHDA residues of an initially trans-rich (top) or cis-rich (bottom) copolyamide, resulting in equilibrium cis/trans ratios of the cycloaliphatic residues, is shown. Legend: a: AA residues; c: *cis*-1,4-CHDA residues; t: *trans*-1,4-CHDA residues.

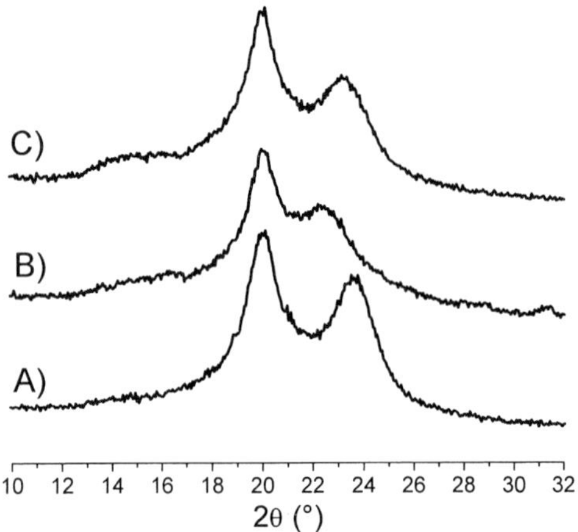

Fig. 4. Room temperature diffraction patterns obtained by WAXS analysis of copolyamides 12.6/12.1,4-CHDA as a function of composition (λ = 1.542 Å). Legend: (A) polyamide 12.6; (B) 16 mole% 1,4-CHDA (cis/trans: 6/94) and (C) 14 mole% 1,4-CHDA (cis/trans: 80/20).

The diffraction pattern of the copolyamide containing 14 mole% 1,4-CHDA with a high cis content is similar to the pattern of the homopolyamide 12.6. This is a clear indication that the cis moieties are not incorporated into the crystals. On the contrary, the copolyamide containing 16 mole % 1,4-CHDA with a high trans content shows a pattern clearly deviating from that of polyamide 12.6, implying that the trans residues are cocrystallising with the adipic acid residues. These X-ray data are fully in agreement with the DSC data shown in Figure 2.

According to the isomerisation mechanism suggested bij Kricheldorf and Schwarz,[1] 1,4-DACH residues are not expected to isomerise. The ^{13}C NMR spectra shown in Figure 5 provide evidence for this. Before and after a 30 min stay at 300 °C, the stereochemistry of the 1,4-DACH-based copolyamides is, within experimental error, the same.

In view of their insensitivity towards isomerisation during thermal analysis, the 1,4-DACH-based partially cycloaliphatic copolyamides are the preferred polymers to study the influence of the stereochemistry of cyclic main chain parts on thermal transitions and crystallisation phenomena. Figure 6 shows that, in analogy with the 1,4-CHDA-based copolyamides, only the trans isomer residues are incorporated into the crystals. The diffraction pattern of the copolyamide containing 20 mole% 1,4-DACH with a high cis content (entry 13 in Table 2) is very similar to the pattern of the homopolyamide 4.14. This once more is a clear indication that the cis moieties are not incorporated into the crystals. On the other hand, the copolyamide containing 20 mole % 1,4-DACH with a high trans content (entry 11 in Table 2) shows a pattern strongly deviating from that of polyamide 4.14, implying that the *trans*-1,4-DACH residues are cocrystallising with the 1,4-diaminobutane residues.

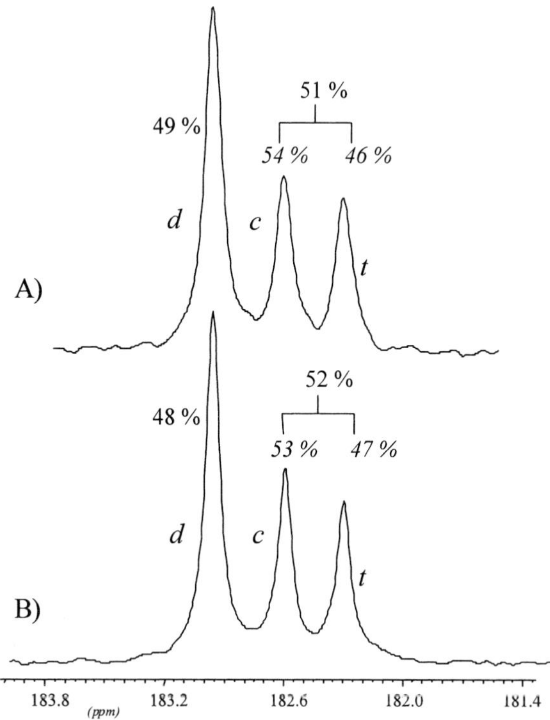

Fig. 5. Solution ^{13}C NMR analysis (TFA-d) of a copolyamide 4.14/1,4-DACH.14, A) after and B) before submission of the sample to 300 °C for 30 min, showing the absence of isomerisation. d: 1,4-DAB residues; c: *cis*-1,4-DACH residues; t: *trans*-1,4-DACH residues.

In view of the WAXS data on the 1,4-DACH based copolyamides, it is not surprising that the $T_{m,end}$ values of polyamide 4.14 and cis-rich 4.14/1,4-DACH.14 80/20 (entry 13, Table 2), all obtained after cooling following a treatment of 30 min at 300 °C, are very similar, *viz.* 228 and 227 °C, respectively, whereas the $T_{m,end}$ of the trans-rich 4.14/1,4-DACH.14 80/20 (entry 11, Table 2) is significantly higher, *viz.* 270 °C. The $T_{m,end}$ of homopolyamide 4.14 corresponds remarkably well with that of its isomer 12.6, being 229 °C.

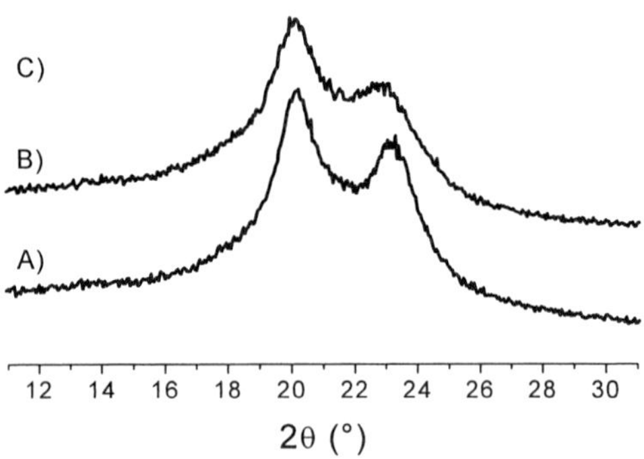

Fig. 6. Room temperature diffraction patterns obtained by WAXD analysis of copolyamides 4.14/1,4-DACH.14 as a function of composition (λ = 1.542 Å). Legend: (A) polyamide 4.14; (B) 20 mole% 1,4-DACH (cis/trans: 75/25) and (C) 20 mole% 1,4-DACH (cis/trans: 0/100).

Copolyamides 4.14 with *trans*-1,4-DACH incorporated show a higher T_m than the isomeric copolyamides 12.6/12.1,4-CHDA containing comparable amounts of initially *trans*-1,4-CHDA. This difference is visualised in a plot of the endset melting point versus composition of the two types of copolyamides (see Fig. 7).

The explanation for this difference is the isomerisation of the 1,4-CHDA residues: after the thermal treatment, part of the *trans*-1,4-CHDA residues has been converted into *cis*-1,4-CHDA residues. In this way, the final content of cocrystallising *trans*-1,4-CHDA units in the copolyamide 12.6/12.1,4-CHDA is lower than the initial content, and accordingly, after partial isomerisation from trans to cis residues, less cycloaliphatic residues are available for incorporation into the crystals, resulting in lower melting temperatures. As shown in Figure 5, the 1,4-DACH residues in copolyamides 4.14/1,4-DACH.14 do not isomerise under the applied conditions. Thus, the content of *trans*-1,4-DACH residues remains constant during thermal analysis.

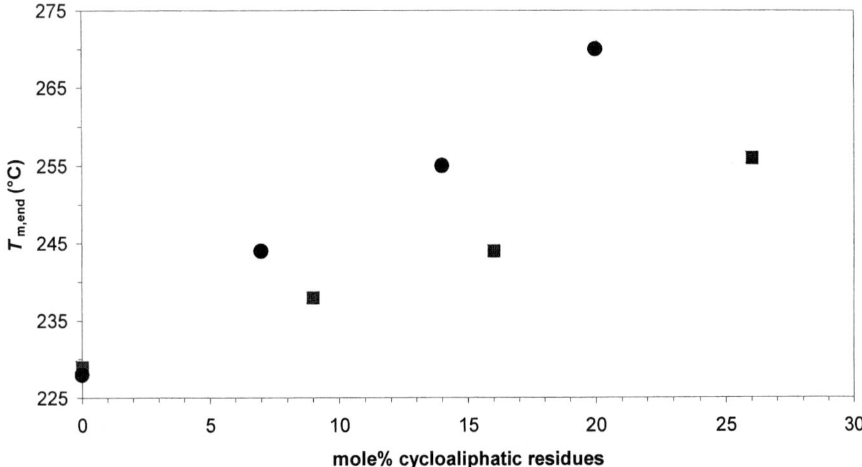

Fig. 7. Melting points of the copolyamides, after cooling following a treatment of 30 min at 300 °C, as a function of composition. Legend: (●) copolyamides 4.14/*trans*-1,4-DACH.14; (■) copolyamides 12.6/12.(initially) *trans*-1,4-CHDA.

Conclusions

A careful DSC and WAXS analysis on two series of isomeric copolyamides, *viz.* copolyamides 12.6/12.1,4-cyclohexanedicarboxylic acid (1,4-CHDA) and copolyamides 4.14/4.1,4-diaminocyclohexane (1,4-DACH), learned that only the trans isomers of both 1,4-DACH and 1,4-CHDA are incorporated into the crystalline phase. During heating, the intitial high trans content is preserved in the case of the non-isomerising 1,4-DACH, whereas during DSC analysis the 1,4-CHDA residues gradually isomerise from a high initial trans content to a significanly lower trans content, which in view of the non-cocrystallising cis residues can explain the lower thermal transition temperatures of the 1,4-CHDA-based copolyamides for similar cycloaliphatic monomer contents.

[1] H. R. Kricheldorf, G. Schwarz, *Makromol Chem* **1987**, *188*, 1281.
[2] A. V. Tenkovtsev, A. B. Rutman, A. Yu. Bilibin, *Makromol Chem* **1992**, *193*, 687.
[3] M. A. Osman, *Macromolecules* **1986**, *19*, 1824.
[4] S. L. Kwolek, R. R. Luise, *Macromolecules* **1986**, *19*, 1789.
[5] B. Reck, H. Ringsdorf, K. Gardner, H. Starkweather, Jr. *Makromol Chem* **1989**, *190*, 2511.
[6] R. Srinivasan, A. Prasad, H. Marand, J. E. McGrath, *Polym Prepr (Am Chem Soc, Div Polym Chem)* **1991**, *32*, 174.
[7] R. Srinivasan, J. E. McGrath, *Polym Prepr (Am Chem Soc, Div Polym Chem)* **1992**, *33*, 503.
[8] R. Srinivasan, T. Moy, J. Saikumar, J. E. McGrath, *Polym Prepr (Am Chem Soc, Div Polym Chem)* **1992**, *33*, 225.
[9] J. S. Ridgway, *J Poly Sci: Part A-1* **1970**, *8*, 3089.
[10] V. D. Kalmykova, M. N. Bogdanov, N. P. Okromchedlidze, I. V. Zhmayeva, V. Ya. Yefremov, *Polymer Science USSR* **1967**, *9*, 2539.
[11] B. Vanhaecht, M. N. Teerenstra, D. R. Suwier, R. Willem, M. Biesemans, C. E. Koning, *J Polym Sci Pol Chem* **2001**, *39*, 833.
[12] B. Vanhaecht, B. Rimez, R. Willem, M. Biesemans, C. E. Koning, *J Polym Sci Pol Chem* **2002**, *40*, 1962.
[13] R. Laatikainen, M. Niemitz, W. J. Malaisse, M. Biesemans, R. Willem, *Magn Res Med* **1996**, *36*, 359.

Vinyl Monomers Containing Active Hydrogen Atoms Are a New Type of Monomer for Polycondensation

Boris A. Rozenberg

Institute of Problems of Chemical Physics Russian Academy of Sciences Chernogolovka, Moscow region, 142432, Russia, E-mail: rozen@icp.ac.ru

Summary: The latest findings in author's laboratory devoted to the understanding kinetic peculiarities and mechanism of vinyl macromonomers formation by the example of anionic polymerization of 2-hydroxyethylacrylate (HEA) and 2-hydroxyethylmethacrylate (HEMA) containing group with mobile hydrogen atoms are discussed. It was shown that polymerization is accompanied by proton-transfer in each elementary propagation step, which leads to isomerization of the backbone and formation of polyester type macromonomers. Polymerization is accompanied by side transesterification reaction with formation of polyester diacrylates and polyester diols. The peculiarities of polymerization mechanism under the action of alkali metals and its alkoholates were established. Expected applications of the polymerization products are discussed.

Keywords: 2-hydroxyethylacrylate, 2-hydroxyethylmethacrylate, macromonomers, proton-transfer anionic polymerization, transesterification reaction

Introduction

Vinyl monomers containing groups with active hydrogen atom like (meth)acrylamides hydroxyalkyl(meth)acrylates, (meth)acryloylalkyl(aryl)urethanes etc. are easily transformed to corresponding heterochain polymers through Michael addition polymerization mechanism under the action of base-type catalysts. For the first time such type of polymer formation was discovered about half a century ago.[1, 2] However, even in the basic work entitled "A Novel Synthesis of Poly-β-alanine from Acrylamide" it was realized that the real mechanism of polymerization is much more complex and it was assumed that one of the propagation steps include a proton transfer reaction. Despite the permanent interest of researchers in such systems[1-12] there is still the obvious lack in understanding kinetic features and mechanism of polymerization and structure of the generated polymers. Brief but practically exhaustive overview of the current status of research on the proton-transfer anionic polymerization of vinyl monomers is given in

 DOI: 10.1002/masy.200350937

reference.[12] A new wave of interest to proton-transfer anionic polymerization of vinyl monomers appeared at the last time[7-15] as to the promising synthetic technique that allows preparing valuable products for technical and biomedical applications. Proton-transfer anionic polymerization can be used for the preparation of a family of macromonomers with different heterochain backbone and related comb-like, superbranched and crosslinked polymers on the basis of vinyl monomers containing groups with mobile hydrogen atom. Anionic polymerization of such a class of vinyl monomers can introduce functional groups into the main chain of polymers and provide necessary terminal functional groups. The objective of the paper is to shorten a gap in understanding the mechanism of the reactions under study.

Experimental

Detailed description of HEA and HEMA purification as well as apparatus for polymerization is described in reference.[12] Analytical techniques IR-, ^{1}H and ^{13}C NMR spectroscopy, ESR, HPLC, GPC, DSC and isothermal microcalorimetry, TMA, elemental analysis, chemical analysis of functional groups used for kinetic investigations and structure characterization of generated polymers are described in the same paper.[12] Special attention was paid to characterization of functional group distribution (FGD) on macromolecules. In order to solve this problem a special HPLC technique in conditions close to the critical ones was worked out for the analysis of synthesized polymers.[13]

Results and Discussion

IR- and NMR spectroscopic data unambiguously indicate that HEA and HEMA are undergone step polymerization through Michael addition reaction under the action of alcoholate type propagating centers formed by any anionic initiator.[12] Step polymerization proceeds via anionic polymerization mechanism accompanying by intra- (path a) or intermolecular (path b) proton-transfer on the ester enolate in each elementary propagation step, which leads to isomerization of the backbone and formation of heterochain (polyester type) polymer. Therefore, such type of polymerization can be called not only proton-transfer but also "isomerization" or "heterochain" polymerization contrary to conventional anionic polymerization of (meth)acrylates resulting in carbochain backbone of polymers.

$$R_jO^- + CH_2{=}CH\overset{O}{\overset{\|}{C}}O(CH_2)_2OH \longrightarrow \left[\begin{array}{c} R_jOCH_2-\bar{C}H\overset{O}{\overset{\|}{C}}O(CH_2)_2OH \\ \updownarrow \\ R_jOCH_2-CH{=}\overset{O^-}{\overset{|}{C}}O(CH_2)_2OH \end{array} \equiv -\overbrace{CH{\cdots}\underset{|}{C}{\cdots}O}^{-}\right]$$

$$\begin{array}{l} \xrightarrow{a} R_jOCH_2-CH_2\overset{O}{\overset{\|}{C}}O(CH_2)_2O^- \\ \xrightarrow[R_iOH]{b} R_jOCH_2-CH_2\overset{O}{\overset{\|}{C}}O(CH_2)_2OH + R_iO^- \end{array}$$

where $R_jO^- = CH_2{=}CH\overset{O}{\overset{\|}{C}}O(CH_2)_2O-(CH_2CH_2\overset{O}{\overset{\|}{C}}OCH_2CH_2O)^-_{j-1}$, $j = 1, 2, 3$ ....

Initiation reactions. The structure of the generated polymers in a great extent depends on a type of initiator used. Formation of alkoxy anion as active propagating center at interaction of hydroxyl groups of monomer or macromonomer with anionic initiator is main mechanism of initiation. However, proton-transfer polymerization of HEMA and HEA under the action of alkali metals is accompanied by conventional vinyl polymerization of methacrylate's double bond.

$$CH_2{=}\underset{CH_3}{\underset{|}{C}}-\overset{O}{\overset{\|}{C}}O(CH_2)_2OH + K \begin{array}{l} \nearrow CH_2{=}\underset{CH_3}{\underset{|}{C}}-\overset{O}{\overset{\|}{C}}O(CH_2)_2O^- K^+ + 1/2\,H_2 \\ \searrow \cdot CH_2-\underset{CH_3}{\underset{|}{\bar{C}}}-\overset{O}{\overset{\|}{C}}O(CH_2)_2OH\ K^+ \end{array}$$

Initiation proceeds with formation of anion-radical, which generates two types of backbone chains (hetero-and carbochain respectively). The final polymers are insoluble in organic solvents and displays network behavior on TMA curve. However, rubbery network polymer based on HEMA is readily dissolved in boiling water due to its alkali hydrolytic decomposition with formation of ethylene glycol, methacrylic, polymethacrylic acids, and 2-hydroxyethyl-β-oxy-α-methylpropionic acids after neutralization, that were identified by ^1H and ^{13}C NMR.[14] The ratio of heterochain and carbochain polymers was measured directly by ^1H NMR and FGD techniques and indirectly calculated using the values of HEMA polymerization enthalpy under the action of potassium and T_g of polymer formed. All methods used are in a good coincidence and give approximately 1:1 molar ratio of hetero- and carbochain polymers formed at HEMA polymerization under the action of potassium at 70-100°C.[14] These polymers are represented a mixture of heterochain homopolymer of HEMA and block copolymer of HEMA with hetero- and carbochain structure of interknot chains.

Addition of inhibitor (2,2,6,6-tetramethyl-N-piperidine oxide) completely suppresses radical polymerization and formation of network polymers. Formation of heterochain polymers is the only result of such inhibited HEMA or HEA anionic polymerization under the action of alkali metals.

Table 1. Characterization of properties and backbone structure of the network polymer formed at HEMA polymerization under the action of potassium.

Backbone structure	ΔH at polymerization / kJ/mol	T_g / °C	^{1}H NMR / fraction	FGD / fraction
Network polymer	-27,7	13		
Carbochain [a)]	-59,0 (0,42)	77 (0,45)	0.50	0.43
Heterochain [a)]	-7,2 (0,58)	-39 (0,55)	0.50	0.57

[a)] Calculated values of ΔH and T_g for the respective backbone structures of HEMA polymers. Calculated fractions of the respective backbone structures are given in brackets. Other values are experimentally measured

One of the most attractive initiators is t-butanol-based alcoholates. This family of initiators with different counterions contrary to the other alcoholates due to unique combination of high basicity and high spatial hindrance initiates HEA and HEMA polymerization via exchange reaction with hydroxyl group of monomer only. There is no direct addition reaction of this initiator to the double bond of monomer that is revealed by analysis of ^{1}H and ^{13}C NMR data. Well-known fact of impossibility to initiate of MMA polymerization under the action of t-butoxides,[16] and what is more, the fact of successful utilization of t-butoxides as ligands for stability and activity control of initiators and active species at conventional anionic polymerization of MMA[17] also confirm this conclusion.

NMR data indicates proceeding of side reactions through acyl-oxygen and alkyl-oxygen scission in the presence of excess of t-butanol, which sometimes are used as a solvent for t-butoxides. There are no side reactions in the absence of t-butanol.[15]

Chain propagation. Establishment of molecular weight relationship as a function of conversion is a key problem in understanding the mechanism of the chain propagation reactions under study. There are two alternative mechanisms of chain growth: chain polymerization ($R_j + M \rightarrow R_{j+1}$) and step growth polymerization ($R_j + R_i \rightarrow R_{j+I}$) in the systems under study.

Chain growth polymerization mechanism is realized at heterogeneous development of the reaction only when the polymerization products R_jOH with $j \geq 2$ are phase separated due to the growth of hydrophobicity of macromonomer formed at increasing chain length. Step growth polymeryzation must be realized in homogeneous system.

All investigated polymerization systems proceed in bulk heterogeneously. The law for living chain growth polymerization is

$$\bar{P}_n = \frac{\alpha[M]_0}{[I]_0} = \alpha \bar{P}_{n,\infty} \qquad (1)$$

where α is conversion of double bonds or hydroxyl groups, $[M]_o$ and $[I]_o$ are initial monomer and initiator concentrations, $\bar{P}_n$ and $\bar{P}_{n,\infty}$ are current and ultimate average polymerization degree, describes satisfactorily our own and literary data[12] for HEA and other hydroxyalkylacrylate polymerizations under the action of different initiators and temperatures. It means that chain growth polymerization mechanism is dominating.

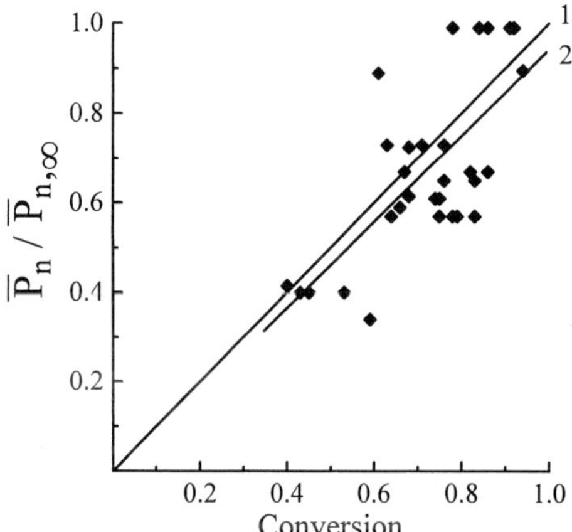

Fig. 1. $\bar{P}_n / \bar{P}_{n,\infty}$ versus conversion at $[I]_o$ / $[M]_o$ ≥0.05. 1 is theoretical line, 2 is linear approximation of experimental data.

Fast exchange equilibrium reaction of propagating alkoxyanion with any terminal hydroxyl group of molecules in the absence of termination reactions makes such polymerizing system “living”.

$$R_jOH + R_iO^- \rightleftharpoons R_jO^- + R_iOH$$

Essential deviation from living chain growth polymerization are observed at initiator to monomer molar ratio < 0.05 that can be related to partial proceeding of the reaction via step growth polymerization under these conditions.

Secondary and side reactions. The polydispersity of polymer increased in the course of polymerization from 1 and tend to 2 at the end of the reaction because of partial parallel proceeding of step growth polymerization and chain exchange reactions.[12, 18] The latter reaction results in redistribution of terminal groups of macromolecules and formation of polyesterdi(meth)acrylates and polyesterdiols in the reaction system. Such types of macromolecules can be observed directly in the reaction product by adsorption chromatography in the conditions close to the critical one[13] (Fig. 2 and 3).

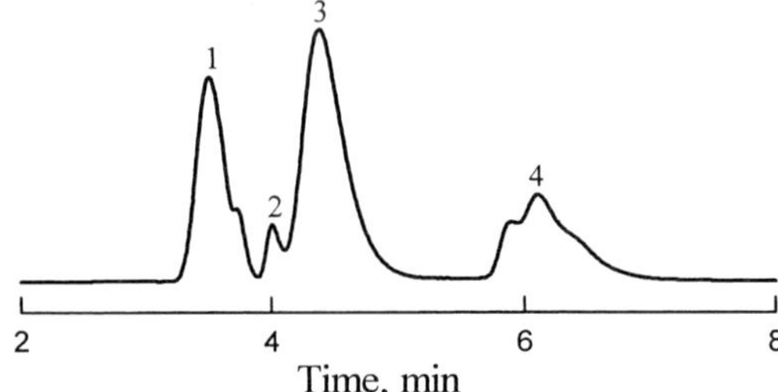

Fig. 2. Functional groups distribution of HEA oligomer produced under the action of BuOLi. Detailed description see in Table 2.Formation of polymers with carbochain backbone is revealed by FGD analysis of polymers produced at HEMA polymerization initiated by alkali metal[13] (Fig. 3).

It is interesting to note that proceeding chain exchange reaction intra-molecularly does not lead to the macrocycles (back-biting cyclization) because the structure of the reaction product is the same as the initial one. Macrocyclization reaction in such systems takes place only when growing alkoxy anion attacks double bond of its own chain.[13] Formation of small cycles is the most probable cyclization reaction.

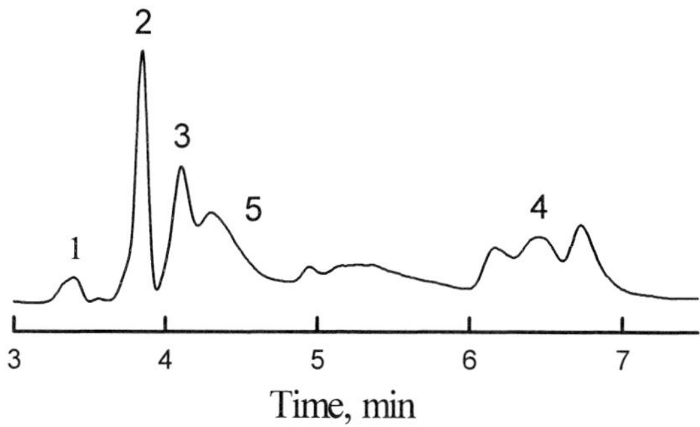

Fig. 3. Functional groups distribution of HEMA oligomer produced under the action of potassium. Detailed description see in Table 2.

Table 2. Molecular characteristics of HEA and HEMA oligomers.

Reaction condition: Monomer; initiator; $[M]_o/[I]_o$; T,°C; conversion	FGD[a)] 1	2	3	4	5	$\bar{M}_n$	MWD $\bar{M}_w$	$\bar{M}_w/\bar{M}_n$
HEA; t-BuOLi; 34,0; 20; 0,80	19.5	9.8	48.4	22.3	–	1120	1800	1.61
HEMA; K; 7,8; 70; 0,93	24.5	4.9	12.9	6.3	42.9	1870		

a)Polyesterdiacrylates (1), macrocycles (2), macromonomers (3), polyesterdiols (4) and carbochain polymer (5).

As expected, heterochain reaction product is readily undergone crosslinking under the action of radical initiators or UV- irradiation.[12] Alkaline hydrolysis of crosslinked polymer results in formation of ethylenglycol, methacrylic, polymethacrylic, and 2-hydroxyethyl-β-oxy-α-methylpropionic acids that were identified by ^1H and ^{13}C NMR spectroscopy.

Chain exchange reaction proceeds via transesterification mechanism with formation of intermediate adduct. Kinetic study of model reaction of ethylacetate with sodium ethanolate in ethanol solution containing traces of water by IR-spectroscopy gives the possibility to estimate the value of rate constant of intermediate adduct formation that is a limiting stage of chain exchange reaction.[15]

Such model system can be described by the following kinetic scheme:

$$RC(=O)-OR_1 + R_1O^- \underset{k_2}{\overset{k_1}{\rightleftharpoons}} RC(O^-)(OR_1)-OR_1$$

$$RC(O^-)(OR_1)-OR_1 + H_2O \xrightarrow{k_3} \left\{RC\langle O,O\rangle\right\}^- + 2R_1OH$$

where R=CH_3, R_1=C_2H_5, k_1, k_2 *u* k_3 are constant rate of the respective reactions.

The rate of ethylacetate (EtOAc) consumption at quasi-stationary approximation on concentration of tetrahedral intermediate reaction product ([Ad]) can be described by the following equation:

$$\frac{d\alpha}{dt} = \frac{k_1 k_3 [EtOAc]_o (1-\alpha\beta)(1-\alpha)[H_2O]}{k_2 + k_3[H_2O]} \quad (2)$$

where α is conversion of EtOAc equal to $\alpha = ([EtOAc]_o - [EtOAc]) / [EtONa]_0$, and $\beta = [EtONa]_o / [EtOAc]_o$. Taking into account that $k_3[H_2O] >> k_2$ solution of this equation gives

$$\frac{\ln(1-\alpha\beta) - \ln(1-\alpha)}{1-\beta} = k_1 [EtOAc]_o t, \quad (3)$$

which quite well describes all kinetic curves at chosen initial conditions (Table 3 and Fig. 4) and the value of k_1=(1,1± 0,1) 10^{-3} l/mol·s determined from initial reaction rate.

Table 3. Rate constant of intermediate adduct formation at transesterification.

№	Reaction condition, T= 25°C				
	$[EtOAc]_o$ / mol/l	$[EtONa]_o$ / mol/l	$[EtOH]_o$ / mol/l	k_1, l/mol·s	
1.	0,089	0,075	16,9	$0{,}93 \cdot 10^{-3}$	
2.	0,10	0,059	16.9	$1{,}17 \cdot 10^{-3}$	k_1=(1,1±0,1)·10^{-3}
3.	0,119	0,041	16.9	$1{,}17 \cdot 10^{-3}$	

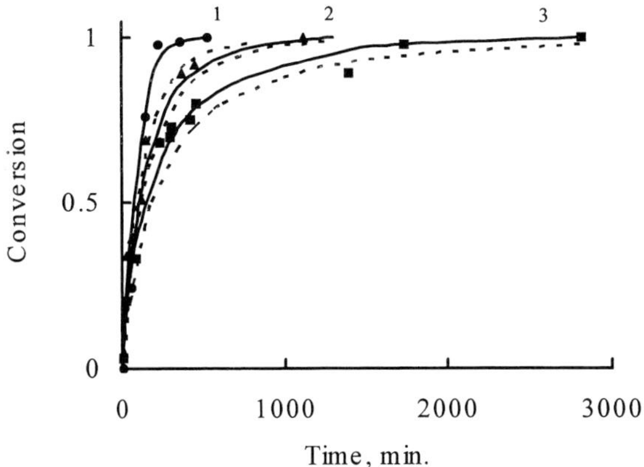

Fig. 4. Kinetics of model reaction of ethylacetate with sodium etoxide in ethanol solution at 25°C (solid lines). Designation of kinetic curves and initial concentration of components are given in Table 3. Calculated kinetic curves are designated by dotted lines.

Conclusion

The monomers studied under the conditions of anionic polymerization display activity towards initiating and propagating anions simultaneously in three points as indicated in the scheme:

$$\overset{\mathbf{1}}{CH_2} = CH\overset{\mathbf{2}}{\underset{\underset{O}{\|}}{C}}-OCH_2CH_2\overset{\mathbf{3}}{OH}$$

Such plural reactivity defines the polymerization mechanism, summarized in the following simplified kinetic scheme:

Initiation $\quad I + M \xrightarrow{k_i} R_1O^-$

Chain propagation $\quad R_1O^- + M \xrightarrow{k_p} R_2O^-$

....

$$R_jO^- + M \xrightarrow{k_p} R_{j+1}O^-$$

$$R_jO^- + R_iOH \xrightarrow{k_p} R_{j+i}OH$$

$$j = 1, 2, 3 \ldots\ldots, \qquad i = 1, 2, 3 \ldots\ldots$$

Proton transfer $R_jO^- + R_iOH \underset{k_{tr2}}{\overset{k_{tr1}}{\rightleftharpoons}} R_iO^- + R_jOH$

$$k_{tr1} >> k_p \ ; \quad K_1 = \frac{k_{tr1}}{k_{tr2}} \sim 1$$

Chain transfer on polymer

$$R_jO^- + R_i\,OH \overset{k_a}{\rightleftharpoons} (\text{Adduct})^- \overset{k_{tr}}{\rightleftharpoons} R_iOCH_2CH_2COOR_{j-k} + {}^-OCH_2CH_2OR_{k-1}OH$$

$$k_a >> k_p; \qquad K_a >> 1$$

Chain termination

$$R_j\,O^- + H_2O \rightleftharpoons R_j\,OH + HO^-$$

$$HO^- + R_j\,OH \xrightarrow{k_t} R_{j-k}OCH_2CH_2COO^- + HOCH_2CH_2OR_{k-1}OH$$

Initiated by alkali metal polymerization, cyclization and chain transfer reactions of t-butoxide anion on polymer with alkyl-oxygen scission discussed ahead are not taken into account in this scheme.

The polymerization mechanism due to the heterogeneous development combines features of chain growth and step growth polymerization.

Acknowledgements

Author is grateful to the Russian Foundation for Basic Research (Project 01–03– 97007) and International Science and Technology Center (Project 02-1918) for the financial support.

[1] A.S. Matlack, Pat. 2672480 USA (1954).
[2] D.S. Breslow, G.E. Hulse, A.S. Matlack, J. Am. Chem. Soc., **1957**, *79,* 3760.
[3] M. Guaita, G. Camino, L. Trossarelli, Makromol. Chem., **1970**, *131,* 309.
[4] T. Saegusa, S. Kobayashi, Y. Kimura, Macromolecules, **1974**, *7,* 256.
[5] T. Saegusa, S. Kobayashi, Y. Kimura, Macromolecules, **1975,** *8*, 950.
[6] Y. Yamada, T. Matsushita, T. Otsu, J. Polym. Sci., Polym. Lett. Ed., **1976**, *14*, 277.
[7] S. Kobayashi, J. Kadokawa, I.-F. Yen, H. Uyama, S. Shoda, Macromolecules, **1992**, *2,5* 6690.
[8] T. Iwamura, I. Tomita, M. Suzuki, T. Endo, J. Polym. Sci., Polym. Chem., **1998**, *36*, 1491.
[9] T. Iwamura, I. Tomita, M. Suzuki, T. Endo, Reactive and Functional Polymers, **1999**, *40*, 115.
[10] J. Kadokawa, Y. Kaneko, S. Yamada, K. Ikuma, H. Tagaya, K. Chiba, Macromol. Rapid Commun., **2000,** *21,* 362.

[11] B.A. Rozenberg, B.A. Komarov, G.N. Boyko, E.A. Dzhavadyan, A.I. Perekhrest, G.A. Estrina, Polym. Sci., **2001**, ***43A.,*** 799.
[12] B.A. Rozenberg, L.M. Bogdanova, E.A. Dzhavadyan, B.A. Komarov, G.N. Boyko, L.L. Gur'eva, G.A. Estrina, Polym. Sci., **2003**, ***45A.,*** ???????.
[13] B.A. Rozenberg, G.A. Estrina, Ya.I. Estrin, Polym. Sci., submitted.
[14] B.A. Rozenberg, G.N. Boyko, L.M. Bogdanova, E.A. Dzhavadyan, B.A. Komarov, P.P. Kusch, G.A. Estrina, Polym. Sci., in press.
[15] B.A. Rozenberg, G.N. Boyko, L.L. Gur'eva, E.A. Dzhavadyan, B.A. Estrina, Polym. Sci., in press.
[16] M. Iijima, Y. Nagasaki, M. Kato, K. Kataoka, Polymer, **1997**, ***38***, 1197.
[17] C.Zune, R.Jerome, Prog. Polym. Sci., **1999,** *24,* 631.
[18] B.A. Rozenberg, V.I. Irzhak, N.S. Enikolopov, Interchain exchange of polymers, Chemistry, Moscow: **1975**.

Synthesis and Characterisation of Polyketanils with Ether Linkages

Danuta Sęk, Agnieszka Iwan, Janusz Kasperczyk, Henryk Janeczek*

Centre of Polymer Chemistry, Polish Academy of Sciences, 34,M. Curie-Sklodowska Str.41-819 Zabrze, Poland
E-mail: polymer@uranos.cto.us.edu.pl

Summary: A series of polyketanils from four diketones with pendant aromatic groups (i.e. dibenzoyl, p-dibenzoylbenzene, p-disebacoylbenzene and trans-1,2-dibenzoylethylene) and two diamines with flexibilizing ether linkages [4,4'-diaminediphenyl ether, 4,4'-(1,3phenylenedioxy)dianiline] has been synthesized using melt polycondensation (180^{0}C, 24h). Spectral and thermal properties of the polyketanils before and after protonation with 1,2-(di-2-ethylhexyl)ester of sulfophthalic acid (DEHEPSA) were investigated.

Keywords: fluorescence, polyketanils (polyketimines), protonation, thermostability

Introduction

In recent years conjugated polymers have been widely studied in regard to their electronic and optoelectronic properties being useful for diverse technological applications. These polymers have various chemical structures of chains and polyenes, polythiophenes, polypyrroles, polyphenylenes, polyphenylenevinylenes and polyaniline can be mention as the main groups. However some interest exhibits also polymers with C=N groups in main chain i. e. polyazomethines (polyimines, poly Schiff-bases) and polyketanils (polyketimines). In contrary to polyazomethines, polyketanils being polycondensation products of diketones and diamines have not been widely studied. However an increasing interest in this group of polymers is observed lastly. Previously, as in the case of the most conjugated polymers, polyketanils were investigated from the point of view of their thermal stability. In several works diacetyl derivatives were used as diketone comonomer. Topchiev et al.[1] condensed 4,4'-diacetyldiphenylmethane and 4,4'-diacetyldiphenylsulphid with 1,4-phenylenediamine. The resulting products melted at about 100^{0}C and their conductivity was in the range of

 DOI: 10.1002/masy.200350938

$10^{-16} – 10^{-11}$ S/cm. The same results were described in [2]. D'Alelio et al.[3] obtained a few series of polyketanils from 1,4-diacetylbenzene, 4,4'-diacetyldiphenyl-methan, -ether, -sulphid and sulphone and various diamines and investigated their thermal stability. 1,4-diacetylbenzene was also used for condensation with 2-methyl-1,4-phenylenediamine by Morgan et al.[4] while Patel et al.[5] condensed 4,4'-diacetyldiphenylether with different diamines. Polyketanils with phenyl pendant groups were synthesized by Volpe et al.[6-8] using para- and meta- dibenzoylobenzene and diamines. Thermal stability of the polyketanils with naphthalene pendant groups were studied by Sęk.[9] Halo-displacement method was used by Yeakel et al.[10] for synthesis of polyketanils with ether linkages. Linear and cyclic oligomers from 4-aminobenzophenone and dendritic oligomers were synthesized by Yamamoto et al.[11-14] Hall et al.[15, 16] investigated polyquinoneimines obtained in polycondensation reaction of antraquinone and its derivatives with aromatic diamines.

But there are not systematic studies of relationships between structure and properties of the polyketanils and to the our best knowledge no influence of the ketimine group protonation on the polyketanils properties was published.

In this work a series of new polyketanils from four diketones with different cores between ketone groups and two aromatic diamines with flexibilizing ether linkages i.e. 4,4'-diaminediphenyl ether (ODA) and 4,4'-(1,3-phenylenedioxy)dianiline (DODA) was synthesized. Melt polycondensation was used for the polyketanils synthesis. Architecturally diverse π-conjugated structures in the polyketanils provides a basis to establish structure – properties relationships.

Experimental

Materials: 4,4'-(1,3-phenylenedioxy)dianiline (Aldrich), trans-1,2-dibenzoylethylene (Aldrich), dibenzoyl (Aldrich) were used as laboratory reagents without further purification.

4,4'-Diaminodiphenyl ether (Merck) was recrystallized from boiling water in the presence of charcoal.

1,2-(di-2-ethylhexyl)ester of sulfophthalic acid (DEHEPSA) was prepared as described in ref.[17]

p-Dibenzoylbenzene, p-disebacoylbenzene were synthesised using the Friedel-Crafts acylation as described in the literature.[18]

Solvents: m-cresol was distilled; dimethylacetamid DMA (Aldrich), methanol were used without purification.

Measurements: Glass transition temperatures of polymers (sample weight about 20 mg, heating rate of 20°C/min) were determined on TA DSC 2010 apparatus using sealed aluminium pen under nitrogen atmosphere (flow rate about 30 ml/min). Thermogravimetric (TG) analyses were performed on a Paulik-Erdey apparatus at a heating rate of 10°C/min under nitrogen. For elemental analysis a 240C Perkin-Elmer analyser was used. Infrared spectra were acquired on a BIO-RAD FTS 40 A spectrometer. Measurements of the IR spectra were done using KBr disc. ^{13}C NMR spectra were recorded on a Varian Inova 300 Spectrometer using $CDCl_3$ as a solvent and TMS as an internal reference. Fluorescence spectra were acquired on a spectrophotometer Fluorolog 3.12 Spex (USA) with a 450 W xenon lamp as a light source. Emission spectra were recorded in DMA solution, at conc. 10^{-4} mol/l. Excitation wavelength was 400 nm. UV-VIS spectra were recorded in m-cresol solution using Hewlett Packard 8452A diode array spectrophotometer. Gel permeation chromatography (GPC) experiments were carried out on Spectra Physics 8800 in temperature 35°C. Differential refractometer Shodex SE61 was used as detector. Tetrahydrofuran was used as the eluent and polystyrene - as standard.

Polymer Synthesis

1 mmol of diketone and 1 mmol of a diamine were heated and stirred at 180°C for 24 hours under nitrogen atmosphere. The mixture was cooled to room temperature, scraped and powdered. The polymers were purified by Soxhlet extraction with methanol and later with acetone and dried at 60°C under vacuum for 24 hours.

Results and Discussion

Structure and Properties of the Polyketanils

Structures of the polyketanils (Fig. 1) were confirmed by spectral and elemental analysis (Table 1).

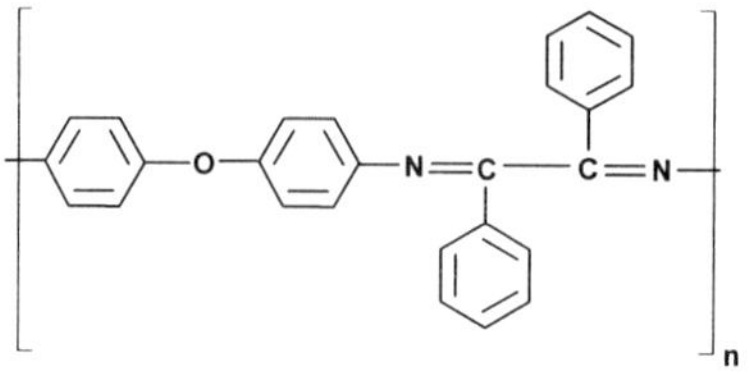

1

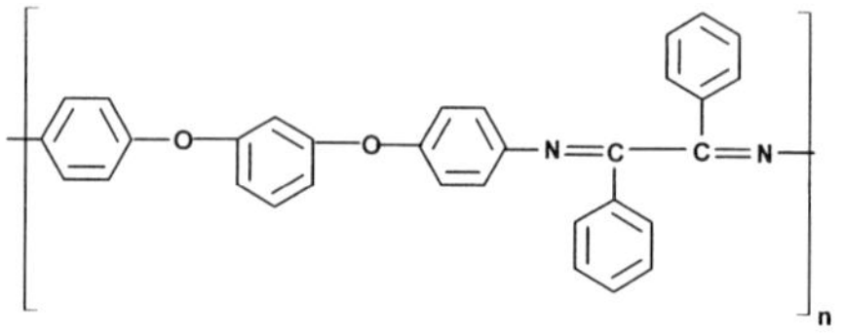

2

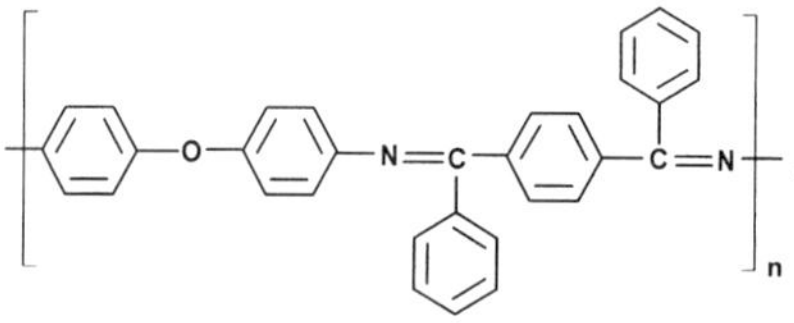

3

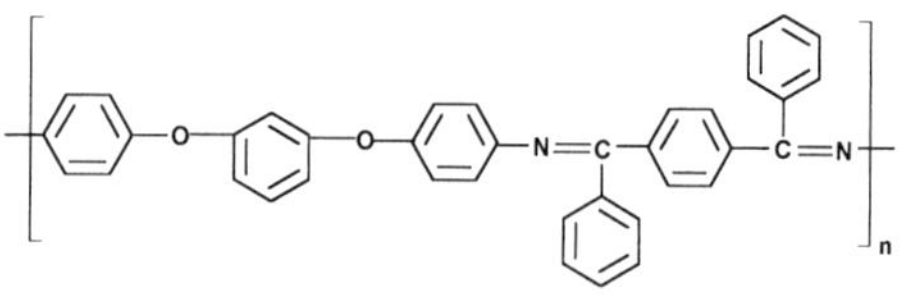

4

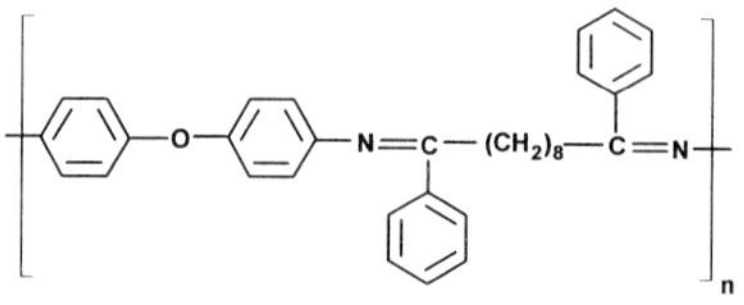

5

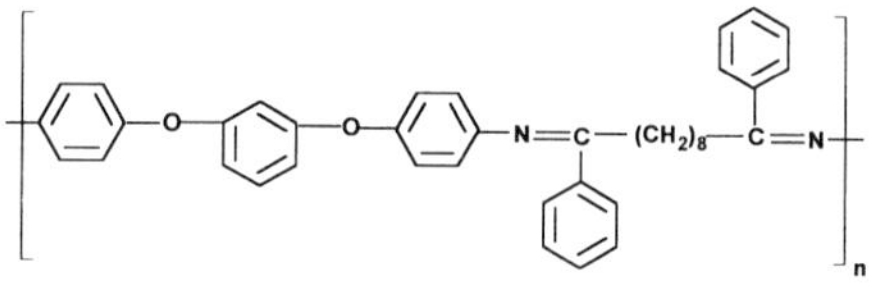

6

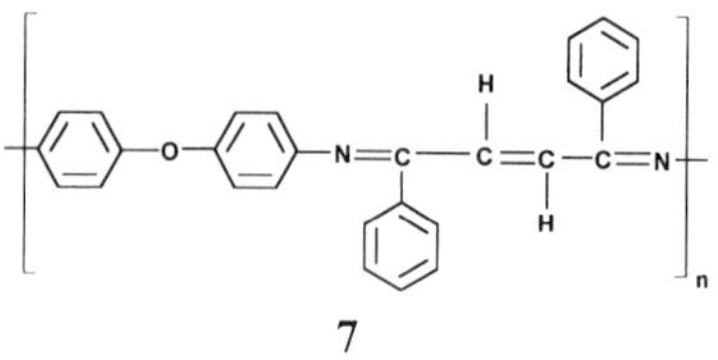

7

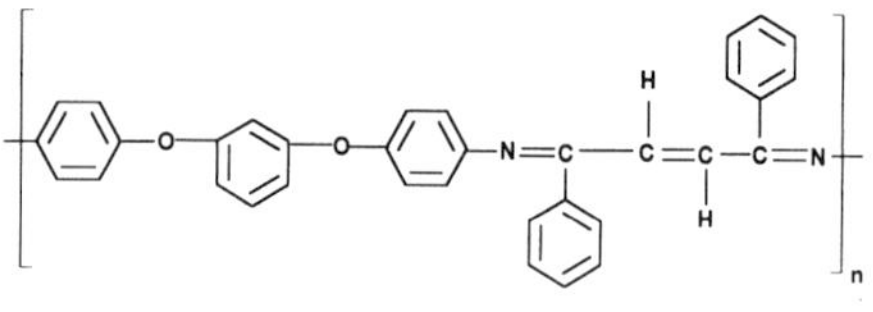

8

Fig. 1. Structure of the polyketanils.

Table 1. Structure and characteristic of the polyketanils.

No	Elemental anal. Found (calcd) [%] C	H	N	FTIR C=N [cm^{-1}]	UV-VIS [nm]	Fluorescence** Emission band [nm]	GPC Mn	GPC Mw (MWD)	Tg (DSC) [^{0}C]	TG in argon Temp of 10% weight loss [°C] (Residue at 1000^{0}C [%])
1	82.33 (83.42)	4.82 (4.81)	7.20 (7.49)	1610	350	**415**	977	1852 (1.9)	145	420 (40)
2	81.33 (82.40)	4.82 (4.72)	5.63 (6.01)	1608	345	425	926	969 (1.0)	70	420 (38)
3	85.07 (85.31)	4.98 (4.92)	6.16 (6.22)	1609	360	**420**	2381	3319 (1.4)	137	500 (58)
4	83.33 (84.11)	4.75 (4.83)	5.05 (5.16)	1606	350	421	3010	3995 (1.3)	118	430 (46)
5	82.55 (83.95)	6.88 (7.00)	5.98 (5.76)	1614	~350*	**453**	1316	2877 (2.1)	124	420 (30)
6	82.90 (83.04)	6.44 (6.57)	4.45 (4.84)	1611	~350*	462	1889	3551 (1.9)	65	390 (29)
7	83.24 (84.00)	4.94 (5.00)	6.61 (7.00)	1610	385	**442**	1091	2312 (2.1)	133	380 (40)
8	82.77 (82.93)	4.86 (4.88)	5.39 (5.69)	1609	385	444	1119	2047 (1.8)	108	420 (38)

*broad with low intensity
**conc. = 10^{-4}mol/l in DMA, exc. = 350 nm

In FTIR spectra an absorption band in the range of 1608 – 1614 cm^{-1} confirmed the presence of C=N bond. At a little shorter wavelength absorb the polyketanils from dioxydianiline than from oxydianiline and the proper diketones. As a matter of fact the C=N absorption band is very closed to the C=C in aromatic ring absorption being observed at 1590 – 1596 cm^{-1} and in some cases these two bands are not good separated but clearly visible. Fig. 2. shows the FTIR spectrum of the polyketanil from oxydianiline and p-dibenzoylbenzene as an example.

An absorption band at 1660 cm^{-1} suggests the presence of ketone groups being at the end of the polymer chains. In ^{13}C NMR spectra of the polyketanils a signal in the range of 166-169 ppm confirms the presence of carbon atom in ketimine group. A low intensity signal observed at 191 ppm (No 7 and 8) and at 198 – 200 ppm (for remain polyketanils) is due to the carboyl carbon atom being end group of the polymer chain.

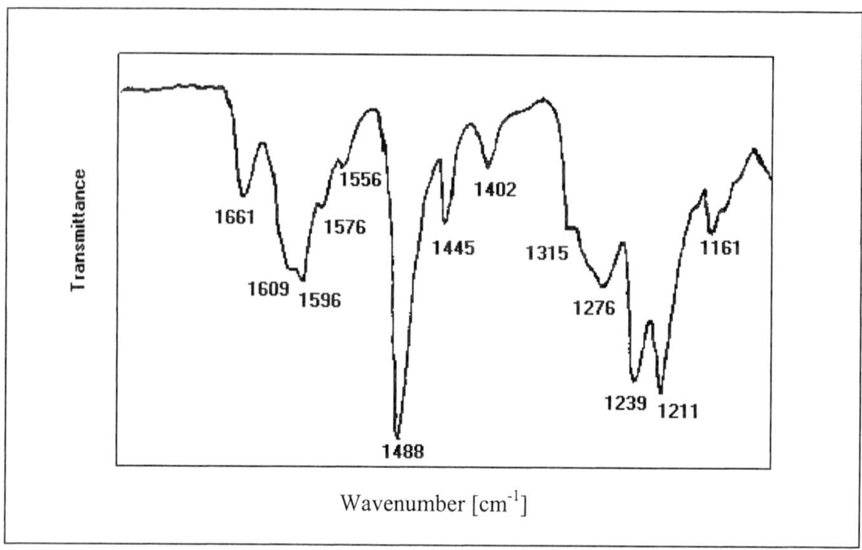

Fig. 2. FTIR spectrum of the polyketanil No 3.

But in the case of ketimine structure the presence of cis and trans isomers is possible even through the trans structure is always supposed to be thermodynamically more stable. A splitting of some signals in ^{13}C NMR spectrum can confirm the presence of isomers. It concerns mainly signals of the quaternary carbon atoms being directly bonded to ketimine carbon. As the diketones used have the different center cores between the carbonyl group the most informative there is the signal from the carbon atom of pendant phenyl rings. Some results are presented in Table 2.

In the polyketanils synthesized from oxydianiline and p-dibenzoylbenzene (No 3 in Table 2) splitting of the signals resulting from the carbon atoms of the pendant benzene rings (4) and the central ring (6) directly connected with ketimine carbon atom (5) confirms the presence of the isomers in the polymer chains. In the spectrum of the polymer from the same diketone and dioxydianiline (No 4) for these carbon atoms the singlets were observed, probably due to existence of the only one isomer. For the other polyketanils investigated the ^{13}C NMR spectra

Table 2. Chemical shift of carbon atoms directly bonded to ketimine carbon atom.

No	Structure	Signal of carbon atom [ppm]				
		4	5	6	7	8
1	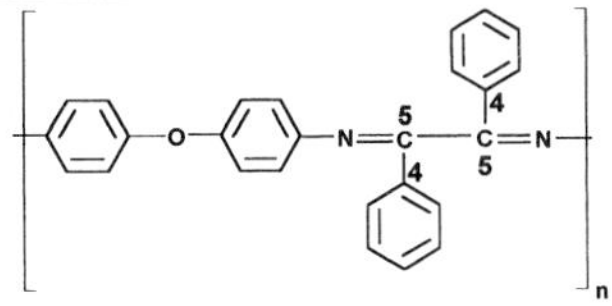	134.56 134.76 135.04 137.01	163.89 163.96 165.75 166.27	-	-	-
2	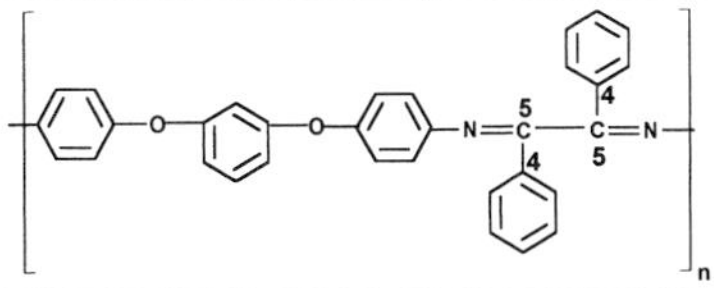	134.32 134.60 135.05	166.46	-	-	-
3	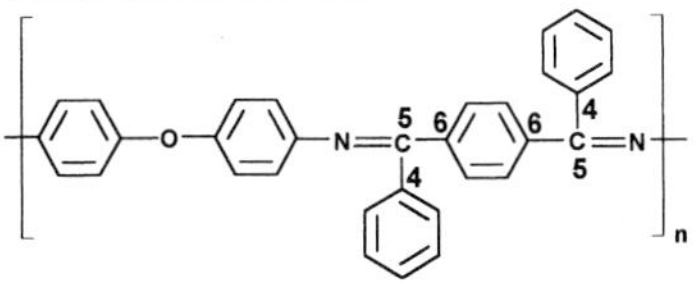	136.02 136.50 137.20 137.50	167.74	138.01 138.60 139.34 139.65	-	-
4	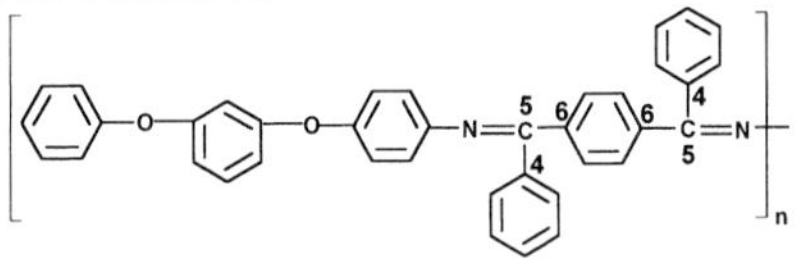	136.53	166.96	137.70	-	-
5	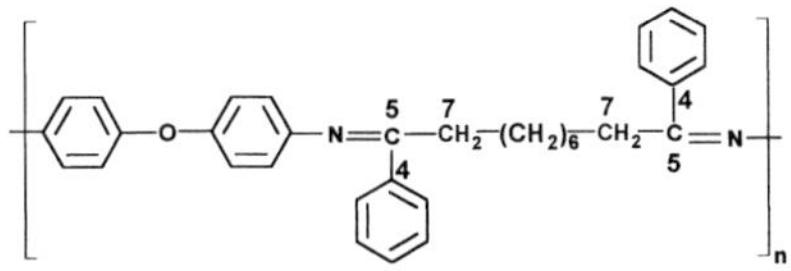	137.05 138.10	170.80 170.78	-	38.58	-
6	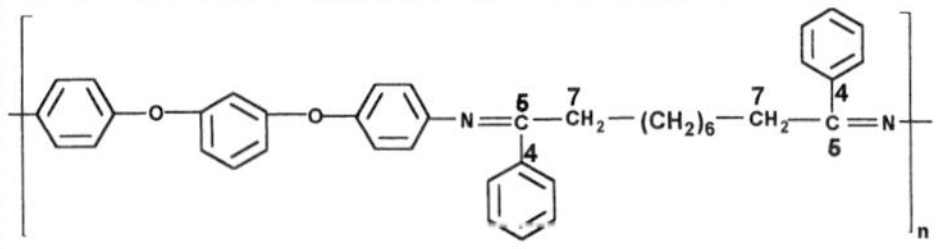	136.97	160.26 160.28	-	38.56	-
7	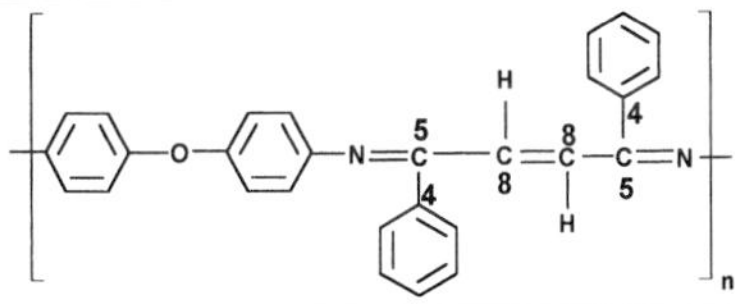	138.94 138.95 139.08 139.11	169.08	-	-	123.70
8	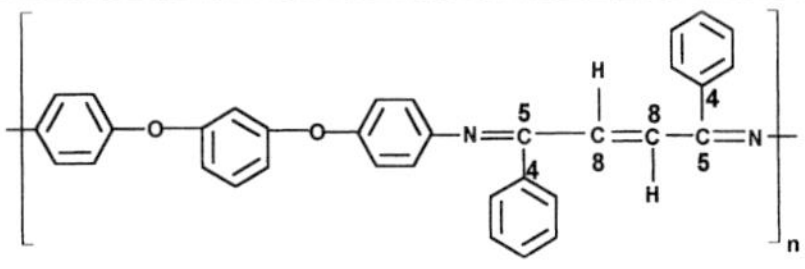	134.32 134.72 135.82 138.42	169.00 169.05 169.20 169.30	-	-	123.74

confirmed the presence of the isomers (detailed study of the whole spectra will be published separately).

Molecular weights of the polyketanils were detected by GPC method. Values of the Mw and Mn are low indicating a formation of oligomers especially in the case of the most stiff diketone used for polycondensation. Never the less these materials form transparent films but some of them are rather brittle. The polyketanils synthesized by melt polycondensation are soluble in many solvents as m-cresol, THF, DMA, chloroform.

Thermal properties of the polyketanils depend on the diketone and the diamine structures as is shown in Table 1. Glass transition temperatures are higher for the polyketanils synthesized from oxydianiline (No 1, 3, 5, 7) than from dioxydianiline (No 2, 4, 6, 8) and the proper diketones. When structures of the diketones are considered one can see that the presence of aliphatic chain caused lowering of the polymers Tg as it was expected. Direct connection of the ketanil groups increases the glass transition temperatures of the polyketanils synthesized from dibenzoyl (No 1 and 2). Comparable values of Tg exhibit polyketanils obtained from oxydianiline and diketones having as a central core phenyl ring (No 3) or carbon-carbon double bond group (No 7). But it is necessary to mention that glass transition temperatures depend also, to some extent, on molecular weight. Our polymers having low molecular weights are probably in the region where Tg may vary with molecular weights. Never the less the found structural relations fit the common rules concerning the influence of stiffness of mer units on glass transition temperatures. And from this point of view the relation for the our materials seems to be true. It is necessary to mention that the polycondensation conditions were not optimized in respect to the diketones used having probably different reactivity.

Results of thermogravimetric analysis show that the polyketanils are thermally stable up to about 400°C.

For the polyketanils photoluminescence was detected in DMA solution. Emission waves were observed in the range of 415 – 460 nm in dependence on the polymer structures (Table 1).

In all the polyketanils synthesized from oxydianiline the emission was observed at a shorter wavelength than for the dioxydianiline independence of the diketone structure. On the other hand if the diketone structure is taking into consideration, the emission at the shortest wavelength exhibits the poliketanil from dibenzoyl while the polyketanils with aliphatic chain in the macromolecules emit light at longer wavelength than the remain ones.

Properties of the Polyketanils after Protonation with 1,2-(di-2-ethylhexyl)ester of Sulfophthalic Acid (DEHEPSA)

Nitrogen atom in ketimine group having free electron pair can be protonated by protonic acids and complexed by Lewis acids. These treatments may influence on electrons delocalization along the polymer chain.

In this work the 1,2-(di-2-ethylhexyl)ester of sulfophthalic acid (DEHEPSA) was used as a protonating agent. But this compound plays a double role i. e. protonates ketimine nitrogen atom and acts as a plasticizer. The protonation effect can be detected as a change of the UV – VIS spectra of the polymers along with the change of color.

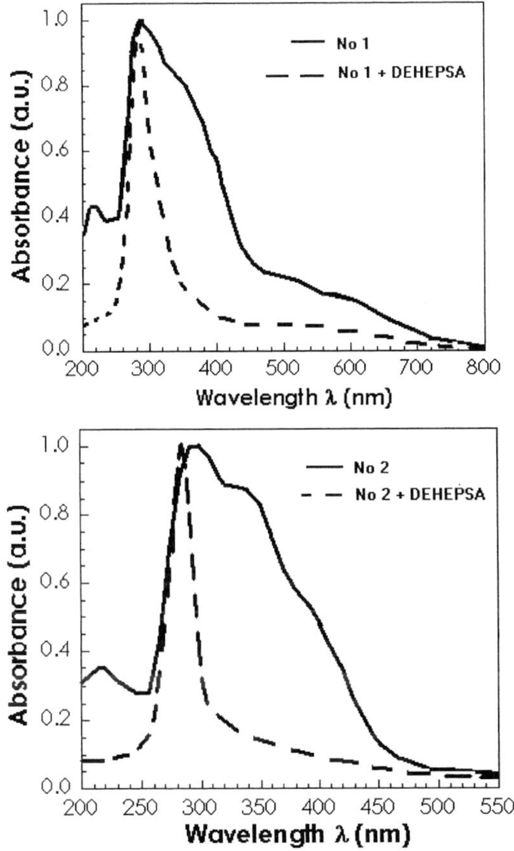

Fig. 3. UV – VIS spectra of the polyketanils No 1 and 2 befor and after protonation.

Fig. 3 shows the UV – VIS spectra of the polyketanils synthesized from dibenzoyl and the diamines (No 1 and 2) before and after protonation. After protonation the absorption band at about 360 nm being responsible for π - π^* transition in ketimine group is not visible. The polymers after protonation change the color to the dipper tone (light brown to brown, yellow-green to brown, respectively).

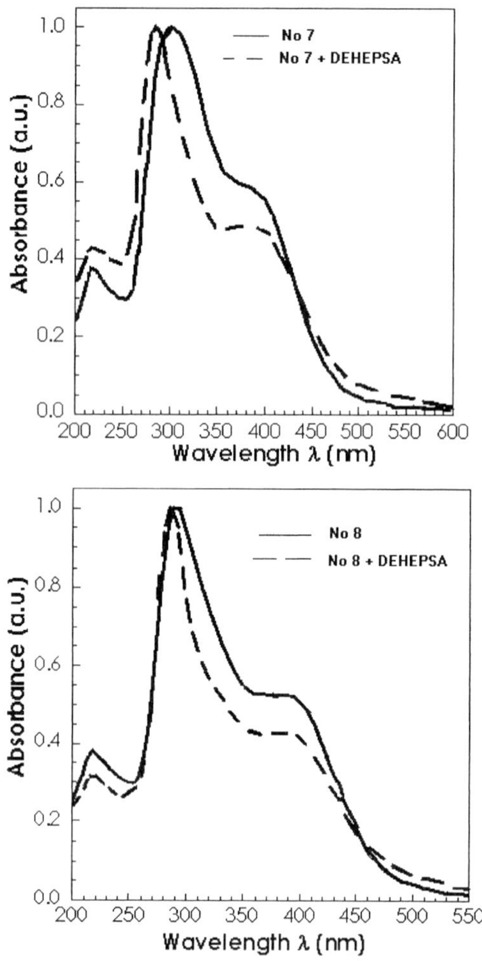

Fig. 4. UV – VIS spectra of the polyketanils No 7 and 8 befor and after protonation.

For the polyketanils from diketone with –C=C- double bond (No 7 and 8), decrease of the absorption intensity at about 385 nm is observed after protonation along with a little higher absorption above 450 nm (change the color from yellow to orange after protonation) (Fig. 4).
In the spectra of the polyketanils with aliphatic chain, asymmetrical absorption band above 300 nm is not structurized. After protonation the narrowing of the band is seen and the absorption is higher at a longer wavelength (about 360 nm). No influence of the diamine structure on the spectra shapes was observed.
Changes in the intensity and shapes of the UV – VIS spectra of the polymers investigated after protonation with DEHEPSA confirm that some delocalization of electrons in the polymers chains took place. Optical spectra before protonation confirmed also the difference in π - electron delocalization in dependence on the polymer structure (Table 1). For example a red (batochromic) shift of 35 nm in λ max was observed while going from the polymer synthesized from oxydianiline and dibenzoyl (No 1) to the one obtained from trans-1,2-dibenzoylethylene (No 7). On the other hand maximum absorption bands of the polyketanils from dioxydianiline and the all diketones exhibited a little blue (hipsochromic) shift in comparison to the ones from oxydianiline.
As it was said before 1,2-(di-2-ethylhexyl)ester of sulfophthalic acid works as a plasticizer. The films after protonation were much more flexible than the virgin ones. Also the glass transitions were observed at lower temperatures. For example the Tg of the polyketanil from oxydianiline and trans-1,2-dibenzoylethylene (No 7) decreased from 133°C to 32°C after protonation while for the polymer from the same diamine and dibenzoyl (No 1) the glass transition temperatures decreased from 145°C to 30°C.

Conclusion

Relationship between structure of a series of soluble polyketanils synthesized from four diketones and diamines containing ether linkages and their thermal and optical properties was investigated.
It was found that the polymers, even if they have rather low molecular weights, exhibit high thermal stability. The polymers emit light in the range of 415 – 460 nm. Protonation of the ketimine groups with 1,2-(di-2-ethylhexyl)ester of sulfophthalic acid (DEHEPSA) caused changes in the UV-VIS spectra that may indicate the changes in delocalization of electrons in

the polymers chains. The protonation compound used acted also as a plasticizing agent and caused the lowering of the glass transition temperatures of the polymers.

Acknowledgements

We thank Prof. A. Proń from CEA in Grenoble for giving us the sample of DEHEPSA.

[1] A. V. Topchiev, V. V. Korshak, U. A. Popov, L. D. Rosenstein, Journal of Polymer Sci., Part C, **1963**, 4, 1305
[2] T. Prot, B. Mendel, Polimery, **1966**, 118
[3] G. F. Alelio, J. M. Hornback, W. F. Strazik, T. R. Dehner, J. Macromol. Sci. – Chem., **1968**, A2(2), 237
[4] P. W. Morgan, S. L. Kwolek, T. C. Pletcher, Macromolecules, **1987**, 20, 729
[5] M. S. Patel, S. R. Patel, J. Polym. Sci.: Polym.Chemis. Editi. **1982**, 20, 1985
[6] A. A. Volpe, L. G. Kaufman, R. G. Dondero, in "Polymers in Space research" Ch. L. Segal, M. Shen, F. N. Kelley, (Eds.), Dekker Marcel, Inc. New York **1970**
[7] A. A. Volpe, L. G. Kaufman, R. G. Dondero, J. Polym. Sci., Chem. **1969**, A3, 1087
[8] L. G. Kaufman, P. T. Funke, A. A. Volpe, Macromolecules **1970**, 3, 358
[9] D. Sęk, Polymer J. **1981**, 13, 13
[10] C. Yeakel, K. Gower, R. S. Mani, Macromol. Chem. **1993**, 194, 2779
[11] M. Higuchi, A. Kimoto, H. Kanazawa, K. Yamamoto, Org. Lett. **1999**, 1, 1881
[12] M. Higuchi, A. Kimoto, S. Shiki, K. Yamamoto, J. Org. Chem. **2000**, 65, 5680
[13] K. Yamamoto, M. Higuchi, S. Shiki, M. Tsuruta, H. Chiba, Nature **2002**, 415, 509
[14] S. Shiki, M. Higuchi, K. Yamamoto, Abstracts of „Polycondensation 2000" Symposium, Tokyo, September, **2000**
[15] J. Dibattista, B. M. Schmidt, A. B. Pachias, H. K. Jr. Hall, J. Polym. Sci. Part A. Polym. Chem. **2002**, 40, 43
[16] H. W. Boone, H. K. Jr. Hall, Macromolecules, **1996**, 29, 5835
[17] T. E. Olinga, J. Fraysse, J. P. Traverers, A. Dufresne, A. Pron, Macromolecules **2000**, 33, 2107
[18] D. Sęk, A. Iwan, H. Janeczek, Polymer Journal, **2002**, Vol. 34, No 12, 924

Macromol. Symp. **2003**, *199*, 467-482

Copolyesters Based on Poly(butylene terephthalate)s Containing Cyclohexyl Groups: Synthesis, Structure and Crystallization

*T. E. Sandhya, C. Ramesh, S. Sivaram**

Division of Polymer Chemistry, National Chemical Laboratory, Pune 411 008, India
E-mail: sivaram@ems.ncl.res.in

Summary: Poly(butylene terephthalate-co-cyclohexylene dimethylene terephthalate) copolymers PBTCT, were synthesized by melt condensation with compositions ranging from 94/08 to 23/77. ^{13}C NMR spectroscopy was used to study the microstructure of the copolyesters and was found to be completely random. The melting temperature, crystallization temperature on cooling, enthalpy of melting and crystallization followed an eutectic behaviour. Thermal and x-ray diffraction studies indicated that the copolyesters in all composition could crystallize. The XRD studies further indicated that PBT rich copolyesters in the range 75 to 100% BT, crystallized in the PBT lattice while the copolyesters rich in PCT having 38 to 100% CT, crystallized in the PCT lattice.

Keywords: cocrystallization, copolyester, copolymerization, sequence analysis, WAXS

Introduction

Polybutylene terephthalate (PBT) is an important thermoplastic polyester widely used in a range of engineering applications.[1] Because of its very easy processing and rapid crystallization, PBT became a rather widely used thermoplastic material.[2] PBT has good electrical and dielectric properties, making it suitable for applications such as electrical and electronic devices. Because of its excellent ensemble of material properties it is a widely used technical polymer in several application fields either alone or in blends with other thermoplastics.[3] The ability to realize many variable and technically potential modifications of PBT as blends and composites has opened a large window for applications in plastics construction, automotive industry, packaging and electrical industry.[4] However PBT does not have the outstanding impact strength which would be desired. It has a low glass transition temperature (T_g) and hence not suitable for applications involving high heat. Poly(1,4-cyclohexylene dimethylene terephthalate) is a semicrystalline polyester having excellent

 DOI: 10.1002/masy.200350939

properties[5,6] like high melting temperature (T_m) 278-318°C and T_g 60-90°C depending on the trans diol content of 50-100%. It requires 300°C for injection molding, too close to its decompostion temperature. The resulting molding parts are brittle and require nucleation as with PET. For this reason this polyester is not used as such as a molding plastic. It is well known that copolymers tend to have better properties corresponding to homopolymers. Copolymerization modifies the crystallization behavior and crystallinity degree which, in turn, is strongly influenced by composition and kind and arrangement of structural units in the chain. Copolyesters of PBT containing 5-20% of 1,4-cyclohexane dimethanol units are reported to possess good impact strength.[7]

Another aspect of the copolyesters is the cocrystallization behavior when the component homopolymers are semicrystalline. Only a few systems have been reported, e.g., poly(3-hydroxybutyrate-co-hydroxyvalerate) where melting temperature and some crystallinity are observed over the entrie range of composition.[8,9,10]

In the present study, poly(butylene terephthalate) (PBT), poly (1,4-cyclohexylene dimethylene terephthalate) (PCT) and poly(butylene terephthalate-co-1,4-cyclohexylene dimethylene terephthalate) (PBTCT) random copolymers were synthesized by melt and solid state polymerization. The sequence analysis was done by ^{13}C NMR spectroscopy. The crystallization behavior of the random copolymers were studied by wide angle X-ray diffraction (WAXD) and differential scanning calorimetry (DSC).

Experimental

Materials

1,4-Dimethyl terephthalate (DMT) (99+%) from Sigma-Aldrich, Inc. was recrystallized from methanol. 1,4-butanediol (BD) (Sigma-Aldrich) was distilled and stored over molecular seives. 1,4-cyclohexanedimethanol (trans/cis 70/30) (CHDM) was dried in vacuum oven before reaction. Tetra isopropyl titanate was distilled under vacuum and used as a solution in dry toluene. Phenol and 1,1,2,2-tetrachloro ethane were used as such.

General Procedure for Synthesis of PBTCT Copolyesters

DMT, BD, CHDM and 0.1 wt% tetraisopropyl titanate were taken in a two neck tube reactor equipped with N_2 gas inlet, short path vacuum distillation adaptor and a spiral trap to collect the distillate. A 1:1.13 ratio of DMT to diol was used in all polymerizations. The transesterification was carried out at 180-210°C for 1 h, followed by another of 4 h at 230-250 °C.

Polycondensation reactions were performed at 250-310 °C and the pressure was slowly reduced to 0.02 mbar over 30 min and isothermally held for 6 h. The reactor was cooled under vacuum and polymer recovered by breaking and cutting into pieces.

Synthesis of PCT by Solid State Polymerization

Dimethyl terephthalate 5.01g (0.026 mol), 1,4-cyclohexane dimethanol 7.06g (0.049 mol) and 0.01 wt% titanium isopropoxide were taken in a tube reactor fitted with N_2 inlet, air condensor and spiral trap to collect the distillate, methanol. The transesterification was done at 190-210°C for 1 h till methanol distillation ceased. Polycondensation was done at 270°C for 30 min and at 290°C under reduced pressure for 15 min. The hot polymer was poured into cold water, filtered and washed several times with hot water. The oligomer was dried in vacuum oven. The oligomer was powdered, coated with titanium isopropoxide in toluene and subjected to solid state polymerization at 220°C for 5 h and at 250°C for 5 h under vacuum(0.02 mbar).

Characterization

Inherent viscosities were measured at 30°C in an automated Schott Gerate AVS 24 viscometer, using an Ubbelohde suspended level viscometer in phenol/1,1`,2,2`-tetrachloroethane(TCE) (60:40 w/w) at a polymer concentration of 0.5 wt%. The PCT prepared by SSP was melted and cooled under vaccum for viscosity measurement as the polymer did not dissolve in the solvent. The X-ray diffraction experiments were performed using a Rigaku Dmax 2500 diffractometer. The system consists of a rotating anode generator and wide-angle powder goniometer and a slit collimated Rigaku SAXS goniometer(model CN2203E.5). The generator was operated at 40 kV and 150 mA. The samples were ground into fine powder and used for the measurements. The samples were scaned between 2θ = 5-35 deg at a speed of 1 deg/min. The calorimetric measurements were done using Perkin-Elmer DSC-7. The samples were heated/cooled at a rate of 10°C/min under nitrogen environment. The melting temperature and heat of fusion were obtained from the heating thermogram and crystallization temperature upon cooling (T_{cc}) from the cooling thermogram.

^{1}H and ^{13}C NMR were performed in a Brucker DRX 500 spectrometer at 25 ± 1°C, operating at 500 and 125 MHz. Polyesters and copolyesters were dissolved in a mixture of deuterated chloroform ($CDCl_3$)/trifluoroacteic acid(TFA-d_1) and spectra were internally referenced to tetramethyl silane. About 12 and 100 mg of sample dissolved in 0.5 ml of solvent for ^{1}H and

^{13}C NMR. For quantitataive ^{13}C NMR analysis of the microstructure, the relaxation delay and spectral width was 2 μs and 250 ppm, respectively., and the relaxation delay was 10s. 1000 FID's were acquired with 3K data points and Fourier transformed.

Results and Discussion

PBT and PBTCT copolymers were prepared by standard melt condensation procedure from dimethyl terephthalate, butanediol and 1,4-cyclohexane dimethanol (trans/cis 70/30) using tetraisopropyl titanate catalyst. The reaction scheme is depicted in Scheme 1. A series of copolyesters were synthesized by changing the ratio of BD/CHDM from 94/6 to 22/78.

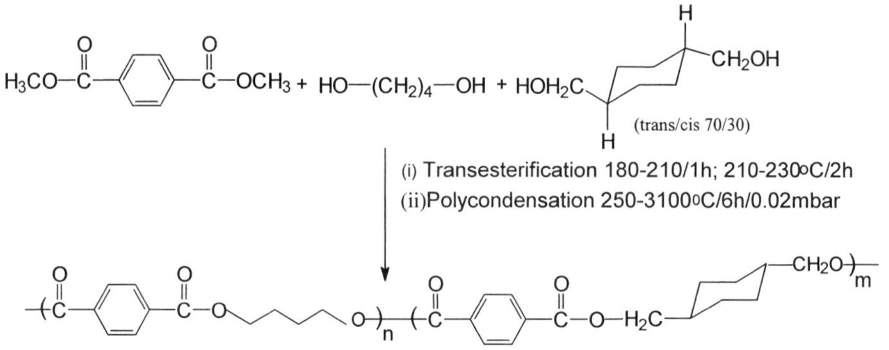

(trans/cis 70/30)

(i) Transesterification 180-210/1h; 210-230oC/2h

(ii)Polycondensation 250-3100oC/6h/0.02mbar

Scheme 1. Synthesis of Poly(butylene terephthalate-co-cyclohexane dimethylene terephthalate).

trans/cis 70/30

(I) Transesterification

(II) Polycondensation

Oligomer

Solid state polymerization

Poly (cyclohexylene dimethylene terephthalate)

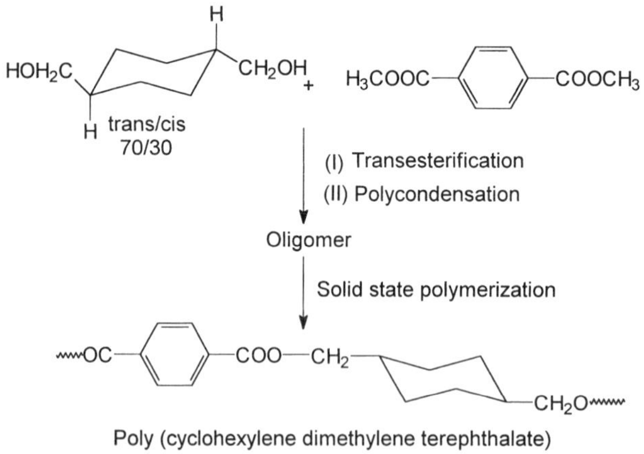

Scheme 2. Synthesis of Poly(cyclohexane dimethlene terephthalate) by melt and SSP.

The copolyesters have viscosities in the range of 0.5-0.9 dL/g and there was no change in the trans/cis ratio of CHDM after the polymerization. However, PCT could not be synthesized by melt method as it underwent degradation. Hence it was synthesised by solid state polymerization[11] (Scheme 2). Oligomer with an inherent viscosity (η_{inh}) of 0.1 dL/g was synthesized by melt condensation and subsequently subjected to SSP to obtain high molecular weight PCT of viscosity 0.66 dL/g. Table 1 gives the SSP conditions and other relevant data.

Table 1. Thermal and viscosity data of PCT samples during SSP.

Sample/conditions	η_{inh} [a] (dL/g)	Melting Peak		Crystallization peak	
		T_m (°C)	[ΔH J/g]	T_c(°C)	[ΔH J/g]
Oligomer	0.10	267	35	258	68
220°C/5.25 h	0.45	297	79	249	55
250°C/5 h	0.66	299	86	246	58

[a]The inherent viscosities were measured after melting the samples under vacuum in 60/40 w/w phenol/TCE mixture

Copolymer Composition and Sequence Analysis

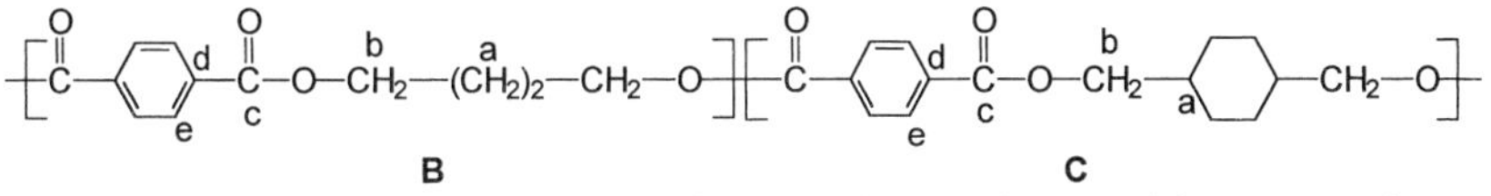

Fig. 1. Chemical structure of PBTCT copolyesters with notations used for NMR assignments.

The chemical structure of PBTCT copolyester with notations used for NMR assignments are shown in Figure 1. The composition of the copolyesters is determined by the ^{1}H NMR and the spectra is shown in Figure 2. The cyclohexylene dimethylene groups of CT unit have two isomers *cis*(equatorial, axial) and *trans* (equatorial, equatorial or axial, axial), the latter (*trans* CT) is more stable than former (*cis* CT). The oxymethylene protons are divided into three groups, viz, BT, *cis* CT and *trans* CT. The copolymer composition of BT, *cis* CT, *trans* CT was estimated from relative peak intensities of oxymethylene proton resonance of butylene and cyclohexylene dimethylene peaks. The ^{1}H and ^{13}C NMR peak assignments of PBT, PCT and PBTCT copolyester is shown in Table 2. The *trans/ cis* ratio determined for all the copolymers are found to be the same as the feed ratio.

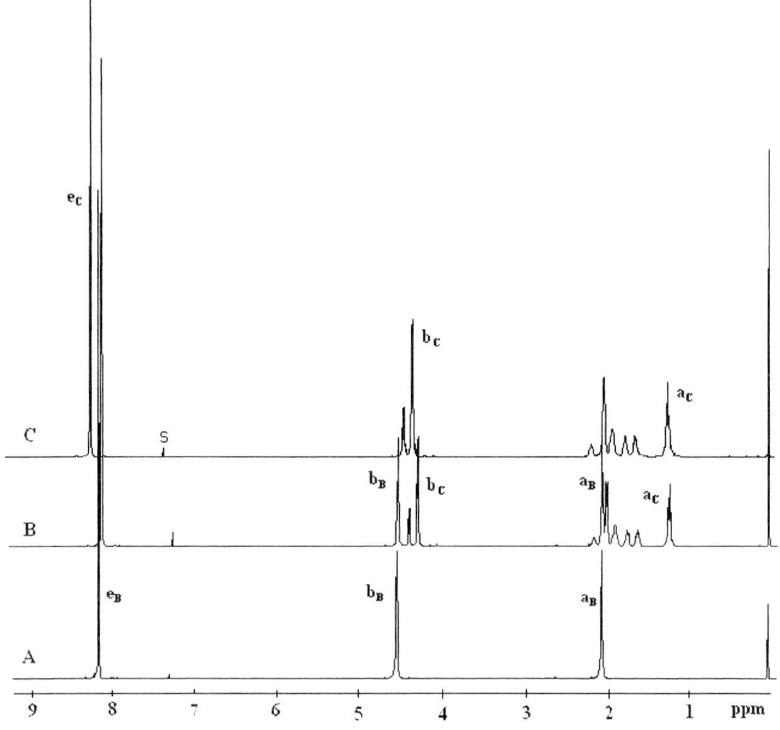

Fig. 2. ^{1}H NMR of (A)BT, (B)$PBT_{45}CT_{55}$ and (C)PCT in $CDCl_3$/TFA-d_1.

Table 2. ^{1}H and ^{13}C NMR Chemical Shifts (δ in ppm) of PBT, PCT and PBTCT copolyesters in $CDCl_3$/TFA-d_1.

				[1]H Chemical shifts		
	^BH_a	^BH_b	^BH_e	^CH_a	^CH_b	^CH_e
PBT	2.02	4.5	8.12			
PCT				1.2-1.99	4.37(d), 4.27(d)	8.15
$PBT_{45}CT_{55}$	2.04	4.52	8.14	1.2-2.00	4.38(d), 4.27(d)	8.15

				[13]C Chemical Shifts				
	^BC_b	^BC_c	^BC_d	^BC_e	^CC_b	^CC_c	^CC_d	^CC_e
PBT	65.9	167.6	133.7	129.8				
PCT					71.3, 69.2	167.9	133.9	129.8
$PBT_{45}CT_{55}$	65.9	167.6	133.7	129.8	71.3, 69.2	167.7	133.9	129.8

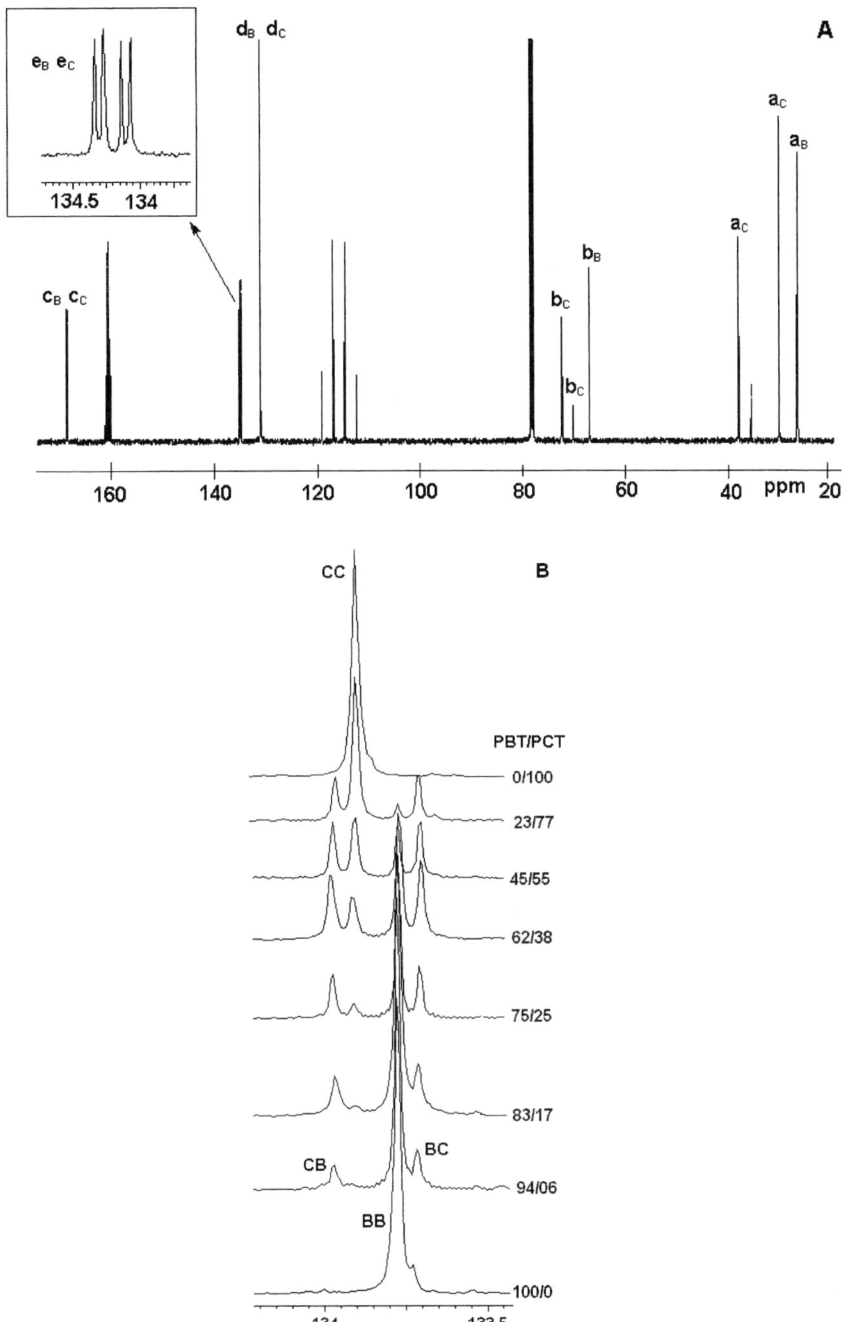

Fig. 3. ^{13}C NMR spectrum of $PBT_{45}CT_{55}$ (A) and quaternary C resonance of PBT, PCT and PBTCT copolyesters (B).

The sequence of the BT and CT units in the copolymer could be determined from the ^{13}C NMR spectra. The quaternary aromatic carbon peak is more sensitive to sequence effects than any other aromatic carbons due to the occurrence of through space and through-bond interactions between neighboring units and was used for sequence quantification.[12,13,14] The spectrum is shown in Figure 3A and the peak assignments is given in the Table 2. The quaternary carbon of the terepthalate (labelled as d) is split into four different signals corresponding to the four dyads, BB(133.8), CC(133.88), BC(133.68), CB(133.95) as shown in Figure 3B and Figure 4. According to Yamadera and Murano,[15] if four kinds of signals due to homolinks and heterolinks are observed in the NMR spectrum of the copolymer, then the average sequence length and the degree of randomness of the copolymer can be determined. The molar fractions of the butylene (B) unit and cyclohexylene (C) units were obtained from integration of the peaks.

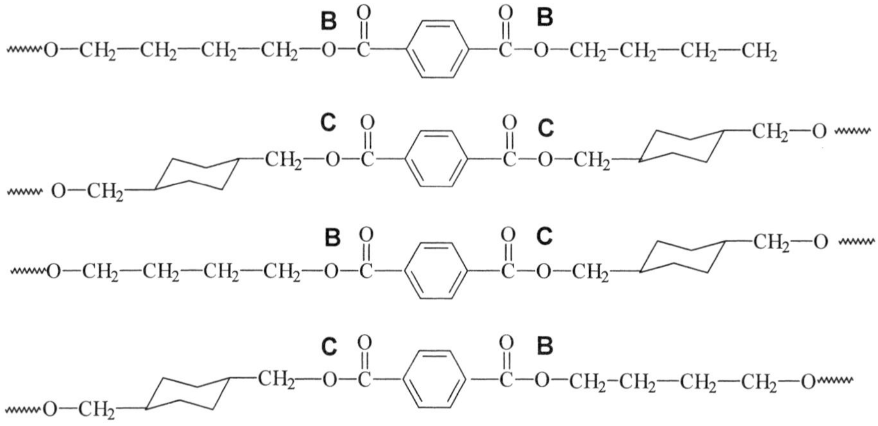

B B
~O—CH2—CH2—CH2—CH2—O—C(=O)—C6H4—C(=O)—O—CH2—CH2—CH2—CH2

C C
~O—CH2—C6H10—CH2—O—C(=O)—C6H4—C(=O)—O—CH2—C6H10—CH2—O~

B C
~O—CH2—CH2—CH2—CH2—O—C(=O)—C6H4—C(=O)—O—CH2—C6H10—CH2—O~

C B
~O—CH2—C6H10—CH2—O—C(=O)—C6H4—C(=O)—O—CH2—CH2—CH2—CH2—O~

Fig. 4. Possible dyad sequences of the quaternary carbon in the PBTCT copolyester.

$$P_B = \frac{f_{BC} + f_{CB}}{2} + f_{BB} \qquad P_C = \frac{f_{BC} + f_{CB}}{2} + f_{CC} \qquad (1)$$

Where P_B is the molar fraction of the butylene unit, P_C the molar fraction of the terephthalate unit and f_{BB}, f_{CC}, f_{BC} and f_{CB} correspond to the proportion of the integrated intensities of BB, CC, BC and CB.

If one could inspect the units along the copolymer chain from one end to the other, the probability of finding a butylene unit placed next to a cyclohexylene unit (or cyclohexylene unit placed next to a butylene unit) would be given by P_{BC} and P_{CB} as given in equation (2)

$$P_{BC} = \frac{f_{BC} + f_{CB}}{2P_B} \qquad P_{CB} = \frac{f_{BC} + f_{BC}}{2} + f_{CC} \qquad (2)$$

The number-average sequence length of butylene terephthalate (BT) and cyclohexylene dimethylene terephthalate (CT) units L_{nB} and L_{nC} repectively and the degree of randomness (B) were calculated using following equations (3) and (4)

$$L_{nB} = \frac{2P_B}{f_{BC} + f_{BC}} \qquad L_{nC} = \frac{2P_C}{f_{BC} + f_{CB}} \qquad (3)$$

$$B = P_{BC} + P_{CB} \qquad (4)$$

For random copolyesters B is unity. A value of B equal to zero indicates a mixture of homopolymers, whilst a value of 2 indicates an alternating distribution.[15] Table 3 shows the average sequence length and degree of randomness. The dyad distribution is also calculated based on Bernoullian statistical model[16] and plotted in Figure 5 along with the experimentally determined values.

Table 3. Sequence distribution and randomness of PBTCT copolyesters determined by ^{13}C NMR.

	Composition[b]		CHDM isomer compostion[b]		Average sequence length		Randomness
Copolyester[a]	X_B	X_C	*cis*	*Trans*	L_{nB}	L_{nC}	B
PBT	100	0	-	-	-	-	-
$PBT_{94}CT_{06}$	93	07	30	70	7.7	1.2	0.95
$PBT_{83}CT_{17}$	84	16	28	72	5.3	1.3	0.98
$PBT_{75}CT_{25}$	74	26	25	75	3.7	1.4	0.99
$PBT_{62}CT_{38}$	61	39	25	75	2.4	1.7	1.01
$PBT_{45}CT_{55}$	45	55	25	75	1.8	2.3	0.98
$PBT_{23}CT_{77}$	22	78	23	77	1.4	5.0	0.93
PCT	0	100	26	74	-	-	-

[a] Experimetnal values were obtained from integration of ^{13}C NMR peaks and ref 16.
[b]calculated from the oxymethylene proton resonance of butanediol and cylohexanedimethanol from ^1H NMR spectra.

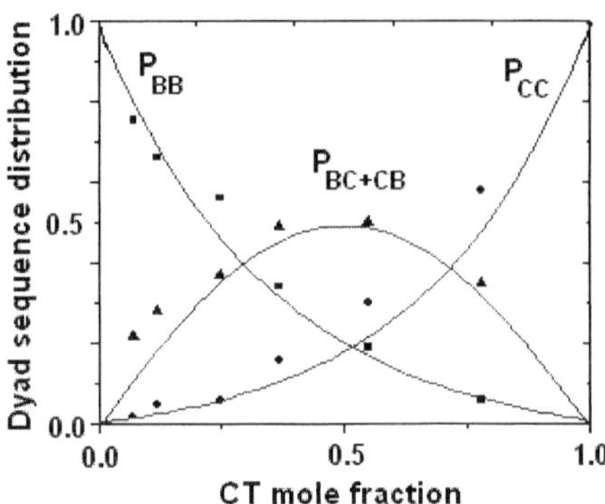

Fig. 5. Dyad sequence distribution as a function of copolymer composition. The solid lines represent the distribution calculated by Bernoullian statistics.

It is observed that the experimentally determined average sequence lengths were in all cases in accordance with that predicted on the basis of ideal copolycondensation statistics with randomness close to unity.[15,17,18]

Crystallization and Melting Behavior of the Copolyesters

The thermal behavior of the copolyesters was studied by differential scanning calorimetry. The copolyesters, in all composition range showed melting endotherm on heating and crystallization exotherm on cooling. The presence of clear melting and crystallization peak indicates cocrystallization behavior of the copolyesters over the entire range of composition. Only few systems show such a behavior.[8,9,10] Figure 6 shows the crystallization exotherms on cooling and the melting endotherms of the copolyesters during the second heating. It may be noted that the as polymerized samples have varying thermal history and hence the thermal properties obtained during the first heating cannot be compared. The various parameters extracted from the thermograms are shown in Table 4. The crystallization, melting and glass transition temperatures of the copolyesters are shown in Figure 7. All the copolyesters have single T_g and show a linear increase with increase in CHDM content. The melting and crystallization temperatures of the copolyesters show eutectic behavior and the eutectic composition is $PBT_{62}CT_{38}$. The typical eutectic behavior indicates that the cocrystallizaion is isodimorphic in nature and monomer units of one type are included in the crystal lattice of the other type.

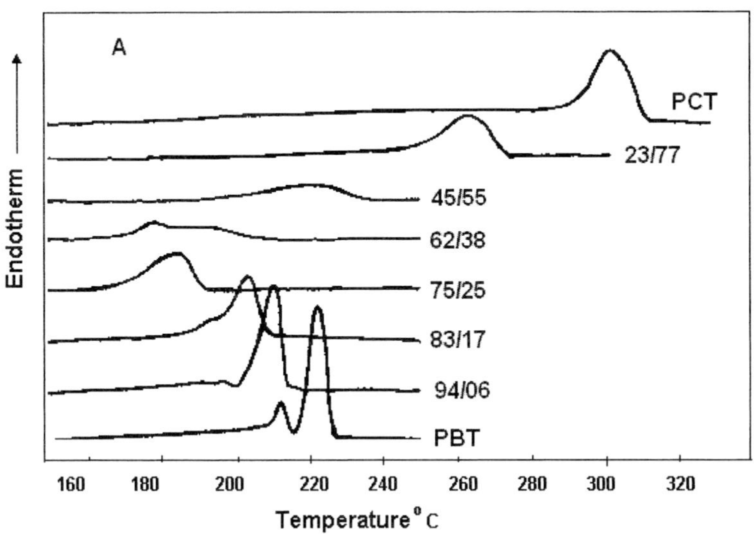

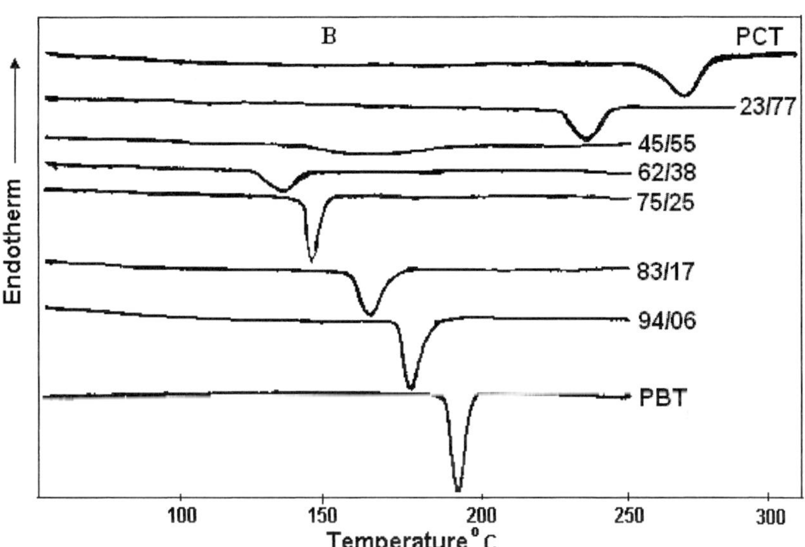

Fig. 6. DSC scans of PBT, PCT and PBTCT copolyesters. (A) second heat (B) first cooling.

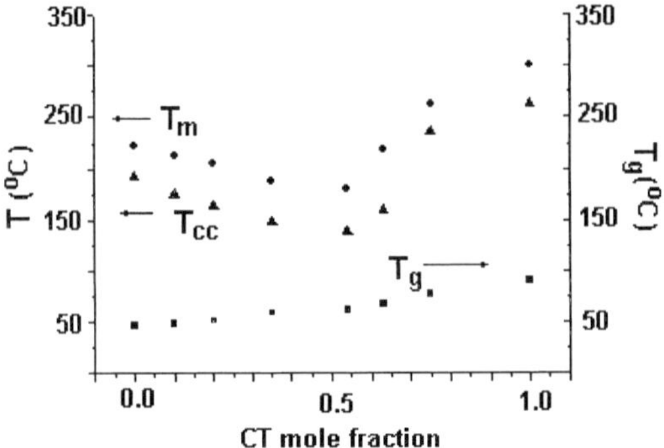

Fig. 7. Melting temperature (T_m), crystallization temperature (T_{cc}) and glass transition temperature (T_g) as a function of comonomer (CHDM) content expressed as molar fraction.

Table 4. Thermal properties of poly(butylene-co-1,4-cyclohexylene dimethyle terephthalate) (PBTCT) copolyesters.

Polyester	η_{inh} dL/g	T_g [a] (°C)	T_m (°C)	ΔH_m (J/g)	T_{cc} (°C)	ΔH_c (J/g)	T_m [b] (°C)	ΔH_m [b] (J/g)
PBT	0.90	40	226	41	193	49	223	54
$PBT_{94}CT_{06}$	0.50	48	212	36	175	45	211	53
$PBT_{83}CT_{17}$	0.57	51	205	34	163	41	206	45
$PBT_{75}CT_{25}$	0.66	59	187	27	149	30	188	37
$PBT_{62}CT_{38}$	0.60	61	180	25	139	26	183	32
$PBT_{45}CT_{55}$	0.75	67	218	19	159	24	210	30
$PBT_{23}CT_{77}$	0.51	77	262	30	237	31	264	40
PCT	0.66	90	296	43	246	58	294	67

[a] Measured by DSC with a heating rate of 10°C/min after quenching from the melt.
[b] Measured from first heating of the annealed samples. Annealing was performed at 15°C below the T_m onset for 2 h.

X-ray Diffraction Studies

The room temperature structure of the samples was analyzed by WAXS. As obtained samples showed diffraction pattern typical of semicrystalline polymers. The diffraction patterns of the samples having compositions close to the eutectic composition are not well resolved. Hence, all the samples were annealed close to their melting temperature for 2 h. After annealing the diffraction peaks are well developed. From the peak position the d-spacings are calculated using Bragg equation. Figure 8 shows the WAXS pattern of the annealed PBT, PCT and PBTCT copolymers. The crystal structure of PBT[4,19] and PCT[20] are reported to be triclinic. PBT shows strong peaks at diffraction angles 16.02, 17.25, 23.28 and 25.16. These peaks are indexed as 010, 101, 100 and 111 planes of triclinic. PCT shows peaks at 15.63, 16.63, 23.41 and 25.6 and are assinged to 011, 010, 100 and 111 planes. The variation of d-spacing with composition shows a break in the region around 30% PCT (Figure 9). The diffraction patterns of copolyesters rich in PBT (74-100% BT) component only show patterns similar to PBT indicating that the copolyesters crystallized in the PBT lattice. When the mole fraction of PCT increases (37-100% CT), the coployesters crystallized in the PCT lattice. The change in the lattice also ocuurs close to the eutectic composition. This shows the ability of PCT to control the crystallization even when it is present in minor proportion. This could arise due to the higher rigidity of PCT when compared to PBT.

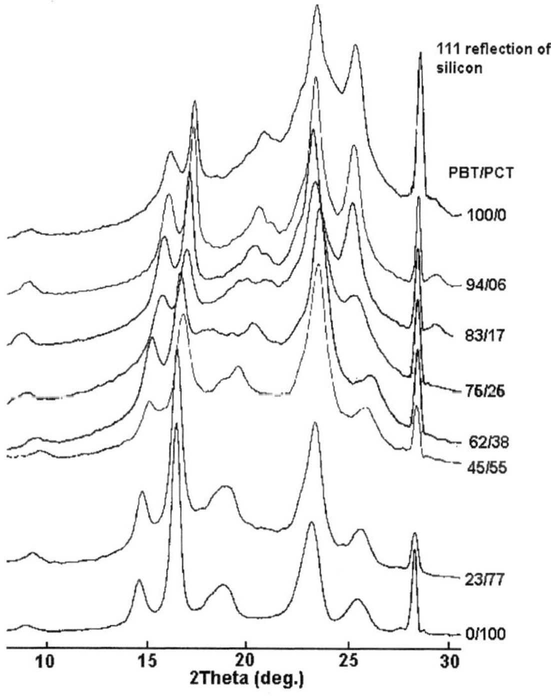

Fig. 8. X-ray diffraction patterns of annealed samples.

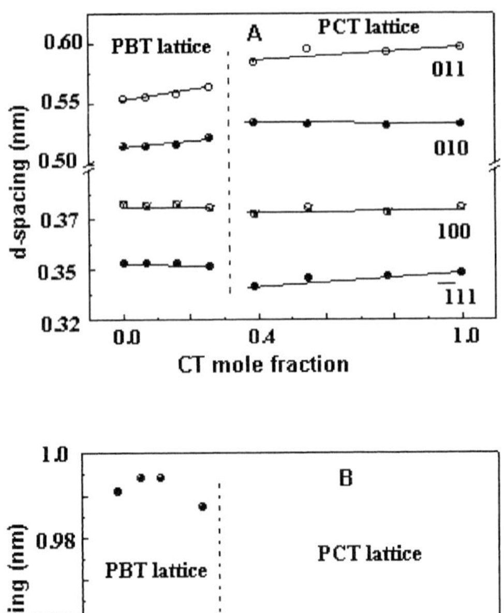

Fig. 9. (A) d-spacigs of 011, 010, 100 and 111 reflections and (B) 001 spacing(along chain direction)as a function of copolymer composition.

According to Jun etal,[21] an average sequence length higher than 3 is required to form crystallites. However, in the present case, copolyesters could crystallize even when the sequence length is less than 3. This indicates similarity in the repeat unit of PBT and PCT. The length of BT and CT units are calculated using conjugate gradient and Newton-Raphson method using Cerius 2 software and found to be similar(1.23 nm for PBT and 1.33 nm for PCT). It appears that similar repeat unit lengths makes the copolyesters to cocrystallize in the same lattice even when the individual sequence length is less than 3.

Conclusions

Poly(butylene terephthalate-co-1,4-cyclohexylene dimethylene terephthalate) (PBTCT) copolymer is synthesized by melt condensation. The NMR spectroscopic analysis indicates that the coploymer is statistically random, irrespective of the composition. The thermal analysis and XRD studies have shown that PBTCT is one of the few copolymers that can be crystallized in all composition. The melting and crystallization from the melt shows typical eutectic behaviour. The eutectic composition is $PBT_{62}CT_{38}$. The T_g shows a linear dependency on the composition and increases with increase in the CHDM proportion in the copolyesters. The copolyesters crystallize in either PBT or PCT lattice depending on the composition, however, only PBT rich compositions favour PBT lattice.

Acknowledgements

TES acknowledges the Council of Scientific and Industrial Research, New Delhi for the award of senior research fellowship.

[1] Jadhav, J. Y.; Kantor, S. W. "*Kirk Othmer Encyclopedia of Chemical Technology*" Vol. 12, 4th edition., John Wiley, NY, **1998.**

[2] East, A. J.; Golden, M. "*Encyclopedia of Chemical Technology*" Vol. 19. John Wiley, **1992**.

[3] Van Berkel, R. W. M.; Van Hartingsveldt, E. A. A.; Van Dersluijs, C. L. "*Handbook of thermoplastics*" Olagoke Okabisi Ed., Marcel Dekker Inc. **1997.**

[4] Radusch, H.-J. "*Handbook of Thermoplastic Polyesters*" Stoyko Fakirov, Ed., Vol.I **2002,** p389.

[5] Schulken, R. M.; Boy, R. E.; Cox, R. H. *J Polym. Sci. Part C* **1964**, 17.

[6] Kibler, C. J.; Bell, A.; Smith, J. G. *J Polym. Sci. Part A*, **1964,** 2, 2115,.

[7] USP 4107150 **(1978)** Philips Petroleum Company, invs.: Campbell, R. W.; Cleary, J. W.

[8] [8a]Bloombergen, S.; Holden, D. A.; Hamer, G. K.; Bluhm, T.L.; Marchessault, R. H. *Macromolecules* **1986**, 19, 2865; [8b] Bluhm, T. L.; Hamer, G. K.; Marchessault, R. H.; Fyfe, C.A.; Veregin, R. P. *Macromolecules* **1986**, 19, 2871; [8c] Kamiya, N.; Sakurai, M.; Inoue, Y.; Chujo, R. *Macromolecules* **1991**, 24, 3888; [8d] Orts, W. J.; Marchessault, R. H.; Bluhm, T. L. *Macromolecules* **1991**, 24, 6435.

[9] [9a] Yoshie, N.; Inoue, Y.; Yoo, H. Y.; Okui, N. *Polymer* **1994**, 34, 1931; [9b] Yoo, H. Y.; Umemoto, S.; Kikutani, T.; Okui, N. *Polymer* **1994**, 34, 117; [9c] Jeong, Y. G.; Jo, W. H.; Lee, S. C. *Macromolecules* **2000** 33, 9705.

[10] [10a]Kunioka, M.; Tamaki, A.; Doi, Y. *Macromolecules* **1989**, 22, 694; [10b] Bloembergen, S.; Holden, D. A.; Bluhm, T. L.; Hamer, G. K.; Marchessault, R. H. *Macromolecules* **1989**, 22, 1663 [10c]VanderHart, D.; Orts, W. J.; Marchessault, R. H. *Macromolecules* **1995**, 28, 6394.

[11] USP 2901466 **(1959)** Eastman Kodak, invs.:Kibler, C. J.; Bell, A.; Smith, J. G.

[12] Martinez de Illarduya, A.; Kint, D P. R.; Munoz-Guerra, S. *Macromolecules* **2000**, 33, 4596.

[13] Abraham, R. J.; Schimperna, G.; Merlo, E. *Macromol. Chem. Phys.* **1990**, 31, 728.

[14] Abis, L.; Pa', R.; Schimperna, G.; Merlo, E. *Macromol. Chem. Phys.* **1994**, 195, 181.

[15] Yamadera, R.; Murano, N. *J. Polym. Sci.* **1967**, 5, 2259.

[16] [16a] Randall, J. C. "*Polymer Sequence Determination*" Academic Press: New York **1977**; p71; [16b] Ibbet, R. N."NMR spectroscopy of polymers'' Blackie Academic & Professional: London, **1993** p50

[17] Korshak, V. V.; Vinogradova, S. V.; Vasner, V. A.; Perfilov, Y. I.; Okulevich, P. O. *J. Polym. Sci., Part A: Polym. Chem.* **1973**, 11, 2209.
[18] Warthem, R.; Schuler, A. S.; Lenz, R. W. *J. Appl. Polym. Sci.* **1979**, 23, 3167.
[19] [19a] Stambaugh, B.; Koenig, J. L.; Lando, J. B. *J. Polym. Sc. Polymer Physics* **1979** 17, 1053; [19b] Kang, H. J.; Park, S.S. *J. Appl. Polym. Sci.* **1999** 72, 593.
[20] Boye, C. A. *J. Polym. Sci.* **1961** 55 275.
[21] H. W.Jun, S. H. Chae, S. S. Park, H. S. Myung, S. S. Im, *Polymer* **1999** 40,1473.

Engineering with Metallo-Supramolecular Polymers: Linear Coordination Polymers and Networks

*S. Schmatloch, U. S. Schubert**

Eindhoven University of Technology and Dutch Polymer Institute, PO Box 513, 5600 MB Eindhoven, The Netherlands
Fax.: 0031 247 4186 E-mail: u.s.schubert@tue.nl

Summary: Here we demonstrate the synthesis of telechelics with different spacer units and different numbers of metal-complexing units, like α-methoxy-ω-(2,2':6',2''-terpyrid-4'-yl)-poly(ethylenoxide)$_{78}$ (**1**), *bis*(2,2':6',2''-terpyrid-4'-yl) di(ethylene glycol) (**2**), *bis*(2,2′:6′,2′′-terpyrid-4'-yl)-poly(ethylene oxide)$_{180}$ (**3**) and *tris*[(2,2′:6′,2′′-terpyrid-4'-yl)-oligo (ethylenoxy-)$_{3.33}$]glycerin (**4**) utilizing 4-chloro-2,2':6',2''-terpyridine. The complexation behaviour of a variety of metal-salts towards the telechelics was studied and different supramolecular architectures were investigated, such as symmetric polymeric complexes and linear coordination polymers. Furthermore, attempts have been undertaken to prepare metallo-supramolecular cross-linked systems.

Keywords: supramolecular structures, telechelics

Introduction

The built-up of supramolecular architectures based on metal-ligand interactions is a very challenging and rapidly growing field of research.[1] Besides the straight forward methodology to create extended macromolecular architectures on the basis of metal complexing telechelics, the metal complex adds a structural as well as a functional benefit to the material. Depending on the utilized metal-salts different association constants and therefore different degrees of polymerization and reversibility are accessible. As a result materials can be synthesized with a tailor made response towards temperature,[2] chelating agents,[3] pH and redox agents.[4] Extended metallo-supramolecular coordination polymers[3, 5-9] can be prepared as well as defined block copolymers.[10] In particular, the utilization of functionalized polymers as supramolecular building blocks is an interesting approach towards high molecular weight polymers.[11-14] The application of *oligo*- or *poly*(ethylene oxides) results in water-soluble coordination polymers. Those systems might serve for applications such as water purification and wastewater treatment,[15] food processing and mining.[16] Water-soluble tailored polymers are also used in the manufacture of cosmetics, the stabilization of colloids and biomedicine.[17] On the basis of amphiphilic block copolymers

 DOI: 10.1002/masy.200350940

metallo-supramolecular micelles[18] and reversible hydrogels can be synthesized. Intelligent microgels from cross-linked, hydrophilic polymers can undergo sudden sol-gel transitions under external influences such as temperature and pH changes.[19] In this way, such systems can release active substances under defined conditions or transmit chemo-mechanical signals.[20] A metallo-supromolecular approach towards cross-linked architectures might lead to enhanced materials for such kind of applications.

Experimental Part

Materials and Instruments

Basic chemicals were obtained from Sigma-Aldrich. Size exclusion chromatography was carried out on BioBeads SX1 columns (CH_2Cl_2). NMR spectra were measured on a Varian Mercury 400 NMR spectrometer. The chemical shifts were calibrated to the residual solvent peaks or TMS. UV/Vis spectra were recorded on a Perkin Elmer Lamda-45 (1 cm cuvettes). The applied solvents are indicated in the experimental section. The used 4'-chloro-2,2':6',2''-terpyridine was donated by the BASF-AG.

Preparation of α-methoxy-ω-(2,2'-6',2''terpyrid-4'-yl)-poly(ethylenoxide)$_{78}$ (1)

534 mg (0.178 mmol) of *α*-methoxy-*ω*-hydroxy-poly(ethylene oxide)$_{78}$ and 49 mg (0.87 mmol) KOH are dissolved in 40 mL of dry DMSO and stirred for 1 h at 60°C. After the addition of 50 mg (0.186 mmol) 4'-chloro-2,2':6',2''-terpyridine the reaction mixture is stirred for additional 24 h at 60°C. The solvent is removed *in vacuo*, 10 mL of water are added and the mixture is extracted four times with 100 mL of chloroform. The combined organic layers were dried over Na_2SO_4. After filtration and evaporation of the solvent, the residue is dissolved in THF and precipitated with diethyl ether. *α*-methoxy-*ω*-(2,2:-6',2''-terpyrid-4'-yl)-poly(ethylene oxide)$_{78}$ is yielded as a colorless solid. Yield: 423 mg, 0.131 mmol, (74%).

^{1}H-NMR (400 MHz, $CDCl_3$): δ_H = 2.60 (t, 3H, J = 6.1 Hz, *H*-PEO), 3.40 (t, 4H, *H*-PEO), 3.64 (m, 410H, *H*-PEO), 4.42 (m, 3H, H-PEO), 7.35 (m, 2H, *H5,5''*), 7.83 (m, 2H, *H4,4''*), 8.07 (m, 2H, *H3',5'*), 8.66 (m, 2H, *H3,3''*), 8.69 (m, 2H, *H6,'6'*).

UV/Vis (CH_3OH): λ_{max}/nm (ε/10^4Lmol^{-1}cm^{-1}) = 239 (2.61), 276 (2.39).

MS (MALDI-TOF, dithranol): $\overline{M}_n$ = 3635, $\overline{M}_w$ = 3659, DPI = 1.01.

$C_{23}H_{27}N_3O$ (361.49): calcd. C 76.45, H 7.48, N 11.63; found C 76.20, H 7.40, N 11.60.

Preparation of bis(2,2':6',2''-terpyrid-4'-yl)-di(ethylene glycol) (2)

500 mg (4.7 mmol) di(ethylene glycol) and 2.60 g (47.11 mmol) KOH are heated in 10 ml dry DMSO 1h at 60°C under argon. 3.03 g (11.31 mmol) 4'-chloro-(2,2':6':2''-terpyridine) are added and heating is continued 50 h at 60°C. After cooling to room temperature the crude product is precipitated by addition of an excess of water. Washing with water and diethyl ether yields 2.60 g (4.57 mmol, 97%) of the product as a colourless powder. m.p. 185°C.

^{1}H-NMR (400 MHz, $CDCl_3$): δ = 8.66 (d, 4H, *J* = 6.00, Hz, *H-6,6''*), 8.56 (d, 4 H, *J* = 8.0 Hz, *H-3,3''*), 8.05 (s, 4H, *H-3',5'*), 7.82 (m, 4H, *H-4,4''*) 7.30 (m, 4H, *H-5,5''*), 4.44 (m, 4H, -C*H*$_2$-O-C*H*$_2$), 4.04 (m, 4H, C_{ter}-O-C*H*$_2$).

UV/Vis (CH_3OH): λ_{max} (ε [10^4 l mol^{-1} cm^{-1})] = 244 nm (4.95), 279 nm (4.71).

Preparation of bis(2,2´:6´,2´´-terpyrid-4'-yl)-poly(ethylene oxide)$_{180}$ (3)

20 g (2.177 mmol) poly(ethylene oxide)$_{180}$ and 1.22 g (21.77 mmol) KOH are dissolved in 200 ml dry DMSO and stirred for 1 h at 60°C. After addition of 1.40 mg (5.225 mmol) 4'-chloro-2,2':6',2''-terpyridin the reaction mixture is stirred additional 72 h at 60°C. The solvent is removed *in vacuo*, 75 mL of water are added and the mixture is extracted four times with 250 mL of chloroform. The combined organic layers are dried over Na_2SO_4. After filtration and evaporation of the solvent, the residue is dissolved in THF and precipitated with diethyl ether. *Bis*(2,2':6',2''-terpyrid-4'-yl)-poly(ethylene oxide)$_{180}$ is yielded as a colorless powder. Yield: 15.92 g, 1.716 mmol, (79%).

^{1}H-NMR (400 MHz, $CDCl_3$): δ_H = 3.48 (m, 4H, *H*-PEO), 3.64 (m, 965H, *H*-PEO), 3.75 (m, 4H, *H*-PEO), 3.82 (m, 4H, *H*-PEO), 3.94 (m, 4H, *H*-PEO), 4.40 (m, 4H, *H*-PEO), 7.33 (m, 4H, *H-5,5´´*), 7.85 (ddd, 4H, *J* = 8.2 Hz, 8.2 Hz, 2.0 Hz, *H-4,4''*), 8.04 (m, 4H, *H-3´,5´*), 8.62 (m, 4H, *H-3,3´´*), 8.63 (m, 4H, H-6,6´´).

UV/Vis (CH_3CN): λ_{max}/nm (ε/10^4 L mol^{-1} cm^{-1}) = 238 (4.58), 274 (4.22).

MS (MALDI-TOF, dithranol): $\overline{M}_n$ = 9280 g/mol, $\overline{M}_w$: = 9380 g/mol, PDI = 1.01.

Preparation of tris[(2,2´:6´,2´´-terpyrid-4'-yl)-oligo(ethylenoxy-)$_{3.33}$]glycerin (4)

1.00 g (1.53 mmol) *tris*[oligo-(ethylenoxy)$_{3.33}$]glycerin and 1.00 g of KOH are dissolved in dry DMSO and stirred for 1 h at 60°C. After the addition of 1.47 g (5.51 mmol) of 4'-chloro-2,2':6',2''-terpyridine the reaction mixture is stirred for further 72 h at 60°C. The solvent is removed *in vacuo* and 10 mL of water are added and the mixture is extracted four times with 50 mL of chloroform. The combined organic layers are dried over Na_2SO_4. After filtration and evaporation of the solvent, the residue is dissolved in THF and precipitated with diethyl ether. *Tris*[(2,2´:6´,2´´-terpyrid-4'-yl)-*oligo*(ethylenoxy)$_{3.33}$]glycerin is yielded as a colorless oil. Yield: 1.94 g, 1.582 mmol, (84%).

^{1}H-NMR (400 MHz, $CDCl_3$): δ_H = 3.61 (m, 28H, *H*-PEO), 3.73 (m, 6H, *H*-PEO), 3.90 (m, 4H, *H*-PEO), 4.37 (m, 6H, *H*-PEO), (m, 6H, *H*-PEO), 7.32 (m, 6H, *H-5,5´´*), 7.84 (m, 6H, *H-4,4''*), 8.04 (m, 6H, *H-3´,5´*), 8.60 (m, 6H, *H-3,3´´*), 8.67 (m, 6H, H-6,6´´).

Preparation of α-methoxy-ω-(2,2':6',2''-terpyrid-4'-yl)-Fe(PF_6)$_2$-poly(ethylene oxide)$_{78}$ (5)

400 mg (116 μmol) of *α*-methoxy-*ω*-(2,2':6',2''-terpyrid-4'-yl)-poly(ethylene oxid)$_{78}$ (**1**) and 10.1 mg (58.2 μmol) Fe(acetat)$_2$ are dissolved in 5 ml MeOH at 60°C under argon and stirred for 48 h. After cooling to room temperature, 37.9 mg (233 μmol) NH_4PF_6 are added. The formed purple precipitate is filtered off and washed consecutively with MeOH. The pure product is obtained after size exclusion chromatography. Yield: 320 mg, 44.3 mmol (on the basis of $\overline{M}_n$ = 7217 g/mol, 76%).

^{1}H-NMR (400 MHz, CD_3CN): δ_H = 3.35 (s, 6H, OC*H$_3$*), 3.63 (m, 546H, *H*-PEO), 4.12 (m, 4H, *H*-PEO), 4.76 (m, 4H, *H*-PEO), 7.09 (dd, 4H, *J* = 6.3 Hz, *H-5,5´´*), 7.18 (d, 4H, *J* = 5.06 Hz, *H-6,6''*), 7.89 (psd. t, 4H, *J* = 7.6, *H-4,4´´*), 8.46 (d 4H, *J* = 8.2 Hz, *H-3,3´´*), 8.50 (m, 4H, *H-3´,5´*).

UV/Vis (CH_3CN): λ_{max}/nm ($\varepsilon/10^4$ L mol^{-1} cm^{-1}) = 244 (3.03), 273 (3.21), 315 (1.92), 360 (0.31), 556 (0.59).

Preparation of bis(2,2':6',2''-terpyrid-4'-yl)-$FeCl_2$-di(ethylene glycol) (6)

250 mg (0.44 mmol) of *bis*(2,2':6',2''-terpyrid-4'-yl)-di(ethylene glycol) are dissolved in 20 mL MeOH under argon and 55.7 mg (0.44 mmol) of $FeCl_2$ are added slowly at room temperature. The reaction mixture is heated 48 h at 60°C, the solvent is removed *in vacuo* and the product is isolated in quantitative yield.

^{1}H-NMR (400 MHz, methanol-d6): δ = 8.88 (s, 4H, *H-3',5'*), 8.75 (d, 4H, J = 8.0 Hz, *H-3,3''*), 7.96 (m, 4H, *H-4,4''*), 7.32 (d, 4H, J = 8.0 Hz, *H-6,6''*) 7.21 (m, 4H, *H-5,5''*), 4.99 (m, 4H, -C*H$_2$*-O-C*H$_2$*), 4.42 (m, 4H, C_{ter}-O-C*H$_2$*).

UV/Vis (CH_3OH): λ_{max} (ε [10^4 l mol-1 cm-1)] = 244 nm (4.39), 275 nm (5.25), 319 nm (3.66), 559 (1.10).

η_{rel} = 1.82 (methanol, c = 20 mg/ml, approx. 2 weight-%).

Preparation of bis(2,2':6',2''-terpyrid-4'-yl)-FeCl$_2$-poly(ethylene oxid)$_{180}$ (7)

300 mg (32.97 µmol) *Bis*(2,2':6',2''-terpyrid-4'-yl) poly(ethylenoxid)$_{180}$ and 4.17 mg (32.97 µmol) FeCl$_2$ were dissolved in 10 mL chloroform and stirred for 14 h at ambient temperature. After the addition of a view drops of methanol the reaction mixture is refluxed for additional 28 h. The solvent is removed *in vacuo* and the product is yielded as a purple solid. Yield: 280 mg, (92%).

^{1}H-NMR (400 MHz, CDCl$_3$): δ = 3.63 (m, 1150H, *H*-PEO), 4.18 (m, 4H, *H*-PEO), 5.82 (m, 4H, *H*-PEO), 7.02-7.90 (m, 8H, *H-5,5'', H-6,6''*), 9.12-9.23 (m, 8H, *H-3,3'', H-3',5'*).

UV/Vis (CH$_3$OH): λ_{max}/nm (ε [10^4 Lmol^{-1}cm^{-1})] = 244 (4.39), 275 (5.25), 319 (3.66), 559 (1.10).

UV/Vis (CH$_3$OH): λ_{max}/nm (ε [10^4 Lmol^{-1}cm^{-1})] = 245 (4.41), 274 (5.07), 559 (0.97).

MS (MALDI-TOF, dithranol): [*bis*(2,2':6',2''-terpyrid-4'-yl) FeCl poly(ethylene oxide)$_{180}$]$^+$, 2x[*bis*(2,2':6',2''-terpyrid-4'-yl) FeCl poly(ethylene oxide)$_{180}$]$^+$, 3x[*bis*(2,2':6',2''-terpyrid-4'-yl) FeCl poly(ethylene oxide)$_{180}$]$^+$, 4x[*bis*(2,2':6',2''-terpyrid-4'-yl) FeCl poly(ethylene oxide)$_{180}$]$^+$, 5x[*bis*(2,2':6',2''-terpyrid-4'-yl) FeCl poly(ethylene oxide)$_{180}$]$^+$, 6x[*bis*(2,2':6',2''-terpyrid-4'-yl) FeCl poly(ethylene oxide)$_{180}$]$^+$.

η_{rel} = 2.41 (methanol, c = 20 mg/ml, approx. 2 weight-%).

Results and Discussion

Synthesis of Telechelics

Herein we demonstrate the preparation of the 2,2':6',2''-terpyridine functionalized telechelics α-methoxy-ω-(2,2'-6',2''terpyrid-4'-yl)-poly(ethylenoxide)$_{78}$ (**1**), *bis*(2,2':6',2''-terpyrid-4'-yl) di(ethylene glycol) (**2**), *bis*(2,2′:6′,2′′-terpyrid-4'-yl)-poly(ethylene oxide)$_{180}$ (**3**) and *tris*[(2,2′:6′,2′′-terpyrid-4'-yl)-oligo(ethylenoxy-)$_{3.33}$]glycerin (**4**). The substitution of several hydroxy-functionalized telechelics with terpyridine moieties was performed applying a literature procedure based on a Williamson type reaction (see Figure 1).[20-24]

The hydroxy-functionalized telechelics were deprotonated applying potassiumhydroxyd in dimethyl sulfoxide. The addition of 4-chloro-2,2':6',2''-terpyridine resulted in a quantitative functionalization. The desired products were characterized via ^{1}H-NMR spectroscopy, UV/Vis and GPC. Complete functionalization could especially be proven by MALDI-TOF-MS.

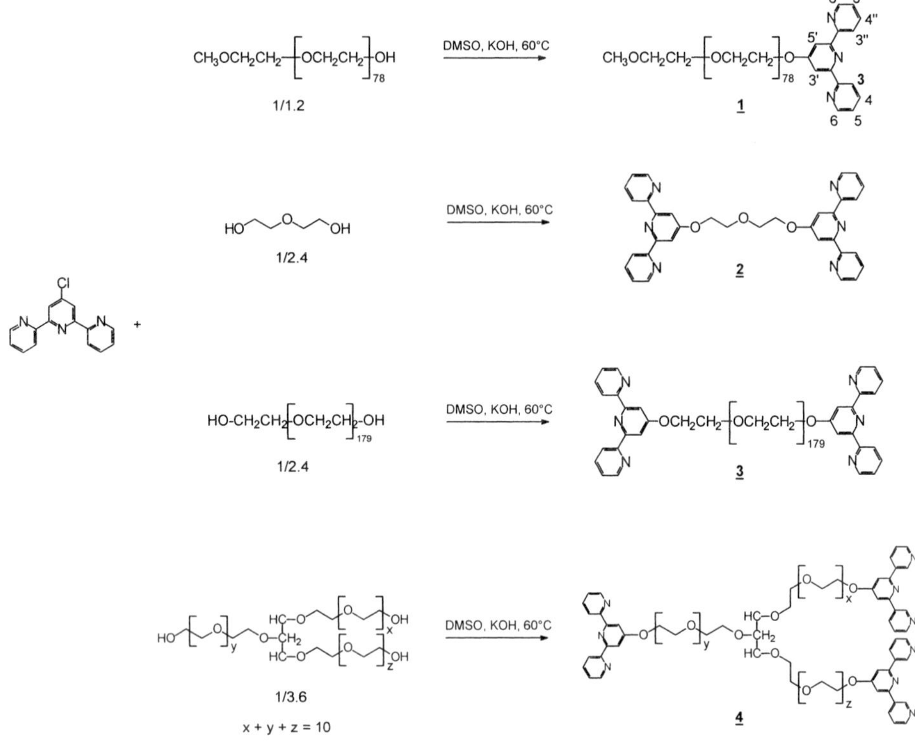

Fig. 1. Synthesis of different 2,2':6',2''-terpyridine-functionalized telechelics: *α-methoxy-ω-(2,2'-6',2''terpyrid-4'-yl)-poly(ethylenoxide)*$_{78}$ (**1**); *bis*(2,2':6',2''-terpyrid-4'-yl)-di (ethylene glycol) (**2**); *bis*(2,2′:6′,2′′-terpyrid-4'-yl) poly (ethylene oxide)$_{180}$ (**3**); *tris*[(2,2′:6′,2′′-terpyrid-4'-yl)-oligo(ethylenoxy-)$_{3.33}$]glycerin (**4**); and *bis*(2,2':6',2''-terpyrid-4'-yl)-poly(propylene oxide)$_{80}$ (**5**).

Synthesis of Model-Complexes

To elaborate suited reaction condition for the formation of metallo-supramolecular coordination polymers on the basis of *bis*(2,2':6',2''-terpyrid-4'-yl)-di(ethylene glycol) and *bis*(2,2′:6′,2′′-terpyrid-4'-yl)-poly(ethylene oxide)$_{180}$ and model complexes were synthesized. In particular iron(II) complexes of the polymeric ligand *α*-methoxy-*ω*-(2,2'-6',2''-terpyrid-4'-yl) poly(ethylene oxid)$_{78}$ (**1**) have been under investigation. Iron(II) was chosen due to the non-toxicity, the low cost, the considerably high binding constants towards 2,2':6',2''-terpyridine and the redox properties of iron(II) that facilitates the "switchability" of the systems.

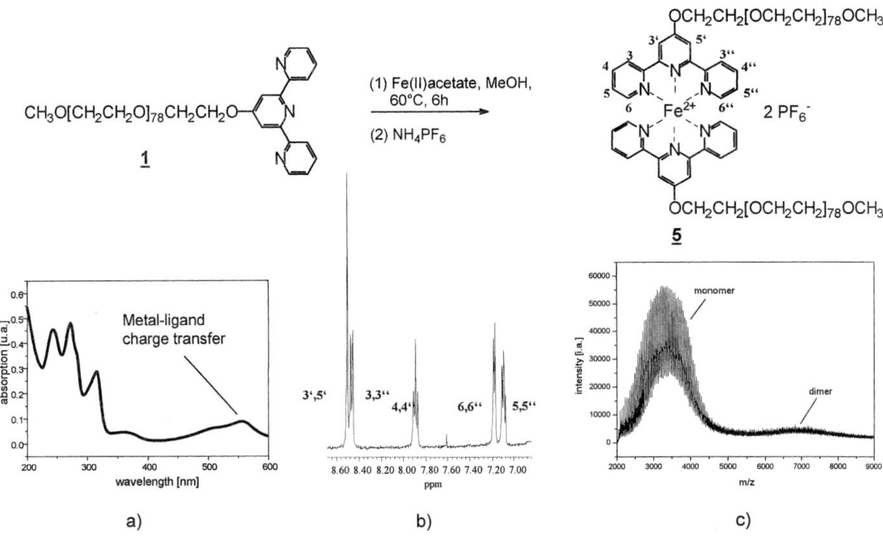

Fig. 2. top: Schematic representation of the formation of the model-complex α-methoxy-ω-(2,2':6',2''-terpyrid-4'-yl)-Fe(PF_6)$_2$-poly(ethylene oxide)78 (5); bottom: Characterization of 5: a) the aromatic region of the ^{1}H-NMR spectrum ($CDCl_3$) shows the typical signals of the complexed terpyridine-unit; b) the UV/Vis spectra shows the typical metal-ligand charge transfer band; c) besides the mass of the monomer, the MALDI-TOF-MS shows the peak with a mass corresponding to the complex.

Addition of $FeCl_2$ under inert conditions to a solution of 1 in methanol resulted in an immediate complex formation, indicated by a colour change. An inert atmosphere was applied in order to avoid an oxidation to Fe^{3+}. After completion of the reaction the chloride counter-ions were exchanged by hexafluorophosphate. The crude product could be successfully purified utilizing size exclusion chromatography (Figure 2).

The formation of the complex could be demonstrated via UV/Vis- and ^{1}H-NMR-spectroscopy and MALDI-TOF-MS. The UV/Vis spectrum of 5 shows a typical metal-ligand-charge transfer band at 556 nm. The bathochromic shift of the π-π* band also indicates the complex formation (Figure 2a). The shift of the terpyridine signals in the aromatic region in the ^{1}H-NMR-spectrum gives further evidence for the formation of the complex. Due to the formation of an octahedral complex geometry the pyridine rings flip from an antiperiplanar into a synperiplanar conformation. Therefore in particular the protons in 6,6'' position are shifted to higher fields and the protons in 3',5' position are shifted to lower fields (Figure 2b). The MALDI-TOF-MS spectrogram of the product shows to different peaks. One corresponds with the molecular weight of the desired complex, the other one with the mass of the uncomplexed ligand (Figure 2c). The latter one is of higher intensity. In particular ^{1}H-NMR-spectroscopy revealed, that there is no free ligand present in the metal complex. Therefore a rupture of the complex under MALDI-TOF-MS measurement conditions seems to be likely.

Synthesis of Linear Coordination Polymers

Reaction at Room Temperature

The complexation behavior of *bis*(2,2′:6′,2′′-terpyrid-4'-yl)-poly(ethylene oxide)$_{180}$ (**3**) towards a wide range of metal-salts were investigated at ambient temperature. Viscosity measurements revealed for all utilized metal-ions (bivalent) a maximum value after the addition of an equivalence of metal-ions. (Figure 3). However, the absolute values for the maximum vary for different ions. The relative viscosity reaches values of η_{rel} = 1.55 with cadmium(II)actetate, η_{rel} = 2.22 with copper(II)actetate, η_{rel} = 2.37 with cobalt(II)acetate, η_{rel} = 2.60 with nickel(II)acetate and η_{rel} = 2.75 with iron(II)acetate. The different maximum values of the viscosities are in agreement with available data for the thermodynamic stability of the formed *bis*complexes. The LogK values for the complex formation of the iron(II)*bis*complex is 13.8, for the nickel(II)*bis*complex is 11.1 and for the Co(II)*bis*complex is 9.9.[25] For other metal-ions there are no data available for the *bis*complexes.

The behaviour of the polymer solutions upon addition of an excess of metal ions are significantly different. It results in an immediate decrease of the viscosity in the cases of cadmium(II)acetate, copper(II)acetate and cobalt(II)acetate. For iron(II)acetate the decrease is much less pronounced and for nickel(II) a plateau is reached, that is almost independent from the amount of added metal-salt. In part, the experimental findings correlate with the thermodynamic stability of the *bis*complexes and available kinetic data. Iron(II)*bis*complexes are known to be kinetically more stable towards ligand exchange (k = 60 min^{-1}), than nickel(II) (k = 780 min^{-1}) and cobalt(II) (k = 9800 min^{-1}) complexes.[26] Therefore, especially the viscosity drop upon addition of an excess of metal-ions for the kinetically very labil cobalt(II)terpyridine coordination polymer is in agreement with the data.

Moreover, for cadmium(II)acetate, copper(II)acetate and cobalt(II)acetate the slope seems to be almost linear, whereas for nickel(II)acetate and iron(II)sulfate a exponential increase of the viscosity can be obtained. Even more striking is the difference between iron(II)chloride and iron(II)sulfate (Figure 4). The complexation or polymerization behaviour respectively seems to be strongly influenced by the utilized counter-ion. For chloride as counter-ion a linear increase of the viscosity with rising metal-ion content can be obtained, whereas for sulfate an exponential increase can be obtained. A strong influence of the counter-ions on the rates of formation as well as the stabilities of *mono-* and *bis*complexes can be derived from that observation. However, those phenomena have to be investigated further in detail.

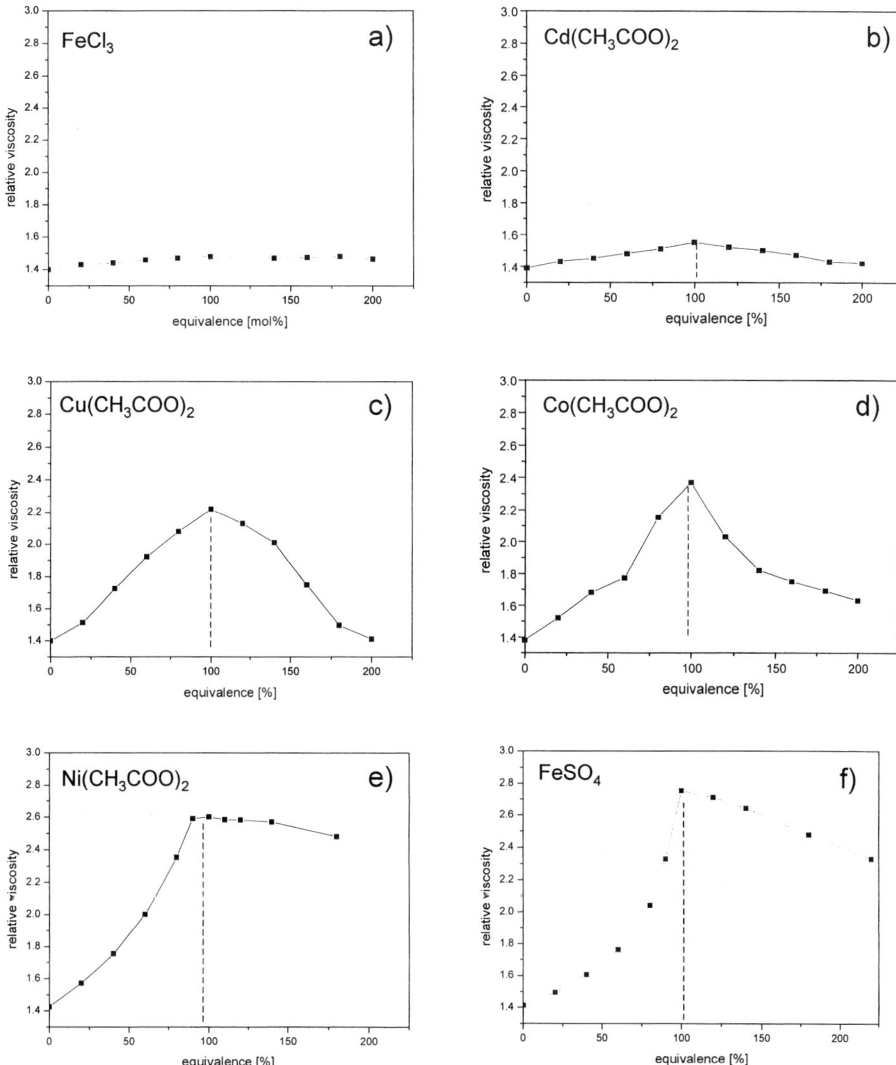

Fig. 3. Viscosimetry experiments: The stepwise addition of a) Cd(II)acetate, b) Cu(II)acetate, c) Co(II)acetate, d) Ni(II)acetate and Fe(II)acetate to a polymer solution (c=100 mg/5 mL methanol) leads to different maximum values of the relative viscosity for an equivalent addition of the metal salt; f) the addition of Fe(III)chloride does not result in a significant increase of the relative viscosity.

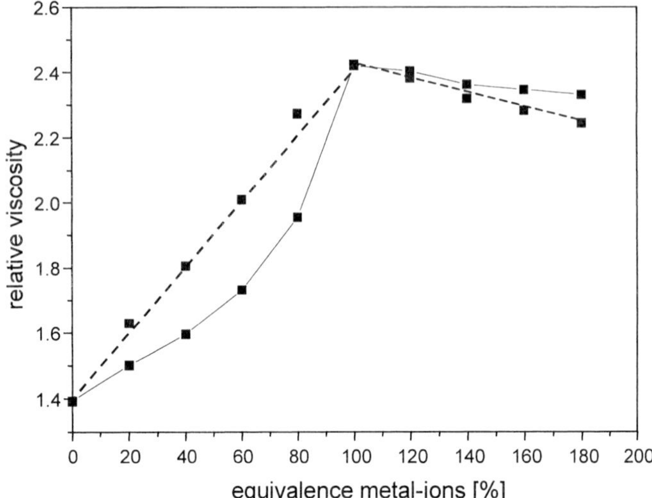

Fig. 4. Influence of the counter-ion onto the complexation/polymerization behaviour of *bis*(2,2′:6′,2′′-terpyrid-4'-yl)-poly(ethylene oxide)$_{180}$: linear increase of the viscosity for chlorine (dotted line) and exponential increase of the viscosity for sulfate (straight line) as counter-ion.

Reaction at Elevated Temperature (Short Spacer and Long Spacer)

The elaborated reaction conditions for the synthesis of the model complex **5** were applied to the synthesis of Fe^{2+} coordination polymers on the basis of the telechelic *bis*(2,2':6',2''-tepyrid-4'-yl) di(ethylene glycol) (**2**) (see Figure 5). In order to obtain water-soluble coordination polymers an exchange of the chloride counter-ions was not undertaken. The quantitative formation of the complex was proven via UV/Vis-, ^{1}H-NMR-spectroscopy and MALDI-TOF-MS. The UV/Vis spectrum shows a metal-ligand-charge transfer-band as well as a bathochromic shift of the π-π* band (see Figure 5a). In addition, the complex formation was monitored by UV/Vis-titration experiments. They show a linear growth of the extinction coefficient of the metal ligand charge transfer (t_{2g}-π*) band at 559 nm with increasing metal-ion content. After addition of 100 mol% Fe^{2+} a maximum for ε is reached, which demonstrates the full conversion. Moreover, that experiment gives evidence of a complete *bis*functionalization of telechelic **1**.

The ^{1}H-NMR-spectrum (Figure 5b) shows the significant shifts of protons 3',5' and 6',6'' after complexation as they are described above. Based on UV/Vis- and ^{1}H-NMR-spectroscopy a quantitative complex formation can be concluded. Attempts were undertaken to determine the molecular weight of the metallo-supramolecular coordination polymer with MALDI-TOF-MS. The main fragments were the free ligand [*bis*(2,2':6',2''-terpyrid-4'-yl) di(ethylene glycole)]$^+$ and a *mono*complex with one counter-ion attached to it [*bis*(2,2':6',2''-terpyrid-4'-yl) FeCl di(ethylene glycole)]$^+$. No higher oligomers could be detected. The observed fragmentation is in agreement with the results of the model-complex α-methoxy-ω-(2,2':6',2''-terpyrid-4'-yl)-Fe$(PF_6)_2$-poly(ethylenoxide)$_{78}$ (**5**).

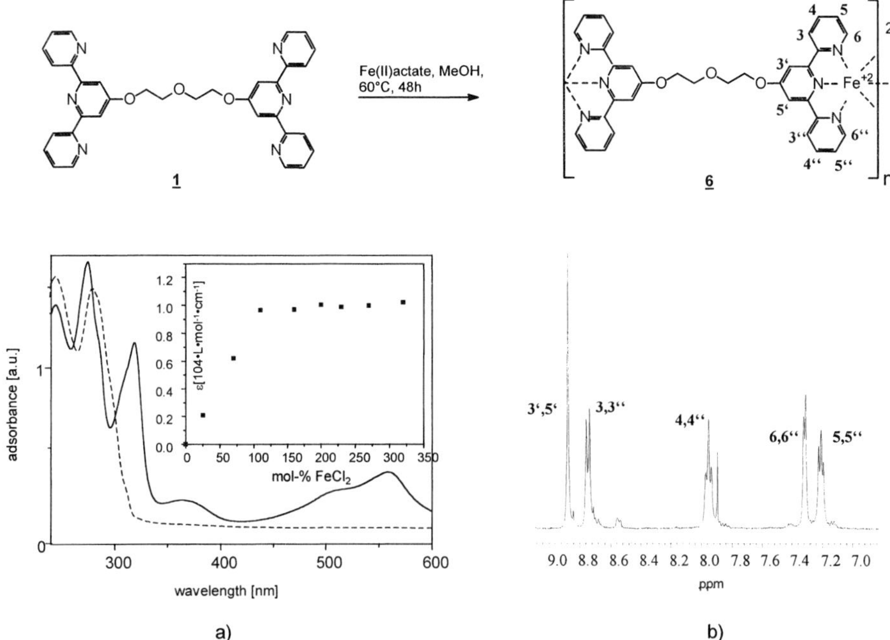

Fig. 5. top: Schematic representation of the complexation of telechelic **1** to form the coordination polymer **6**; a) the UV/Vis spectra shows the typical metal-ligand charge transfer band; inset: UV/Vis titration experiments reveal a maximum value for the extinction coefficient after an equimolar addition of $FeCl_2$; b) the aromatic region of the ^{1}H-NMR spectrum (CD_3OD) shows the typical signals of the complexed terpyridine-unit.

In order to investigate the capability of opening and closing the metal-ligand bond and therefore the reversibility of the coordination polymers several parameters were investigated, such as temperature and the influence of chelating agents. The experimental outcome was monitored via ^{1}H-NMR, UV/Vis spectroscopy and AFM measurements.

First investigations on the temperature sensitivity of Fe^{2+} *bis*(2,2':6',2''-tepyrid-4'-yl) di(ethylene glycol) coordination polymers were undertaken. Upon heating the purple colour of a drop-casted film of **2** slightly fades at about 210°C. Heating to 270°C does not results in a complete disappearance of the colour, but it leads to the degradation of the polymer. For other terpyridine-containing polymers a cleavage of the Fe^{2+}-ligand bond could be already obtained at temperatures of about 160°C.[2] The higher stability of the coordination polymer 6 might be due to the high charge density that might lead to very stable, net ionic structures in the bulk material. Heating up to 210°C and cooling down to room temperature within 30 min, does not lead to the reappearance of the purple colour of the metal-complexes. After 12 h at room temperature the opened complexes reform, which can be monitored by the reappearance of the purple colour. Experiments with a competitive ligand HEDTA revealed the full reversibility of the polymer formation.[3]

A similar reaction procedure as for **5** and **6** has been applied for the complexation/polymerization of *bis*(2,2′:6′,2′′-terpyrid-4'-yl)-poly(ethylene oxide)$_{180}$ (Figure 6, top image). Chloroform was used as solvent to achieve an exposed position of the complexing moieties in solution rather than an integration into the coiled polymer matrix. To avoid oxidation to iron(III) the reaction was performed under inert conditions. The complexation/polymerization could be demonstrated by UV/Vis measurements and ^{1}H-NMR spectroscopy. Besides the typical chemical shifts of the octahedral terpyridine complexes the broadening of the signals reveal the formation of extended coordination polymers (Figure 6a). Moreover, the MALDI-TOF-MS spectrogram shows peaks matching to the masses of oligomers up to six repeating units (Figure 6b). Higher oligomers could not be obtained for similar reason as stated for the compounds 5 and 6. The low values of the relative viscosities of $\eta = 2.41$ for 7 in comparison with $\eta = 1.85$ for the coordination polymer *bis*(2,2':6',2''-terpyrid-4'-yl)-iron(II)-chloride-di(ethylene glycol) (6) are in agreement with other randomly coiled polymers. The film forming properties of the material from methanol solution (in contrary to the free telechelic **3**) reveals as well its polymeric character.

As known from literature and indicated by viscosity measurements (see Figure 4a), iron(III) tends to form *mono*complexes. Oxidation of the metal centers of 7 therefore should result in a degradation of the formed polymer and reduction should on the other hand lead to anew complexation/polymerization. Attemps were undertaken to switch between the metallo-supramolecular coordination polymer and monomers units purely by oxidizing the complex bound iron(II) at air upon heating. The process was monitored by ^{1}H-NMR spectroscopy (Figure 6c). After fourteen hour of refluxing under air clearly the signals of the free ligand (approx δ – values 8.65 and 8.10) are approaching which corresponds to the breaking-up of the iron(II) terpyridine *bis*complexes upon oxidation. Utilizing *N*-ethylmorpholine as a reducing agent the reformation of the complexes/polymers can be achieved (see dissapearance of the free ligand peaks in the ^{1}H-NMR). The experimental data demonstrate the feasibility of the concept. However, different methodologies for the redox processes have to be investigated in order to achieve a quantitative "switching" under milder conditions.

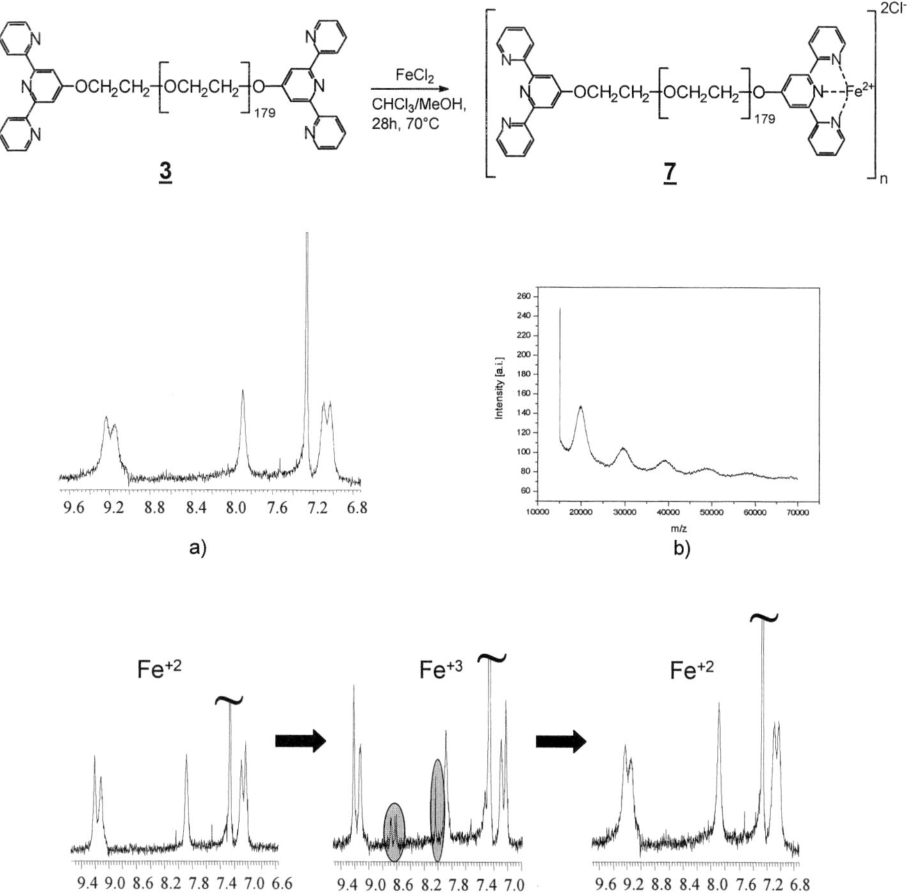

Fig. 6. top: Schematic representation of the complexation of telechelic 3 to form the coordination polymer 7; a) the aromatic region of the ^{1}H-NMR spectrum ($CDCl_3$) shows the typical signals of the complexed terpyridine-unit; the broadening of the peaks reveal the polymeric character of the material; d) MALDI-TOF-MS spectrogram shows oligomers up to six repeating units; c) appearance/disappearance of the signals of the free ligand in the aromatic region of the ^{1}H-NMR spectrum ($CDCl_3$) upon oxidation/reduction of the central metal-ion.

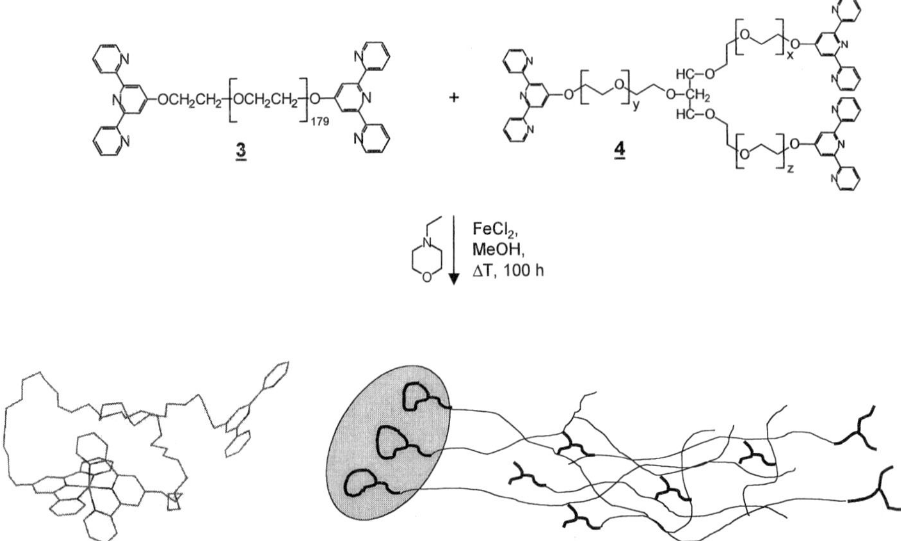

Fig. 7. Schematic representation of the formation of a metallo-supramolecular cross-linked architecture (bottom right) utilizing *bis*(2,2′:6′,2′′-terpyrid-4'-yl) poly(ethylene oxide)$_{180}$ (3) and *tris*[(2,2′:6′,2′′-terpyrid-4'-yl) oligo (ethylenoxy-)$_{3.33}$]glycerin (**4**). Molecular modeling of the intramolecular complexed cross-linker (bottom left).

In order to create metallo-supramolecular cross-linked architectures, attempts were undertaken to incorporate the *tris*functionalized cross-linker 4 into the polymer-matrix of *bis*(2,2':6',2''-terpyrid-4'-yl)-iron(II)-chloride-poly(ethylene oxid)$_{180}$. Initial experiments revealed that a built-in at room temperature is not feasible. However, a quantitative cross-linking can neither be obtained at elevated temperatures (Figure 7). During the reaction only a small quantity of cross-linked material precipitates from the methanol solution. From molecular modeling data a strong tendency of the cross-linker to form intramolecular complexes could be derived (Figure 7c). Therefore they act as chain stopper rather than cross-linker. Stiffer *multi*functional ligands might lead to enhanced cross-linking.

Conclusion

We have demonstrated the synthesis and full characterization of several 2,2':6',2''-terpyridine functionalized metallo-supramolecular building units and could prove their feasibility to form polymeric complexes and coordination polymers. Applying different metal-ions, metallo-supramolecular polymers with different stability ratios between monomeric and polymeric complexes could be synthesized in an self-assembly manner. Therefore, depending on the metal-ion and the metal ion content, solutions with tailor made and tuneable viscosities are easily accessible. However, due to the randomly coiled character of the polymers the range of viscosity values seems to be limited. Furthermore, the reversibility of iron(II) polymers *bis*(2,2':6',2''-terpyrid-4'-yl)-$FeCl_2$-di(ethylene glycol) (**6**) and *bis*(2,2':6',2''-terpyrid-4'-yl)-$FeCl_2$-poly(ethylene oxid)$_{180}$ (**7**) upon the increase of temperature, addition of chelating ligands (HEDTA) and the application of oxidizing/reducing conditions could be demonstrated.

[1] U. S. Schubert, C. Eschbaumer, *Angew. Chem.* **2002**, *114*, 3016-3050; *Angew. Chem. Int. Ed.* **2002**, *41*, 2892 – 2926.
[2] M. Heller, U. S. Schubert, *Macromol. Rapid Commun.* **2001**, *22*, 1359 – 1363.
[3] S. Schmatloch, U. S. Schubert, *Macromol. Rapid Commun.* **2002**, *in press*.
[4] Y. Chujo, K. Sada, T. Saegusa, *Macromolecules*, **1993**, *26*, 6315 – 6319.
[5] [5a] S. Kelch, M. Rehahn, *Macromolecules* **1998**, *31*, 4102 – 4106; [5b] S. Kelch, M. Rehahn, *Macromolecules* **1997**, *30*, 6185 – 6193.
[6] S. Bernard, K. Takada, D. J. Díaz, H. D. Abruña, H. Mürner, *J. Am. Chem Soc.* **2001**, *123*, 10265 – 10271.
[7] M. Kimura, M. Sano, T. Muto, K. Hanabusa, H. Shirai, *Macromolecules*, **1999**, *32*, 7951 – 7953.
[8] M. Schütte, D. G. Kurth, M. R. Linford, H. Cölfen, H. Möhwald, *Angew. Chem.* **1998**, *110*, 3058 – 3061; *Angew. Chem. Int. Ed.* **1998**, *37*, 2891 – 2893.
[9] R. Dobrawa, F. Würthner, *Chem. Commun.* **2002**, 1878 – 1879.
[10] B. G. G. Lohmeijer, U. S. Schubert, *Angew. Chem.* **2002**, *114*, 3980 – 3984.
[11] U. S. Schubert, C. Eschbaumer, *Polym. Preprints* **2000**, *41(1)*, 542 – 543.
[12] U. S. Schubert, O. Hien, C. Eschbaumer. *Macromol. Rapid Commun.* **2000**, *21*, 1156 – 1161.
[13] U. S. Schubert, C. Eschbaumer, *Macromol. Symp.* **2001**, *163*, 177 – 187.
[14] S. Schmatloch, U. S. Schubert, *Polym. Preprints* **2001**, *42(2)*, 395 – 396.
[15] V. Kislik, A. Eyal, *Water Science & Technology: Water Supply* **2001**, *1*, 119 – 129.
[16] K.O. Havelka, C.L. McCormick, Eds. *Specialty Monomers and Polymers*, ACSSymp. Ser. 755, Washington, **2000**.
[17] S. W. Shalaby, C. L. McCormick, G. B. Butler, Eds. *"Water-soluble polymers: synthesis, solution properties and applications"*, ACS Symp. Ser. 467, Washington, **1991,** 350 –502.
[18] [18a] J.-F. Gohy, B. G. G. Lohmeijer, U. S. Schubert, *Macromol. Rapid Commun.* **2002**, *23*, 555 – 560; [18b] J.-F. Gohy, B. G. G. Lohmeijer, S.K. Varshney, U. S. Schubert, *Macromolecules* **2002**, *35*, 7427 – 7435.
[19] K. Kobayashi, H. Suzuki, *Chemical Sensors* **2000**, *16*, 70 – 72.
[20] U. S. Schubert, C. Eschbaumer, *Polym. Preprints* **1999**, *40*(2), 1070 – 1071.
[25] R. H. Hoyler, C. D. Hubbard, S. F. A. Kettle, R. G. Wilkins, *Inorg. Chem.* **1966**, *5*, 622 – 625.
[26] R. Hogg, R. G. Wilkins, *J. Chem. Soc.* **1962**, 341 – 350.

Condensation Polymers Containing Silicon and Germanium in the Main Chain

Luis H. Tagle

Facultad de Química, Pontificia Universidad Católica de Chile, P.O. Box 306, Santiago, Chile
E-mail: ltagle@puc.cl

Summary: Poly(carbonates), poly(thiocarbonates) and poly(esteres) containing silicon and/or germanium in the main chain were obtained under phase transfer conditions. Polymers were synthesized in a biphasic system $NaOH/CH_2Cl_2$ at 20ºC using several phase transfer catalysts, and characterized by IR and ^1H and ^{13}C NMR. The results were evaluated by the yields and the inherent viscosity values. The process was effective observing an increase of both parameters in comparison with the essays without catalyst. The increases depended of the nature of the polymer and the catalyst. In poly(ester) synthesis there was an increase of these parameters when the NaOH concentration was increased due to a salting out effect of the diphenolate from the aqueous phase to the organic one. Also poly(amides) containing silicon or germanium were synthesized by solution polymerization.

Keywords: condensation polymers, germanium, phase transfer catalysis, silicon

Introduction

Organosilicon condensation polymers such as poly(amides), poly(esters), poly(imides) and others, in which the silicon atom is in the main chain, and bonded to aromatic or aliphatic groups, have been described and their properties studied.[1]

However, condensation polymers containing a germanium atom in the main chain and bonded to four carbon atoms have not been described. Only some very special polymers with the germanium atom forming part of the side groups have been described.[2-3] Also inorganic poly(germanes) with this structure $(GeR_2)_n$, have been described and their properties studied.[4]

DOI: 10.1002/masy.200350941

For many years we have focussed our attention on the synthesis of condensation polymers, using phase transfer catalysis as the polymerization method. In this sense we have described and studied the synthesis of several kinds of polymers, such as poly(carbonates), poly(thiocarbonates) and poly(esters). This technique offers advantages compared to solution or interphase polymerization.[5]

In the classical scheme, the catalyst, normally a quaternary onium salt has the function of transferring the diphenolate in the form of an ionic pair from the aqueous phase, to the organic one, in which the polymerization takes place.[6-7] We have studied the influence of the nature of the catalysts in the yields and in the inherent viscosity values of the obtained polymers. The principal limitation of this technique can be the insolubility of the polymeric chain in the reaction media, because this factor has an important effect on the molecular weight.

In this paper we summarise our works with respect to the synthesis of poly(carbonates) and poly(thiocarbonates) derived from diphenols or acid dichlorides, containing silicon or germanium bonded to four organic groups. The objective is to obtain polymers containing one or both heteroatoms in the main chain using the phase transfer catalysis as the polymerization method. Also poly(amides) from diamines containing these heteroatoms but by solution polymerization were obtained.

Experimental

Monomers were obtained according to described procedures from dichloro-dimethyl or dichloro-diphenyl-silane or -germane, and 4-bromo-phenol for the diphenols,[7] 4-bromo-toluene for the acid dichlorides,[8-10] and 4-bromo-N,N-bis(trimethylsilyl)-aniline for the diamines.[11]

In general the polymerizations were carried out in a biphasic system. Normally CH_2Cl_2 was used as organic solvent in which the acid dichloride, phosgene or thiophosgene were dissolved. The diphenol and the catalyst were dissolved in a sodium hidroxide aqueous solution. In all the polymerizations essays without catalysts were made, in order to evaluate the behaviour of the interphase of the system. Monomers and polymers were characterized by IR, ^{1}H and ^{13}C NMR spectroscopy.

Results and Discussion

Poly(carbonates) and poly(thiocarbonates) with the following structures derived from four diphenols with silicon or germanium and phosgene or thiophosgene respectively, were obtained and characterized.[12] Table 1 shows the results obtained for poly(carbonates) **Ia** and **IIa** and poly(thiocarbonates) **Ib** and **IIb**.

Ia: Y = Ge; R = $-CH_3$; X = O **Ib**: Y = Ge; R = $-CH_3$; X = S
IIa: Y = Ge; R = $-C_6H_5$; X = O **IIb**: Y = Ge; R = $-C_6H_5$; X = S
IIIa: Y = Si; R = $-CH_3$; X = O **IIIb**: Y = Si; R = $-CH_3$; X = S
IVa: Y = Si; R = $-C_6H_5$; X = O **IVb**: Y = Si; R = $-C_6H_5$; X = S

Table 1. Yields and inherent viscosities obtained for poly(carbonates) **Ia** and **IIa** and poly(thiocarbonates) **Ib** and **IIb** (Ref. 12).

Polymer	Ia		Ib		IIa		IIb	
Catalyst	%	$\eta^{a)}$	%	$\eta^{a)}$	%	$\eta^{a)}$	%	$\eta^{a)}$
none	38	0.04	20	0.16	45	0.08	29	0.04
TBAB	48	0.24	61	0.24	27	0.17	39	0.21
ALIQUAT	30	0.04	56	0.12	37	0.13	18	0.13
BTEAC	57	0.20	52	0.12	36	0.25	37	0.17
HDTMAB	35	0.04	42	0.16	46	0.13	19	0.08
HDTBPB	7	0.20	55	0.16	41	0.13	20	0.04

a) Inherent viscosity in $CHCl_3$ at 25°C (conc. = 0.3 g/dl).

In general, there is an increase of the yields when the catalysts were used, showing the effectiveness of the transfer process, and in less proportion of the inherent viscosity values. For polymers in which the heteroatom is bonded to mehtyl groups, good results were obtained with tetrabutylammonium bromide (TBAB), which due to its symmetrical structure, has been described as effective for the transfer process of the diphenolate derived from bisphenol A. Benzyltriethylammonium chloride (BTEAC) due to its more hydrophilic character, showed a relatively good behaviour for polymers in which the heretoatom is bonded to phenyl groups.

Table 2 shows the results obtained for poly(carbonates) **IIIa** and **IVa** and poly(thiocarbonates) **IIIb** and **IVb**.

Table 2. Yields and inherent viscosities obtained for poly(carbonates) **IIIa** and **IVa** and poly(thiocarbonates) **IIIb** and **IVb** (Ref. 12).

Polymer	**IIIa**		**IIIb**		**IVa**		**IVb**	
Catalyst	%	η[a]	%	η[a]	%	η[a]	%	η[a]
none	---	-----	---	-----	32	0.04	13	0.04
TBAB	38	0.16	14	0.08	48	0.08	33	0.08
ALIQUAT	17	0.13	16	0.08	41	0.08	31	0.08
BTEAC	35	0.16	7	0.04	42	0.08	34	0.08
HDTMAB	23	0.13	21	0.08	54	0.13	46	0.13
HDTBPB	7	0.10	4	0.04	40	0.08	19	0.04

[a] Inherent viscosity in $CHCl_3$ at 25°C (conc. = 0.3 g/dl).

It is possible to see an increase of the yields with respect to the value obtained without catalyst. However the increase of the inherent viscosity values was low, the behaviour of the catalysts being very similar. In general the yields and the inherent viscosity values obtained for these polymers were low, due probably to a hydrolytic process of both, the phosgene or thiophosgene, and the polymeric chain, promoted by lypophilic catalysts, which has been described for other poly(carbonates) and poly(thiocarbonates).[5] Due to the low values of inherent viscosity we have principally oligomeric species.

On the other hand, in this work we used a stoichiometric amount of sodium hydroxide in order to avoid the decomposition of the monomer. In another work of poly(carbonates) and poly(thiocarbonates) synthesis, in which we used a higher concentration of sodium hydroxide, we obtained high yields, probably due to a salting out effect.

Poly(carbonates) derived from the bischloroformate of bisphenol A and the same diphenols with silicon or germanium were obtained. The resulting polymers contain two carbonate groups and silicon or germanium in the main chain.[13] In this synthesis we used two catalysts: tetrabutylammonium (TBAB) bromide and benzyltriethylammonium chloride (BTEAC), and three sodium hydroxide concentrations: stoichiometric, twice the stoichiometric and three times the stoichiometric. The results are summarised in Table 3.

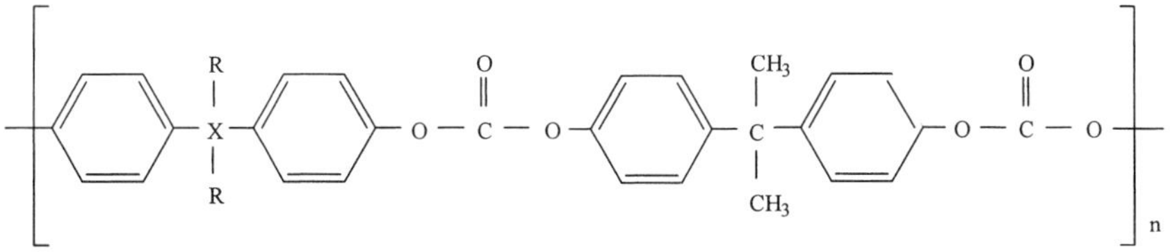

V: X = Si R = CH_3 **VII**: X = Si R = C_6H_5

VI: X = Ge R = CH_3 **VIII**: X = Ge R = C_6H_5

In the results it is possible to see that in several cases there is an increase of the yields when the sodium hydroxide concentration is increased and also when the catalysts are used. In general, there are no important differences between the use of twice or three times the sodium hydroxide concentration with respect to the stoichiometric. With all the diphenols the effect was similar. In general both parameters have the tendency to increase when the sodium hydroxide concentration is increased and when the catalysts are used. These effects were attributed to a salting out effect of the diphenolate from the aqueous phase to the organic one when the sodium hydroxide concentration is increased.

Table 3. Yields and inherent viscosities obtained for the poly(carbonates) **V** and **VI** derived from the diphenols bis(4-hydroxyphenyl)-dimethyl-silane and bis(4-hydroxyphenyl)-dimethyl-germane, poly(carbonates) **VII** and **VIII** derived from the diphenols bis(4-hydroxyphenyl)-diphenyl-silane and bis(4-hydroxyphenyl)-diphenyl-germane (Ref. 13).

		Catalyst					
		---		TBAB		BTEAC	
Poly(carbonate)	NaOH/phenol[a]	%	$\eta^{b)}$	%	$\eta^{b)}$	%	$\eta^{b)}$
V	2 / 1	32	0.04	56	0.08	58	0.08
V	4 / 1	49	0.11	63	0.11	58	0.11
V	6 / 1	44	0.11	55	0.11	54	0.11
VI	2 / 1	13	0.08	56	0.12	55	0.08
VI	4 / 1	72	0.07	95	0.18	88	0.14
VI	6 / 1	73	0.11	91	0.18	83	0.14
VII	2 / 1	31	0.0459	0.04	55	0.08	
VII	4 / 1	42	0.0850	0.12	86	0.20	
VII	6 / 1	48	0.1272	0.16	71	0.16	
VIII	2 / 1	37	0.0869	0.08	60	0.12	
VIII	4 / 1	45	0.0862	0.12	67	0.12	
VIII	6 / 1	39	0.0867	0.12	62	0.12	

[a] Molar ratio.
[b] Inherent viscosity, in $CHCl_3$ at 25°C (c = 0.3 g/dL).

Poly(esters) of two different type were also synthesized. The first group is derived from the same diphenols with silicon or germanium and methyl of phenyl groups bonded to the heteroatoms, and terephthaloyl or isophthaloyl dichlorides.[14] The results are summarised in

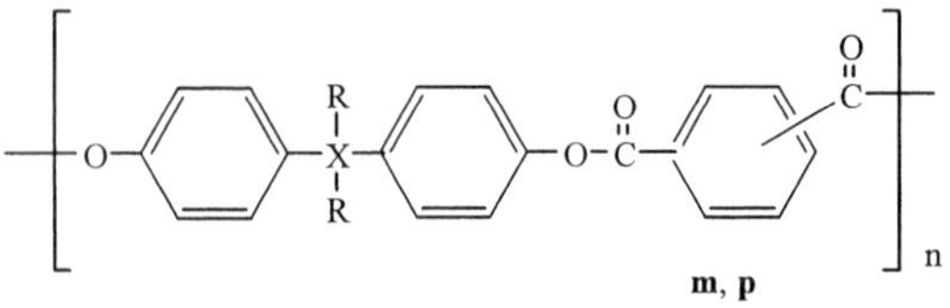

m, p

Ia : R = $-CH_3$; X = Si **Ib** : R = $-CH_3$; X = Ge
IIa: R = $-C_6H_5$; X = Si **IIb**: R = $-C_6H_5$; X = Ge

Table 4.

Table 4. Yields and inherent viscosities obtained for poly(esters) derived from the diphenols **I-a**, **I-b**, **II-a** and **II-b** (Ref. 14).

		Catalyst					
		---		TBAB		TEBAC	
Poly(ester)	NaOH/phenol [a)]	%	$\eta^{b)}$	%	$\eta^{b)}$	%	$\eta^{b)}$
I-a-m	2 / 1	43	0.25	47	0.12	15	0.12
I-a-m	3 / 1	52	0.25	77	0.24	58	0.12
I-a-m	4 / 1	83	0.30	90	0.23	64	0.24
I-a-p	2 / 1	22	0.20	17	0.10	16	0.10
I-a-p	3 / 1	40	0.20	46	0.20	40	0.30
I-a-p	4 / 1	49	0.30	54	0.20	61	0.44
I-b-m	2 / 1	25	0.10	14	0.24	37	0.05
I-b-m	3 / 1	45	0.10	17	0.24	75	0.10
I-b-m	4 / 1	73	0.15	26	0.24	81	0.10
I-b-p	2 / 1	43	0.10	24	0.15	60	0.05
I-b-p	3 / 1	76	0.10	80	0.15	83	0.05
I-b-p	4 / 1	87	0.15	80	0.25	77	0.15
Poly(ester)	NaOH/phenol	%	$\eta^{c)}$	%	$\eta^{c)}$	%	$\eta^{c)}$
II-a-m	2 / 1	27	0.06	40	0.08	30	0.09
II-a-m	3 / 1	49	0.10	84	0.10	72	0.08
II-a-m	4 / 1	65	0.13	94	0.12	80	0.09
II-a-p	2 / 1	33	0.15	50	0.07	26	0.12
II-a-p	3 / 1	48	0.12	62	0.10	84	0.12
II-a-p	4 / 1	65	0.16	88	0.16	92	0.18
II-b-m	2 / 1	8	0.08	28	0.20	23	0.10
II-b-m	3 / 1	37	0.08	77	0.23	82	0.08
II-b-m	4 / 1	73	0.24	98	0.28	97	0.09
II-b-p	2 / 1	31	0.08	20	0.10	25	0.10
II-b-p	3 / 1	42	0.11	83	0.20	59	0.08
II-b-p	4 / 1	55	0.23	95	0.23	83	0.15

[a)] Molar ratio
[b)] Inherent viscosity, in N-methyl-pirrolidone, at 25°C (c = 0.3 g/dL)
[c)] Inherent viscosity, in $CHCl_3$, at 25°C (c = 0.3 g/dL)

Two catalysts (TBAB and BTEAC) and three sodium hydroxide concentrations were used and in all cases there was an increase of the yields when this concentration was increased, with and without catalysts. This tendency was also attributed to a salting out effect, which increases the transfer process of the ionic pair from the aqueous phase to the organic one.

Also in some cases there was an increase of the inherent viscosity values. It is difficult to explain the decrease of the inherent viscosity values when the catalysts were used. Probably there is a hydrolytic process of the polymeric chains. Neither were there any differences between the diphenols containing silicon or germanium atoms.

The second group of poly(esters), are the following sixteen polymers derived from four acid dichlorides with germanium or silicon and methyl or phenyl groups and the same diphenols shown previously. The resulting poly(esters) contain two heteroatoms and in the synthesis we used twice the stoichiometric amount of sodium hydroxide.[15] The results are shown in Table 5.

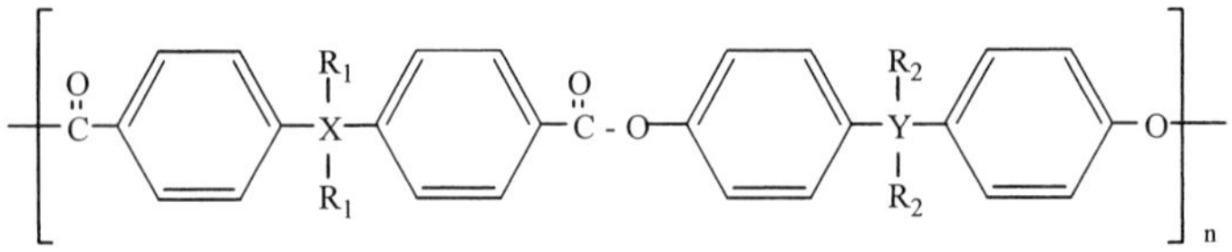

I : X = Si , Y = Si , R_1 = $-C_6H_5$, R_2 = $-CH_3$
II : X = Si , Y = Ge , R_1 = $-C_6H_5$, R_2 = $-CH_3$
III: X = Ge , Y = Si , R_1 = $-C_6H_5$, R_2 = $-CH_3$
IV: X = Ge , Y = Ge , R_1 = $-C_6H_5$, R_2 = $-CH_3$

V : X = Si , Y = Si , R_1 = R_2 = $-C_6H_5$
VI : X = Si , Y = Ge, R_1 = R_2 = $-C_6H_5$
VII : X = Ge , Y = Si , R_1 = R_2 = $-C_6H_5$
VIII : X = Ge , Y = Ge, R_1 = R_2 = $-C_6H_5$

IX : X = Si , Y = Si , R_1 = R_2 = $-CH_3$
X : X = Si , Y = Ge, R_1 = R_2 = $-CH_3$
XI : X = Ge , Y = Si , R_1 = R_2 = $-CH_3$
XII : X = Ge , Y = Ge, R_1 = R_2 = $-CH_3$

XIII : X = Si , Y = Si , R_1 = $-CH_3$, R_2 = $-C_6H_5$
XIV : X = Si , Y = Ge, R_1 = $-CH_3$, R_2 = $-C_6H_5$
XV : X = Ge , Y = Si , R_1 = $-CH_3$, R_2 = $-C_6H_5$
XVI : X = Ge , Y = Ge, R_1 = $-CH_3$, R_2 = $-C_6H_5$

The results for poly(esters) **I - XVI** showed that when the catalysts were used, it is possible to see very low increases of the yields and in some cases, of the inherent viscosity values. Neither are there important differences in the behaviour of the catalysts. A very negative effect was the insolubility of these poly(esters) in the reaction media, which hindered the growth of the polymeric chain and implies in the future the use of other organic solvents. In spite of thcsc results, the phase transfer process had certain effectiveness in this polyester synthesis. With respect to these polymers, at this moment we are working on the effect of lower polymerization temperatures in the inherent viscosity values.

Table 5. Yields and inherent viscosities obtained for poly(esters) **I** – **XVI** (Ref. 15).

	Catalyst					
	---		TBAB		BTEAC	
Poly(ester)	%	η[a)]	%	η[a)]	%	η[a)]
I	75	0.12	85	0.16	80	0.20
II	55	0.08	62	0.12	67	0.12
III	72	0.16	74	0.20	76	0.28
IV	57	0.08	66	0.12	67	0.12
V	70	0.16	78	0.24	89	0.28
VI	80	0.08	87	0.20	89	0.24
VII	57	0.12	72	0.24	89	0.24
VIII	71	0.16	78	0.20	87	0.24
IX	35	0.08	50	0.12	89	0.16
X	47	0.12	69	0.20	60	0.16
XI	40	0.12	51	0.16	74	0.12
XII	35	0.08	50	0.12	48	0.12
XIII	36	0.12	50	0.12	89	0.16
XIV	67	0.12	75	0.16	70	0.12
XV	71	0.12	89	0.20	68	0.16
XVI	78	0.16	79	0.20	64	0.12

a) Inherent viscosity, in $CHCl_3$ at 25°C (c = 0.3 g/dL).

Poly(amides) derived from two aromatic diamines with silicon or germanium and isophthaloyl or terephthaloyl dichlorides were synthesized, according to the following structures, and the results are summarised in Table 6.[16]

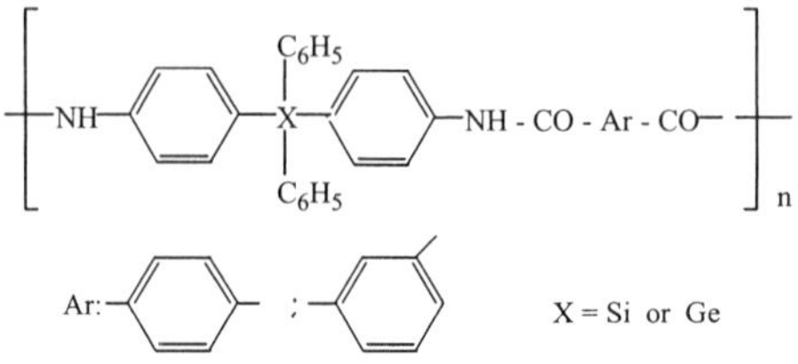

$$\left[-NH-C_6H_4-X(C_6H_5)_2-C_6H_4-NH-CO-Ar-CO- \right]_n$$

Ar: I (1,4-phenylene) ; II (1,3-phenylene) X = Si or Ge

Table 6. Yields, inherent viscosity, thermogravimetric data and glass transition temperatures obtained for the poly(amides) (Ref. 16).

Poly(amide)	Yield (%)	η_{inh} [a)]	TDT$^{10\%}$ (°C)	Tg (°C)
I-Si	91	0.45	400	178
I-Ge	95	0.50	410	133
II-Si	99	0.42	200	138
II-Ge	84	0.49	225	127

a) inherent in o-chloro-phenol at 25°C (c = 0.3 g/dL)

In general the yields and the inherent viscosity values were good and similar between the poly(amides).

In this case the thermal properties were studied, showing that the poly(amides) derived from terephthalic acid had higher thermal stability attributed to the more symmetrical structures. On the other hand those with germanium showed a little more stability than those with silicon. The bond carbon-silicon has a little higher polarity than the bond carbon-germanium, and this has higher bond enthalpy. This difference would explain the higher stability of the germanium containing poly(amides).

The higher Tg values were obtained for the more symmetrical structures. Poly(amides) with silicon showed higher Tg values, which can be due to the larger size of the germanium atom, which increases the flexibility of the polymeric chain and consequently lower values of Tg are obtained.

Acknowledgements

The authors acknowledge the financial support of FONDECYT (Fondo Nacional de Investigación Científica y Tecnológica) and DIPUC (Dirección de Investigación y Postgrado de la Universidad Católica de Chile).

[1] S.F. Thames, K.G. Panjnani, *J. Inorg. and Organomet. Chem.* **1996, 6**, 59.
[2] T.J. Peckham, J.A. Massey, M. Edwards, I. I. Manners, D.A. Foucher, *Macromolecules*. **1996**, 29, 2396.
[3] H. Ito, T. Masuda, T. Higashimura, *J. Polym. Sci., Part A, Polym. Chem.* **1996**, 34, 2925.
[4] M. Okano, H. Fukai, M. Arakawa, H. Hamano, *Electrochem. Commun.* **1999**, 1, 223.
[5] L.H. Tagle, *"Phase transfer catalysis in polymer synthesis"*. In *"Handbook of phase transfer catalysis"*, Eds. Y. Sasson, R. Neumann, Blackie Academic and Professional, 1997, p. 200, and references therein.
[6] C.M. Starks, C. Liotta, *"Phase transfer catalysis. Principles and Techniques"*, Academic Press, New York, 1978, p. 42.
[7] W. Davidson, B.R. Laliberte, C.M. Goddard, M.C. Henry, *J. Organomet. Chem.* **1972**, 36, 283.
[8] M. Maienthal, M. Hellmann, C.P. Haber, L.A. Hymo, S. Carpenter, J. Carr, *J. Am. Chem. Soc.* **1954**, 76, 6392.
[9] H.N. Kovacs, A.D. Delman, B.B. Simms, *J. Polym. Sci., Part A-1*. **1968**, 6, 2103.
[10] J. Zhang, Q. Sun, X. Hou, *Macromolecules*. **1993**, 26, 7176.
[11] J.R. Pratt, W.D. Massey, F.H. Pinkerton, S.F. Thames, *J. Org. Chem.* **1975**, 40, 1090.
[12] L.H. Tagle, J.C. Vega, F.R. Diaz, D. Radic, L. Gargallo, P. Valenzuela, *J. Macromol. Sci., Pure Appl. Chem.* **2000**, A-37, 997.
[13] L.H. Tagle, unpublished results.
[14] L.H. Tagle, F.R. Diaz, M. Nuñez, F. Canario, *Intern. J. Polymeric Mat.* (in press).
[15] L.H. Tagle, F.R. Diaz, J.C. Vega, P. Valenzuela, *Eur. Polym. J.* (in press).
[16] L.H. Tagle, F.R. Diaz, D. Radic, A Opazo, J.M. Espinoza, *J. Inorg. and Organomet. Chem.* **2000**, 10, 73.

Aromatic Polymers with Side Oxadiazole Rings as Luminescent Materials in LEDs

Maria Bruma,[*1] *Elena Hamciuc,*[1] *Burkhard Schulz,*[2] *Thomas Köpnick,*[3] *Yvette Kaminorz,*[4] *Jenifer Robison*[5]

[1] Institute of Macromolecular Chemistry, Aleea Ghica Voda 41A, 6600 Iasi, Romania
Email: mbruma@icmpp.tuiasi.ro; ehamciuc@icmpp.tuiasi.ro
[2] University of Potsdam, FZDOBS, Am Neuen Palais 10, 14469 Potsdam, Germany
[3] Institute of Thin Film Technology and Microsensors, Kanntstr 55, 14513 Teltow, Germany
[4] University of Potsdam, Institute of Physics, Am Neuen Palais 10, 14415 Potsdam, Germany
[5] Tyco Electronics Corporation, Menlo Park, California 94025, USA

Summary: Aromatic polyamides and polyazomethines with side oxadiazole rings have been synthesized by using aromatic diamines containing pendent substituted oxadiazole groups and a diacid chloride having diphenylsilane or hexafluoroisopropylidene, or an aromatic dialdehyde with fluorene unit, respectively. These polymers were easily soluble in amidic solvents. Very thin films which were deposited from polymer solutions onto silicon wafers exhibited smooth, pinhole-free surface in atomic force microscopy investigations. The polymers showed high thermal stability with decomposition temperature being above 400°C. Some of them exhibited blue photoluminescence, in the range of 450-480 nm, making them promising candidates for future use as high performance materials in the construction of light emitting devices.

Keywords: oxadiazole, photoluminescence, polyamides, polyazomethines, thin films

Introduction

Aromatic polymers containing 1,3,4-oxadiazole rings in the main chain are well-known for their high thermal resistance in oxidative atmosphere, good hydrolitic stability, low dielectric constant and tough mechanical properties.[1] There is currently much research directed towards the discovery of new blue light-emitting polymers, with characteristics of high efficiency and high reliability. For such a purpose polyoxadiazoles are of great interest because due to the electron-withdrawing character of the 1,3,4-oxadiazole rings they can facilitate the injection and transport of electrons.[2] But, aromatic polyoxadiazoles are rigid, rod-like molecules and are insoluble in organic solvents and do not have a glass transition (T_g) which makes their processing quite difficult. To improve the solubility and lower the T_g, various approaches have been undertaken such as introduction of flexible side groups on the aromatic rings[3] or bulky moities, such as „cardo“ groups[4] in the main chain. Another way would be the incorporation of oxadiazole rings as pendent groups on a polymer chain.[5]

 DOI: 10.1002/masy.200350942

Therefore, we considered interesting to make polymers in which the oxadiazole rings are attached as side groups to an aromatic polyamide or polyazomethine backbone. Here we present the synthesis of certain aromatic polyamides and polyazomethines containing the 1,3,4-oxadiazole rings in the side groups; polyamides have the diphenyl silane or 6F units in the main chain, while polyazomethines have fluorene units in the main chain. The properties of these polymers, such as solubility, thermal stability, glass transition, film forming ability and quality of thin films, as well as their photoluminescence ability are discussed.

Experimental

Synthesis of the Monomers

Aromatic diamines, **1,** having an oxadiazole ring in the side group, namely 2-(4-dimethylaminophenyl)-5-(3,5-diaminophenyl)-1,3,4-oxadiazole, 2-(4-tert-butylamino)-5-(3,5-diaminophenyl)-1,3,4-oxadiazole and 2-(4-dimethylaminophenyl)-5-[2,5-bis(*p*-aminophenoxy)-phenyl]-1,3,4-oxadiazole respectively,[6] two diacid chlorides such as bis(*p*-chloro-carbonyl-phenylene)-diphenylsilane,[7] **2**, and hexafluoroisopropylidene-bis(*p*-benzoyl chloride),[8] **3**, respectively, and an aromatic dialdehyde, **4**, having fluorene unit, namely 9,9-bis(p-formyl-phenoxy-4-phenylene) fluorine,[9] have been prepared by published procedures and thoroughly purified, and their structures are shown in Scheme 1.

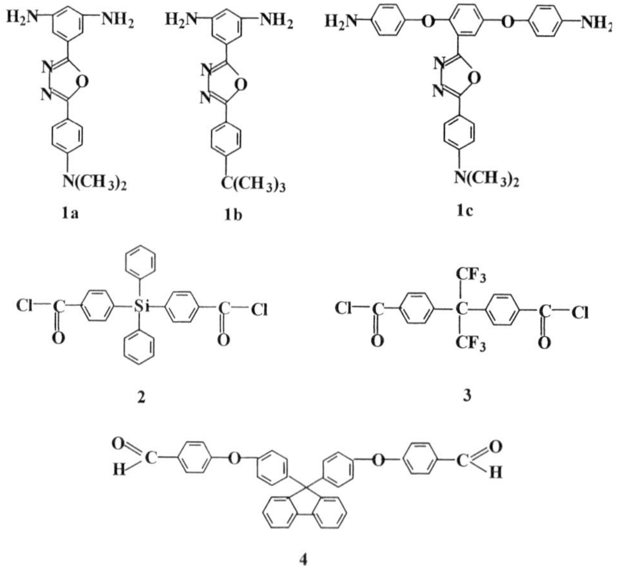

H_2N NH_2 H_2N NH_2 H_2N O O NH_2

N O N N O N N O N

$N(CH_3)_2$ $C(CH_3)_3$ $N(CH_3)_2$

1a **1b** **1c**

CF_3

Cl C Si C Cl Cl C C C Cl

O O O CF_3 O

2 **3**

O C O O C O

H H

4

Scheme 1. Structures of the monomers **1**, **2**, **3** and **4**.

Synthesis of the Polymers

Six polyamides containing 1,3,4-oxadiazole ring in the side group, three of them having silicon in the main chain, **5**, and the other three having 6F in the main chain, **6**, have been synthesized by low temperature solution polycondensation reaction of equimolar amounts of a diamino-oxadiazole **1** with bis(*p*-chlorocarbonyl-phenylene)-diphenylsilane, **2**, or with hexafluoroisopropylidene-bis(*p*-benzoyl chloride), **3**, in NMP as a solvent and with pyridine as an acid acceptor, and their structures are shown in Scheme 2.

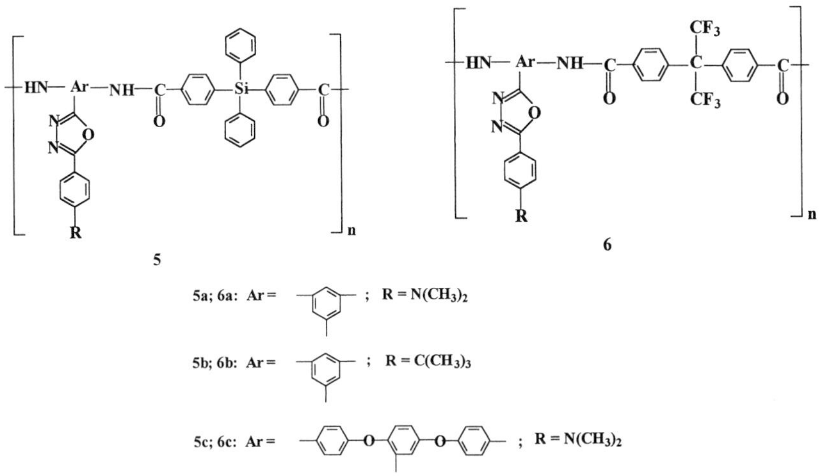

5a; 6a: Ar = ; $R = N(CH_3)_2$

5b; 6b: Ar = ; $R = C(CH_3)_3$

5c; 6c: Ar = ; $R = N(CH_3)_2$

Scheme 2. Structures of polyamides with oxadiazole units in the side group, **5** and **6**.

Similarly, three polyazomethines having 1,3,4-oxadiazole ring in the side group, **7**, have been prepared by polycondensation at 20-80°C of equimolar quantities of a diamino-oxadiazole **1** with 9,9-bis(p-formylphenoxy-4-phenylene) fluorene **4**, by using NMP as a solvent. The structures of these polyazomethines are shown in Scheme 3. The resulting polymer was separated from solution by precipitation in water, followed by washing with water and ethanol and drying in oven at 120°C.

Preparation of Polymer Films

1g of a polymer was dissolved in 10 mL NMP. Half of the solution was cast onto glass plates and were dried in an oven at 80°C, 110°C, 140°C, 170°C and 210°C for 15 minutes each. Transparent homogeneous films resulted, which were stripped off the plates by immersion in hot water. These films had a thickness of 20-30 μm. The other part of polymer solution was dilluted with NMP to 1% concentration and was used to deposit very thin films, in the range of tens of nanometres, onto silicon wafers, by a spin-coating technique, at a speed of 5000 rotation/min. These films, as-deposited, were gradually heated to 210°C in the same way as described earlier to remove the solvent, and were used for atomic force microscopy (AFM) investigations.

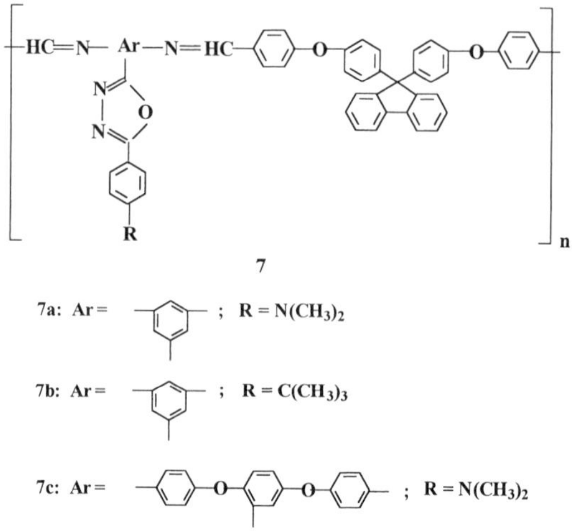
HC=N– Ar—N=HC
N
O
N
R
n
7
7a: Ar =
; R = N(CH3)2
7b: Ar =
; R = C(CH3)3
7c: Ar =
; R = N(CH3)2

Scheme 3. Structures of polyazomethines with side oxadiazole rings, **7**.

Measurements

Infrared spectra were recorded with a Nicolet Magna FTIR spectrometer in transmission mode. The thermogravimetric analysis (TGA) of the precipitated polymers was performed with a Seiko RTG 220 thermobalance, operating at a heating rate of 5°C/min, in air. The glass transition temperature (T_g) of the precipitated polymers was determined with a Seiko differential scanning calorimeter DSC 6200. Model molecules for a polymer fragment were obtained by molecular mechanics (MM^+) by means of the Hyperchem program, version 4.0.[10] The surfaces of the very thin films as-deposited on silicon wafers were studied by atomic force microscopy (AFM) with a SA1/BD2 apparatus (Park Scientific Instruments) in the contact mode. For photoluminescence (PL) measurements the polymer was spin-coated on a silicon substrate from NMP solution and heated afterwards to remove the solvent. A UV lamp peaking at 365 nm was used as excitation source. For recording the PL spectra, the emitted light was transfered by an optical fiber to an Insta Spec CCD detector used with an

MS 257 monochromator from L.O.T. Oriel Instruments. All PL spectra were corrected for the sensitivity of the detector.

Results and Discussion

Aromatic polyamides **5** and **6** containing the 1,3,4-oxadiazole ring in the side group and silicon or 6F units in the main chain were prepared by polycondensation reaction of aromatic diamines having pendent oxadiazole ring with a diacid chloride incorporating the diphenyl silane or 6F group (Scheme 2). Aromatic polyazomethines having the 1,3,4-oxadiazole ring in the side group **7** and fluorene unit in the main chain were prepared by polycondensation reaction of the diamines containing pendent oxadiazole ring with a fluorene-containing dialdehyde (Scheme 3). The expected structures of the polymers **5**, **6** and **7** were confirmed by IR spectra as follows. The strong absorption bands which appeared in the spectra of polymers **5** and **6** at 3400-3440 cm^{-1} and 1660-1670 cm^{-1} were attributed to amide groups. The absorption band at 1600 cm^{-1} in the spectra of the polymers **7** was assigned to –CH=N- groups. In the spectra of polymers **V** the absorption peaks at 1425-1430, 1105-1110 and 700 cm^{-1} were attributed to silicon-phenyl bonds. In the spectra of polymers **6** the absorption peaks at 1210 cm^{-1} were assigned to 6F groups. All the polymers exhibited broad absorption at 3060 cm^{-1} which was characteristic to CH aromatic bonds; the IR bands at 960-970 cm^{-1} and 1020 cm^{-1} were assigned to oxadiazole rings. The absorptions at 2920-2960 cm^{-1} were due to methyl groups.

All these polymers are soluble in polar amidic solvents such as N-methylpyrolidinone NMP, dimethylformamide (DMF) and dimethylacetamide (DMAc). Their improved solubility as compared with that of conventional aromatic polyamides, poly(arylene-oxadiazole)s and aromatic polyazomethines can be explained by the presence of pendent groups which disturbe the packing of macromolecular chains and thus facilitate the diffusion of small molecules of solvent which leads to better solubility. In addition, the voluminous diphenyl silane, 6F or fluorene units in the main chain introduce more flexibility and consequently make the shape of the macromolecules to be far from a „rigid rod“, as evidenced by molecular modelling (Figure 1 and Figure 2). The good solubility makes the present polymers potential candidates for practical applications in spin-coating and casting processes.

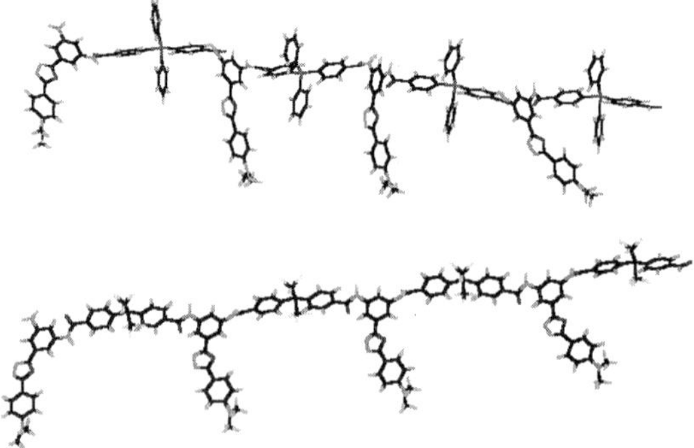

Fig. 1. Models of two polyamides **5a** (top) and **6a** (bottom).

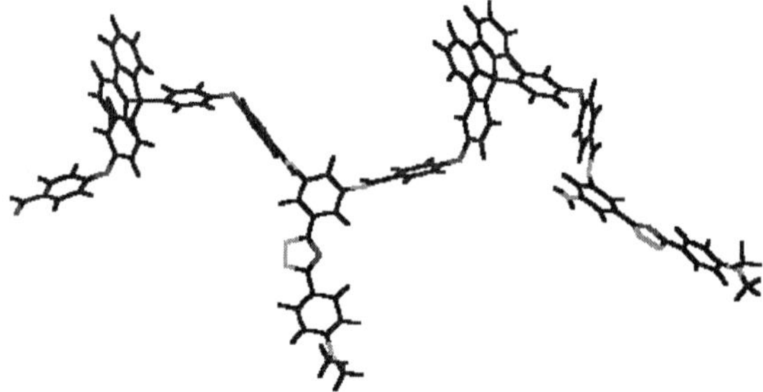

Fig. 2. Model of the polyazomethine **7a**.

All these polymers possess film-forming ability. Thin transparent films having a thickness of tens of micrometres which were prepared by casting technique were tough and flexible. Also, very thin coatings having thicknesses of tens of nanometres have been deposited onto silicon wafers. The quality of such films as-deposited on silicon substrates was studied by AFM. The films exhibited very smooth surfaces over large scanning ranges (1-100 μm). The values of root mean square (rms) roughness calculated from the AFM data lie in the range of 6-12 Å being of the same order of magnitude as that of the highly polished silicon wafers which were used as substrates. This means that the deposited films are very smooth and homogeneous. They do not show any pinholes or cracks and are practically deffectless. The films had strong adhesion to the silicon wafers. These qualities are very much required when such films are used in microelectronic devices.[11] A typical AFM image is shown in Figure 3.

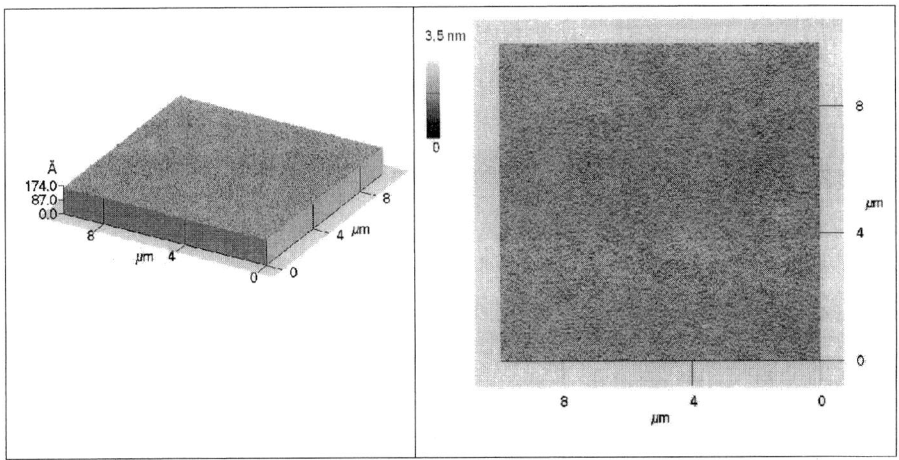

Fig. 3. AFM image of a film made from polymer **5a** (left: side view; right: top view).

The thermal stability of the polymers was evaluated by thermogravimetric analysis (TGA). All these polymers exhibited high thermal stability with insignificant weight loss up to 350°C. The polyamides **5** and **6** begin to decompose (IDT) in the range of 409-425°C. The temperature of maximum decomposition rate is in the domain of 426-522°C (Table 1).

Table 1. Properties of the polyamides containing side oxadiazole groups.

Polymer	IDT [a)] (°C)	T_{max} [b)] (°C)	T_g (°C)	Photoluminescence	
				Maximum (nm)	Half width (nm)
5a	413.7	479.4	275.8	475	102
5b	425.2	454.5	Not detected		
5c	409.3	429.8	151.6	460	114
6a	420.3	521.7	Not detected	480	108
6b	417.2	428.1	Not detected		
6c	411.2	425.9	153.6	465	106

[a)] Initial Decomposition Temperature = Temperature of 10% weight loss.
[b)] Temperature of maximum rate of decomposition.

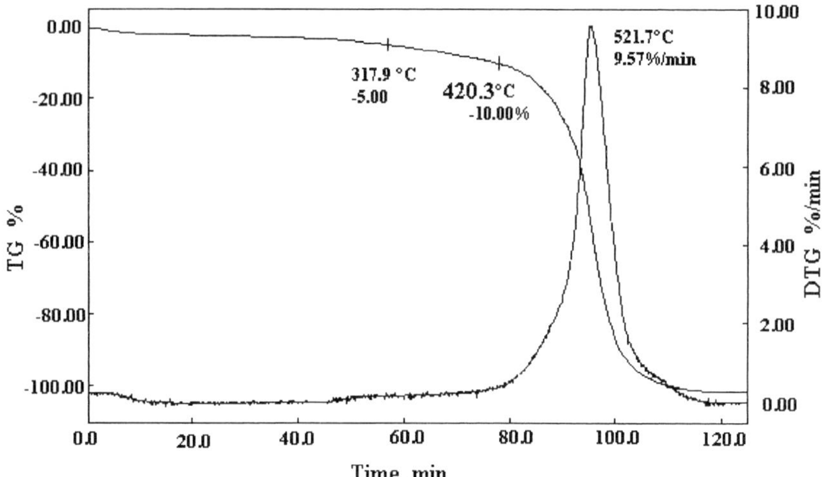

Fig. 4. TG and DTG curves of the polymer **6a**.

The polyamides **5b**, **6a** and **6b** did not exhibit a glass transition when heated to 320°C in DSC experiments. The other polyamides did show a glass transition, in the temperature range of 151-276°C. As expected, the polymers **5c** and **6c**, which contain some ether linkages in the main chain, have lower T_g, in the range of 151-154°C, due to the increase in flexibility of polymer backbone determined by these ether bridges. It can be noticed that there is a large „window" between the glass transition and decomposition temperature of these two polymers (**5c** and **6c**) which makes them attractive for thermoforming processing.

The light emitting ability of these polymers was evaluated on the basis of photoluminescence spectra. The polyamides **5a**, **5c**, **6a** and **6c** containing the dimethylamino substituent in the *para*-position of the chromophoric diphenyl-1,3,4-oxadiazole unit show intensive blue emission with the maximum between 460 and 480 nm (Table 1). Also, the polyazomethines **7a** and **7c**, containing fluorene units and dimethylamino substituents in *para* – position of the diphenyloxadiazole moiety show blue photoluminescence, in the range 450-480 nm. In contrast to other diphenyloxadiazole side chain polymers such as polymethylmethacrylate derivatives,[5,12] in the present polymers parts of the main chain are conjugated and therefore influence the luminescent behavior. The coupling bond of the side to the main chain influences the luminescence behavior significantly. The half width of the luminescence is around 100 nm, being in the usual range observed for other polymeric emission materials, and does not change significantly with the chemical structure of the present polymers. Typical photoluminescence spectra of polyamides with side oxadiazole groups are shown in Figure 5, while those of polyazomethines with side oxadiazole groups are shown in Figure 6.

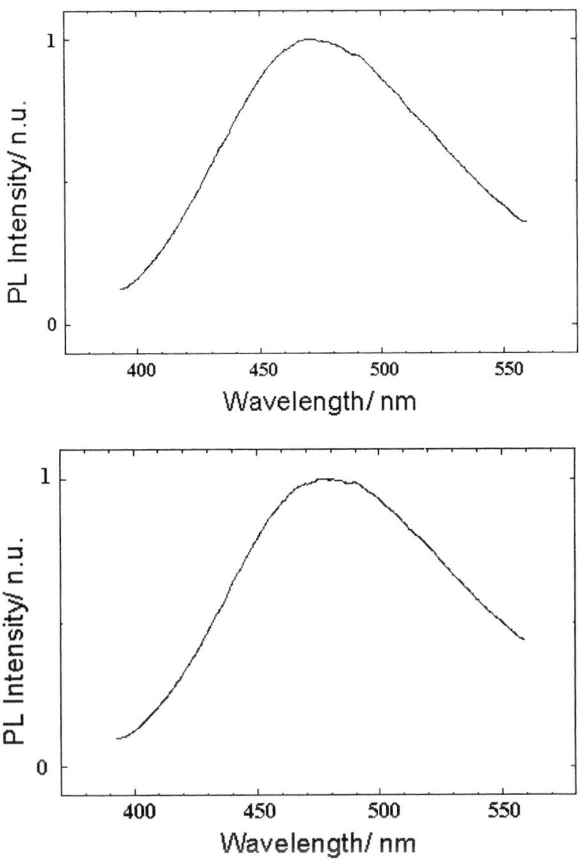

Fig. 5. Photoluminescence spectra of polyamides containing side oxadiazole rings, **5c** (top) and **6c** (bottom).

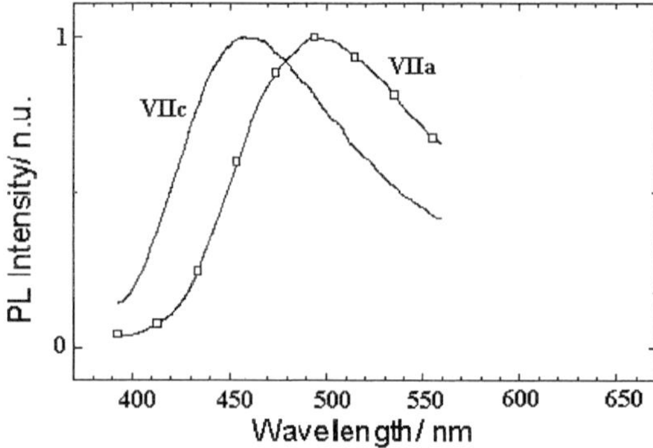

Fig. 6. Photoluminescence spectra of polyazomethines containing side oxadiazole rings, **7a** and **7c.**

Very thin films deposited from polyazomethines containing fluorene units **7** were investigated for use as negative resist materials in electron-beam litography. Fine structures could be designed by electron-beam irradiation follwed by development. A typical AFM image of a such structured film is shown in Figure 7.

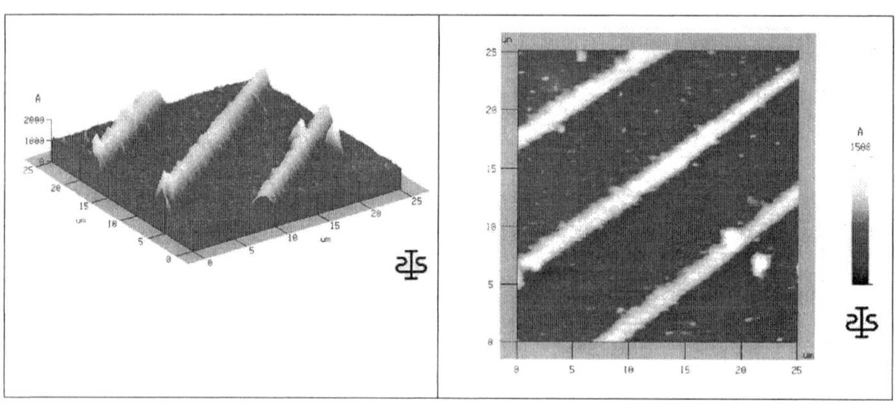

Spin-coating:	5 % polymer solution in NMP; 5.000 rpm/30 s
Substrate:	Silicon, hydrophile (H_2SO_4/H_2O_2 4/1, 120 °C, 30 min)
Pre-bake:	80, 110, 140, 170, 200 °C (20 min each temperature)
Electron-beam:	700 $\mu C/cm^2$; 10 kV (30 s)
Development:	30 s NMP, 30 s iPrOH, 1 min H_2O_2 (ultrasonic bath)

Fig. 7. AFM image of a structured film deposited from polyazomethine **7a** (left: side view; right: top-view).

Conclusions

The incorporation of 1,3,4-oxadiazole rings as side groups into an aromatic polyamide or polyazomethine backbone, together with certain flexible bridges such as diphenyl silane or hexafluoroisopropylidene in the case of polyamides, or together with voluminous fluorene units, in the case of polyazomethines, in the main chain, gave soluble polymers which could be easily processed into thin films from solutions, while maintaining a high thermal stability. Some of these polymers did exhibit a glass transition, with a large interval between the glass transition and decomposition temperature, which may be advantageous for their processing by thermoforming techniques, as well. The very thin coatings which were deposited onto silicon wafers showed a strong adhesion to the substrates and a smooth, homogeneous surface, practically deffectless. The polyamides containing the dimethylamino substituent in the para position of the diphenyl-oxadiazole side groups, as well as related polyazomethines containing fluorene unit in the main chain showed intensive blue photoluminescence, being promissing candidates for future use in light emitting devices. Potential applications in optoelectronics, microelectronics or other related advanced fields are forseen.

Acknowledgements

It is a pleasure to acknowledge the financial support provided to MB and EH by the AiF (Arbeitsgemeinschaft industrieller Forschungsvereinigungen) in Germany and Ministry of Education and Research in Romania. Our warm thanks also go to Burkhard Stiller for performing the AFM investigations.

[1] B. Schulz, M. Bruma, L. Brehmer, *Adv. Mater.*, **1997**, *9*, 601.
[2] Q. Pei, Y. Yang, *Adv. Mater.*, **1995**, *7*, 559.
[3] Z. Peng, Z. Bao, M. Galvin, *Adv. Mater.*, **1998**, *10*, 680.
[4] A.T. Shermukhamedov, A. Oktai, *Uzb. Htm. Zh.*, **1989**, *3*, 69; *Chem. Abst.*, **1989**, *111*, 165357 v.
[5] E. Greczmiel, P. Posch, H. W. Schmidt, P. Strohriegel, *Macromol. Symp.*, **1996**, *102*, 371.
[6] M. Bruma, E. Hamciuc, B. Schulz, T. Kopnick, Y. Kaminorz, J. Robison, *J. Appl. Polym. Sci.*, in press.
[7] M. Bruma, B. Schulz, T. Kopnick, R. Dietel, B. Stiller, F. Mercer, V. N. Reddy, *High Perform. Polym.*, **1998**, *10*, 207.
[8] M. Kane, L. A. Wells, P. E. Cassidy, *High Perform. Polym.*, **1991**, *3*, 191.
[9] K. H. Park, T. Tani, M. Kakimoto, Y. Imai, *Macromol. Chem. Phys.*, **1998**, *199*, 1029.
[10] Hypercube Inc. (Ontario), *Hyperchem,* Version **1994**, 4.0.
[11] N. Xu, M. R. Coleman, *J. Appl. Polym. Sci.*, **1997**, *66*, 459.
[12] Y. Kaminorz, B. Schulz, L. Brehmer, *Synth. Met.*, **2000**, *111/112*, 75.